Intermediate Algebra and Geometry

Charles P. McKeague

CENGAGE Learning

Australia • Brazil • Japan • Korea • Mexico • Singapore • Spain • United Kingdom • United States

CENGAGE Learning

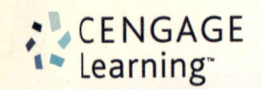

Intermediate Algebra and Geometry

Intermediate Algebra: A Text/Workbook, 8th Edition
Charles P. McKeague

© 2010 Cengage Learning. All rights reserved.

Trigonometry, 5th Edition
Charles P. McKeague | Mark Turner

© 2004 Cengage Learning. All rights reserved.

Executive Editors:
 Maureen Staudt
 Michael Stranz

Senior Project Development Manager:
 Linda deStefano

Marketing Specialist:
 Courtney Sheldon

Senior Production/Manufacturing Manager:
 Donna M. Brown

PreMedia Manager:
 Joel Brennecke

Sr. Rights Acquisition Account Manager:
 Todd Osborne

Cover Image:
 Getty Images*

*Unless otherwise noted, all cover images used by Custom Solutions, a part of Cengage Learning, have been supplied courtesy of Getty Images with the exception of the Earthview cover image, which has been supplied by the National Aeronautics and Space Administration (NASA).

ALL RIGHTS RESERVED. No part of this work covered by the copyright herein may be reproduced, transmitted, stored or used in any form or by any means graphic, electronic, or mechanical, including but not limited to photocopying, recording, scanning, digitizing, taping, Web distribution, information networks, or information storage and retrieval systems, except as permitted under Section 107 or 108 of the 1976 United States Copyright Act, without the prior written permission of the publisher.

For product information and technology assistance, contact us at
Cengage Learning Customer & Sales Support, 1-800-354-9706
For permission to use material from this text or product,
submit all requests online at **cengage.com/permissions**
Further permissions questions can be emailed to
permissionrequest@cengage.com

This book contains select works from existing Cengage Learning resources and was produced by Cengage Learning Custom Solutions for collegiate use. As such, those adopting and/or contributing to this work are responsible for editorial content accuracy, continuity and completeness.

Compilation © 2011 Cengage Learning
ISBN-13: 978-1-133-06668-2

ISBN-10: 1-133-06668-2
Cengage Learning
5191 Natorp Boulevard
Mason, Ohio 45040
USA

Cengage Learning is a leading provider of customized learning solutions with office locations around the globe, including Singapore, the United Kingdom, Australia, Mexico, Brazil, and Japan. Locate your local office at:
international.cengage.com/region.
Cengage Learning products are represented in Canada by Nelson Education, Ltd.
For your lifelong learning solutions, visit **www.cengage.com/custom.**
Visit our corporate website at **www.cengage.com.**

Printed in the United States of America

Brief Contents

Chapter 1   Basic Concepts and Properties   1

Chapter 2   Exponents and Polynomials   75

Chapter 3   Rational Expressions and Rational Equations   157

Chapter 4   Functions   235

Chapter 5   Geometry   301

Answers to Odd-Numbered Problems   A-1

Solutions to Selected Practice Problems   S-1

Index   I-1

Contents

1 Basic Concepts and Properties   1

Introduction   1

1.1 Real Numbers   3
1.2 Arithmetic with Real Numbers   15
1.3 Scientific Notation   31
1.4 Solving First Degree Equations   39
1.5 Problem Solving Involving First Degree Equations   53

Summary   65
Review   69
Test   71
Projects   73

2 Exponents and Polynomials   75

Introduction   75

2.1 Properties of Exponents   77
2.2 Polynomials, Sums, and Differences   89
2.3 Multiplication of Polynomials   99
2.4 The Greatest Common Factor   107
2.5 Special Factoring   113
2.6 Factoring Trinomials   121
2.7 A General Review of Factoring   129
2.8 Solving Equations by Factoring   135

Summary   147
Review   151
Cumulative Review   153
Test   154
Projects   155

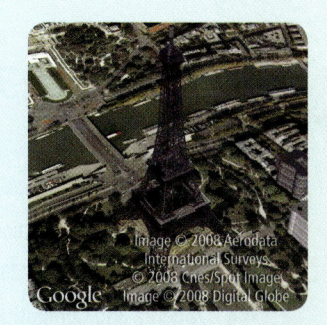

3 Rational Expressions and Rational Equations 157

Introduction 157

3.1 Reducing to Lowest Terms 159
3.2 Multiplication and Division of Rational Expressions 171
3.3 Addition and Subtraction of Rational Expressions 181
3.4 Equations Involving Rational Expressions 193
3.5 Simplified Form for Radicals 203
3.6 The Quadratic Formula 215

Summary 227
Review 229
Cumulative Review 231
Test 232
Projects 233

4 Functions 235

Introduction 235

4.1 Paired Data and Graphing Ordered Pairs 237
4.2 Introduction to Functions 253
4.3 Function Notation 269
4.4 Algebra and Composition with Functions 283

Summary 293
Review 295
Cumulative Review 297
Test 298
Projects 299

5 Geometry   301

Introduction   301

5.1 Angles   303

5.2 Parallel Lines   315

5.3 Classification of Triangles   327

5.4 Congruent Triangles   337

5.5 Parallelograms   347

5.6 Similar Figures and Proportions   355

5.7 The Pythagorean Theorem   365

5.8 Right Triangle Trigonometry   377

5.9 Solving Right Triangles   385

5.10 Area and Perimeter   399

Summary   409

Review   413

Cumulative Review   415

Test   416

Projects   417

Answers to Odd-Numbered Problems   A-1

Solutions to Selected Practice Problems   S-1

Index   I-1

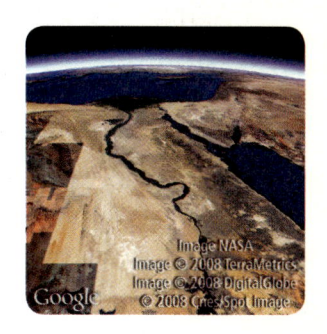

Preface to the Instructor

I have a passion for teaching mathematics. That passion carries through to my textbooks. My goal is a textbook that is user-friendly for both students and instructors. For students, this book forms a bridge to college algebra with clear, concise writing, continuous review, and interesting applications. For the instructor, I build features into the text that reinforce the habits and study skills we know will bring success to our students.

Our Proven Commitment to Student Success

Here are some important success features of the book.

Chapter Pretest These are meant as a diagnostic test taken before the starting work in the chapter. Much of the material here is learned in the chapter so proficiency on the pretests is not necessary.

Getting Ready for Chapter X This is a set of problems that students need in order to be successful in the current chapter. These are review problems intended to reinforce the idea that all topics in the course are built on topics they have learned already.

Practice Problems The Practice Problems, with their answers and solutions, are the key to moving students through the material in this book. We call it the EPAS system for Example, Practice Problem, Answer, Solution. Students begin by reading through the text, stopping after they have read through an example. Then they work the Practice Problem in the margin next to the example. When they are finished, they check the answers to the Practice Problem at the bottom of the page. If they have made a mistake, they try the problem again. If they still cannot get it right, they look at the Solutions to Selected Practice Problems in the back of the book. After working their way through the section in this manner, they are ready to start on the problems in the Problem Set.

Getting Ready for Class Just before each problem set is a list of four questions under the heading Getting Ready for Class. These problems require written responses from students and are to be done before students come to class. The answers can be found by reading the preceding section. These questions reinforce the importance of reading the section before coming to class.

Blueprint for Problem Solving Found in the main text, this feature is a detailed outline of steps required to successfully attempt application problems. Intended as a guide to problem solving in general, the blueprint takes the student through the solution process to various kinds of applications.

Organization of the Problem Sets

The problem sets begin with drill problems that are linked to the section objectives. Following the drill problems we have the following categories of problems.

Applying the Concepts Students are always curious about how the mathematics they are learning can be applied, so we have included applied problems in many of the problem sets in the book and have labeled them to show students the array of uses of mathematics. These applied problems are written in an inviting way, many times accompanied by new interesting illustrations to help students overcome some of the apprehension associated with application problems.

Getting Ready for the Next Section Many students think of mathematics as a collection of discrete, unrelated topics. Their instructors know that this is not the case. The Getting Ready for the Next Section problems reinforce the cumulative, connected nature of this course by showing how the concepts and techniques flow one from another throughout the course. These problems review the material that students will need in order to be successful in the next section, gently preparing students to move forward.

Maintaining Your Skills One of the major themes of our book is continuous review. We strive to continuously hone techniques learned earlier by keeping the important concepts in the forefront of the course. The Maintaining Your Skills problems review material from the previous chapter, or they review problems that form the foundation of the course—the problems that you expect students to be able to solve when they get to the next course.

End-of-Chapter Summary, Review, and Assessment

We have learned that students are more comfortable with a chapter that sums up what they have learned thoroughly and accessibly, and reinforces concepts and techniques well. To help students grasp concepts and get more practice, each chapter ends with the following features that together give a comprehensive reexamination of the chapter.

Chapter Summary The Chapter Summary recaps all main points from the chapter in a visually appealing grid. In the margin is an example that illustrates the type of problem associated with the topic being reviewed. When students prepare for a test, they can use the chapter summary as a guide to the main concepts of the chapter.

Chapter Review Following the Chapter Summary in each chapter is the Chapter Review. It contains an extensive set of problems that review all the main topics in the chapter. This feature can be used flexibly, as assigned review, as a recommended self-test for students as they prepare for examinations, or as an in-class quiz or test.

Cumulative Review Starting in Chapter 2, following the Chapter Review in each chapter, is a set of problems that reviews material from preceding chapters. This keeps students current with past topics and helps them retain the information they study.

Chapter Test This is a set of problems representative of all the main points of the chapter. These don't contain as many problems as the Chapter Review, and should be completed in 50 minutes.

Chapter Projects Each chapter closes with a pair of projects. One is a Group Project, suitable for students to work on in class. The second project is a Research Project for students to do outside of class and tends to be open ended.

Additional Feature of the Book

Using Technology Scattered throughout the book is material that shows how graphing calculators can be used to enhance the topics being covered.

Acknowledgments

I would like to thank my editor at Cengage Learning, Marc Bove, for his help and encouragement with this project. Many thanks also to Rich Jones, my developmental editor, for his suggestions on content, and his availability for consulting. Devin Christ, the head of production at our office, was a tremendous help in organizing and planning the details of putting this book together. Mary Gentilucci, Michael Landrum and Tammy Fisher-Vasta assisted with error checking and proofreading. Special thanks to my other friends at Cengage Learning: Sam Subity and Shaun Williams for handling the media and ancillary packages on this project, and Hal Humphrey, my project manager, who did a great job of coordinating everyone and everything in order to publish this book.

Finally, I am grateful to the following instructors for their suggestions and comments: Patricia Clark, Sinclair CC; Matthew Hudock, St. Phillip's College; Bridget Young, Suffolk County CC; Bettie Truitt, Black Hawk College; Armando Perez, Laredo CC; Diane Allen, College of Technology Idaho State; Jignasa Rami, CCBC Catonsville; Yon Kim, Passaic Community College; Elizabeth Chu, Suffolk County CC, Ammerman; Marilyn Larsen, College of the Mainland; Sherri Ucravich, University of Wisconsin; Scott Beckett, Jacksonville State University; Nimisha Raval, Macon Technical Institute; Gary Franchy, Davenport University, Warren; Debbi Loeffler, CC of Baltimore County; Scott Boman, Wayne County CC; Dayna Coker, Southwestern Oklahoma State; Annette Wiesner, University of Wisconsin; Anne Kmet, Grossmont College; Mary Wagner-Krankel, St. Mary's University; Joseph Deguzman, Riverside CC, Norco; Deborah McKee, Weber State University; Gail Burkett, Palm Beach CC; Lee Ann Spahr, Durham Technical CC; Randall Mills, KCTCS Big Sandy CC/Tech; Jana Bryant, Manatee CC; Fred Brown, University of Maine, Augusta; Jeff Waller, Grossmont College; Robert Fusco, Broward CC, FL; Larry Perez, Saddleback College, CA; Victoria Anemelu, San Bernardino Valley, CA; John Close, Salt Lake CC, UT; Randy Gallaher, Lewis and Clark CC; Julia Simms, Southern Illinois U; Julianne Labbiento, Lehigh Carbon CC; Joanne Kendall, Cy-Fair College; Ann Davis, Northeastern Tech.

Pat McKeague
July 2011

Preface to the Student

I often find my students asking themselves the question "Why can't I understand this stuff the first time?" The answer is "You're not expected to." Learning a topic in mathematics isn't always accomplished the first time around. There are many instances when you will find yourself reading over new material a number of times before you can begin to work problems. That's just the way things are in mathematics. If you don't understand a topic the first time you see it, that doesn't mean there is something wrong with you. Understanding mathematics takes time. The process of understanding requires reading the book, studying the examples, working problems, and getting your questions answered.

How to Be Successful in Mathematics

1. If you are in a lecture class, be sure to attend all class sessions on time. You cannot know exactly what goes on in class unless you are there. Missing class and then expecting to find out what went on from someone else is not the same as being there yourself.

2. Read the book and work the Practice Problems. It is best to read the section that will be covered in class beforehand. Reading in advance, even if you do not understand everything you read, is still better than going to class with no idea of what will be discussed. As you read through each section, be sure to work the Practice Problems in the margin of the text. Each Practice Problem is similar to the example with the same number. Look over the example and then try the corresponding Practice Problem. The answers to the Practice Problems are at the bottom of the page. If you don't get the correct answer, see if you can rework the problem correcly. If you miss it a second time, check your solution with the Solution to the Selected Practice Problems in the back of the book.

3. Work problems every day and check your answers. The key to success in mathematics is working problems. The more problems you work, the better you will become at working them. The answers to the odd-numbered problems are given in the back of the book. When you have finished an assignment, be sure to compare your answers with those in the book. If you have made a mistake, find out what it is, and correct it.

4. Do it on your own. Don't be misled into thinking someone else's work is your own. Having someone else show you how to work a problem is not the same as working the same problem yourself. It is okay to get help when you are stuck. As a matter of fact, it is a good idea. Just be sure you do the work yourself.

5. Review every day. After you have finished the problems your instructor has assigned, take another 15 minutes and review a section you have already completed. The more you review, the longer you will retain the material you have learned.

6. Don't expect to understand every new topic the first time you see it. Sometimes you will understand everything you are doing, and sometimes you won't. That's just the way things are in mathematics. Expecting to understand each new topic the first time you see it can lead to disappointment and frustration. The process of understanding takes time. It requires that you read the book, work problems, and get your questions answered.

7. Spend as much time as it takes for you to master the material. No set formula exists for the exact amount of time you need to spend on mathematics to master it. You will find out as you go along what is or isn't enough time for you. If you end up spending two or more hours on each section in order to master the material there, then that's how much time it takes; trying to get by with less will not work.

8. Relax. It's probably not as difficult as you think.

Basic Concepts and Properties

1

Chapter Outline

1.1 Real Numbers

1.2 Arithmetic with Real Numbers

1.3 Scientific Notation

1.4 Solving First Degree Equations

1.5 Problem Solving Involving First Degree Equations

Introduction

The Google Earth image is of the Leaning Tower of Pisa. Pisa is the birthplace of the Italian mathematician Fibonacci. It was in Pisa that Fibonacci wrote the first European algebra book *Liber Abaci*. It was published in 1202. Below is the cover of an English translation of that book, published in 2003, along with a quote from the Mathematics Association of America.

MAA Online, March 2003:
"The *Liber abaci* of Leonardo Pisano (today commonly called Fibonacci) is one of the fundamental works of European mathematics. No other book did more to establish the basic framework of arithmetic and algebra as they developed in the Western world."

In his book, Fibonacci shows how to solve linear equations in one variable, which is the main topic in this chapter.

Chapter Pretest

The numbers in brackets indicate the section(s) to which the problems correspond.

Simplify each expression using the rule for order of operations. [1.2]

1. $7 \cdot 2^2 + 3 \cdot 5^2$
2. $5 + 3(2 \cdot 9 - 8)$
3. $6 - 12 \div 3 + 5 \cdot 2$
4. $18 - 2[3^2 - 4(2^3 - 8)]$

Simplify each of the following as much as possible. [1.2]

5. $\dfrac{-3(-2) - (-6)}{1 - (-3)}$
6. $-5 - 3\left[\dfrac{-4(-6) + 1}{-6(2) - 3}\right]$
7. $-\dfrac{5}{6} + \dfrac{3}{2} - \left(-\dfrac{4}{7}\right)$
8. $-\dfrac{1}{3}(9x)$

9. $-5(2x + 7) - 4x$
10. $3(4y + 1) - (7y + 6)$
11. $-1 + 5(3x - 2) - 3x$
12. $2 + 4a + 4(3a - 6)$

13. Subtract $\dfrac{1}{2}$ from the product of -6 and $\dfrac{3}{12}$.
14. Subtract $-\dfrac{5}{8}$ from the product of 2 and $\dfrac{7}{16}$.

Solve the following equations. [1.4]

15. $x + 4 = 3$
16. $5y = -2$

17. $4(3x - 1) + 3(x - 2) = 7x + 2$
18. $-.07x - .02 = .05 - .03(4x + 2)$

Getting Ready for Chapter 1

To get started in this book, we assume that you can do simple addition and multiplication problems with whole numbers and decimals. To check to see that you are ready for this chapter, work each of the problems below without using a calculator.

1. $5.7 - 1.9$
2. $10 - 9.5$
3. $0.3(6)$
4. $39 \div 13$
5. $28 \div 7$

6. $125 - 81$
7. $630 \div 63$
8. $210 \div 10$
9. $2 \cdot 3 \cdot 5 \cdot 7$
10. $3 \cdot 7 \cdot 11$

11. $2 \cdot 3 \cdot 3 \cdot 5 \cdot 7$
12. $18{,}000 - 4{,}500$
13. $1.6 + 2.0$
14. $7(0.05)$
15. $7{,}546 \div 35$

16. $1.2052 - 1$
17. $250(3.14)$
18. $109 - 36 + 14$
19. $0.08(4{,}000)$
20. $3.9 \div 1.3$

Simplify each expression.

21. $-6\left(\dfrac{2}{3}\right)$
22. $-24\left(\dfrac{3}{8}\right)$
23. $\dfrac{4}{3}(-9)$
24. $\dfrac{2}{3}\left(-\dfrac{21}{16}\right)$
25. $-\dfrac{3}{8} + \left(-\dfrac{1}{2}\right)$

26. $-\dfrac{7}{8} + \dfrac{1}{2}$
27. $\dfrac{1}{2}(2x)$
28. $\dfrac{4}{3}\left(\dfrac{3}{4}x\right)$
29. $3(2y - 1) + y$
30. $8 - 3(4x - 2) + 5x$

Chapter 1 Basic Concepts and Properties

Real Numbers

1.1

Objectives
- **A** Find the opposite of a real number.
- **B** Multiply fractions.
- **C** Find the reciprocal of a real number.
- **D** Simplify expressions involving absolute value.
- **E** Recognize and apply the properties of real numbers.
- **F** Recognize and apply the distributive property.
- **G** Add and subtract fractions.
- **H** Simplify algebraic expressions.

Introduction

The area of the large rectangle shown here can be found in two ways: We can multiply its length a by its width $b + c$, or we can find the areas of the two smaller rectangles and add those areas to find the total area.

Area of large rectangle: $a(b + c)$

Sum of the areas of two smaller rectangles: $ab + ac$

Because the area of the large rectangle is the sum of the areas of the two smaller rectangles, we can write:

$$a(b + c) = ab + ac$$

This equation is called the *distributive property*. It is one of the properties we will be discussing in this section. Before we arrive at the distributive property, we need to review some basic definitions and vocabulary.

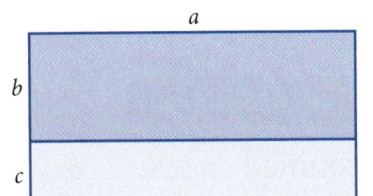

A Opposites

Definition
Any two real numbers the same distance from 0, but in opposite directions from 0 on the number line, are called **opposites**, or **additive inverses**.

EXAMPLE 1 The numbers -3 and 3 are opposites. So are π and $-\pi$, $\frac{3}{4}$ and $-\frac{3}{4}$, and $\sqrt{2}$ and $-\sqrt{2}$.

The negative sign in front of a number can be read in a number of different ways. It can be read as "negative" or "the opposite of." We say -4 is the opposite of 4, or negative 4. The one we use will depend on the situation. For instance, the expression $-(-3)$ is best read "the opposite of negative 3." Because the opposite of -3 is 3, we have $-(-3) = 3$. In general, if a is any positive real number, then

$$-(-a) = a \quad \text{The opposite of a negative is a positive.}$$

B Review of Multiplication with Fractions

Before we go further with our study of real numbers, we need to review multiplication with fractions. Recall that for the fraction $\frac{a}{b}$, a is called the numerator and b is called the denominator. To multiply two fractions, we simply multiply numerators and multiply denominators.

EXAMPLE 2 Multiply $\frac{3}{5} \cdot \frac{7}{8}$.

SOLUTION $\frac{3}{5} \cdot \frac{7}{8} = \frac{3 \cdot 7}{5 \cdot 8} = \frac{21}{40}$

EXAMPLE 3 Multiply $8 \cdot \frac{1}{5}$.

SOLUTION $8 \cdot \frac{1}{5} = \frac{8}{1} \cdot \frac{1}{5} = \frac{8 \cdot 1}{1 \cdot 5} = \frac{8}{5}$

PRACTICE PROBLEMS

1. Give the opposite of each of the following numbers.
   - a. 5
   - b. $\frac{1}{4}$
   - c. -3
   - d. $-\sqrt{5}$

Note In past math classes you may have written fractions like $\frac{8}{5}$ (improper fractions) as mixed numbers, such as $1\frac{3}{5}$. In algebra, it is usually better to leave them as improper fractions.

2. Multiply $\frac{3}{7} \cdot \frac{2}{5}$.

3. Multiply $9 \cdot \frac{1}{4}$.

Answers
1. a. -5 b. $-\frac{1}{4}$ c. 3 d. $\sqrt{5}$
2. $\frac{6}{35}$
3. $\frac{9}{4}$

4. Multiply $\left(\frac{3}{4}\right)^3$.

Chapter 1 Basic Concepts and Properties

EXAMPLE 4 Multiply $\left(\frac{2}{3}\right)^4$.

SOLUTION $\left(\frac{2}{3}\right)^4 = \frac{2}{3} \cdot \frac{2}{3} \cdot \frac{2}{3} \cdot \frac{2}{3} = \frac{16}{81}$

C Reciprocals

The idea of multiplication of fractions is useful in understanding the concept of the reciprocal of a number. Here is the definition.

> **Definition**
> Any two real numbers whose product is 1 are called **reciprocals**, or **multiplicative inverses**.

5. Give the reciprocal of 5.

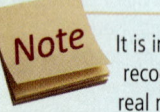

EXAMPLE 5 Give the reciprocal of 3.

SOLUTION
Number	Reciprocal	
3	$\frac{1}{3}$	Because $3 \cdot \frac{1}{3} = \frac{3}{1} \cdot \frac{1}{3} = \frac{3}{3} = 1$

6. Give the reciprocal of $\frac{1}{4}$.

EXAMPLE 6 Give the reciprocal of $\frac{1}{6}$.

SOLUTION
Number	Reciprocal	
$\frac{1}{6}$	6	Because $\frac{1}{6} \cdot 6 = \frac{1}{6} \cdot \frac{6}{1} = \frac{6}{6} = 1$

7. Give the reciprocal of $\frac{3}{4}$.

EXAMPLE 7 Give the reciprocal of $\frac{4}{5}$.

SOLUTION
Number	Reciprocal	
$\frac{4}{5}$	$\frac{5}{4}$	Because $\frac{4}{5} \cdot \frac{5}{4} = \frac{20}{20} = 1$

8. Give the reciprocal of x if $x \neq 0$.

EXAMPLE 8 Give the reciprocal of a.

SOLUTION
Number	Reciprocal	
a	$\frac{1}{a}$	Because $a \cdot \frac{1}{a} = \frac{a}{1} \cdot \frac{1}{a} = \frac{a}{a} = 1 \quad (a \neq 0)$

Although we will not develop multiplication with negative numbers until later in this chapter, you should know that the reciprocal of a negative number is also a negative number. For example, the reciprocal of -5 is $-\frac{1}{5}$.

D The Absolute Value of a Real Number

Note It is important to recognize that if x is a real number, $-x$ is not necessarily negative. For example, if x is 5, then $-x$ is -5. However, if x were -5, then $-x$ would be $-(-5)$, which is 5.

> **Definition**
> The **absolute value** of a number (also called its **magnitude**) is the distance the number is from 0 on the number line. If x represents a real number, then the absolute value of x is written $|x|$.

Answers
4. $\frac{27}{64}$
5. $\frac{1}{5}$
6. 4
7. $\frac{4}{3}$
8. $\frac{1}{x}$

1.1 Real Numbers

This definition of absolute value is geometric in form since it defines absolute value in terms of the number line. Here is an alternative definition of absolute value that is algebraic in form since it involves only symbols.

Alternative Definition
If x represents a real number, then the **absolute value** of x is written $|x|$, and is given by

$$|x| = \begin{cases} x & \text{if } x \geq 0 \\ -x & \text{if } x < 0 \end{cases}$$

If the original number is positive or 0, then its absolute value is the number itself. If the number is negative, its absolute value is its opposite (which must be positive). The absolute value of a number is always positive because absolute value descrives a distance, and distance is always a positive number.

EXAMPLE 9 Write $|5|$ without absolute value symbols.

SOLUTION $|5| = 5$

EXAMPLE 10 Write $|-2|$ without absolute value symbols.

SOLUTION $|-2| = 2$

EXAMPLE 11 Write $\left|-\frac{1}{2}\right|$ without absolute value symbols.

SOLUTION $\left|-\frac{1}{2}\right| = \frac{1}{2}$

EXAMPLE 12 Write $-|-3|$ without absolute value symbols.

SOLUTION $-|-3| = -3$

EXAMPLE 13 Write $-|5|$ without absolute value symbols.

SOLUTION $-|5| = -5$

EXAMPLE 14 Write $-|-\sqrt{2}|$ without absolute value symbols.

SOLUTION $-|-\sqrt{2}| = -\sqrt{2}$

E Properties of Real Numbers

We know that adding 3 and 7 gives the same answer as adding 7 and 3. The order of two numbers in an addition problem can be changed without changing the result. This fact about numbers and addition is called the *commutative property of addition*.

For all the properties listed in this section, a, b, and c represent real numbers.

Write the following without absolute value symbols.

9. $|7|$

10. $|-4|$

11. $\left|-\frac{2}{3}\right|$

12. $-|-6|$

13. $-|7|$

14. $-|-\sqrt{3}|$

Answers
9. 7
10. 4
11. $\frac{2}{3}$
12. −6
13. −7
14. $-\sqrt{3}$

Complete each statement so it is an example of the commutative property.

15. $5 + x = x + \underline{}$

16. $7 \cdot 3 = 3 \cdot \underline{}$

> **Note** The other two basic operations (subtraction and division) are not commutative. If we change the order in which we are subtracting or dividing two numbers, we change the result.

17. Simplify $5 + (7 + y)$.

18. Simplify $3(2x)$.

19. Simplify $\frac{1}{3}(3a)$.

Answers
15. $5 + x = x + 5$
16. $7 \cdot 3 = 3 \cdot 7$
17. $12 + y$
18. $6x$
19. a

Chapter 1 Basic Concepts and Properties

Commutative Property of Addition
In symbols: $a + b = b + a$
In words: The *order* of the numbers in a sum does not affect the result.

Commutative Property of Multiplication
In symbols: $a \cdot b = b \cdot a$
In words: The *order* of the numbers in a product does not affect the result.

EXAMPLE 15 The statement $3 + 7 = 7 + 3$ is an example of the commutative property of addition.

EXAMPLE 16 The statement $3 \cdot x = x \cdot 3$ is an example of the commutative property of multiplication.

Another property of numbers you have used many times has to do with grouping. When adding $3 + 5 + 7$, we can add the 3 and 5 first and then the 7, or we can add the 5 and 7 first and then the 3. Mathematically, it looks like this: $(3 + 5) + 7 = 3 + (5 + 7)$. Operations that behave in this manner are called *associative* operations.

Associative Property of Addition
In symbols: $a + (b + c) = (a + b) + c$
In words: The *grouping* of the numbers in a sum does not affect the result.

Associative Property of Multiplication
In symbols: $a(bc) = (ab)c$
In words: The *grouping* of the numbers in a product does not affect the result.

The following examples illustrate how the associative properties can be used to simplify expressions that involve both numbers and variables.

EXAMPLE 17 Simplify $2 + (3 + y)$ by using the associative property.

SOLUTION
$2 + (3 + y) = (2 + 3) + y$ Associative property
$ = 5 + y$ Addition

EXAMPLE 18 Simplify $5(4x)$ by using the associative property.

SOLUTION
$5(4x) = (5 \cdot 4)x$ Associative property
$ = 20x$ Multiplication

EXAMPLE 19 Simplify $\frac{1}{4}(4a)$ by using the associative property.

SOLUTION
$\frac{1}{4}(4a) = \left(\frac{1}{4} \cdot 4\right)a$ Associative property
$\phantom{\frac{1}{4}(4a)} = 1a$ Multiplication
$\phantom{\frac{1}{4}(4a)} = a$

1.1 Real Numbers

EXAMPLE 20 Simplify $2\left(\frac{1}{2}x\right)$ by using the associative property.

SOLUTION
$2\left(\frac{1}{2}x\right) = \left(2 \cdot \frac{1}{2}\right)x$ Associative property
$= 1x$ Multiplication
$= x$

EXAMPLE 21 Simplify $6\left(\frac{1}{3}x\right)$ by using the associative property.

SOLUTION
$6\left(\frac{1}{3}x\right) = \left(6 \cdot \frac{1}{3}\right)x$ Associative property
$= 2x$ Multiplication

F Distributive Property

Our next property involves both addition and multiplication. It is called the *distributive property* and is stated as follows.

> **Distributive Property**
> In symbols: $a(b + c) = ab + ac$
> In words: Multiplication *distributes* over addition.

You will see as we progress through the book that the distributive property is used very frequently in algebra. To see that the distributive property works, compare the following:

$3(4 + 5)$ $3(4) + 3(5)$
$3(9)$ $12 + 15$
27 27

In both cases the result is 27. Because the results are the same, the original two expressions must be equal, or $3(4 + 5) = 3(4) + 3(5)$.

EXAMPLE 22 Apply the distributive property to $5(4x + 3)$ and then simplify the result.

SOLUTION
$5(4x + 3) = 5(4x) + 5(3)$ Distributive property
$= 20x + 15$ Multiplication

EXAMPLE 23 Apply the distributive property to $6(3x + 2y)$ and then simplify the result.

SOLUTION
$6(3x + 2y) = 6(3x) + 6(2y)$ Distributive property
$= 18x + 12y$ Multiplication

EXAMPLE 24 Apply the distributive property to $\frac{1}{2}(3x + 6)$ and then simplify the result.

SOLUTION
$\frac{1}{2}(3x + 6) = \frac{1}{2}(3x) + \frac{1}{2}(6)$ Distributive property
$= \frac{3}{2}x + 3$ Multiplication

---

20. Simplify $5\left(\frac{1}{5}x\right)$.

21. Simplify $9\left(\frac{2}{3}x\right)$.

Apply the distributive property and then simplify.

22. $7(6x + 8)$

23. $9(7x + 11y)$

24. $\frac{1}{3}(3x + 6)$

Answers
20. x
21. $6x$
22. $42x + 56$
23. $63x + 99y$
24. $x + 2$

Chapter 1 Basic Concepts and Properties

25. $4(5y + 2) + 8$

EXAMPLE 25 Apply the distributive property to $2(3y + 4) + 2$ and then simplify the result.

SOLUTION
$$\begin{aligned} 2(3y + 4) + 2 &= 2(3y) + 2(4) + 2 & \text{Distributive property}\\ &= 6y + 8 + 2 & \text{Multiplication}\\ &= 6y + 10 & \text{Addition} \end{aligned}$$

We can combine our knowledge of the distributive property with multiplication of fractions to manipulate expressions involving fractions. Here are some examples that show how we do this.

Apply the distributive property, then simplify if possible.

26. $a\left(2 - \dfrac{1}{a}\right)$

EXAMPLE 26 Apply the distributive property to $a\left(1 + \dfrac{1}{a}\right)$, and then simplify if possible.

SOLUTION $a\left(1 + \dfrac{1}{a}\right) = a \cdot 1 + a \cdot \dfrac{1}{a} = a + 1$

27. $5\left(\dfrac{1}{5}x + 8\right)$

EXAMPLE 27 Apply the distributive property to $3\left(\dfrac{1}{3}x + 5\right)$, and then simplify if possible.

SOLUTION $3\left(\dfrac{1}{3}x + 5\right) = 3 \cdot \dfrac{1}{3}x + 3 \cdot 5 = x + 15$

28. $12\left(\dfrac{2}{3}x + \dfrac{3}{4}y\right)$

EXAMPLE 28 Apply the distributive property to $6\left(\dfrac{1}{3}x + \dfrac{1}{2}y\right)$, and then simplify if possible.

SOLUTION $6\left(\dfrac{1}{3}x + \dfrac{1}{2}y\right) = 6 \cdot \dfrac{1}{3}x + 6 \cdot \dfrac{1}{2}y = 2x + 3y$

Combining Similar Terms

The distributive property can also be used to combine similar terms. (For now, a *term* is a number, or the product of a number with one or more variables.) Similar terms are terms with the same variable part. The terms $3x$ and $5x$ are similar, as are $2y$, $7y$, and $-3y$, because the variable parts are the same.

Combine similar terms.

29. $4x + 9x$

EXAMPLE 29 Use the distributive property to combine similar terms.

SOLUTION
$$\begin{aligned} 3x + 5x &= (3 + 5)x & \text{Distributive property}\\ &= 8x & \text{Addition} \end{aligned}$$

30. $y + 5y$

EXAMPLE 30 Use the distributive property to combine similar terms.

SOLUTION
$$\begin{aligned} 3y + y &= (3 + 1)y & \text{Distributive property}\\ &= 4y & \text{Addition} \end{aligned}$$

G Review of Addition with Fractions

To add fractions, each fraction must have the same denominator.

Answers
25. $20y + 16$
26. $2a - 1$
27. $x + 40$
28. $8x + 9y$
29. $13x$
30. $6y$

> **Definition**
> The **least common denominator** (LCD) for a set of denominators is the smallest number divisible by *all* the denominators.

The first step in adding fractions is to find a common denominator for all the denominators. We then rewrite each fraction (if necessary) as an equivalent fraction with the common denominator. Finally, we add the numerators and reduce to lowest terms if necessary.

EXAMPLE 31 Add $\dfrac{5}{12} + \dfrac{7}{18}$.

SOLUTION The least common denominator for the denominators 12 and 18 must be the smallest number divisible by both 12 and 18. We can factor 12 and 18 completely and then build the LCD from these factors.

$$\left.\begin{array}{l}12 = 2 \cdot 2 \cdot 3 \\ 18 = 2 \cdot 3 \cdot 3\end{array}\right\} \text{LCD} = 2 \cdot 2 \cdot 3 \cdot 3 = 36$$

12 divides the LCD
18 divides the LCD

Next, we rewrite our original fractions as equivalent fractions with denominators of 36. To do so, we multiply each original fraction by an appropriate form of the number 1.

$$\dfrac{5}{12} + \dfrac{7}{18} = \dfrac{5}{12} \cdot \dfrac{\mathbf{3}}{\mathbf{3}} + \dfrac{7}{18} \cdot \dfrac{\mathbf{2}}{\mathbf{2}} = \dfrac{15}{36} + \dfrac{14}{36}$$

Finally, we add numerators and place the result over the common denominator, 36.

$$\dfrac{15}{36} + \dfrac{14}{36} = \dfrac{15 + 14}{36} = \dfrac{29}{36}$$

H Simplifying Expressions

We can use the commutative, associative, and distributive properties together to simplify expressions.

EXAMPLE 32 Simplify $7x + 4 + 6x + 3$.

SOLUTION We begin by applying the commutative and associative properties to group similar terms.

$7x + 4 + 6x + 3 = (7x + 6x) + (4 + 3)$ Commutative and associative properties
$ = (7 + 6)x + (4 + 3)$ Distributive property
$ = 13x + 7$ Addition

EXAMPLE 33 Simplify $4 + 3(2y + 5) + 8y$.

SOLUTION As we will see, the rule for order of operations indicates that we are to multiply before adding, so we must distribute the 3 across $2y + 5$ first.

$4 + 3(2y + 5) + 8y = 4 + 6y + 15 + 8y$ Distributive property
$ = (6y + 8y) + (4 + 15)$ Commutative and associative properties
$ = (6 + 8)y + (4 + 15)$ Distributive property
$ = 14y + 19$ Addition

---

31. Add $\dfrac{3}{14} + \dfrac{7}{30}$.

32. Simplify $10x + 3 + 7x + 12$.

33. Simplify $9 + 5(4y + 8) + 10y$.

Answers
31. $\dfrac{47}{105}$
32. $17x + 15$
33. $30y + 49$

Chapter 1 Basic Concepts and Properties

The remaining properties of real numbers have to do with the numbers 0 and 1.

> **Additive Identity Property**
> There exists a unique number 0 such that
> In symbols: $a + 0 = a$ and $0 + a = a$
>
> **Multiplicative Identity Property**
> There exists a unique number 1 such that
> In symbols: $a(1) = a$ and $1(a) = a$
>
> **Additive Inverse Property**
> In symbols: $a + (-a) = 0$
> In words: Opposites add to 0.
>
> **Multiplicative Inverse Property**
> For every real number a, except 0, there exists a unique real number $\frac{1}{a}$ such that
> In symbols: $a\left(\frac{1}{a}\right) = 1$
> In words: Reciprocals multiply to 1.

State the property of real numbers that justifies each statement.

34. $2 + 0 = 2$

35. $3(1) = 3$

36. $-5 + 5 = 0$

37. $\frac{3}{4} \cdot \frac{4}{3} = 1$

EXAMPLES State the property of real numbers that justifies each statement.

34. $7(1) = 7$ Multiplicative identity property

35. $4 + (-4) = 0$ Additive inverse property

36. $6\left(\frac{1}{6}\right) = 1$ Multiplicative inverse property

37. $(5 + 0) + 2 = 5 + 2$ Additive identity property

Getting Ready for Class

After reading through the preceding section, respond in your own words and in complete sentences.

1. Describe the commutative property of multiplication.
2. Give definitions for each of the following:
   a. The opposite of a number
   b. The absolute value of a number
3. Explain why subtraction and division are not commutative operations.
4. Explain why zero does not have a multiplicative inverse.

Answers
34. Additive identity
35. Multiplicative identity
36. Additive inverse
37. Multiplicative inverse

Problem Set 1.1

A C Complete the following table. [Examples 1, 5–8]

	NUMBER	OPPOSITE	RECIPROCAL
1.	4		
2.	−3		
3.	$-\frac{1}{2}$		
4.	$\frac{5}{6}$		
5.		−5	
6.		7	
7.		$-\frac{3}{8}$	
8.		$\frac{1}{2}$	
9.			−6
10.			−3
11.			$\frac{1}{3}$
12.			$-\frac{1}{4}$

13. Name two numbers that are their own reciprocals.

14. Give the number that has no reciprocal.

15. Name the number that is its own opposite.

16. The reciprocal of a negative number is negative—true or false?

D Write each of the following without absolute value symbols. [Examples 9–14]

17. $|-2|$ 18. $|-7|$ 19. $\left|-\frac{3}{4}\right|$ 20. $\left|\frac{5}{6}\right|$

21. $|\pi|$ 22. $|-\sqrt{2}|$ 23. $-|4|$ 24. $-|5|$

25. $-|-2|$ 26. $-|-10|$ 27. $-\left|-\frac{3}{4}\right|$ 28. $-\left|\frac{7}{8}\right|$

B Multiply the following. [Examples 2–4]

29. $\frac{3}{5} \cdot \frac{7}{8}$ 30. $\frac{6}{7} \cdot \frac{9}{5}$ 31. $\frac{1}{3} \cdot 6$ 32. $\frac{1}{4} \cdot 8$

33. $\left(\frac{2}{3}\right)^3$ 34. $\left(\frac{4}{5}\right)^2$ 35. $\left(\frac{1}{10}\right)^4$ 36. $\left(\frac{1}{2}\right)^5$

37. $\frac{3}{5} \cdot \frac{4}{7} \cdot \frac{6}{11}$ 38. $\frac{4}{5} \cdot \frac{6}{7} \cdot \frac{3}{11}$ 39. $\frac{4}{3} \cdot \frac{3}{4}$ 40. $\frac{5}{8} \cdot \frac{8}{5}$

E Use the associative property to rewrite each of the following expressions and then simplify the result. [Examples 17–21]

41. $4 + (2 + x)$
42. $6 + (5 + 3x)$
43. $(a + 3) + 5$
44. $(4a + 5) + 7$

45. $5(3y)$
46. $7(4y)$
47. $\frac{1}{3}(3x)$
48. $\frac{1}{5}(5x)$

49. $4\left(\frac{1}{4}a\right)$
50. $7\left(\frac{1}{7}a\right)$
51. $\frac{2}{3}\left(\frac{3}{2}x\right)$
52. $\frac{4}{3}\left(\frac{3}{4}x\right)$

F Apply the distributive property to each expression. Simplify when possible. [Examples 22–24]

53. $3(x + 6)$
54. $5(x + 9)$
55. $2(6x + 4)$
56. $3(7x + 8)$

57. $5(3a + 2b)$
58. $7(2a + 3b)$
59. $\frac{1}{3}(4x + 6)$
60. $\frac{1}{2}(3x + 8)$

61. $\frac{1}{5}(10 + 5y)$
62. $\frac{1}{6}(12 + 6y)$
63. $(5t + 1)8$
64. $(3t + 2)5$

F The problems below are problems you will see later in the book. Apply the distributive property, then simplify if possible. [Examples 26–28]

65. $3(3x + y - 2z)$
66. $2(2x - y + z)$
67. $10(0.3x + 0.7y)$
68. $10(0.2x + 0.5y)$

69. $100(0.06x + 0.07y)$
70. $100(0.09x + 0.08y)$
71. $3\left(x + \frac{1}{3}\right)$
72. $5\left(x - \frac{1}{5}\right)$

73. $2\left(x - \frac{1}{2}\right)$
74. $7\left(x + \frac{1}{7}\right)$
75. $x\left(1 + \frac{2}{x}\right)$
76. $x\left(1 - \frac{1}{x}\right)$

77. $a\left(1 - \frac{3}{a}\right)$
78. $a\left(1 + \frac{1}{a}\right)$
79. $8\left(\frac{1}{8}x + 3\right)$
80. $4\left(\frac{1}{4}x - 9\right)$

81. $6\left(\frac{1}{2}x - \frac{1}{3}y\right)$
82. $12\left(\frac{1}{4}x - \frac{1}{6}y\right)$
83. $12\left(\frac{1}{4}x + \frac{2}{3}y\right)$
84. $12\left(\frac{2}{3}x - \frac{1}{4}y\right)$

85. $20\left(\frac{2}{5}x + \frac{1}{4}y\right)$
86. $15\left(\frac{2}{3}x + \frac{2}{5}y\right)$

F Apply the distributive property to each expression. Simplify when possible. [Example 25]

87. $3(5x + 2) + 4$
88. $4(3x + 2) + 5$
89. $4(2y + 6) + 8$
90. $6(2y + 3) + 2$

91. $5(1 + 3t) + 4$
92. $2(1 + 5t) + 6$
93. $3 + (2 + 7x)4$
94. $4 + (1 + 3x)5$

G Add the following fractions. [Example 31]

95. $\frac{2}{5} + \frac{1}{15}$

96. $\frac{5}{8} + \frac{1}{4}$

97. $\frac{17}{30} + \frac{11}{42}$

98. $\frac{19}{42} + \frac{13}{70}$

99. $\frac{9}{48} + \frac{3}{54}$

100. $\frac{6}{28} + \frac{5}{42}$

101. $\frac{25}{84} + \frac{41}{90}$

102. $\frac{23}{70} + \frac{29}{84}$

G The problems below are problems you will see later in the book. Simplify each expression.

103. $\frac{3}{14} + \frac{7}{30}$

104. $\frac{3}{10} + \frac{11}{42}$

105. $32\left(\frac{3}{4}\right) - 16\left(\frac{3}{4}\right)^2$

106. $32\left(\frac{3}{2}\right) - 16\left(\frac{3}{2}\right)^2$

H Use the commutative, associative, and distributive properties to simplify the following. [Example 32–33]

107. $5a + 7 + 8a + a$

108. $6a + 4 + a + 4a$

109. $3y + y + 5 + 2y + 1$

110. $4y + 2y + 3 + y + 7$

111. $2(5x + 1) + 2x$

112. $3(4x + 1) + 9x$

113. $7 + 2(4y + 2)$

114. $6 + 3(5y + 2)$

115. $3 + 4(5a + 3) + 4a$

116. $8 + 2(4a + 2) + 5a$

117. $5x + 2(3x + 8) + 4$

118. $7x + 3(4x + 1) + 7$

E Identify the property or properties of real numbers that justifies each of the following. [Examples 15–16, 34–37]

119. $3 + 2 = 2 + 3$

120. $3(ab) = (3a)b$

121. $5 \cdot x = x \cdot 5$

122. $2 + 0 = 2$

123. $4 + (-4) = 0$

124. $1(6) = 6$

125. $x + (y + 2) = (y + 2) + x$

126. $(a + 3) + 4 = a + (3 + 4)$

127. $4(5 \cdot 7) = 5(4 \cdot 7)$

128. $6(xy) = (xy)6$

129. $4 + (x + y) = (4 + y) + x$

130. $(r + 7) + s = (r + s) + 7$

131. $3(4x + 2) = 12x + 6$

132. $5\left(\frac{1}{5}\right) = 1$

1.1 Problem Set   13

Applying the Concepts

Area and the Distributive Property Find the area of each of the following rectangles in two ways: first by multiplying length and width, and then by adding the areas of the two smaller rectangles together.

133.

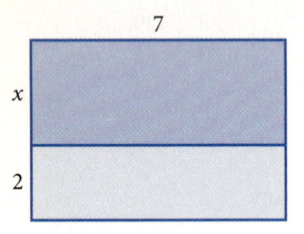

134.

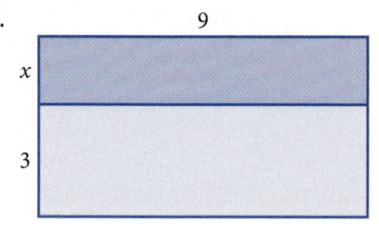

135.

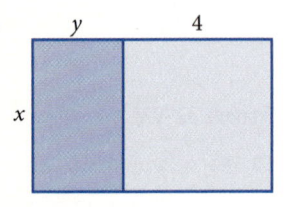

136.

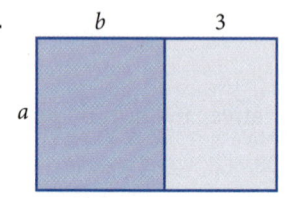

137. **Rhind Papyrus** In approximately 1650 b.c., a mathematical document called the *Rhind Papyrus* (part of which is shown here) was written in ancient Egypt. An "exercise" in this document asked the reader to find "a quantity such that when it is added to one fourth of itself results in 15." Verify this quantity must be 12.

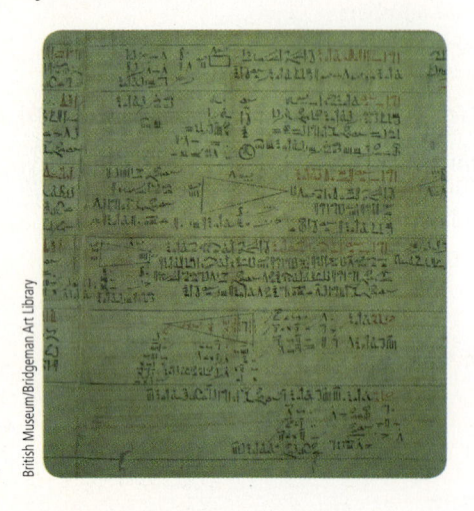

138. **Clock Arithmetic** In a normal clock with 12 hours on its face, 12 is the additive identity because adding 12 hours to any time on the clock will not change the hands of the clock. Also, if we think of the hour hand of a clock, the problem $10 + 4$ can be taken to mean: The hour hand is pointing at 10; if we add 4 more hours, it will be pointing at what number? Reasoning this way, we see that in clock arithmetic $10 + 4 = 2$ and $9 + 6 = 3$. Find the following in clock arithmetic:

$10(\text{o'clock}) + 4(\text{hours}) = 2(\text{o'clock})$

a. $10 + 5$
b. $10 + 6$
c. $10 + 1$
d. $10 + 12$
e. $x + 12$

Arithmetic with Real Numbers

1.2

Objectives
- **A** Add and subtract real numbers.
- **B** Multiply and divide real numbers.
- **C** Apply the rule for order of operations.
- **D** Simplify algebraic expressions.
- **E** Divide fractions.
- **F** Find the value of an expression.

Introduction

The temperature at the airport is 70°F. A plane takes off and reaches its cruising altitude of 28,000 feet, where the temperature is −40°F. Find the difference in the temperatures at takeoff and at cruising altitude.

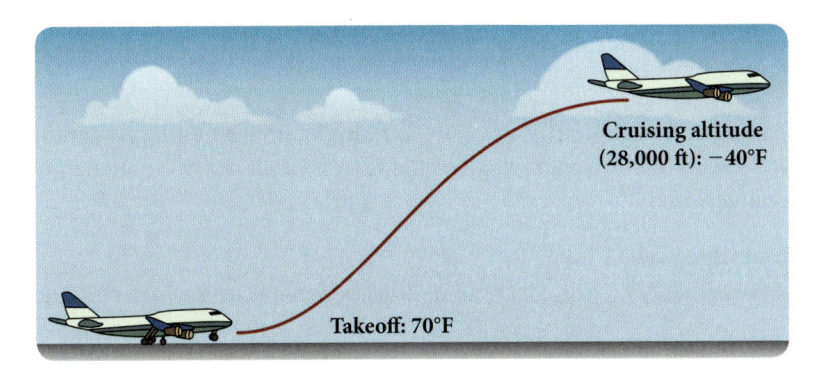

Cruising altitude (28,000 ft): −40°F

Takeoff: 70°F

We know intuitively that the difference in temperature is 110°F. If we write this problem using symbols, we have

$$70 - (-40) = 110$$

In this section we review the rules for arithmetic with real numbers, which will include problems such as this one.

A Adding Real Numbers

The purpose of this section is to review the rules for arithmetic with real numbers and the justification for those rules. We can justify the rules for addition of real numbers geometrically by use of the real number line. Consider the sum of −5 and 3:

$$-5 + 3$$

We can interpret this expression as meaning "start at the origin and move 5 units in the negative direction and then 3 units in the positive direction." With the aid of a number line we can visualize the process.

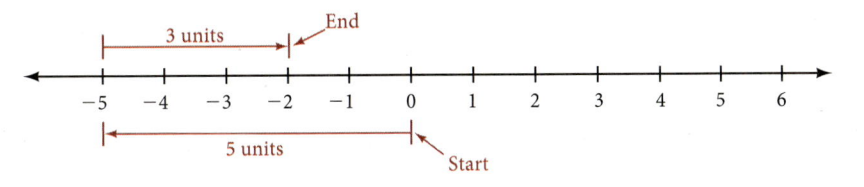

Because the process ends at −2, we say the sum of −5 and 3 is −2:

$$-5 + 3 = -2$$

We can use the real number line in this way to add any combination of positive and negative numbers.

Note We are showing addition of real numbers on the number line to justify the rule we will write for addition of positive and negative numbers. You may want to skip ahead and read the rule on the next page first, and then come back and read through this discussion again. The discussion here is the "why" behind the rule.

Chapter 1 Basic Concepts and Properties

The sum of −4 and −2, −4 + (−2), can be interpreted as starting at the origin, moving 4 units in the negative direction, and then 2 more units in the negative direction:

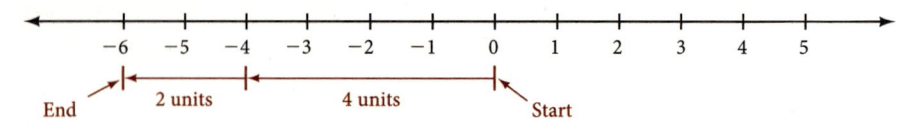

Because the process ends at −6, we say the sum of −4 and −2 is −6.

$$-4 + (-2) = -6$$

We can eliminate actually drawing a number line by simply visualizing it mentally. The following example gives the results of all possible sums of positive and negative 5 and 7.

EXAMPLE 1 Add all combinations of positive and negative 5 and 7.

SOLUTION
$$5 + 7 = 12$$
$$-5 + 7 = 2$$
$$5 + (-7) = -2$$
$$-5 + (-7) = -12$$

Looking closely at the relationships in Example 1 (and trying other similar examples if necessary), we can arrive at the following rule for adding two real numbers.

To Add Two Real Numbers
With the *same* sign:
Step 1: Add their absolute values.
Step 2: Attach their common sign. If both numbers are positive, their sum is positive; if both numbers are negative, their sum is negative.

With *opposite* signs:
Step 1: Subtract the smaller absolute value from the larger.
Step 2: Attach the sign of the number whose absolute value is larger.

Subtracting Real Numbers

In order to have as few rules as possible, we will not attempt to list new rules for the *difference* of two real numbers. We will instead define it in terms of addition and apply the rule for addition.

Definition
If a and b are any two real numbers, then the **difference** of a and b, written $a - b$, is given by
$$a - b = a + (-b)$$
To subtract b, add the opposite of b.

We define the process of subtracting b from a as being equivalent to adding the opposite of b to a. In short, we say, "subtraction is addition of the opposite."

PRACTICE PROBLEMS
1. Add.
   $3 + 5 =$
   $-3 + 5 =$
   $3 + (-5) =$
   $-3 + (-5) =$

Note This rule is the most important rule we have had so far. It is very important that you use it exactly the way it is written. Your goal is to become fast and accurate at adding positive and negative numbers. When you have finished reading this section and working the problems in the problem set, you should have attained that goal.

Answer
1. 8, 2, −2, −8

1.2 Arithmetic with Real Numbers

EXAMPLE 2 Subtract 5 − 3.

SOLUTION
$$5 - 3 = 5 + (-3) \quad \text{Subtracting 3 is equivalent to adding } -3$$
$$= 2$$

EXAMPLE 3 Subtract −7 − 6.

SOLUTION
$$-7 - 6 = -7 + (-6) \quad \text{Subtracting 6 is equivalent to adding } -6$$
$$= -13$$

EXAMPLE 4 Subtract 9 − (−2).

SOLUTION
$$9 - (-2) = 9 + 2 \quad \text{Subtracting } -2 \text{ is equivalent to adding 2}$$
$$= 11$$

EXAMPLE 5 Subtract −6 − (−5).

SOLUTION
$$-6 - (-5) = -6 + 5 \quad \text{Subtracting } -5 \text{ is equivalent to adding 5}$$
$$= -1$$

EXAMPLE 6 Subtract −9 −(−3).

SOLUTION Because subtraction is not commutative, we must be sure to write the numbers in the correct order. Because we are subtracting −3, the problem looks like this when translated into symbols:

$$-9 - (-3) = -9 + 3 \quad \text{Change to addition of the opposite}$$
$$= -6 \quad \text{Add}$$

EXAMPLE 7 Add −4 to the difference of −2 and 5.

SOLUTION The difference of −2 and 5 is written −2 − 5. Adding −4 to that difference gives us

$$(-2 - 5) + (-4) = -7 + (-4) \quad \text{Simplify inside parentheses}$$
$$= -11 \quad \text{Add}$$

B Multiplying Real Numbers

Multiplication with whole numbers is simply a shorthand way of writing repeated addition.

For example, 3(−2) can be evaluated as follows:

$$3(-2) = -2 + (-2) + (-2) = -6$$

We can evaluate the product −3(2) in a similar manner if we first apply the commutative property of multiplication:

$$-3(2) = 2(-3) = -3 + (-3) = -6$$

From these results it seems reasonable so say that the product of a positive and a negative is a negative number.

---

2. Subtract 7 − 4.

3. Subtract −6 − 3.

4. Subtract 8 − (−2).

5. Subtract −4 − (−6).

6. Subtract −5 from −3.

7. Add −8 to the difference of −7 and 4.

Note This discussion is to show why the rule for multiplication of real numbers is written the way it is. Even if you already know how to multiply positive and negative numbers, it is a good idea to review the "why" that is behind it all.

Answers
2. 3
3. −9
4. 10
5. 2
6. 2
7. −19

Chapter 1 Basic Concepts and Properties

The last case we must consider is the product of two negative numbers, such as −3(−2). To evaluate this product we will look at the expression −3[2 + (−2)] in two different ways. First, since 2 + (−2) = 0, we know the expression −3[2 + (−2)] is equal to 0. On the other hand, we can apply the distributive property to get

$$-3[2 + (-2)] = -3(2) + (-3)(-2) = -6 + ?$$

Because we know the expression is equal to 0, it must be true that our ? is 6 because 6 is the only number we can add to −6 to get 0. Therefore, we have

$$-3(-2) = 6$$

Here is a summary of what we have so far:

ORIGINAL NUMBERS HAVE		THE ANSWER IS
The same sign	3(2) = 6	Positive
Different signs	3(−2) = −6	Negative
Different signs	−3(2) = −6	Negative
The same sign	−3(−2) = 6	Positive

To Multiply Two Real Numbers
Step 1: Multiply their absolute values.
Step 2: If the two numbers have the *same* sign, the product is positive. If the two numbers have *opposite* signs, the product is negative.

EXAMPLE 8 Multiply all combinations of positive and negative 7 and 3.

SOLUTION
$7(3) = 21$
$7(-3) = -21$
$-7(3) = -21$
$-7(-3) = 21$

Dividing Real Numbers

Definition
If a and b are any two real numbers, where $b \neq 0$, then the **quotient** of a and b, written $\frac{a}{b}$ is given by

$$\frac{a}{b} = a \cdot \left(\frac{1}{b}\right)$$

Dividing a by b is equivalent to multiplying a by the reciprocal of b. In short, we say, "division is multiplication by the reciprocal."

Because division is defined in terms of multiplication, the same rules hold for assigning the correct sign to a quotient as held for assigning the correct sign to a product; that is, *the quotient of two numbers with like signs is positive, while the quotient of two numbers with unlike signs is negative.*

---

8. Multiply.
4(5)
4(−5)
−4(5)
−4(−5)

Answer
8. 20, −20, −20, 20

1.2 Arithmetic with Real Numbers

EXAMPLE 9 Divide each of the following.

a. $\dfrac{6}{3}$ b. $\dfrac{6}{-3}$ c. $\dfrac{-6}{3}$ d. $\dfrac{-6}{-3}$

SOLUTION

a. $\dfrac{6}{3} = 6 \cdot \left(\dfrac{1}{3}\right) = 2$

b. $\dfrac{6}{-3} = 6 \cdot \left(-\dfrac{1}{3}\right) = -2$

c. $\dfrac{-6}{3} = -6 \cdot \left(\dfrac{1}{3}\right) = -2$

d. $\dfrac{-6}{-3} = -6 \cdot \left(-\dfrac{1}{3}\right) = 2$

Notice these examples indicate that if a and b are positive real numbers then

$$\dfrac{-a}{b} = \dfrac{a}{-b} = -\dfrac{a}{b}$$

and

$$\dfrac{-a}{-b} = \dfrac{a}{b}$$

The second step in the preceding examples is written only to show that each quotient can be written as a product. It is not actually necessary to show this step when working problems.

C Order of Operations

When evaluating a mathematical expression, we will perform the operations in the following order, beginning with the expression in the innermost parentheses or brackets and working our way out.

1. Simplify all numbers with exponents, working from left to right if more than one of these numbers is present.
2. Then, do all multiplications and divisions left to right.
3. Finally, perform all additions and subtractions left to right.

In the examples that follow, we find a combination of operations. In each case we use the rule for order of operations.

EXAMPLE 10 Simplify $(-2 - 3)(5 - 9)$.

SOLUTION $(-2 - 3)(5 - 9) = (-5)(-4)$ Simplify inside parentheses
$= 20$ Multiply

EXAMPLE 11 Simplify $2 - 5(7 - 4) - 6$.

SOLUTION $2 - 5(7 - 4) - 6 = 2 - 5(3) - 6$ Simplify inside parentheses
$= 2 - 15 - 6$ Then, multiply
$= -19$ Finally, subtract, left to right

EXAMPLE 12 Simplify $2(4 - 7)^3 + (-2 - 3)^2$.

SOLUTION $2(4 - 7)^3 + 3(-2 - 3)^2 = 2(-3)^3 + 3(-5)^2$ Simplify inside parentheses
$= 2(-27) + 3(25)$ Evaluate numbers with exponents
$= -54 + 75$ Multiply
$= 21$ Add

---

9. Divide.

a. $\dfrac{12}{4}$

b. $\dfrac{12}{-4}$

c. $\dfrac{-12}{4}$

d. $\dfrac{-12}{-4}$

10. Simplify $(-6 - 4)(8 - 12)$

11. $3 - 7(9 - 5) - 2$

12. $3(5 - 8)^3 - 2(-1 - 1)^2$

Answers
9. a. 3   b. −3   c. −3   d. 3
10. 40
11. −27
12. −89

13. $2\left(\dfrac{x}{2}\right)$

14. Simplify $2\left(\dfrac{x}{2} - 3\right)$.

15. Simplify $\dfrac{-6-6}{-5-3}$.

16. Simplify $\dfrac{5(-6) + 3(-2)}{4(-3) + 3}$.

Chapter 1 Basic Concepts and Properties

D Simplifying Expressions

We can combine our knowledge of the properties of multiplication with our definition of division to simplify more expressions involving fractions. Here are two examples:

EXAMPLE 13 Simplify $6\left(\dfrac{t}{3}\right)$

SOLUTION
$6\left(\dfrac{t}{3}\right) = 6\left(\dfrac{1}{3}t\right)$ Dividing by 3 is the same as multiplying by $\dfrac{1}{3}$

$= \left(6 \cdot \dfrac{1}{3}\right)t$ Associative property

$= 2t$ Multiplication

EXAMPLE 14 Simplify $3\left(\dfrac{t}{3} - 2\right)$

SOLUTION $3\left(\dfrac{t}{3} - 2\right) = 3 \cdot \dfrac{t}{3} - 3 \cdot 2$ Distributive property

$= t - 6$ Multiplication

Our next examples involve more complicated fractions. The fraction bar works like parentheses to separate the numerator from the denominator. Although we don't write expressions this way, here is one way to think of the fraction bar:

$$\dfrac{-8 - 8}{-5 - 3} = (-8 - 8) \div (-5 - 3)$$

As you can see, if we apply the rule for order of operations to the expression on the right, we would work inside each set of parentheses first, then divide. Applying this to the expression on the left, we work on the numerator and denominator separately, then we divide or reduce the resulting fraction to lowest terms.

EXAMPLE 15 Simplify $\dfrac{-8-8}{-5-3}$ as much as possible.

SOLUTION $\dfrac{-8-8}{-5-3} = \dfrac{-16}{-8}$ Simplify numerator and denominator separately

$= 2$ Divide

EXAMPLE 16 Simplify $\dfrac{-5(-4) + 2(-3)}{2(-1) - 5}$ as much as possible.

SOLUTION $\dfrac{-5(-4) + 2(-3)}{2(-1) - 5} = \dfrac{20 - 6}{-2 - 5}$

$= \dfrac{14}{-7}$

$= -2$

Answers
13. x
14. $x - 6$
15. $\dfrac{3}{2}$
16. 4

1.2 Arithmetic with Real Numbers

EXAMPLE 17 Simplify $\dfrac{2^3 + 3^3}{2^2 - 3^2}$ as much as possible.

SOLUTION
$$\dfrac{2^3 + 3^3}{2^2 - 3^2} = \dfrac{8 + 27}{4 - 9}$$
$$= \dfrac{35}{-5}$$
$$= -7$$

Remember, since subtraction is defined in terms of addition, we can restate the distributive property in terms of subtraction; that is, if a, b, and c are real numbers, then $a(b - c) = ab - ac$.

EXAMPLE 18 Simplify $3(2y - 1) + y$.

SOLUTION We begin by multiplying the 3 and $2y - 1$. Then, we combine similar terms:

$3(2y - 1) + y = 6y - 3 + y$ Distributive property
$\qquad\qquad\qquad = 7y - 3$ Combine similar terms

EXAMPLE 19 Simplify $8 - 3(4x - 2) + 5x$.

SOLUTION First we distribute the -3 across the $4x - 2$.

$8 - 3(4x - 2) + 5x = 8 - 12x + 6 + 5x$
$\qquad\qquad\qquad\qquad = -7x + 14$

EXAMPLE 20 Simplify $5(2a + 3) - (6a - 4)$.

SOLUTION We begin by applying the distributive property to remove the parentheses. The expression $-(6a - 4)$ can be thought of as $-1(6a - 4)$. Thinking of it in this way allows us to apply the distributive property.

$-1(6a - 4) = -1(6a) - (-1)(4) = -6a + 4$

Here is the complete problem:

$5(2a + 3) - (6a - 4) = 10a + 15 - 6a + 4$ Distributive property
$\qquad\qquad\qquad\qquad\quad = 4a + 19$ Combine similar terms

E Dividing Fractions

Next we review division with fractions and division with the number 0.

EXAMPLE 21 Divide and reduce to lowest terms: $\dfrac{3}{4} \div \dfrac{6}{11}$

SOLUTION
$\dfrac{3}{4} \div \dfrac{6}{11} = \dfrac{3}{4} \cdot \dfrac{11}{6}$ Definition of division
$\qquad\qquad = \dfrac{33}{24}$ Multiply numerators, multiply denominators
$\qquad\qquad = \dfrac{11}{8}$ Divide numerator and denominator by 3

---

17. Simplify $\dfrac{3^3 - 4^3}{3^2 + 4^2}$.

18. Simplify $2(5y - 1) - y$.

19. Simplify $6 - 2(5x + 1) + 4x$.

20. Simplify $4(3a + 1) - (7a - 6)$.

Divide and reduce to lowest terms.

21. $\dfrac{3}{5} \div \dfrac{6}{7}$

Answers
17. $-\dfrac{37}{25}$
18. $9y - 2$
19. $-6x + 4$
20. $5a + 10$
21. $\dfrac{7}{10}$

22. $12 \div \frac{3}{4}$

23. $-\frac{5}{6} \div 10$

Chapter 1 Basic Concepts and Properties

EXAMPLE 22 Divide and reduce to lowest terms: $10 \div \frac{5}{6}$

SOLUTION
$10 \div \frac{5}{6} = \frac{10}{1} \cdot \frac{6}{5}$ Definition of division

$= \frac{60}{5}$ Multiply numerators, multiply denominators

$= 12$ Divide

EXAMPLE 23 Divide and reduce to lowest terms: $-\frac{3}{8} \div 6$

SOLUTION
$-\frac{3}{8} \div 6 = -\frac{3}{8} \cdot \frac{1}{6}$ Definition of division

$= -\frac{3}{48}$ Multiply numerators, multiply denominators

$= -\frac{1}{16}$ Divide numerator and denominator by 3

Division with the Number 0

For every division problem an associated multiplication problem involving the same numbers exists. For example, the following two problems say the same thing about the numbers 2, 3, and 6:

Division	Multiplication
$\frac{6}{3} = 2$	$6 = 2(3)$

We can use this relationship between division and multiplication to clarify division involving the number 0.

First of all, dividing 0 by a number other than 0 is allowed and always results in 0. To see this, consider dividing 0 by 5. We know the answer is 0 because of the relationship between multiplication and division. This is how we write it:

$\frac{0}{5} = 0$ because $0 = 0(5)$

On the other hand, dividing a nonzero number by 0 is not allowed in the real numbers. Suppose we were attempting to divide 5 by 0. We don't know whether there is an answer to this problem, but if there is, let's say the answer is a number that we can represent with the letter n. If 5 divided by 0 is a number n, then

$\frac{5}{0} = n$ and $5 = n(0)$

But this is impossible because no matter what number n is, when we multiply it by 0 the answer must be 0. It can never be 5. In algebra, we say expressions like $\frac{5}{0}$ are undefined because there is no answer to them; that is, division by 0 is not allowed in the real numbers.

Answers
22. 16
23. $-\frac{1}{12}$

F Finding the Value of an Algebraic Expression

As we mentioned earlier in this chapter, an algebraic expression is a combination of numbers, variables, and operation symbols. Each of the following is an algebraic expression

$$7a \qquad x^2 - y^2 \qquad 2(3t - 4) \qquad \frac{2x - 5}{6}$$

An expression such as $2(3t - 4)$ will take on different values depending on what number we substitute for t. For example, if we substitute -8 for t, then the expression $2(3t - 4)$ becomes $2[3(-8) - 4)]$, which simplifies to -56. If we apply the distributive property to $2(3t - 4)$, we have

$$2(3t - 4) = 6t - 8$$

Substituting -8 for t in the simplified expression gives us $6(-8) - 8 = -56$, which is the same result we obtained previously. As you would expect, substituting the same number into an expression, and any simplified form of that expression, will yield the same result.

EXAMPLE 24 Evaluate the expressions $(a + 4)^2$, $a^2 + 16$, and $a^2 + 8a + 16$ when a is -2, 0, and 3.

SOLUTION Organizing our work with a table, we have

a	$(a + 4)^2$	$a^2 + 16$	$a^2 + 8a + 16$
-2	$(-2 + 4)^2 = 4$	$(-2)^2 + 16 = 20$	$(-2)^2 + 8(-2) + 16 = 4$
0	$(0 + 4)^2 = 16$	$0^2 + 16 = 16$	$0^2 + 8(0) + 16 = 16$
3	$(3 + 4)^2 = 49$	$3^2 + 16 = 25$	$3^2 + 8(3) + 16 = 49$

When we study polynomials later in the book, you will see that the expressions $(a + 4)^2$ and $a^2 + 8a + 16$ are equivalent, and that neither one is equivalent to $a^2 + 16$.

STUDY SKILLS
Don't Let Your Intuition Fool You
As you become more experienced and more successful in mathematics, you will be able to trust your mathematical intuition. For now, though, it can get in the way of success. For example, if you ask a beginning algebra student to "subtract 3 from -5" many will answer -2 or 2. Both answers are incorrect, even though they may seem intuitively true.

---

24. Evaluate each expression when x is -1.
    a. $7(4x - 6)$
    b. $28x - 42$
    c. $28x - 6$

Answer
24. a. -70 b. -70 c. -34

Getting Ready for Class

After reading through the preceding section, respond in your own words and in complete sentences.

1. For each of the following expressions, give an example of an everyday situation that is modeled by the expression.
   a. $35 − $12 = $23
   b. −$35 − $12 = −$47

2. For each of the following expressions, give an example of an everyday situation that is modeled by the expression.
   a. 3(−$25) = −$75
   b. (−$100) ÷ 5 = −$20

3. Why is division by 0 not allowed?

4. Why isn't the statement "two negatives make a positive" true?

Problem Set 1.2

A Find each of the following sums. [Example 1]

1. $6 + (-2)$
2. $11 + (-5)$
3. $-6 + 2$
4. $-11 + 5$

A Find each of the following differences. [Examples 2–7]

5. $-7 - 3$
6. $-6 - 9$
7. $-7 - (-3)$
8. $-6 - (-9)$

9. $\frac{3}{4} - \left(-\frac{5}{6}\right)$
10. $\frac{2}{3} - \left(-\frac{7}{5}\right)$
11. $\frac{11}{42} - \frac{17}{30}$
12. $\frac{13}{70} - \frac{19}{42}$

13. Subtract 5 from -3.
14. Subtract -3 from 5.

15. Find the difference of -4 and 8.
16. Find the difference of 8 and -4.

17. Subtract $4x$ from $-3x$.
18. Subtract $-5x$ from $7x$.

19. What number do you subtract from 5 to get -8?
20. What number do you subtract from -3 to get 9?

21. Add -7 to the difference of 2 and 9.
22. Add -3 to the difference of 9 and 2.

23. Subtract $3a$ from the sum of $8a$ and a.
24. Subtract $-3a$ from the sum of $3a$ and $5a$.

Chapter 1 Basic Concepts and Properties

B Find the following products. [Example 8]

25. $3(-5)$
26. $-3(5)$
27. $-3(-5)$
28. $4(-6)$

29. $2(-3)(4)$
30. $-2(3)(-4)$
31. $-2(5x)$
32. $-5(4x)$

33. $-\dfrac{1}{3}(-3x)$
34. $-\dfrac{1}{6}(-6x)$
35. $-\dfrac{2}{3}\left(-\dfrac{3}{2}y\right)$
36. $-\dfrac{2}{5}\left(-\dfrac{5}{2}y\right)$

37. $-2(4x - 3)$
38. $-2(-5t + 6)$
39. $-\dfrac{1}{2}(6a - 8)$
40. $-\dfrac{1}{3}(6a - 9)$

C Simplify each expression as much as possible. [Examples 10–12]

41. $3(-4) - 2$
42. $-3(-4) - 2$
43. $4(-3) - 6(-5)$

44. $-6(-3) - 5(-7)$
45. $2 - 5(-4) - 6$
46. $3 - 8(-1) - 7$

47. $4 - 3(7 - 1) - 5$
48. $8 - 5(6 - 3) - 7$
49. $2(-3)^2 - 4(-2)^3$

50. $5(-2)^2 - 2(-3)^3$
51. $(2 - 8)^2 - (3 - 7)^2$
52. $(5 - 8)^2 - (4 - 8)^2$

53. $7(3 - 5)^3 - 2(4 - 7)^3$
54. $3(-7 + 9)^3 - 5(-2 + 4)^3$
55. $-3(2 - 9) - 4(6 - 1)$

56. $-5(5 - 6) - 7(2 - 8)$
57. $2 - 4[3 - 5(-1)]$
58. $6 - 5[2 - 4(-8)]$

59. $(8 - 7)[4 - 7(-2)]$
60. $(6 - 9)[15 - 3(-4)]$
61. $-3 + 4[6 - 8(-3 - 5)]$

62. $-2 + 7[2 - 6(-3 - 4)]$
63. $5 - 6[-3(2 - 9) - 4(8 - 6)]$
64. $9 - 4[-2(4 - 8) - 5(3 - 1)]$

65. $1(-2) - 2(-16) + 1(9)$
66. $6(1) - 1(-5) + 1(2)$
67. $1(1) - 3(-2) + (-2)(-2)$

1.2 Problem Set

C **D** The problems below are problems you will see later in the book. Simplify. [Examples 10–12, 15]

68. $-2(-14) + 3(-4) - 1(-10)$

69. $-4(0)(-2) - (-1)(1)(1) - 1(2)(3)$

70. $1(0)(1) + 3(1)(4) + (-2)(2)(-1)$

71. $1[0 - (-1)] - 3(2 - 4) + (-2)(-2 - 0)$

72. $-3(-1 - 1) + 4(-2 + 2) - 5[2 - (-2)]$

73. $3(-2)^2 + 2(-2) - 1$

74. $4(-1)^2 + 3(-1) - 2$

75. $2(-2)^3 - 3(-2)^2 + 4(-2) - 8$

76. $5 \cdot 2^3 - 3 \cdot 2^2 + 4 \cdot 2 - 5$

77. $\dfrac{0 - 4}{0 - 2}$

78. $\dfrac{0 + 6}{0 - 3}$

79. $\dfrac{-4 - 4}{-4 - 2}$

80. $\dfrac{6 + 6}{6 - 3}$

81. $\dfrac{-6 + 6}{-6 - 3}$

82. $\dfrac{4 - 4}{4 - 2}$

83. $\dfrac{2 - 4}{2 - 2}$

84. $\dfrac{3 + 6}{3 - 3}$

85. $\dfrac{3 - (-1)}{-3 - 3}$

86. $\dfrac{-1 - 3}{3 - (-3)}$

D The problems below are problems you will see later in the book. Apply the distributive property, then simplify if possible. [Examples 13–14]

87. $-1(5 - x)$

88. $-1(a - b)$

89. $-1(7 - x)$

90. $-1(6 - y)$

91. $-3(2x - 3y)$

92. $-1(x - 2z)$

93. $6\left(\dfrac{x}{2} - 3\right)$

94. 94. $6\left(\dfrac{x}{3} + 1\right)$

95. $12\left(\dfrac{a}{4} + \dfrac{1}{2}\right)$

96. $15\left(\dfrac{a}{3} + 2\right)$

97. $15\left(\dfrac{x}{5} + 4\right)$

98. $10\left(\dfrac{x}{2} - 9\right)$

99. $12\left(\dfrac{y}{2} + \dfrac{y}{4} + \dfrac{y}{6}\right)$

100. $12\left(\dfrac{y}{3} - \dfrac{y}{6} + \dfrac{y}{2}\right)$

D Simplify each expression. [Examples 18–20]

101. $3(5x + 4) - x$

102. $4(7x + 3) - x$

103. $6 - 7(3 - m)$

104. $3 - 5(5 - m)$

105. $7 - 2(3x - 1) + 4x$

106. $8 - 5(2x - 3) + 4x$

107. $5(3y + 1) - (8y - 5)$

108. $4(6y + 3) - (6y - 6)$

109. $4(2 - 6x) - (3 - 4x)$

110. $7(1 - 2x) - (4 - 10x)$

111. $10 - 4(2x + 1) - (3x - 4)$

112. $7 - 2(3x + 5) - (2x - 3)$

B **E** Use the definition of division to write each division problem as a multiplication problem, then simplify. [Examples 9, 21–23]

113. $\dfrac{8}{-4}$

114. $\dfrac{-8}{4}$

115. $\dfrac{4}{0}$

116. $\dfrac{-7}{0}$

117. $\dfrac{0}{-3}$

118. $\dfrac{0}{5}$

119. $-\dfrac{3}{4} \div \dfrac{9}{8}$

120. $-\dfrac{2}{3} \div \dfrac{4}{9}$

121. $-8 \div \left(-\dfrac{1}{4}\right)$

122. $-12 \div \left(-\dfrac{2}{3}\right)$

123. $-40 \div \left(-\dfrac{5}{8}\right)$

124. $-30 \div \left(-\dfrac{5}{6}\right)$

125. $\dfrac{4}{9} \div (-8)$

126. $\dfrac{3}{7} \div (-6)$

D Simplify as much as possible. [Examples 16–17]

127. $\dfrac{3(-1) - 4(-2)}{8 - 5}$

128. $\dfrac{6(-4) - 5(-2)}{7 - 6}$

129. $\dfrac{4(-3) - 5(-2)}{8 - 6}$

130. $-9 - 5\left[\dfrac{11(-1) - 9}{4(-3) + 2(5)}\right]$

131. $6 - (-3)\left[\dfrac{2 - 4(3 - 8)}{1 - 5(1 - 3)}\right]$

132. $8 - (-7)\left[\dfrac{6 - 1(6 - 10)}{4 - 3(5 - 7)}\right]$

F Complete each of the following tables. [Example 24]

133.

a	b	SUM $a + b$	DIFFERENCE $a - b$	PRODUCT ab	QUOTIENT $\frac{a}{b}$
3	12				
−3	12				
3	−12				
−3	−12				

134.

a	b	SUM $a + b$	DIFFERENCE $a - b$	PRODUCT ab	QUOTIENT $\frac{a}{b}$
8	2				
−8	2				
8	−2				
−8	−2				

135.

x	$3(5x - 2)$	$15x - 6$	$15x - 2$
−2			
−1			
0			
1			
2			

136.

x	$(x + 1)^2$	$x^2 + 1$	$x^2 + 2x + 1$
−2			
−1			
0			
1			
2			

137. Find the value of $-\dfrac{b}{2a}$ when

a. $a = 3, b = -6$
b. $a = -2, b = 6$
c. $a = -1, b = -2$
d. $a = -0.1, b = 27$

138. Find the value of $b^2 - 4ac$ when

a. $a = 3, b = -2$, and $c = 4$
b. $a = 1, b = -3$, and $c = -28$
c. $a = 1, b = -6$, and $c = 9$
d. $a = 0.1, b = -27$, and $c = 1{,}700$

Use a calculator to simplify each expression. If rounding is necessary, round your answers to the nearest ten thousandth (4 places past the decimal point). You will see these problems later in the book.

139. $\dfrac{1.3802}{0.9031}$

140. $\dfrac{1.0792}{0.6990}$

141. $\dfrac{1}{2}(-0.1587)$

142. $\dfrac{1}{2}(-0.7948)$

143. $\dfrac{1}{2}\left(\dfrac{1.2}{1.4} - 1\right)$

144. $\dfrac{1}{2}\left(\dfrac{1.3}{1.1} - 1\right)$

145. $\dfrac{(6.8)(3.9)}{7.8}$

146. $\dfrac{(2.4)(1.8)}{1.2}$

147. $\dfrac{0.0005(200)}{(0.25)^2}$

148. $\dfrac{0.0006(400)}{(0.25)^2}$

149. $-500 + 27(100) - 0.1(100)^2$

150. $-500 + 27(170) - 0.1(170)^2$

151. $-0.05(130)^2 + 9.5(130) - 200$

152. $-0.04(130)^2 + 8.5(130) - 210$

Applying the Concepts

153. Time Zones The continental United States is divided into four time zones. When it is 4:00 in the Pacific zone, it is 5:00 in the Mountain zone, 6:00 in the Central zone, and 7:00 in the Eastern zone.

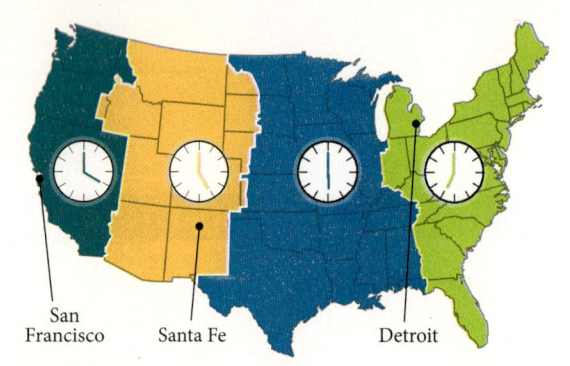

You board a plane at 6:55 p.m. in California (Pacific zone) and take 2 hours and 15 minutes to arrive in Santa Fe, New Mexico (Mountain zone) and another 3 hours and 20 minutes to arrive at your final destination, Detroit, Michigan (Eastern zone). At what local times did you arrive at Santa Fe and Detroit?

154. Oceans and Mountains The deepest ocean depth is 35,840 feet, found in the Pacific Ocean's Mariana Trench. The tallest mountain is Mount Everest, with a height of 29,028 feet. What is the difference between the highest point on earth and the lowest point on earth?

Scientific Notation 1.3

Objectives
- **A** Write numbers in scientific notation.
- **B** Convert numbers written in scientific notation to standard form.
- **C** Multiply and divide numbers written in scientific notation.

There are many disciplines that deal with very large numbers and others that deal with very small numbers. For example, in astronomy, distances commonly are given in light-years. A light-year is the distance that light will travel in one year. It is approximately

$$5,880,000,000,000 \text{ miles}$$

It can be difficult to perform calculations with numbers in this form because of the number of zeros present. Scientific notation provides a way of writing very large, or very small, numbers in a more manageable form.

A Scientific Notation

> **Definition**
> A number is in scientific notation when it is written as the product of a number between 1 and 10 and an integer power of 10. A number written in scientific notation has the form
>
> $$n \times 10^r$$
>
> where $1 \leq n < 10$ and $r =$ an integer.

EXAMPLE 1 The speed of light is 186,000 miles per second. Write 186,000 in scientific notation.

SOLUTION To write this number in scientific notation, we rewrite it as the product of a number between 1 and 10 and a power of 10. To do so, we move the decimal point 5 places to the left so that it appears between the 1 and the 8, giving us 1.86. Then we multiply this number by 10^5. The number that results has the same value as our original number but is written in scientific notation. Here is our result:

$$186,000 = 1.86 \times 10^5$$

Both numbers have exactly the same value. The number on the left is written in *standard form,* while the number on the right is written in scientific notation.

B Converting to Standard Form

EXAMPLE 2 If your pulse rate is 60 beats per minute, then your heart will beat 8.64×10^4 times each day. Write 8.64×10^4 in standard form.

SOLUTION Because 10^4 is 10,000, we can think of this as simply a multiplication problem. That is,

$$8.64 \times 10^4 = 8.64 \times 10,000 = 86,400$$

Looking over our result, we can think of the exponent 4 as indicating the number of places we need to move the decimal point to write our number in standard form. Because our exponent is positive 4, we move the decimal point from its original position, between the 8 and the 6, four places to the right. If we need to add any zeros on the right we do so. The result is the standard form of our number, 86,400.

PRACTICE PROBLEMS
1. Write 27,500 in scientific notation.

2. Write 7.89×10^5 in standard form.

Answers
1. 2.75×10^4
2. 789,000

Chapter 1 Basic Concepts and Properties

Next, we turn our attention to writing small numbers in scientific notation. To do so, we use negative exponents. For example, the number 0.00075, when written in scientific notation, is equivalent to 7.5×10^{-4}. Here's why:

$$7.5 \times 10^{-4} = 7.5 \times \frac{1}{10^4} = 7.5 \times \frac{1}{10,000} = \frac{7.5}{10,000} = 0.00075$$

The table below lists some other numbers both in scientific notation and in standard form.

EXAMPLE 3   Each pair of numbers in the table below is equal.

Standard Form		Scientific Notation
376,000	=	3.76×10^5
49,500	=	4.95×10^4
3,200	=	3.2×10^3
591	=	5.91×10^2
46	=	4.6×10^1
8	=	8×10^0
0.47	=	4.7×10^{-1}
0.093	=	9.3×10^{-2}
0.00688	=	6.88×10^{-3}
0.0002	=	2×10^{-4}
0.000098	=	9.8×10^{-5}

As we read across the table, for each pair of numbers, notice how the decimal point in the number on the right is placed so that the number containing the decimal point is always a number between 1 and 10. Correspondingly, the exponent on 10 keeps track of how many places the decimal point was moved in converting from standard form to scientific notation. In general, when the exponent is positive, we are working with a large number. On the other hand, when the exponent is negative, we are working with a small number. (By small number, we mean a number that is less than 1, but larger than 0.)

We end this section with a diagram that shows two numbers, one large and one small, that are converted to scientific notation.

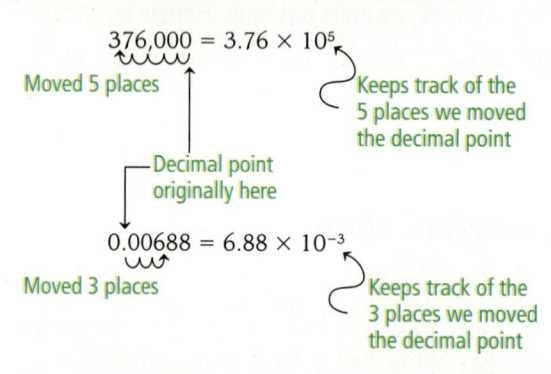

$376,000 = 3.76 \times 10^5$

Moved 5 places — Keeps track of the 5 places we moved the decimal point

Decimal point originally here

$0.00688 = 6.88 \times 10^{-3}$

Moved 3 places — Keeps track of the 3 places we moved the decimal point

C Multiplication and Division with Numbers Written in Scientific Notation

In this section, we extend our work with scientific notation to include multiplication and division with numbers written in scientific notation. To work the problems in this section, we use the material presented in the previous two sections, along with the commutative and associative properties of multiplication and the rule for multiplication with fractions. Here is our first example.

---

3. Fill in the missing numbers in the table below:

	Standard Form		Scientific Notation
a.	24,500	=	
b.		=	5.6×10^5
c.	0.000789	=	
d.		=	4.8×10^{-3}

Answer
3. a. 2.45×10^4 b. 560,000
   c. 7.89×10^{-4} d. 0.0048

1.3 Scientific Notation

EXAMPLE 4   Multiply: $(3.5 \times 10^8)(2.2 \times 10^{-5})$

SOLUTION   First we apply the commutative and associative properties to rearrange the numbers, so that the decimal numbers are grouped together and the powers of 10 are also.

$$(3.5 \times 10^8)(2.2 \times 10^{-5}) = (3.5)(2.2) \times (10^8)(10^{-5})$$

Next, we multiply the decimal numbers together and then the powers of ten. To multiply the powers of ten, we add exponents.

$$= 7.7 \times 10^{8+(-5)}$$
$$= 7.7 \times 10^3$$

4. Multiply: $(2.5 \times 10^6)(1.4 \times 10^2)$

EXAMPLE 5   Find the product of 130,000,000 and 0.000005. Write your answer in scientific notation.

SOLUTION   We begin by writing both numbers in scientific notation. Then we proceed as we did in Example 4: We group the numbers between 1 and 10 separately from the powers of 10.

$$(130{,}000{,}000)(0.000005) = (1.3 \times 10^8)(5 \times 10^{-6})$$
$$= (1.3)(5) \times (10^8)(10^{-6})$$
$$= 6.5 \times 10^2$$

5. Find the product of 2,200,000 and 0.00015.

Our next examples involve division with numbers in scientific notation.

EXAMPLE 6   Divide: $\dfrac{8 \times 10^3}{4 \times 10^{-6}}$

SOLUTION   To separate the numbers between 1 and 10 from the powers of 10, we "undo" the multiplication and write the problem as the product of two fractions. Doing so looks like this:

$$\frac{8 \times 10^3}{4 \times 10^{-6}} = \frac{8}{4} \times \frac{10^3}{10^{-6}} \quad \text{Write as two separate fractions}$$

Next, we divide 8 by 4 to obtain 2. Then we divide 10^3 by 10^{-6} by subtracting exponents.

$$= 2 \times 10^{3-(-6)} \quad \text{Divide}$$
$$= 2 \times 10^9$$

6. Divide: $\dfrac{6 \times 10^5}{2 \times 10^{-4}}$

EXAMPLE 7   Divide: $\dfrac{0.00045}{1{,}500{,}000}$

SOLUTION   To begin, write each number in scientific notation.

$$\frac{0.00045}{1{,}500{,}000} = \frac{4.5 \times 10^{-4}}{1.5 \times 10^6} \quad \text{Write numbers in scientific notation}$$

Next, as in the previous example, we write the problem as two separate fractions in order to group the numbers between 1 and 10 together, as well as the powers of 10.

$$= \frac{4.5}{1.5} \times \frac{10^{-4}}{10^6} \quad \text{Write as two separate fractions}$$
$$= 3 \times 10^{-4-6} \quad \text{Divide}$$
$$= 3 \times 10^{-10} \quad -4 - 6 = -4 + (-6) = -10$$

7. Divide: $\dfrac{0.0038}{19{,}000{,}000}$

Answers
4. 3.5×10^8
5. 3.3×10^2
6. 3.0×10^9
7. 2.0×10^{-10}

8. Simplify: $\dfrac{(6.8 \times 10^{-4})(3.9 \times 10^2)}{7.8 \times 10^{-6}}$

9. Simplify: $\dfrac{(0.000035)(45,000)}{0.000075}$

Chapter 1 Basic Concepts and Properties

EXAMPLE 8 Simplify: $\dfrac{(6.8 \times 10^5)(3.9 \times 10^{-7})}{7.8 \times 10^{-4}}$

SOLUTION We group the numbers between 1 and 10 separately from the powers of 10:

$$\dfrac{(6.8 \times 10^5)(3.9 \times 10^{-7})}{7.8 \times 10^{-4}} = \dfrac{(6.8)(3.9)}{7.8} \times \dfrac{(10^5)(10^{-7})}{10^{-4}}$$

$$= 3.4 \times 10^{5+(-7)-(-4)}$$
$$= 3.4 \times 10^2$$

EXAMPLE 9 Simplify: $\dfrac{(35,000)(0.0045)}{7,500,000}$

SOLUTION We write each number in scientific notation, and then we proceed as we have in the examples above.

$$\dfrac{(35,000)(0.0045)}{7,500,000} = \dfrac{(3.5 \times 10^4)(4.5 \times 10^{-3})}{7.5 \times 10^6}$$

$$= \dfrac{(3.5)(4.5)}{7.5} \times \dfrac{(10^4)(10^{-3})}{10^6}$$

$$= 2.1 \times 10^{4+(-3)-6}$$
$$= 2.1 \times 10^{-5}$$

Getting Ready for Class

After reading through the preceding section, respond in your own words and in complete sentences.

1. What is scientific notation?
2. What types of numbers are frequently written with scientific notation?
3. The following problem is not complete. Why not?

    $(6.0 \times 10^3)(5.2 \times 10^4) = 31.2 \times 10^7$

4. In writing the weight of a paper clip, in kilograms, would we use a positive power of ten or a negative power of ten?

Answers
8. 3.4×10^4
9. 2.1×10^4

Problem Set 1.3

A Write each number in scientific notation. [Examples 1, 3]

1. 425,000
2. 635,000
3. 6,780,000
4. 5,490,000

5. 11,000
6. 29,000
7. 89,000,000
8. 37,000,000

B Write each number in standard form. [Examples 2, 3]

9. 3.84×10^4
10. 3.84×10^7
11. 5.71×10^7
12. 5.71×10^5

13. 3.3×10^3
14. 3.3×10^2
15. 8.913×10^7
16. 8.913×10^5

A Write each number in scientific notation. [Examples 1, 3]

17. 0.00035
18. 0.0000035
19. 0.0007
20. 0.007

21. 0.06035
22. 0.0006035
23. 0.1276
24. 0.001276

B Write each number in standard form. [Examples 2, 3]

25. 8.3×10^{-4}
26. 8.3×10^{-7}
27. 6.25×10^{-2}
28. 6.25×10^{-4}

29. 3.125×10^{-1}
30. 3.125×10^{-2}
31. 5×10^{-3}
32. 5×10^{-5}

Applying the Concepts

Super Bowl Advertising and Viewers The cost of a 30-second television ad along with the approximate number of viewers for four different Super Bowls is shown below. Complete the table by writing the ad cost in scientific notation, and the number of viewers in standard form.

	Year	Super Bowl	Ad Cost	Ad Cost in Scientific Notation	Viewers in Scientific Notation	Number of Viewers
33.	1967	I	$40,000		5.12×10^6	
34.	1982	XVI	$345,000		8.52×10^7	
35.	1997	XXXI	$1,200,000		8.79×10^7	
36.	2011	XLV	$3,000,000		1.11×10^8	

Galilean Moons The planet Jupiter has about 60 known moons. In the year 1610 Galileo first discovered the four largest moons of Jupiter, Io, Europa, Ganymede, and Callisto. These moons are known as the Galilean moons. Each moon has a unique period, or the time it takes to make a trip around Jupiter. Fill in the tables below.

37.

Jupiter's Moon	Period (seconds)
Io	153,000
Europa	3.07×10^5
Ganymede	618,000
Callisto	1.44×10^6

38.

Jupiter's Moon	Distance from Jupiter (kilometers)
Io	422,000
Europa	6.17×10^5
Ganymede	1,070,000
Callisto	1.88×10^6

Computer Science The smallest amount of data that a computer can hold is measured in bits. A byte is the next largest unit and is equal to 8, or 2^3, bits. Fill in the table below.

	Unit	Number of Bytes — Exponential Form	Scientific Notation
39.	Kilobyte	$2^{10} = 1,024$	
40.	Megabyte	$2^{20} \approx 1,048,000$	
41.	Gigabyte	$2^{30} \approx 1,074,000,000$	
42.	Terabyte	$2^{40} \approx 1,099,500,000,000$	

Maintaining Your Skills

Simplify the following expressions.

43. $7x - 3x$

44. $11y + 6y$

45. $5x + 3x - 11x$

46. $3x - 7x + 2x$

47. $7x + 3x - x$

48. $12x - 8x + 5x$

49. $3a - 14a + 5a$

50. $2a - 7a - 10a$

1.3 Problem Set

C Find each product. Write all answers in scientific notation. [Examples 4, 5]

51. $(2 \times 10^4)(3 \times 10^6)$

52. $(3 \times 10^3)(1 \times 10^5)$

53. $(2.5 \times 10^7)(6 \times 10^3)$

54. $(3.8 \times 10^6)(5 \times 10^3)$

55. $(7.2 \times 10^3)(9.5 \times 10^{-6})$

56. $(8.5 \times 10^5)(4.2 \times 10^{-9})$

57. $(36,000)(450,000)$

58. $(25,000)(620,000)$

59. $(4,200)(0.00009)$

60. $(0.0000065)(86,000)$

Find each quotient. Write all answers in scientific notation. [Examples 3, 4]

61. $\dfrac{3.6 \times 10^5}{1.8 \times 10^2}$

62. $\dfrac{9.3 \times 10^{15}}{3.0 \times 10^5}$

63. $\dfrac{8.4 \times 10^{-6}}{2.1 \times 10^3}$

64. $\dfrac{6.0 \times 10^{-10}}{1.5 \times 10^3}$

65. $\dfrac{3.5 \times 10^5}{7.0 \times 10^{-10}}$

66. $\dfrac{1.6 \times 10^7}{8.0 \times 10^{-14}}$

67. $\dfrac{540,000}{9,000}$

68. $\dfrac{750,000,000}{250,000}$

69. $\dfrac{0.00092}{46,000}$

70. $\dfrac{0.00000047}{235,000}$

Simplify each expression, and write all answers in scientific notation. [Examples 5, 6]

71. $\dfrac{(3 \times 10^7)(8 \times 10^4)}{6 \times 10^5}$

72. $\dfrac{(4 \times 10^9)(6 \times 10^5)}{8 \times 10^3}$

73. $\dfrac{(2 \times 10^{-3})(6 \times 10^{-5})}{3 \times 10^{-4}}$

74. $\dfrac{(4 \times 10^{-5})(9 \times 10^{-10})}{6 \times 10^{-6}}$

75. $\dfrac{(3.5 \times 10^{-4})(4.2 \times 10^5)}{7 \times 10^3}$

76. $\dfrac{(2.4 \times 10^{-6})(3.6 \times 10^3)}{9 \times 10^5}$

77. $\dfrac{(0.00087)(40,000)}{1,160,000}$

78. $\dfrac{(0.0045)(24,000)}{270,000}$

79. $\dfrac{(525)(0.0000032)}{0.0025}$

80. $\dfrac{(465)(0.000004)}{0.0093}$

Applying the Concepts

Super Bowl Advertising and Viewers The cost of a 30-second television ad along with the approximate number of viewers for four different Super Bowls is shown below. Complete the table by finding the cost per viewer by dividing the ad cost by the number of viewers.

	Year	Super Bowl	Ad Cost	Number of Viewers	Cost per Viewer
81.	1967	I	4.0×10^4	5.12×10^6	
82.	1982	XVI	3.45×10^5	8.52×10^7	
83.	1997	XXXI	1.2×10^6	8.79×10^7	
84.	2011	XLV	3.0×10^6	1.11×10^8	

85. **Pyramids and Scientific Notation** The Great Pyramid at Giza is one of the largest and oldest man-made structures in the world. It weighs over 10^{10} kilograms. If each stone making up the pyramid weighs approximately 4,000 kilograms, how many stones make up the structure? Write your answer in scientific notation.

86. **Technology** A CD-ROM holds about 700 megabytes, or 7.0×10^8 bytes, of data. A four-and-a-half minute song downloaded from the Internet uses 5 megabytes, or 5.0×10^6 bytes, of storage space. How many four-and-a-half minute songs can fit on a CD?

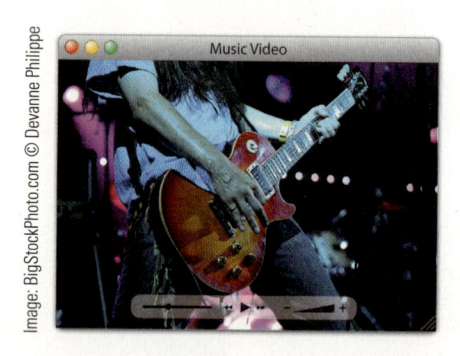

Maintaining Your Skills

Find the value of each of the following expressions when $x = 4$.

87. $3x - 5$
88. $-2 - 2x$
89. $-5x + 6$
90. $4x + 7$

Find the value of each of the following expressions when $x = -3$.

91. $3x - 5$
92. $-2 - 2x$
93. $-5x + 6$
94. $4x + 7$

Solving First Degree Equations

1.4

Objectives

A Solve a linear equation in one variable.

A Solving Linear Equations

A *linear equation in one variable* is any equation that can be put in the form

$$ax + b = c$$

where a, b, and c are constants and $a \neq 0$. For example, each of the equations

$$5x + 3 = 2 \qquad 2x = 7 \qquad 2x + 5 = 0$$

is linear because they can be put in the form $ax + b = c$. In the first equation, $5x$, 3, and 2 are called *terms* of the equation: $5x$ is a variable term; 3 and 2 are constant terms.

> **Definition**
> The **solution set** for an equation is the set of all numbers that, when used in place of the variable, make the equation a true statement.

EXAMPLE 1 The solution set for $2x - 3 = 9$ is $\{6\}$ since replacing x with 6 makes the equation a true statement.

$$\begin{aligned}
\text{If} \qquad & x = 6 \\
\text{then} \qquad & 2x - 3 = 9 \\
\text{becomes} \qquad & 2(6) - 3 = 9 \\
& 12 - 3 = 9 \\
& 9 = 9 \quad \text{A true statement}
\end{aligned}$$

PRACTICE PROBLEMS

1. Show that $x = 3$ is a solution to the equation $4x - 2 = 10$.

> **Definition**
> Two or more equations with the same solution set are called **equivalent equations**.

EXAMPLE 2 The equations $2x - 5 = 9$, $x - 1 = 6$, and $x = 7$ are all equivalent equations because the solution set for each is $\{7\}$.

2. Show that the equations $3x + 1 = 16$ and $4x - 6 = 14$ both have solution set $\{5\}$.

Properties of Equality

The first property of equality states that adding the same quantity to both sides of an equation preserves equality. Or, more importantly, adding the same amount to both sides of an equation *never changes* the solution set. This property is called the *addition property of equality* and is stated in symbols as follows.

> **Addition Property of Equality**
> For any three algebraic expressions, A, B, and C,
> $$\begin{aligned} \text{if} \quad & A = B \\ \text{then} \quad & A + C = B + C \end{aligned}$$
> *In words:* Adding the same quantity to both sides of an equation will not change the solution set.

Answers
1. See Solutions to Selected Practice Problems.
2. See Solutions to Selected Practice Problems.

Chapter 1 Basic Concepts and Properties

Our second new property is called the *multiplication property of equality* and is stated as follows.

Multiplication Property of Equality
For any three algebraic expressions A, B, and C, where $C \neq 0$,
$$\text{if} \quad A = B$$
$$\text{then} \quad AC = BC$$

In words: Multiplying both sides of an equation by the same nonzero quantity will not change the solution set.

Note Because subtraction is defined in terms of addition and division is defined in terms of multiplication, we do not need to introduce separate properties for subtraction and division. The solution set for an equation will never be changed by subtracting the same amount from both sides or by dividing both sides by the same nonzero quantity.

EXAMPLE 3 Solve $\frac{3}{4}x + 5 = -4$.

SOLUTION We begin by adding -5 to both sides of the equation. Once this has been done, we multiply both sides by the reciprocal of $\frac{3}{4}$, which is $\frac{4}{3}$.

$$\frac{3}{4}x + 5 = -4$$

$$\frac{3}{4}x + 5 + (-5) = -4 + (-5) \quad \text{Add } -5 \text{ to both sides}$$

$$\frac{3}{4}x = -9$$

$$\frac{4}{3}\left(\frac{3}{4}x\right) = \frac{4}{3}(-9) \quad \text{Multiply both sides by } \frac{4}{3}$$

$$x = -12 \qquad \frac{4}{3}(-9) = \frac{4}{3}\left(-\frac{9}{1}\right) = -\frac{36}{3} = -12$$

3. Solve $\frac{2}{3}x + 4 = -8$.

EXAMPLE 4 Find the solution set for $3a - 5 = -6a + 1$.

SOLUTION To solve for *a* we must isolate it on one side of the equation. Let's decide to isolate *a* on the left side by adding $6a$ to both sides of the equation.

$$3a - 5 = -6a + 1$$

$$3a + 6a - 5 = -6a + 6a + 1 \quad \text{Add } 6a \text{ to both sides}$$

$$9a - 5 = 1$$

$$9a - 5 + 5 = 1 + 5 \quad \text{Add 5 to both sides}$$

$$9a = 6$$

$$\frac{1}{9}(9a) = \frac{1}{9}(6) \quad \text{Multiply both sides by } \frac{1}{9}$$

$$a = \frac{2}{3} \qquad \frac{1}{9}(6) = \frac{6}{9} = \frac{2}{3}$$

4. Find the solution set for $3a - 3 = -5a + 9$.

Note From the previous chapter we know that multiplication by a number and division by its reciprocal always produce the same result. Because of this fact, instead of multiplying each side of our equation by $\frac{1}{9}$, we could just as easily divide each side by 9. If we did so, the last two lines in our solution would look like this:
$$\frac{9a}{9} = \frac{6}{9}$$
$$a = \frac{2}{3}$$

The solution set is $\left\{\frac{2}{3}\right\}$.

We can check our solution in Example 4 by replacing *a* in the original equation with $\frac{2}{3}$.

When $a = \frac{2}{3}$

the equation $3a - 5 = -6a + 1$

becomes $3\left(\frac{2}{3}\right) - 5 \stackrel{?}{=} -6\left(\frac{2}{3}\right) + 1$

$2 - 5 \stackrel{?}{=} -4 + 1$

$-3 = -3$ A true statement

Note We are placing a question mark over the equal sign because we don't know yet if the expression on the left will be equal to the expression on the right.

Answers
3. -18
4. $\left\{\frac{3}{2}\right\}$

1.4 Solving First Degree Equations

There will be times when we solve equations and end up with a negative sign in front of the variable. The next example shows how to handle this situation.

EXAMPLE 5 Solve each equation.

a. $-x = 4$ b. $-y = -8$

SOLUTION Neither equation can be considered solved because of the negative sign in front of the variable. To eliminate the negative signs, we simply multiply both sides of each equation by -1.

a. $\quad -x = 4 \qquad\qquad$ b. $\ -y = -8$

$\quad -\mathbf{1}(-x) = -\mathbf{1}(4) \qquad -\mathbf{1}(-y) = -\mathbf{1}(-8) \qquad$ Multiply each side by -1

$\qquad x = -4 \qquad\qquad\quad y = 8$

5. Solve each equation.
   a. $-x = \dfrac{2}{3}$
   b. $-y = -4$

EXAMPLE 6 Solve $\dfrac{2}{3}x + \dfrac{1}{2} = -\dfrac{3}{8}$.

SOLUTION We can solve this equation by applying our properties and working with fractions, or we can begin by eliminating the fractions. Let's use both methods.

METHOD 1 Working with the fractions.

$\dfrac{2}{3}x + \dfrac{1}{2} + \left(-\dfrac{\mathbf{1}}{\mathbf{2}}\right) = -\dfrac{3}{8} + \left(-\dfrac{\mathbf{1}}{\mathbf{2}}\right) \qquad$ Add $-\dfrac{1}{2}$ to each side

$\dfrac{2}{3}x = -\dfrac{7}{8} \qquad\qquad -\dfrac{3}{8} + \left(-\dfrac{1}{2}\right) = -\dfrac{3}{8} + \left(-\dfrac{4}{8}\right)$

$\dfrac{\mathbf{3}}{\mathbf{2}}\left(\dfrac{2}{3}x\right) = \dfrac{\mathbf{3}}{\mathbf{2}}\left(-\dfrac{7}{8}\right) \qquad$ Multiply each side by $\dfrac{3}{2}$

$x = -\dfrac{21}{16}$

6. Solve $\dfrac{3}{5}x + \dfrac{1}{3} = -\dfrac{5}{6}$.

METHOD 2 Eliminating the fractions in the beginning.

Our original equation has denominators of 3, 2, and 8. The least common denominator, abbreviated LCD, for these three denominators is 24, and it has the property that all three denominators will divide it evenly. If we multiply both sides of our equation by 24, each denominator will divide into 24, and we will be left with an equation that does not contain any denominators other than 1.

$\mathbf{24}\left(\dfrac{2}{3}x + \dfrac{1}{2}\right) = \mathbf{24}\left(-\dfrac{3}{8}\right) \qquad$ Multiply each side by the LCD 24

$24\left(\dfrac{2}{3}x\right) + 24\left(\dfrac{1}{2}\right) = 24\left(-\dfrac{3}{8}\right) \qquad$ Distributive property on the left side

$16x + 12 = -9 \qquad$ Multiply

$16x = -21 \qquad$ Add -12 to each side

$x = -\dfrac{21}{16} \qquad$ Multiply each side by $\dfrac{1}{16}$

CHECK To check our solution, we substitute $x = -\dfrac{21}{16}$ back into our original equation to obtain

$\dfrac{2}{3}\left(-\dfrac{21}{16}\right) + \dfrac{1}{2} \stackrel{?}{=} -\dfrac{3}{8}$

Answers
5. a. $-\dfrac{2}{3}$ b. 4
6. $-\dfrac{35}{18}$

Chapter 1 Basic Concepts and Properties

$$-\frac{7}{8} + \frac{1}{2} \stackrel{?}{=} -\frac{3}{8}$$

$$-\frac{7}{8} + \frac{4}{8} \stackrel{?}{=} -\frac{3}{8}$$

$$-\frac{3}{8} = -\frac{3}{8} \quad \text{A true statement}$$

EXAMPLE 7 Solve the equation $0.06x + 0.05(10{,}000 - x) = 560$.

SOLUTION We can solve the equation in its original form by working with the decimals, or we can eliminate the decimals first by using the multiplication property of equality and solve the resulting equation. Here are both methods.

METHOD 1 Working with the decimals.

$0.06x + 0.05(10{,}000 - x) = 560$	Original equation
$0.06x + 0.05(10{,}000) - 0.05x = 560$	Distributive property
$0.01x + 500 = 560$	Simplify the left side
$0.01x + 500 + (-\mathbf{500}) = 560 + (-\mathbf{500})$	Add -500 to each side
$0.01x = 60$	
$\dfrac{0.01x}{\mathbf{0.01}} = \dfrac{60}{\mathbf{0.01}}$	Divide each side by 0.01
$x = 6{,}000$	

METHOD 2 Eliminating the decimals in the beginning: To move the decimal point two places to the right in $0.06x$ and 0.05, we multiply each side of the equation by 100.

$0.06x + 0.05(10{,}000 - x) = 560$	Original equation
$0.06x + 500 - 0.05x = 560$	Distributive property
$\mathbf{100}(0.06x) + \mathbf{100}(500) - \mathbf{100}(0.05x) = \mathbf{100}(560)$	Multiply each side by 100
$6x + 50{,}000 - 5x = 56{,}000$	
$x + 50{,}000 = 56{,}000$	Simplify the left side
$x = 6{,}000$	Add $-50{,}000$ to each side

Using either method, the solution to our equation is 6,000.

CHECK We check our work (to be sure we have not made a mistake in applying the properties or in arithmetic) by substituting 6,000 into our original equation and simplifying each side of the result separately, as the following shows.

$$0.06(\mathbf{6{,}000}) + 0.05(10{,}000 - \mathbf{6{,}000}) \stackrel{?}{=} 560$$

$$0.06(6{,}000) + 0.05(4{,}000) \stackrel{?}{=} 560$$

$$360 + 200 \stackrel{?}{=} 560$$

$$560 = 560 \quad \text{A true statement}$$

Here is a list of steps to use as a guideline for solving linear equations in one variable.

---

7. Solve.
   $0.08x + 0.10(8{,}000 - x) = 680$

Answer
7. 6,000

1.4 Solving First Degree Equations

> **Strategy** Solving Linear Equations in One Variable
>
> **Step 1:** **a.** Use the distributive property to separate terms, if necessary.
>
> **b.** If fractions are present, consider multiplying both sides by the LCD to eliminate the fractions. If decimals are present, consider multiplying both sides by a power of 10 to clear the equation of decimals.
>
> **c.** Combine similar terms on each side of the equation.
>
> **Step 2:** Use the addition property of equality to get all variable terms on one side of the equation and all constant terms on the other side. A **variable term** is a term that contains the variable (for example, 5x). A **constant term** is a term that does not contain the variable (the number 3, for example).
>
> **Step 3:** Use the multiplication property of equality to get the variable by itself on one side of the equation.
>
> **Step 4:** Check your solution in the original equation to be sure that you have not made a mistake in the solution process.

As you will see as you work through the problems in the problem set, it is not always necessary to use all four steps when solving equations. The number of steps used depends on the equation. In Example 8 there are no fractions or decimals in the original equation, so Step 1b will not be used.

EXAMPLE 8 Solve $3(2y - 1) + y = 5y + 3$.

SOLUTION Applying the steps outlined in the preceding strategy, we have

Step 1: a. $3(2y - 1) + y = 5y + 3$ Original equation

$6y - 3 + y = 5y + 3$ Distributive property

c. $7y - 3 = 5y + 3$ Simplify

Step 2: $7y + (-5y) - 3 = 5y + (-5y) + 3$ Add $-5y$ to both sides

$2y - 3 = 3$

$2y - 3 + 3 = 3 + 3$ Add $+3$ to both sides

$2y = 6$

Step 3: $\frac{1}{2}(2y) = \frac{1}{2}(6)$ Multiply by $\frac{1}{2}$

$y = 3$

Step 4: Check

When $y = 3$

the equation $3(2y - 1) + y = 5y + 3$

becomes $3(2 \cdot 3 - 1) + 3 \stackrel{?}{=} 5 \cdot 3 + 3$

$3(5) + 3 \stackrel{?}{=} 15 + 3$

$18 = 18$ A true statement

8. Solve: $2(5y - 3) + 4 = 9y - 6$

Answer
8. -4

9. Solve for x.
   $6 - 2(5x - 1) + 4x = 20$

Note It would be a mistake to subtract 3 from 8 first because the rule for order of operations indicates we are to do multiplication before subtraction.

10. Solve: $3(5x + 1) = 10 + 15x$

11. Solve: $-4 + 8x = 2(4x - 2)$

Answers
9. -2
10. No solution
11. Identity

Chapter 1 Basic Concepts and Properties

EXAMPLE 9 Solve the equation $8 - 3(4x - 2) + 5x = 35$.

SOLUTION We must begin by distributing the -3 across the quantity $4x - 2$.

Step 1: a. $8 - 3(4x - 2) + 5x = 35$ Original equation
$8 - 12x + 6 + 5x = 35$ Distributive property
c. $-7x + 14 = 35$ Simplify
Step 2: $-7x = 21$ Add -14 to each side
Step 3: $x = -3$ Multiply by $-\frac{1}{7}$

Step 4: When x is replaced by -3 in the original equation, a true statement results. Therefore, -3 is the solution to our equation.

Identities and Equations with No Solution

Two special cases are associated with solving linear equations in one variable, each of which is illustrated in the following examples.

EXAMPLE 10 Solve for x: $2(3x - 4) = 3 + 6x$.

SOLUTION Applying the distributive property to the left side gives us

$6x - 8 = 3 + 6x$ Distributive property

Now, if we add $-6x$ to each side, we are left with the following

$-8 = 3$

which is a false statement. This means that there is no solution to our equation. Any number we substitute for x in the original equation will lead to a similar false statement.

EXAMPLE 11 Solve for x: $-15 + 3x = 3(x - 5)$.

SOLUTION We start by applying the distributive property to the right side.

$-15 + 3x = 3x - 15$ Distributive property

If we add $-3x$ to each side, we are left with the true statement

$-15 = -15$

In this case, our result tells us that any number we use in place of x in the original equation will lead to a true statement. Therefore, all real numbers are solutions to our equation. We say the original equation is an *identity* because the left side is always identically equal to the right side.

Getting Ready for Class

After reading through the preceding section, respond in your own words and in complete sentences.

1. What is a solution to an equation?
2. What are equivalent equations?
3. Describe how to eliminate fractions in an equation.
4. Suppose that when you solve an equation your result is the statement "$3 = -3$." What would you conclude about the solution to the equation?

Problem Set 1.4

A Solve each of the following equations. [Examples 1–11]

1. $x - 5 = 3$
2. $x + 2 = 7$
3. $2x - 4 = 6$
4. $3x - 5 = 4$

5. $7 = 4a - 1$
6. $10 = 3a - 5$
7. $3 - y = 10$
8. $5 - 2y = 11$

9. $-3 - 4x = 15$
10. $-8 - 5x = -6$
11. $-3 = 5 + 2x$
12. $-12 = 6 + 9x$

13. $-300y + 100 = 500$
14. $-20y + 80 = 30$
15. $160 = -50x - 40$
16. $110 = -60x - 50$

17. $-x = 2$
18. $-x = \frac{1}{2}$
19. $-a = -\frac{3}{4}$
20. $-a = -5$

21. $\frac{2}{3}x = 8$
22. $\frac{3}{2}x = 9$
23. $-\frac{3}{5}a + 2 = 8$
24. $-\frac{5}{3}a + 3 = 23$

25. $8 = 6 + \frac{2}{7}y$
26. $1 = 4 + \frac{3}{7}y$
27. $2x - 5 = 3x + 2$
28. $5x - 1 = 4x + 3$

29. $-3a + 2 = -2a - 1$

30. $-4a - 8 = -3a + 7$

31. $5 - 2x = 3x + 1$

32. $7 - 3x = 8x - 4$

33. $11x - 5 + 4x - 2 = 8x$

34. $2x + 7 - 3x + 4 = -2x$

35. $6 - 7(m - 3) = -1$

36. $3 - 5(2m - 5) = -2$

37. $7 + 3(x + 2) = 4(x - 1)$

38. $5 + 2(4x - 4) = 3(2x - 1)$

39. $5 = 7 - 2(3x - 1) + 4x$

40. $20 = 8 - 5(2x - 3) + 4x$

1.4 Problem Set

41. $\dfrac{1}{2}x + \dfrac{1}{4} = \dfrac{1}{3}x + \dfrac{5}{4}$

42. $\dfrac{2}{3}x - \dfrac{3}{4} = \dfrac{1}{6}x + \dfrac{21}{4}$

43. $-\dfrac{2}{5}x + \dfrac{2}{15} = \dfrac{2}{3}$

44. $-\dfrac{1}{6}x + \dfrac{2}{3} = \dfrac{1}{4}$

45. $\dfrac{3}{4}(8x - 4) = \dfrac{2}{3}(6x - 9)$

46. $\dfrac{3}{5}(5x + 10) = \dfrac{5}{6}(12x - 18)$

47. $\dfrac{1}{4}(12a + 1) - \dfrac{1}{4} = 5$

48. $\dfrac{2}{3}(6x - 1) + \dfrac{2}{3} = 4$

49. $0.35x - 0.2 = 0.15x + 0.1$

50. $0.25x - 0.05 = 0.2x + 0.15$

51. $0.42 - 0.18x = 0.48x - 0.24$

52. $0.3 - 0.12x = 0.18x + 0.06$

Chapter 1 Basic Concepts and Properties

Solve each equation, if possible.

53. $3x - 6 = 3(x + 4)$

54. $7x - 14 = 7(x - 2)$

55. $4y + 2 - 3y + 5 = 3 + y + 4$

56. $7y + 5 - 2y - 3 = 6 + 5y - 4$

57. $2(4t - 1) + 3 = 5t + 4 + 3t$

58. $5(2t - 1) + 1 = 2t - 4 + 8t$

Now that you have practiced solving a variety of equations, we can turn our attention to the types of equations you will see as you progress through the book. Each equation appears later in the book exactly as you see it below.

Solve each equation.

59. $3x + 2 = 0$

60. $5x - 4 = 0$

61. $0 = 6{,}400a + 70$

62. $0 = 6{,}400a + 60$

63. $x + 2 = 2x$

64. $x + 2 = 7x$

65. $0.07x = 1.4$

66. $0.02x = 0.3$

67. $5(2x + 1) = 12$ **68.** $4(3x - 2) = 21$ **69.** $50 = \dfrac{K}{48}$ **70.** $50 = \dfrac{K}{24}$

71. $100P = 2{,}400$ **72.** $3.5d = 16(3.5)^2$ **73.** $x + (3x + 2) = 26$ **74.** $2(1) + y = 4$

75. $2x - 3(3x - 5) = -6$ **76.** $2(2y + 6) + 3y = 5$ **77.** $2(2x - 3) + 2x = 45$ **78.** $2(4x - 10) + 2x = 12.5$

79. $3x + (x - 2) \cdot 2 = 6$ **80.** $2x - (x + 1) = -1$ **81.** $15 - 3(x - 1) = x - 2$ **82.** $4x - 4(x - 3) = x + 3$

83. $2(x + 3) + x = 4(x - 3)$ **84.** $5(y + 2) - 4(y + 1) = 3$ **85.** $6(y - 3) - 5(y + 2) = 8$ **86.** $2(x + 3) + 3(x + 5) = 2x$

87. $2(20 + x) = 3(20 - x)$ **88.** $6(7 + x) = 5(9 + x)$ **89.** $2x + 1.5(75 - x) = 127.5$ **90.** $x + 0.06x = 954$

91. $0.08x + 0.09(9{,}000 - x) = 750$

92. $0.08x + 0.09(9{,}000 - x) = 500$

93. $0.12x + 0.10(15{,}000 - x) = 1{,}600$

94. $0.09x + 0.11(11{,}000 - x) = 1{,}150$

95. $5\left(\dfrac{19}{15}\right) + 5y = 9$

96. $4\left(\dfrac{19}{15}\right) - 2y = 4$

97. $2\left(\dfrac{29}{22}\right) - 3y = 4$

98. $2x - 3\left(-\dfrac{5}{11}\right) = 4$

99. Work each problem according to the instructions given:
   a. Solve: $8x - 5 = 0$
   b. Solve: $8x - 5 = -5$
   c. Add: $(8x - 5) + (2x - 5)$
   d. Solve: $8x - 5 = 2x - 5$
   e. Multiply: $8(x - 5)$
   f. Solve: $8(x - 5) = 2(x - 5)$

100. Work each problem according to the instructions given:
   a. Solve: $3x + 6 = 0$
   b. Solve: $3x + 6 = 4$
   c. Add: $(3x + 6) + (7x + 4)$
   d. Solve: $3x + 6 = 7x + 4$
   e. Multiply: $3(x + 5)$
   f. Solve: $3(x + 6) = 7(x + 2)$

Applying the Concepts

101. Cost of a Taxi Ride The taximeter was invented in 1891 by Wilhelm Bruhn. The city of Chicago charges $1.80 plus $0.40 per mile for a taxi ride.

   a. A woman paid a fare of $6.60. Write an equation that connects the fare the woman paid, the miles she traveled, n, and the charges the taximeter computes.

   b. Solve the equation from part (a) to determine how many miles the woman traveled.

102. Coughs and Earaches In 2007, twice as many people visited their doctor because of a cough than an earache. The total number of doctor's visits for these two ailments was reported to be 45 million.

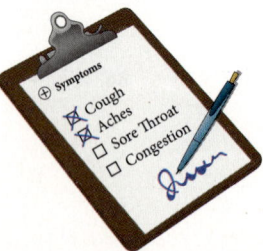

   a. Let x represent the number of earaches reported in 2007, then write an expression using x for the number of coughs reported in 2007.

   b. Write an equation that relates 45 million to the variable x.

   c. Solve the equation from part (b) to determine the number of people who visited their doctor in 2007 to report an earache.

103. **Population Density** In July 2001 the population of Puerto Rico was estimated to be 3,937,000 people, with a population density of 1,125 people per square mile.

   a. Let A represent the area of Puerto Rico in square miles, and write an equation that shows that the population is equal to the product of the area and the population density.

   b. Solve the equation from part (a), rounding your solution to the nearest square mile.

104. **Solving Equations by Trial and Error** Sometimes equations can be solved most easily by trial and error. Solve the following equations by trial and error.

   a. Find x and y if $x \cdot y + 1 = 36$, and both x and y are prime.

   b. Find w, t, and z if $w + t + z + 10 = 52$, and w, t, and z are consecutive terms of a Fibonacci sequence.

   c. Find x and y if $x \neq y$ and $x^y = y^x$.

105. **Cars** The chart shows the fastest cars in America. The sum of five times the speed of a car and 23 is the maximum speed of the SSC Ultimate Aero. What is the speed of the car?

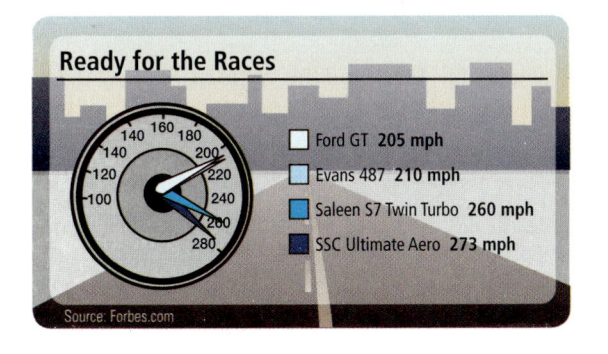

Ready for the Races
- Ford GT 205 mph
- Evans 487 210 mph
- Saleen S7 Twin Turbo 260 mph
- SSC Ultimate Aero 273 mph

Source: Forbes.com

106. **Sound** The chart shows the decibel level of various sounds. The difference of 3 times the decibel level of city traffic heard from the inside of a car and 67 is the decibel level of a blue whale. What is the decibel level of city traffic?

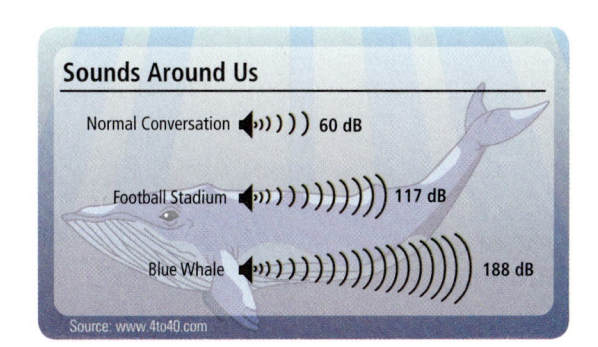

Sounds Around Us
- Normal Conversation 60 dB
- Football Stadium 117 dB
- Blue Whale 188 dB

Source: www.4to40.com

Maintaining Your Skills

From this point on, each problem set will contain a number of problems under the heading Maintaining Your Skills. These problems cover the most important skills you have learned in previous sections and chapters. By working these problems regularly, you will keep yourself current on all the topics we have covered and, possibly, need less time to study for tests and quizzes.

Identify the property (or properties) that justifies each of the following statements.

107. $ax = xa$

108. $5\left(\dfrac{1}{5}\right) = 1$

109. $3 + (x + y) = (3 + x) + y$

110. $3 + (x + y) = (x + y) + 3$

111. $3 + (x + y) = (3 + y) + x$

112. $7(3x - 5) = 21x - 35$

113. $5(1) = 5$

114. $5 + 0 = 5$

115. $4(xy) = 4(yx)$

116. $4(xy) = (4y)x$

117. $2 + 0 = 2$

118. $2 + (-2) = 0$

Getting Ready for the Next Section

Problems under the heading, *Getting Ready for the Next Section*, are problems that you must be able to work in order to understand the material in the next section. In this case, the problems below are variations on the types of problems you have already worked in this problem set. They are exactly the types of problems you will see in explanations and examples in the next section.

Solve each equation.

119. $x \cdot 42 = 21$

120. $x \cdot 84 = 21$

121. $25 = 0.4x$

122. $35 = 0.4x$

123. $12 - 4y = 12$

124. $-6 - 3y = 6$

125. $525 = 900 - 300p$

126. $375 = 900 - 300p$

127. $486.7 = 78.5 + 31.4h$

128. $486.7 = 113.0 + 37.7h$

129. Find the value of $2x - 1$ when x is
    a. 2   b. 3   c. 5

130. Find the value of $\dfrac{1}{x + 1}$ when x is
    a. 1   b. 2   c. 3

Extending the Concepts

131. $\dfrac{x + 4}{5} - \dfrac{x + 3}{3} = -\dfrac{7}{15}$

132. $\dfrac{x + 1}{7} - \dfrac{x - 2}{2} = \dfrac{1}{14}$

133. $\dfrac{1}{x} - \dfrac{2}{3} = \dfrac{2}{x}$

134. $\dfrac{1}{x} - \dfrac{3}{5} = \dfrac{2}{x}$

135. $\dfrac{x + 3}{2} - \dfrac{x - 4}{4} = -\dfrac{1}{8}$

136. $\dfrac{x - 3}{5} - \dfrac{x + 1}{10} = -\dfrac{1}{10}$

137. $\dfrac{x - 1}{2} - \dfrac{x + 2}{3} = \dfrac{x + 3}{6}$

138. $\dfrac{x + 2}{4} - \dfrac{x - 1}{3} = \dfrac{x + 2}{6}$

1.5 Problem Solving Involving First Degree Equations

Objectives

A Apply the Blueprint for Problem Solving to a variety of application problems.

B Use a formula to construct a table of paired data.

A Blueprint for Problem Solving

In this section we use the skills we have developed for solving equations to solve problems written in words. You may find that some of the examples and problems are more realistic than others. Since we are just beginning our work with application problems, even the ones that seem unrealistic are good practice. What is important in this section is the *method* we use to solve application problems, not the applications themselves. The method, or strategy, that we use to solve application problems is called the *Blueprint for Problem Solving*. It is an outline that will overlay the solution process we use on all application problems.

> **Blueprint for Problem Solving**
>
> **Step 1: Read** the problem, and then mentally **list** the items that are known and the items that are unknown.
>
> **Step 2: Assign a variable** to one of the unknown items. (In most cases this will amount to letting x = the item that is asked for in the problem.) Then **translate** the other **information** in the problem to expressions involving the variable.
>
> **Step 3: Reread** the problem, and then **write an equation,** using the items and variable listed in steps 1 and 2, that describes the situation.
>
> **Step 4: Solve the equation** found in step 3.
>
> **Step 5: Write your answer** using a complete sentence.
>
> **Step 6: Reread** the problem, and **check** your solution with the original words in the problem.

A number of substeps occur within each of the steps in our blueprint. For instance, with steps 1 and 2 it is always a good idea to draw a diagram or picture if it helps you visualize the relationship between the items in the problem.

EXAMPLE 1 The length of a rectangle is 3 inches less than twice the width. The perimeter is 45 inches. Find the length and width.

SOLUTION When working problems that involve geometric figures, a sketch of the figure helps organize and visualize the problem.

Step 1: **Read and list.**
Known items: The figure is a rectangle. The length is 3 inches less than twice the width. The perimeter is 45 inches.
Unknown items: The length and the width

Step 2: **Assign a variable and translate information.**
Since the length is given in terms of the width (the length is 3 less than twice the width), we let x = the width of the rectangle. The length is 3 less than twice the width, so it must be $2x - 3$. The diagram in Figure 1 is a visual description of the relationships we have listed so far.

Step 3: **Reread and write an equation.**
The equation that describes the situation is

Twice the length + twice the width = perimeter
$2(2x - 3)$ + $2x$ = 45

PRACTICE PROBLEMS

1. The width of a rectangle is 10 feet less than 4 times the length. If the perimeter is 12.5 feet, find the length and width.

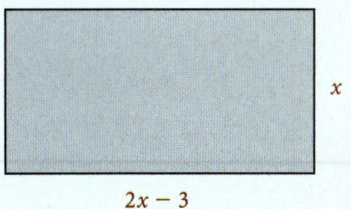

FIGURE 1

Step 4: Solve the equation.
$$2(2x - 3) + 2x = 45$$
$$4x - 6 + 2x = 45$$
$$6x - 6 = 45$$
$$6x = 51$$
$$x = 8.5$$

Step 5: Write the answer.
The width is 8.5 inches. The length is $2x - 3 = 2(8.5) - 3 = 14$ inches.

Step 6: Reread and check.
If the length is 14 inches and the width is 8.5 inches, then the perimeter must be $2(14) + 2(8.5) = 28 + 17 = 45$ inches. Also, the length, 14, is 3 less than twice the width.

Remember as you read through the steps in the solutions to the examples in this section that step 1 is done mentally. Read the problem and then *mentally* list the items that you know and the items that you don't know. The purpose of step 1 is to give you direction as you begin to work application problems. Finding the solution to an application problem is a process; it doesn't happen all at once. The first step is to read the problem with a purpose in mind. That purpose is to mentally note the items that are known and the items that are unknown.

EXAMPLE 2 Suppose Pat bought a new Ford Mustang with a 5.0-liter engine. The total price, which includes the price of the car plus sales tax, was $17,481.75. If the sales tax rate is 7.25%, what was the price of the car?

SOLUTION

Step 1: Read and list.
Known items: The total price is $17,481.75. The sales tax rate is 7.25%, which is 0.0725 in decimal form.
Unknown item: The price of the car

Step 2: Assign a variable and translate information.
If we let x = the price of the car, then to calculate the sales tax, we multiply the price of the car x by the sales tax rate:
$$\text{Sales tax} = (\text{sales tax rate})(\text{price of the car})$$
$$= 0.0725x$$

Step 3: Reread and write an equation.
$$\text{Car price} + \text{sales tax} = \text{total price}$$
$$x + 0.0725x = 17{,}481.75$$

Step 4: Solve the equation.
$$x + 0.0725x = 17{,}481.75$$
$$1.0725x = 17{,}481.75$$
$$x = \frac{17{,}481.75}{1.0725}$$
$$= 16{,}300.00$$

Step 5: Write the answer.
The price of the car is $16,300.00.

Step 6: Reread and check.
The price of the car is $16,300.00. The tax is $0.0725(16,300) = \$1{,}181.75$. Adding the retail price and the sales tax we have a total bill of $17,481.75.

---

2. If Pat had purchased the convertible version of the Ford Mustang discussed in Example 2, the total price (the price of the car plus sales tax) would have been $23,466.30. Find the price of the convertible if the sales tax rate is 7.25%.

Answer
1. length = 3.25 feet, width = 3 feet
2. $21,880.00

Interest Problem

EXAMPLE 3 Suppose a person invests a total of $10,000 in two accounts. One account earns 5% annually, and the other earns 6% annually. If the total interest earned from both accounts in a year is $560, how much is invested in each account?

SOLUTION

Step 1: *Read and list.*
Known items: Two accounts. One pays interest of 5%, and the other pays 6%. The total invested is $10,000.
Unknown items: The number of dollars invested in each individual account

Step 2: *Assign a variable and translate information.*
If we let x equal the amount invested at 6%, then $10,000 - x$ is the amount invested at 5%. The total interest earned from both accounts is $560. The amount of interest earned on x dollars at 6% is $0.06x$, whereas the amount of interest earned on $10,000 - x$ dollars at 5% is $0.05(10,000 - x)$.

	Dollars at 6%	Dollars at 5%	Total
Number of	x	$10,000 - x$	$10,000$
Interest on	$0.06x$	$0.05(10,000 - x)$	560

Step 3: *Reread and write an equation.*
The last line gives us the equation we are after:

$$0.06x + 0.05(10,000 - x) = 560$$

Step 4: *Solve the equation.*
To make this equation a little easier to solve, we begin by multiplying both sides by 100 to move the decimal point two places to the right:

$$6x + 5(10,000 - x) = 56,000$$
$$6x + 50,000 - 5x = 56,000$$
$$x + 50,000 = 56,000$$
$$x = 6,000$$

Step 5: *Write the answer.*
The amount of money invested at 6% is $6,000. The amount of money invested at 5% is $10,000 - $6,000 = $4,000.

Step 6: *Reread and check.*
To check our results, we find the total interest from the two accounts:

The interest on $6,000 at 6% is $0.06(6,000) = 360$
The interest on $4,000 at 5% is $0.05(4,000) = 200$
_____
The total interest $= \$560$

B Table Building

We can use our knowledge of formulas to build tables of paired data. As you will see, equations or formulas that contain exactly two variables produce pairs of numbers that can be used to construct tables.

---

3. Howard invests a total of $8,000 in two accounts. One account earns 8% annually, and the other earns 10% annually. If the total interest earned from both accounts in a year is $680, how much is invested in each account?

Answer
3. $6,000 at 8%, $2,000 at 10%

4. Rework Example 4 if the string is 14 inches long and the rectangle has a width of 1, 2, 3, 4, 5, or 6 inches.

12 inches

Chapter 1 Basic Concepts and Properties

EXAMPLE 4

A piece of string 12 inches long is to be formed into a rectangle. Build a table that gives the length of the rectangle if the width is 1, 2, 3, 4, or 5 inches. Then find the area of each of the rectangles formed.

SOLUTION Because the formula for the perimeter of a rectangle is $P = 2l + 2w$ and our piece of string is 12 inches long, the formula we will use to find the lengths for the given widths is $12 = 2l + 2w$. To solve this formula for l, we divide each side by 2 and then subtract w. The result is $l = 6 - w$. Table 1 organizes our work so that the formula we use to find l for a given value of w is shown, and we have added a last column to give us the areas of the rectangles formed. The units for the first three columns are inches, and the units for the numbers in the last column are square inches.

TABLE 1
LENGTH, WIDTH, AND AREA

Width (in.) w	Length (in.) $l = 6 - w$	l	Area (in.²) $A = lw$
1	$l = 6 - 1$	5	5
2	$l = 6 - 2$	4	8
3	$l = 6 - 3$	3	9
4	$l = 6 - 4$	2	8
5	$l = 6 - 5$	1	5

Figures 2 and 3 show two bar charts constructed from the information in Table 1.

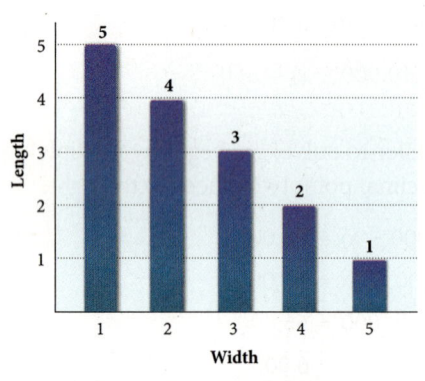

FIGURE 2 Length and width of rectangles with perimeters fixed at 12 inches

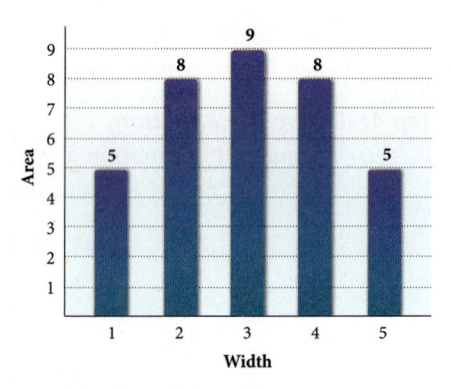

FIGURE 3 Area and width of rectangles with perimeters fixed at 12 inches

Answers
4. See Solutions to Selected Practice Problems.

1.5 Problem Solving Involving First Degree Equations

USING TECHNOLOGY

Calculators

A number of graphing calculators have table-building capabilities. We can let the calculator variable X represent the widths of the rectangles in Example 4. To find the lengths, we set variable Y_1 equal to $6 - X$. The area of each rectangle can be found by setting variable Y_2 equal to $X * Y_1$. To have the calculator produce the table automatically, we use a table minimum of 0 and a table increment of 1. Here is a summary of how the graphing calculator is set up:

Table Setup	Y Variables Setup
Table minimum = 0	$Y_1 = 6 - X$
Table increment = 1	$Y_2 = X * Y_1$
Independent variable: Auto	
Dependent variable: Auto	

The table will look like this:

X	Y_1	Y_2
0	6	0
1	5	5
2	4	8
3	3	9
4	2	8
5	1	5
6	0	0

STUDY SKILLS

Begin to Develop Confidence with Word Problems

The main difference between people who are good at working word problems and those who are not seems to be confidence. People with confidence know that no matter how long it takes them, they will eventually be able to solve the problem they are working on. Those without confidence begin by saying to themselves, "I'll never be able to work this problem." If you are in this second group, then instead of telling yourself that you can't do word problems, that you don't like them, or that they're not good for anything anyway, decide to do whatever it takes to master them.

Getting Ready for Class

After reading through the preceding section, respond in your own words and in complete sentences.

1. What is the first step in solving an application problem?
2. What is the biggest obstacle between you and success in solving application problems?
3. Write an application problem for which the solution depends on solving the equation $2x + 2 \cdot 3 = 18$.
4. What is the last step in solving an application problem? Why is this step important?

Left Blank Intentionally

Problem Set 1.5

Solve each application problem. Be sure to follow the steps in the Blueprint for Problem Solving. [Example 1]

A Geometry Problems

1. **Rectangle** A rectangle is twice as long as it is wide. The perimeter is 60 feet. Find the dimensions.

2. **Rectangle** The length of a rectangle is 5 times the width. The perimeter is 48 inches. Find the dimensions.

3. **Square** A square has a perimeter of 28 feet. Find the length of each side.

4. **Square** A square has a perimeter of 36 centimeters. Find the length of each side.

5. **Triangle** A triangle has a perimeter of 23 inches. The medium side is 3 inches more than the shortest side, and the longest side is twice the shortest side. Find the shortest side.

6. **Triangle** The longest side of a triangle is two times the shortest side, whereas the medium side is 3 meters more than the shortest side. The perimeter is 27 meters. Find the dimensions.

7. **Rectangle** The length of a rectangle is 3 meters less than twice the width. The perimeter is 18 meters. Find the width.

8. **Rectangle** The length of a rectangle is 1 foot more than twice the width. The perimeter is 20 feet. Find the dimensions.

9. **Livestock Pen** A livestock pen is built in the shape of a rectangle that is twice as long as it is wide. The perimeter is 48 feet. If the material used to build the pen is $1.75 per foot for the longer sides and $2.25 per foot for the shorter sides (the shorter sides have gates, which increase the cost per foot), find the cost to build the pen.

10. **Garden** A garden is in the shape of a square with a perimeter of 42 feet. The garden is surrounded by two fences. One fence is around the perimeter of the garden, whereas the second fence is 3 feet from the first fence, as Figure 4 indicates. If the material used to build the two fences is $1.28 per foot, what was the total cost of the fences?

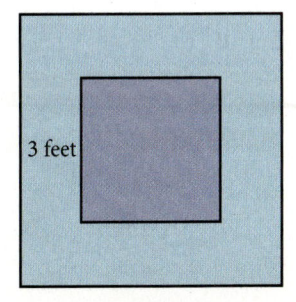

FIGURE 4

A Percent Problems [Example 2]

11. Money Shane returned from a trip to Las Vegas with $300.00, which was 50% more money than he had at the beginning of the trip. How much money did Shane have at the beginning of his trip?

12. Items Sold Every item in the Just a Dollar store is priced at $1.00. When Mary Jo opens the store, there is $125.50 in the cash register. When she counts the money in the cash register at the end of the day, the total is $1,058.60. If the sales tax rate is 8.5%, how many items were sold that day?

13. Monthly Salary An accountant earns $3,440 per month after receiving a 5.5% raise. What was the accountant's monthly income before the raise? Round your answer to the nearest cent.

14. Textbook Price Suppose a college bookstore buys a textbook from a publishing company and then marks up the price they paid for the book 33% and sells it to a student at the marked-up price. If the student pays $75.00 for the textbook, what did the bookstore pay for it? Round your answer to the nearest cent.

15. Movies *Batman Forever* grossed $52.8 million on its opening weekend and had one of the most successful movie launches in history. If *Batman Forever* accounted for approximately 53% of all box office receipts that weekend, what were the total box office receipts? Round your answer to the nearest tenth of a million dollars.

16. Hourly Wage A sheet metal worker earns $26.80 per hour after receiving a 4.5% raise. What was the sheet metal worker's hourly pay before the raise? Round your answer to the nearest cent.

A Interest Problems [Example 4]

17. Investing A woman has a total of $9,000 to invest. She invests part of the money in an account that pays 8% per year and the rest in an account that pays 9% per year. If the interest earned in the first year is $750, how much did she invest in each account?

	Dollars at 8%	Dollars at 9%	Total
Number of			
Interest on			

18. Investing A man invests $12,000 in two accounts. If one account pays 10% per year and the other pays 7% per year, how much was invested in each account if the total interest earned in the first year was $960?

	Dollars at 10%	Dollars at 7%	Total
Number of			
Interest on			

19. **Investing** A total of $15,000 is invested in two accounts. One of the accounts earns 12% per year, and the other earns 10% per year. If the total interest earned in the first year is $1,600, how much was invested in each account?

20. **Investing** A total of $11,000 is invested in two accounts. One of the two accounts pays 9% per year, and the other account pays 11% per year. If the total interest paid in the first year is $1,150, how much was invested in each account?

21. **Investing** Stacy has a total of $6,000 in two accounts. The total amount of interest she earns from both accounts in the first year is $500. If one of the accounts earns 8% interest per year and the other earns 9% interest per year, how much did she invest in each account?

22. **Investing** Travis has a total of $6,000 invested in two accounts. The total amount of interest he earns from the accounts in the first year is $410. If one account pays 6% per year and the other pays 8% per year, how much did he invest in each account?

A Miscellaneous Problems

23. **Ticket Prices** Miguel is selling tickets to a barbecue. Adult tickets cost $6.00 and children's tickets cost $4.00. He sells six more children's tickets than adult tickets. The total amount of money he collects is $184. How many adult tickets and how many children's tickets did he sell?

	Adult	Child
Number	x	$x + 6$
Income	$6(x)$	$4(x + 6)$

24. **Working Two Jobs** Maggie has a job working in an office for $10 an hour and another job driving a tractor for $12 an hour. One week she works in the office twice as long as she drives the tractor. Her total income for that week is $416. How many hours did she spend at each job?

	Office	Tractor
Hours Worked	$2x$	x
Wages Earned	$10(2x)$	$12x$

25. **Sales Tax** A woman owns a small, cash-only business in a state that requires her to charge 6% sales tax on each item she sells. At the beginning of the day, she has $250 in the cash register. At the end of the day, she has $1,204 in the register. How much money should she send to the state government for the sales tax she collected?

26. **Sales Tax** A store is located in a state that requires 6% tax on all items sold. If the store brings in a total of $3,392 in one day, how much of that total was sales tax?

B Table Building [Example 5]

27. Use $h = 32t - 16t^2$ to complete the table.

t	0	$\frac{1}{4}$	1	$\frac{7}{4}$	2
h					

28. Use $s = \dfrac{60}{t}$ to complete the table.

t	4	6	8	10
s				

29. Horse Racing The graph shows the total amount of money wagered on the Kentucky Derby. Use the information to fill in the table.

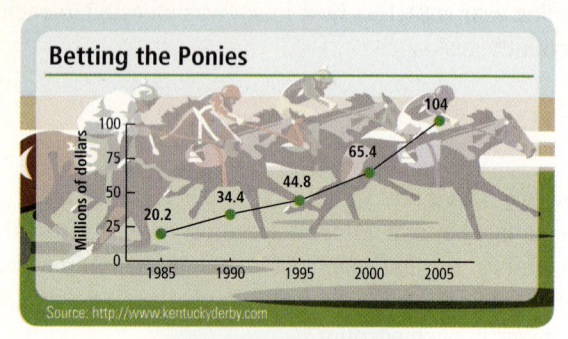

Betting the Ponies
- 1985: 20.2
- 1990: 34.4
- 1995: 44.8
- 2000: 65.4
- 2005: 104

Source: http://www.kentuckyderby.com

Year	Bets (millions of dollars)
1985	
1990	
1995	
2000	
2005	

30. BMI Sometimes body mass index is used to determine if a person is underweight, overweight, or normal. The equation used to find BMI is found by dividing a persons weight in kilograms by their height squared in meters. So if w is weight and h is height then your BMI is BMI $= \left(\frac{w}{h^2}\right)$. Use the formula to fill in the table. Round to the nearest tenth.

Weight (kg)	Height (m)	BMI (kg/m²)
38	1.25	
49	1.5	
53	1.75	
75	2	

Coffee Sales Use the information to complete the tables below. Round to the nearest tenth of a billion dollars.

31. HOT COFFEE SALES

Year	Sales (billions of dollars)
2005	
2006	
2007	
2008	
2009	

32. TOTAL COFFEE SALES

Year	Sales (billions of dollars)
2005	
2006	
2007	
2008	
2009	

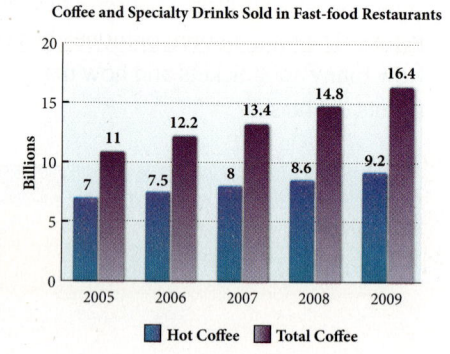

Java Sales
Coffee and Specialty Drinks Sold in Fast-food Restaurants

- 2005: Hot Coffee 7, Total Coffee 11
- 2006: Hot Coffee 7.5, Total Coffee 12.2
- 2007: Hot Coffee 8, Total Coffee 13.4
- 2008: Hot Coffee 8.6, Total Coffee 14.8
- 2009: Hot Coffee 9.2, Total Coffee 16.4

33. Distance A search is being conducted for someone guilty of a hit-and-run felony. In order to set up roadblocks at appropriate points, the police must determine how far the guilty party might have traveled during the past half-hour. Use the formula $d = rt$ with $t = 0.5$ hour to complete the following table.

Speed (miles per hour)	Distance (miles)
20	
30	
40	
50	
60	
70	

34. Speed To determine the average speed of a bullet when fired from a rifle, the time is measured from when the gun is fired until the bullet hits a target that is 1,000 feet away. Use the formula $d = rt$ with $d = 1,000$ feet to complete the following table.

Time (seconds)	Rate (feet per second)
1.00	
0.80	
0.64	
0.50	
0.40	
0.32	

35. Current A boat that can travel 10 miles per hour in still water is traveling along a stream with a current of 4 miles per hour. The distance the boat will travel upstream is given by the formula $d = (r - c) \cdot t$, and the distance it will travel downstream is given by the formula $d = (r + c) \cdot t$. Use these formulas with $r = 10$ and $c = 4$ to complete the following table.

Time (hours)	Distance Upstream (miles)	Distance Downstream (miles)
1		
2		
3		
4		
5		
6		

36. Wind A plane that can travel 300 miles per hour in still air is traveling in a wind stream with a speed of 20 miles per hour. The distance the plane will travel against the wind is given by the formula $d = (r - w) \cdot t$, and the distance it will travel with the wind is given by the formula $d = (r + w) \cdot t$. Use these formulas with $r = 300$ and $w = 20$ to complete the following table.

Time (hours)	Distance Against Wind (miles)	Distance With Wind (miles)
.5		
1		
1.5		
2		
2.5		
3		

Maximum Heart Rate In exercise physiology, a person's maximum heart rate, in beats per minute, is found by subtracting his age in years from 220. So, if A represents your age in years, then your maximum heart rate is

$$M = 220 - A$$

Use this formula to complete the following tables.

37.

Age (years)	Maximum Heart Rate (beats per minute)
18	
19	
20	
21	
22	
23	

38.

Age (years)	Maximum Heart Rate (beats per minute)
15	
20	
25	
30	
35	
40	

Problems 38–39 may be solved using a graphing calculator.

39. Livestock Pen A farmer buys 48 feet of fencing material to build a rectangular livestock pen. Fill in the second column of the table to find the length of the pen if the width is 2, 4, 6, 8, 10, or 12 feet. Then find the area of each of the pens formed.

w	l	A
2		
4		
6		
8		
10		
12		

40. Model Rocket A small rocket is projected straight up into the air with a velocity of 128 feet per second. The formula that gives the height h of the rocket t seconds after it is launched is

$$h = -16t^2 + 128t$$

Use this formula to find the height of the rocket after 1, 2, 3, 4, 5, and 6 seconds.

Time (seconds)	Height (feet)
1	
2	
3	
4	
5	
6	

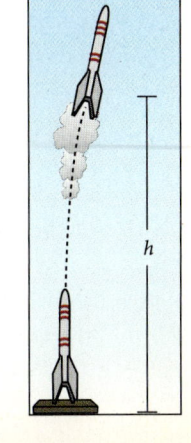

Left Blank Intentionally

Chapter 1 Summary

The numbers in brackets refer to the section(s) in which the topic can be found.

■ Opposites [1.1]

Any two real numbers the same distance from 0 on the number line, but in opposite directions from 0, are called *opposites*, or *additive inverses*. Opposites always add to 0.

EXAMPLES

1. The numbers 5 and −5 are opposites; their sum is 0.
   $5 + (−5) = 0$

■ Reciprocals [1.1]

Any two real numbers whose product is 1 are called *reciprocals*. Every real number has a reciprocal except 0. The number and its reciprocal must have the same sign.

2. The numbers 3 and $\frac{1}{3}$ are reciprocals; their product is 1.
   $3\left(\frac{1}{3}\right) = 1$

■ Absolute Value [1.1]

The *absolute value* of a real number is its distance from 0 on the number line. If $|x|$ represents the absolute value of x, then

$$|x| = \begin{cases} x & \text{if } x \geq 0 \\ -x & \text{if } x < 0 \end{cases}$$

The absolute value of a real number is never negative.

3. $|5| = 5$
   $|-5| = 5$

■ Properties of Real Numbers [1.1]

	For Addition	For Multiplication
Commutative	$a + b = b + a$	$ab = ba$
Associative	$a + (b + c) = (a + b) + c$	$a(bc) = (ab)c$
Identity	$a + 0 = a$	$a \cdot 1 = a$
Inverse	$a + (-a) = 0$	$a\left(\frac{1}{a}\right) = 1$
Distributive	$a(b + c) = ab + ac$	

■ Addition [1.2]

To add two real numbers with
1. *The same sign:* Simply add absolute values and use the common sign.
2. *Different signs:* Subtract the smaller absolute value from the larger absolute value. The answer has the same sign as the number with the larger absolute value.

4. $5 + 3 = 8$
   $5 + (-3) = 2$
   $-5 + 3 = -2$
   $-5 + (-3) = -8$

Chapter 1 Summary

Subtraction [1.2]

5. $6 - 2 = 6 + (-2) = 4$
$6 - (-2) = 6 + 2 = 8$

If a and b are real numbers,
$$a - b = a + (-b)$$
To subtract b, add the opposite of b.

Multiplication [1.2]

6. $5(4) = 20$
$5(-4) = -20$
$-5(4) = -20$
$-5(-4) = 20$

To multiply two real numbers, simply multiply their absolute values. Like signs give a positive answer. Unlike signs give a negative answer.

Division [1.2]

7. $\frac{12}{-3} = -4$
$\frac{-12}{-3} = 4$

If a and b are real numbers and $b \neq 0$, then
$$\frac{a}{b} = a \cdot \left(\frac{1}{b}\right)$$
To divide by b, multiply by the reciprocal of b.

Order of Operations [1.2]

8. $10 + (2 \cdot 3^2 - 4 \cdot 2)$
$= 10 + (2 \cdot 9 - 4 \cdot 2)$
$= 10 + (18 - 8)$
$= 10 + 10$
$= 20$

When evaluating a mathematical expression, we will perform the operations in the following order, beginning with the expression in the innermost parentheses or brackets and working our way out.

1. Simplify all numbers with exponents, working from left to right if more than one of these numbers is present.
2. Then, do all multiplications and divisions left to right.
3. Finally, perform all additions and subtractions left to right.

Scientific Notation [1.3]

9. $768,000 = 7.68 \times 10^5$
$0.00039 = 3.9 \times 10^{-4}$

A number is in scientific notation when it is written as the product of a number between 1 and 10 and an integer power of 10.

Addition Property of Equality [1.4]

10. We can solve
$$x + 3 = 5$$
by adding -3 to both sides:
$$x + 3 + (-3) = 5 + (-3)$$
$$x = 2$$

For algebraic expressions A, B, and C,

if $A = B$

then $A + C = B + C$

This property states that we can add the same quantity to both sides of an equation without changing the solution set.

Multiplication Property of Equality [1.4]

For algebraic expressions A, B, and C,

    if $A = B$

    then $AC = BC$ $(C \neq 0)$

Multiplying both sides of an equation by the same nonzero quantity never changes the solution set.

11. We can solve $3x = 12$ by multiplying both sides by $\frac{1}{3}$.

$$3x = 12$$
$$\frac{1}{3}(3x) = \frac{1}{3}(12)$$
$$x = 4$$

Strategy for Solving Linear Equations in One Variable [1.4]

Step 1: a. Use the distributive property to separate terms, if necessary.

           b. If fractions are present, consider multiplying both sides by the LCD to eliminate the fractions. If decimals are present, consider multiplying both sides by a power of 10 to clear the equation of decimals.

           c. Combine similar terms on each side of the equation.

Step 2: Use the addition property of equality to get all variable terms on one side of the equation and all constant terms on the other side. A variable term is a term that contains the variable (for example, $5x$). A constant term is a term that does not contain the variable (the number 3, for example).

Step 3: Use the multiplication property of equality to get the variable by itself on one side of the equation.

Step 4: Check your solution in the original equation to be sure that you have not made a mistake in the solution process.

12. Solve: $3(2x - 1) = 9$.

$$3(2x - 1) = 9$$
$$6x - 3 = 9$$
$$6x - 3 + 3 = 9 + 3$$
$$6x = 12$$
$$\frac{1}{6}(6x) = \frac{1}{6}(12)$$
$$x = 2$$

Blueprint for Problem Solving [1.5]

Step 1: **Read** the problem, and then mentally **list** the items that are known and the items that are unknown.

Step 2: **Assign a variable** to one of the unknown items. (In most cases this will amount to letting x = the item that is asked for in the problem.) Then **translate** the other **information** in the problem to expressions involving the variable.

Step 3: **Reread** the problem, and then **write an equation,** using the items and variables listed in steps 1 and 2, that describes the situation.

Step 4: **Solve the equation** found in step 3.

Step 5: **Write your answer** using a complete sentence.

Step 6: **Reread** the problem, and **check** your solution with the original words in the problem.

13. The perimeter of a rectangle is 32 inches. If the length is 3 times the width, find the dimensions.

Step 1: This step is done mentally.

Step 2: Let x = the width. Then the length is $3x$.

Step 3: The perimeter is 32; therefore
$$2x + 2(3x) = 32$$

Step 4: $8x = 32$
                  $x = 4$

Step 5: The width is 4 inches. The length is $3(4) = 12$ inches.

Step 6: The perimeter is $2(4) + 2(12)$, which is 32. The length is 3 times the width.

Left Blank Intentionally

Chapter 1   Review

Give the opposite and reciprocal of each number. [1.1]

1. 2

2. $-\dfrac{2}{5}$

Simplify. [1.1]

3. $|-3|$
4. $-|-5|$
5. $|-4|$
6. $|10-16|$

Match each expression on the left with the letter of the appropriate property (or properties) on the right. [1.1]

7. $x + 3 = 3 + x$
8. $(x + 2) + 3 = x + (2 + 3)$
9. $3(x + 4) = 3(4 + x)$
10. $(5x)y = x(5y)$
11. $(x + 2) + y = (x + y) + 2$
12. $3(1) = 3$
13. $5 + 0 = 5$
14. $5 + (-5) = 0$

a. Commutative property of addition
b. Commutative property of multiplication
c. Associative property of addition
d. Associative property of multiplication
e. Additive identity
f. Multiplicative identity
g. Additive inverse
h. Multiplicative inverse

Apply the distributive property. [1.1]

15. $-2(3x - 5)$
16. $-3(2x - 7)$
17. $-\dfrac{1}{2}(2x - 6)$
18. $-3(5x - 1)$

Combine similar terms. [1.1]

19. $-2y + 4y$
20. $-3x - x + 7x$
21. $3x - 2 + 5x + 7$
22. $2y + 4 - y - 2$

Simplify. [1.1]

23. $7 - 2(3y - 1) + 4y$
24. $4(3x - 1) - 5(6x + 2)$
25. $4(2a - 5) - (3a + 2)$

Find the following sums and differences. [1.2]

26. $5 - 3$
27. $-5 - (-3)$
28. $7 + (-2) - 4$
29. $6 - (-3) + 8$
30. $|-4| - |-3| + |-2|$
31. $|7 - 9| - |-3 - 5|$
32. $6 - (-3) - 2 - 5$
33. $2 \cdot 3^2 - 4 \cdot 2^3 + 5 \cdot 4^2$
34. $-\dfrac{1}{12} - \dfrac{1}{6} - \dfrac{1}{4} - \dfrac{1}{3}$
35. $-\dfrac{1}{3} - \dfrac{1}{4} - \dfrac{1}{6} - \dfrac{1}{12}$

Multiply. [1.2]

36. $\dfrac{3}{4} \cdot \dfrac{8}{5} \cdot \dfrac{5}{6}$
37. $\left(\dfrac{3}{4}\right)^3$
38. $\dfrac{1}{4} \cdot 8$
39. $6(-7)$
40. $-3(5)(-2)$
41. $7(3x)$
42. $-3(2x)$

Divide. [1.2]

43. $-\dfrac{5}{8} \div \dfrac{3}{4}$
44. $-12 \div \dfrac{1}{3}$
45. $\dfrac{3}{5} \div 6$
46. $\dfrac{4}{7} \div (-2)$

Simplify each expression. [1.2]

47. $2 + 3 \cdot 5$

48. $10 - 2 \cdot 3$

49. $20 \div 2 + 3$

50. $30 \div 6 + 4 \div 2$

51. $3 + 2(5 - 2)$

52. $(10 - 2)(7 - 3)$

53. $3 \cdot 4^2 - 2 \cdot 3^2$

54. $3 + 5(2 \cdot 3^2 - 10)$

55. $2(-5) - 3$

56. $3(-4) - 5$

57. $6 + 3(-2)$

58. $7 + 2(-4)$

59. $-3(2) - 5(6)$

60. $-4(3)^2 - 2(-1)^3$

61. $8 - 2(6 - 10)$

62. $(8 - 2)(6 - 10)$

63. $\dfrac{3(-4) - 8}{-5 - 5}$

64. $\dfrac{9(-1)^3 - 3(-6)^2}{6 - 9}$

65. $4 - (-2)\left[\dfrac{6 - 3(-4)}{1 + 5(-2)}\right]$

Write each number in scientific notation. [1.3]

66. 73,800,000

67. 0.002935

Write each number in standard form. [1.3]

68. 4.4×10^9

69. 1.6×10^{-5}

Solve each equation [1.4]

70. $x - 3 = 7$

71. $5x - 2 = 8$

72. $400 - 100a = 200$

73. $5 - \dfrac{2}{3}a = 7$

74. $4x - 2 = 7x + 7$

75. $\dfrac{3}{2}x - \dfrac{1}{6} = -\dfrac{7}{6}x - \dfrac{1}{6}$

76. $7y - 5 - 2y = 2y - 3$

77. $\dfrac{3y}{4} - \dfrac{1}{2} + \dfrac{3y}{2} = 2 - y$

78. $3(2x + 1) = 18$

79. $-\dfrac{1}{2}(4x - 2) = -x$

80. $8 - 3(2t + 1) = 5(t + 2)$

81. $8 + 4(1 - 3t) = -3(t - 4) + 2$

Solve each application. In each case, be sure to show the equation that describes the situation. [1.5]

82. Geometry The length of a rectangle is 3 times the width. The perimeter is 32 feet. Find the length and width.

83. Geometry The three sides of a triangle are given by three consecutive integers. If the perimeter is 12 meters, find the length of each side.

Chapter 1 Test

The numbers in brackets indicate the section(s) to which the problems correspond.

1. Simplify each of the following. [1.1]
   a. $-(-3)$ b. $-|-2|$

State the property or properties that justify each of the following. [1.1]

2. $4 + x = x + 4$

3. $5(1) = 5$

4. $3(xy) = (3x) \cdot y$

5. $(a + 1) + b = a + (1 + b)$

Simplify each of the following as much as possible. [1.1, 1.2]

6. $\dfrac{-4(-1) - (-10)}{5 - (-2)}$

7. $3 - 2\left[\dfrac{8(-1) - 7}{-3(2) - 4}\right]$

8. $-\dfrac{3}{8} + \dfrac{5}{12} - \left(-\dfrac{7}{9}\right)$

9. $-\dfrac{1}{2}(8x)$

10. $-4(3x + 2) + 7x$

11. $5(2y - 3) - (6y - 5)$

12. $3 + 4(2x - 5) - 5x$

13. $2 + 5a + 3(2a - 4)$

14. Add $-\dfrac{2}{3}$ to the product of -2 and $\dfrac{5}{6}$.

15. Subtract $\dfrac{3}{4}$ from the product of -4 and $\dfrac{7}{16}$.

Simplify each expression using the rule for order of operations. [1.2]

16. $3 \cdot 2^2 + 5 \cdot 3^2$

17. $6 + 2(4 \cdot 3 - 10)$

18. $12 - 8 \div 4 + 2 \cdot 3$

19. $20 - 4[3^2 - 2(2^3 - 6)]$

20. Write 0.0000385 in scientific notation. [1.3]

21. Write 6.75×10^6 in standard form. [1.3]

Simplify each expression completely. Write your answer using scientific notation. [1.3]

22. $(2.7 \times 10^6)(6.3 \times 10^{-4})$

23. $\dfrac{36{,}500{,}000}{73{,}000}$

24. $\dfrac{6.8 \times 10^{-8}}{1.7 \times 10^{-4}}$

25. $\dfrac{(640)(0.000000032)}{0.008}$

Solve the following equations. [1.4]

26. $x - 5 = 7$

27. $3y = -4$

28. $5(x - 1) - 2(2x + 3) = 5x - 4$

29. $0.07 - 0.02(3x + 1) = -0.04x + 0.01$

Solve each geometry problem. [1.5]

30. Find the dimensions of the rectangle below if its length is four times the width and the perimeter is 60 cm.

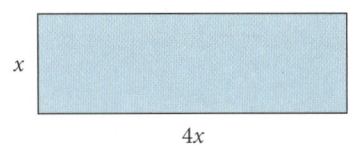

31. The perimeter of the triangle below is 35 meters. Find the value of x.

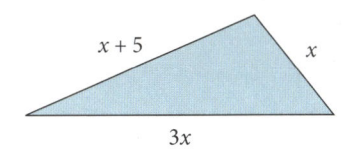

32. **Technology** A broad band cable connection can download information from the Internet at 3 megabits per second, or 3.0×10^6 bits per second. A music video streams over the Internet at 250 kilobits per second, or 2.5×10^5 bits per second. How many music videos can be played at the same time over this connection?

Left Blank Intentionally

Chapter 1 Projects

BASIC CONCEPTS AND PROPERTIES

GROUP PROJECT

Finding the Maximum Height of a Model Rocket

Number of People 3

Time Needed 20 minutes

Equipment Paper and pencil

Background In this chapter, we used formulas to do some table building. Once we have a table, it is sometimes possible to use just the table information to extend what we know about the situation described by the table. In this project, we take some basic information from a table and then look for patterns among the table entries. Once we have established the patterns, we continue them and, in so doing, solve a realistic application problem.

Procedure A model rocket is launched into the air. Table 1 gives the height of the rocket every second after takeoff for the first 5 seconds. Figure 1 is a graphical representation of the information in Table 1.

TABLE 1
HEIGHT OF A MODEL ROCKET

Time (seconds)	Height (feet)
0	0
1	176
2	320
3	432
4	512
5	560

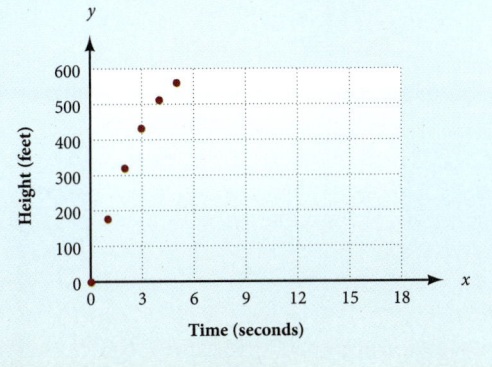

FIGURE 1

TABLE 1
HEIGHT OF A MODEL ROCKET

Time (seconds)	Height (feet)	First Differences	Second Differences
0	0		
1	176		
2	320		
3	432		
4	512		
5	560		

1. Table 1 is shown again above with two new columns. Fill in the first five entries in the First Differences column by finding the difference of consecutive heights. For example, the second entry in the First Differences column will be the difference of 320 and 176, which is 144.

2. Start filling in the Second Differences column by finding the differences of the First Differences.

3. Once you see the pattern in the Second Differences table, fill in the rest of the entries.

4. Now, using the results in the Second Differences table, go back and complete the First Differences table.

5. Now, using the results in the First Differences table, go back and complete the Heights column in the original table.

6. Plot the rest of the points from Table 1 on the graph in Figure 1.

7. What is the maximum height of the rocket?

8. How long was the rocket in the air?

RESEARCH PROJECT

THE EQUAL SIGN

We have been using the equal sign, =, for some time now. The first published use of the symbol was in 1557, with the publication of *The Whetstone of Witte* by the English mathematician and physician Robert Recorde. Research the first use of the symbols we use for addition, subtraction, multiplication, and division and then write an essay on the subject from your results.

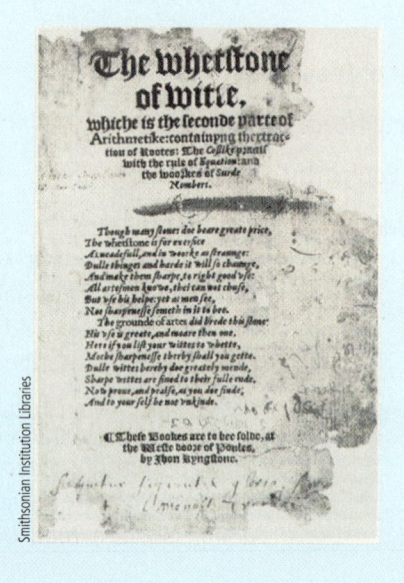

Exponents and Polynomials

Chapter Outline

2.1 Properties of Exponents
2.2 Polynomials, Sums, and Differences
2.3 Multiplication of Polynomials
2.4 The Greatest Common Factor
2.5 Special Factoring
2.6 Factoring Trinomials
2.7 A General Review of Factoring
2.8 Solving Equations by Factoring

Introduction

The French mathematician and philosopher, Blaise Pascal, was born in France in 1623. Both a scientist and a mathematician, he is credited for defining the scientific method. In the image below, Pascal carries a barometer to the top of the bell tower at the church of Saint-Jacques-de-la-Boucherie, overlooking Paris, to test a scientific theory.

The triangular array of numbers to the right of the painting is called Pascal's triangle. Pascal's triangle is connected to the work we will do in this chapter when we find increasing powers of binomials.

Chapter Pretest

Simplify. All answers should contain positive exponents only. (Assume all variables are nonzero.) [2.1]

1. $\left(\dfrac{2}{5}\right)^{-3}$
2. 3^{-4}
3. $a^{-8} \cdot a^9 \cdot a^3$
4. $\dfrac{x^{-3}}{x^{-5}}$
5. $(4ab^2)^2 \cdot (2a^3b)^3$
6. $\dfrac{(5x^2y)^{-2}}{(2x^2y^{-3})^{-1}}$

Simplify the following expressions. [2.2]

7. $\left(\dfrac{6}{5}x^3 - 2x - \dfrac{3}{5}\right) - \left(\dfrac{6}{5}x^2 - \dfrac{2}{5}x + \dfrac{3}{5}\right)$
8. $5 - 7[9(2x + 1) - 16x]$

Multiply. [2.3]

9. $(x + 7)(-5x + 4)$
10. $(3x - 2)(2x^2 + 6x - 5)$
11. $(3a^4 - 7)^2$
12. $(2x + 3)(2x - 3)$
13. $x(x - 7)(3x + 4)$
14. $\left(2x - \dfrac{1}{7}\right)\left(7x + \dfrac{1}{2}\right)$

Factor the following expressions. [2.4, 2.5, 2.6, 2.7]

15. $x^2 - 6x + 5$
16. $15x^4 + 33x^2 - 36$
17. $81x^4 - 16y^4$
18. $6ax - ay + 18b^2x - 3b^2y$
19. $y^3 + \dfrac{1}{27}$
20. $3x^4y^4 + 15x^3y^5 - 72x^2y^6$
21. $a^2 - 2ab - 36 + b^2$
22. $16 - x^4$

Solve each equation [2.8]

23. $\dfrac{1}{4}x^2 = -\dfrac{21}{8}x - \dfrac{5}{4}$
24. $243x^3 = 81x^4$
25. $(x + 5)(x - 2) = 8$
26. $x^3 + 5x^2 - 9x - 45 = 0$

Let $f(x) = x^2 - 2x - 15$. Find all values for the variable x for which $f(x) = g(x)$. [2.8]

27. $g(x) = 0$
28. $g(x) = 5 - 3x$

Getting Ready for Chapter 2

1. $-6 - (-9)$
2. $-5 - (-7)$
3. $2 - 18$
4. $3 - 10$
5. $-8(4)$
6. $-3(2)$
7. $\dfrac{6}{12}$
8. $\dfrac{18}{36}$
9. 3^4
10. 2^3
11. 10^3
12. 10^4
13. $(-2)^3$
14. $(-3)^3$
15. $\dfrac{1}{5^2}$
16. $\dfrac{1}{3^2}$
17. $\dfrac{1}{(-2)^3}$
18. $\dfrac{1}{(-5)^3}$
19. $\dfrac{1}{\left(\dfrac{2}{3}\right)^2}$
20. $\dfrac{1}{\left(\dfrac{3}{4}\right)^2}$
21. $4.52 \times 1{,}000$
22. $3.91 \times 10{,}000$
23. $376{,}000 \div 100{,}000$
24. $27{,}400 \div 10{,}000$

Chapter 2 Exponents and Polynomials

2.1 Properties of Exponents

Objectives
A Simplify expressions using the properties of exponents.

The figure shows a square and a cube, each with a side of length 1.5 centimeters. To find the area of the square, we raise 1.5 to the second power: 1.5^2. To find the volume of the cube, we raise 1.5 to the third power: 1.5^3.

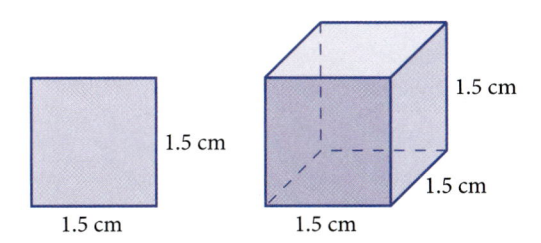

Because the area of the square is 1.5^2, we say second powers are *squares;* that is, x^2 is read "x squared." Likewise, since the volume of the cube is 1.5^3, we say third powers are *cubes,* that is, x^3 is read "x cubed." Exponents and the vocabulary associated with them are topics we will study in this section.

A Properties of Exponents

In this section, we will be concerned with the simplification of expressions that involve exponents. We begin by making some generalizations about exponents, based on specific examples.

EXAMPLE 1 Write the product $x^3 \cdot x^4$ with a single exponent.

SOLUTION
$$x^3 \cdot x^4 = (x \cdot x \cdot x)(x \cdot x \cdot x \cdot x)$$
$$= (x \cdot x \cdot x \cdot x \cdot x \cdot x \cdot x)$$
$$= x^7 \quad \text{Notice: } 3 + 4 = 7$$

We can generalize this result into the first property of exponents.

Property 1 for Exponents
If a is a real number and r and s are integers, then
$$a^r \cdot a^s = a^{r+s}$$

EXAMPLE 2 Write $(5^3)^2$ with a single exponent.

SOLUTION
$$(5^3)^2 = 5^3 \cdot 5^3$$
$$= 5^6 \quad \text{Notice: } 3 \cdot 2 = 6$$

Generalizing this result, we have a second property of exponents.

Property 2 for Exponents
If a is a real number and r and s are integers, then
$$(a^r)^s = a^{r \cdot s}$$

A third property of exponents arises when we have the product of two or more numbers raised to an integer power.

PRACTICE PROBLEMS
1. Write the product $x^5 \cdot x^4$ with a single exponent.

2. Simplify $(4^5)^2$ by writing an equivalent expression with a single exponent.

Answers
1. x^9
2. 4^{10}

2.1 Properties of Exponents

77

Chapter 2 Exponents and Polynomials

3. Expand $(2x)^3$ and then multiply.

EXAMPLE 3 Expand $(3x)^4$ and then multiply.

SOLUTION
$$(3x)^4 = (3x)(3x)(3x)(3x)$$
$$= (3 \cdot 3 \cdot 3 \cdot 3)(x \cdot x \cdot x \cdot x)$$
$$= 3^4 \cdot x^4 \quad \text{Notice: The exponent 4 distributes over the product } 3x$$
$$= 81x^4$$

Generalizing Example 3 we have property 3 for exponents.

> **Property 3 for Exponents**
> If a and b are any two real numbers and r is an integer, then
> $$(ab)^r = a^r \cdot b^r$$

Here are some examples that use combinations of the first three properties of exponents to simplify expressions involving exponents.

Simplify each expression.
4. $(-2x^3)(4x^5)$

EXAMPLE 4 Simplify $(-3x^2)(5x^4)$ using the properties of exponents.

SOLUTION
$$(-3x^2)(5x^4) = -3(5)(x^2 \cdot x^4) \quad \text{Commutative and associative}$$
$$= -15x^6 \quad \text{Property 1 for exponents}$$

5. $(-3y^3)^3(2y^6)$

EXAMPLE 5 Simplify $(-2x^2)^3(4x^5)$ using the properties of exponents.

SOLUTION
$$(-2x^2)^3(4x^5) = (-2)^3(x^2)^3(4x^5) \quad \text{Property 3}$$
$$= -8x^6 \cdot (4x^5) \quad \text{Property 2}$$
$$= (-8 \cdot 4)(x^6 \cdot x^5) \quad \text{Commutative and associative}$$
$$= -32x^{11} \quad \text{Property 1}$$

6. $(a^2)^3(a^3b^4)^2(b^5)^2$

EXAMPLE 6 Simplify $(x^2)^4(x^2y^3)^2(y^4)^3$ using the properties of exponents.

SOLUTION
$$(x^2)^4(x^2y^3)^2(y^4)^3 = x^8 \cdot x^4 \cdot y^6 \cdot y^{12} \quad \text{Properties 2 and 3}$$
$$= x^{12}y^{18} \quad \text{Property 1}$$

Note This property is actually a definition; that is, we are defining negative-integer exponents as indicating reciprocals. Doing so gives us a way to write an expression with a negative exponent as an equivalent expression with a positive exponent.

The next property of exponents deals with negative integer exponents.

> **Property 4 for Exponents**
> If a is any nonzero real number and r is a positive integer, then
> $$a^{-r} = \frac{1}{a^r}$$

Write with positive exponents, then simplify.
7. 3^{-2}

EXAMPLE 7 Write 5^{-2} with positive exponents, then simplify.

SOLUTION $\quad 5^{-2} = \dfrac{1}{5^2} = \dfrac{1}{25}$

8. $(-5)^{-3}$

Answers
3. $8x^3$
4. $-8x^8$
5. $-54y^{15}$
6. $a^{12}b^{18}$
7. $\dfrac{1}{9}$
8. $-\dfrac{1}{125}$

EXAMPLE 8 Write $(-2)^{-3}$ with positive exponents, then simplify.

SOLUTION $\quad (-2)^{-3} = \dfrac{1}{(-2)^3} = \dfrac{1}{-8} = -\dfrac{1}{8}$

2.1 Properties of Exponents

EXAMPLE 9 Write $\left(\dfrac{3}{4}\right)^{-2}$ with positive exponents, then simplify.

SOLUTION $\left(\dfrac{3}{4}\right)^{-2} = \dfrac{1}{\left(\dfrac{3}{4}\right)^{2}} = \dfrac{1}{\dfrac{9}{16}} = \dfrac{16}{9}$

If we generalize the result in Example 9, we have the following extension of property 4,

$$\left(\dfrac{a}{b}\right)^{-r} = \left(\dfrac{b}{a}\right)^{r}$$

which indicates that raising a fraction to a negative power is equivalent to raising the reciprocal of the fraction to the positive power.

Property 3 indicated that exponents distribute over products. Since division is defined in terms of multiplication, we can expect that exponents will distribute over quotients as well. Property 5 is the formal statement of this fact.

> **Property 5 for Exponents**
> If a and b are any two real numbers with $b \neq 0$, and r is an integer, then
> $$\left(\dfrac{a}{b}\right)^{r} = \dfrac{a^{r}}{b^{r}}$$

PROOF OF PROPERTY 5

$$\left(\dfrac{a}{b}\right)^{r} = \underbrace{\left(\dfrac{a}{b}\right)\left(\dfrac{a}{b}\right)\left(\dfrac{a}{b}\right) \cdots \left(\dfrac{a}{b}\right)}_{r \text{ factors}}$$

$$= \dfrac{a \cdot a \cdot a \cdots a}{b \cdot b \cdot b \cdots b} \begin{array}{l} \leftarrow r \text{ factors} \\ \leftarrow r \text{ factors} \end{array}$$

$$= \dfrac{a^{r}}{b^{r}}$$

Since multiplication with the same base resulted in addition of exponents, it seems reasonable to expect division with the same base to result in subtraction of exponents.

> **Property 6 for Exponents**
> If a is any nonzero real number, and r and s are any two integers, then
> $$\dfrac{a^{r}}{a^{s}} = a^{r-s}$$

Notice again that we have specified r and s to be any integers. Our definition of negative exponents is such that the properties of exponents hold for all integer exponents, whether positive or negative integers. Here is proof of property 6.

PROOF OF PROPERTY 6
Our proof is centered on the fact that division by a number is equivalent to multiplication by the reciprocal of the number.

$\dfrac{a^{r}}{a^{s}} = a^{r} \cdot \dfrac{1}{a^{s}}$ Dividing by a^{s} is equivalent to multiplying by $\dfrac{1}{a^{s}}$

$\phantom{\dfrac{a^{r}}{a^{s}}} = a^{r} a^{-s}$ Property 4

$\phantom{\dfrac{a^{r}}{a^{s}}} = a^{r+(-s)}$ Property 1

$\phantom{\dfrac{a^{r}}{a^{s}}} = a^{r-s}$ Definition of subtraction

---

Write with positive exponents, then simplify.

9. $\left(\dfrac{2}{3}\right)^{-2}$

Answer
9. $\dfrac{9}{4}$

Chapter 2 Exponents and Polynomials

10. Apply property 6 to each expression, and then simplify. All answers should contain positive exponents only.

a. $\dfrac{3^7}{3^4}$

b. $\dfrac{x^3}{x^{10}}$

c. $\dfrac{a^5}{a^{-7}}$

d. $\dfrac{m^{-3}}{m^{-5}}$

EXAMPLE 10 Apply property 6 to each expression, and then simplify the result. All answers that contain exponents should contain positive exponents only.

SOLUTION

a. $\dfrac{2^8}{2^3} = 2^{8-3} = 2^5 = 32$

b. $\dfrac{x^2}{x^{18}} = x^{2-18} = x^{-16} = \dfrac{1}{x^{16}}$

c. $\dfrac{a^6}{a^{-8}} = a^{6-(-8)} = a^{14}$

d. $\dfrac{m^{-5}}{m^{-7}} = m^{-5-(-7)} = m^2$

Let's complete our list of properties by looking at how the numbers 0 and 1 behave when used as exponents.

We can use the original definition for exponents when the number 1 is used as an exponent.

$$a^1 = \underbrace{a}_{1\text{ factor}}$$

For 0 as an exponent, consider the expression $\dfrac{3^4}{3^4}$. Since $3^4 = 81$, we have

$$\dfrac{3^4}{3^4} = \dfrac{81}{81} = 1$$

However, because we have the quotient of two expressions with the same base, we can subtract exponents.

$$\dfrac{3^4}{3^4} = 3^{4-4} = 3^0$$

Hence, 3^0 must be the same as 1.

Summarizing these results, we have our last property for exponents.

> **Property 7 for Exponents**
> If a is any real number, then
> $$a^1 = a$$
> and
> $$a^0 = 1 \quad \text{(as long as } a \neq 0\text{)}$$

11. Simplify.

a. $(3xy^5)^0$

b. $(3xy^5)^1$

EXAMPLE 11 Simplify.

SOLUTION

a. $(2x^2y^4)^0 = 1$

b. $(2x^2y^4)^1 = 2x^2y^4$

Here are some examples that use many of the properties of exponents. There are a number of ways to proceed on problems like these. You should use the method that works best for you.

Simplify.

12. $\dfrac{(x^2)^{-4}(x^3)^6}{(x^{-3})^5}$

EXAMPLE 12 Simplify $\dfrac{(x^3)^{-2}(x^4)^5}{(x^{-2})^7}$.

SOLUTION

$\dfrac{(x^3)^{-2}(x^4)^5}{(x^{-2})^7} = \dfrac{x^{-6}x^{20}}{x^{-14}}$ Property 2

$= \dfrac{x^{14}}{x^{-14}}$ Property 1

$= x^{28}$ Property 6: $x^{14-(-14)} = x^{28}$

Answers
10. a. 27   b. $\dfrac{1}{x^7}$ c. a^{12} d. m^2
11. a. 1   b. $3xy^5$
12. x^{25}

2.1 Properties of Exponents

EXAMPLE 13 Simplify $\dfrac{6a^5b^{-6}}{12a^3b^{-9}}$.

SOLUTION
$$\dfrac{6a^5b^{-6}}{12a^3b^{-9}} = \dfrac{6}{12} \cdot \dfrac{a^5}{a^3} \cdot \dfrac{b^{-6}}{b^{-9}} \quad \text{Write as separate fractions}$$
$$= \dfrac{1}{2}a^2b^3 \quad \text{Property 6}$$

NOTE: This last answer also can be written as $\dfrac{a^2b^3}{2}$. Either answer is correct.

EXAMPLE 14 Simplify $\dfrac{(4x^{-5}y^3)^2}{(x^4y^{-6})^{-3}}$.

SOLUTION
$$\dfrac{(4x^{-5}y^3)^2}{(x^4y^{-6})^{-3}} = \dfrac{16x^{-10}y^6}{x^{-12}y^{18}} \quad \text{Properties 2 and 3}$$
$$= 16x^2y^{-12} \quad \text{Property 6}$$
$$= 16x^2 \cdot \dfrac{1}{y^{12}} \quad \text{Property 4}$$
$$= \dfrac{16x^2}{y^{12}} \quad \text{Multiplication}$$

Question: in what way are $(-5)^2$ and -5^2 different?
Answer: In the first case, the base is -5. In the second case, the base is 5. The answer to the first is 25. The answer to the second is -25. Can you tell why? Would there be a difference in the answers if the exponent in each case were changed to 3?

> **STUDY SKILLS**
> **Test Properties About Which You Are Unsure**
> From time to time you will be in a situation in which you would like to apply a property or rule, but you are not sure if it is true. You can always test a property or statement by substituting numbers for variables. For instance, I always have students that rewrite $(x + 3)^2$ as $x^2 + 9$, thinking the two expressions are equivalent. The fact that the two expressions are not equivalent becomes obvious when we substitute 10 for x in each one.
>
> When $x = 10$, the expression $(x + 3)^2$ is $(10 + 3)^2 = 13^2 = 169$
>
> When $x = 10$, the expression $x^2 + 9 = 10^2 + 9 = 100 + 9 = 109$
>
> It is not unusual, nor is it wrong, to try occasionally to apply a property that doesn't exist. If you have any doubt about generalizations you are making, test them by replacing variables with numbers and simplifying.

> ## Getting Ready for Class
> *After reading through the preceding section, respond in your own words and in complete sentences.*
> 1. Explain the difference between -2^4 and $(-2)^4$.
> 2. Explain the difference between 2^5 and 2^{-5}.
> 3. If a positive base is raised to a negative exponent, can the result be a negative number?
> 4. State Property 1 for exponents in your own words.

Simplify.
13. $\dfrac{18a^7b^{-4}}{36a^2b^{-8}}$

Simplify.
14. $\dfrac{(3x^{-2}y^7)^2}{(x^5y^{-2})^{-4}}$

Answers
13. $\dfrac{a^5b^4}{2}$
14. $9x^{16}y^6$

Left Blank Intentionally

Problem Set 2.1

Evaluate each of the following.

1. 4^2
2. $(-4)^2$
3. -4^2
4. $-(-4)^2$
5. -0.3^3
6. $(-0.3)^3$
7. 2^5
8. 2^4
9. $\left(\dfrac{1}{2}\right)^3$
10. $\left(\dfrac{3}{4}\right)^2$
11. $-\left(\dfrac{5}{6}\right)^2$
12. $-\left(\dfrac{7}{8}\right)^2$

A Use the properties of exponents to simplify each of the following as much as possible. [Examples 1–6]

13. $x^5 \cdot x^4$
14. $x^6 \cdot x^3$
15. $(2^3)^2$
16. $(3^2)^2$
17. $\left(-\dfrac{2}{3}x^2\right)^3$
18. $\left(-\dfrac{3}{5}x^4\right)^3$
19. $-3a^2(2a^4)$
20. $5a^7(-4a^6)$
21. $a^5 \cdot a^9$
22. $y^2 \cdot y^5$
23. $(4^2)^4$
24. $(6^3)^5$
25. $(-3x^4)^2$
26. $\left(-\dfrac{5}{2}x^5\right)^3$
27. $-4b^2(-2b^5)$
28. $6a^3(-4a^4)$
29. $-\left(\dfrac{1}{2}a^3\right)^4$
30. $-(-5x)^3$

A Write each of the following with positive exponents. Then simplify as much as possible. [Examples 7–9]

31. 3^{-2}
32. $(-5)^{-2}$
33. $(-2)^{-5}$
34. 2^{-5}
35. $\left(\dfrac{3}{4}\right)^{-2}$
36. $\left(\dfrac{3}{5}\right)^{-2}$
37. $\left(\dfrac{1}{3}\right)^{-2} + \left(\dfrac{1}{2}\right)^{-3}$
38. $\left(\dfrac{1}{2}\right)^{-2} + \left(\dfrac{1}{3}\right)^{-3}$

A Simplify each expression. Write all answers with positive exponents only. (Assume all variables are nonzero.)

39. $x^{-4}x^7$
40. $x^{-3}x^8$
41. $(a^2b^{-5})^3$

42. $(a^4b^{-3})^3$

43. $(5y^4)^{-3}(2y^{-2})^3$

44. $(3y^5)^{-2}(2y^{-4})^3$

45. $\left(\dfrac{1}{2}x^3\right)\left(\dfrac{2}{3}x^4\right)\left(\dfrac{3}{5}x^{-7}\right)$

46. $\left(\dfrac{1}{7}x^{-3}\right)\left(\dfrac{7}{8}x^{-5}\right)\left(\dfrac{8}{9}x^8\right)$

47. $(4a^5b^2)(2b^{-5}c^2)(3a^7c^4)$

48. $(3a^{-2}c^3)(5b^{-6}c^5)(4a^6b^{-2})$

49. $(2x^2y^{-5})^3(3x^{-4}y^2)^{-4}$

50. $(4x^{-4}y^9)^{-2}(5x^4y^{-3})^2$

A Use the properties of exponents to simplify each expression. Write all answers with positive exponents only. (Assume all variables are nonzero.) [Examples 10–14]

51. $\dfrac{x^{-1}}{x^9}$

52. $\dfrac{x^{-3}}{x^5}$

53. $\dfrac{a^4}{a^{-6}}$

54. $\dfrac{a^5}{a^{-2}}$

55. $\dfrac{t^{-10}}{t^{-4}}$

56. $\dfrac{t^{-8}}{t^{-5}}$

57. $\left(\dfrac{x^5}{x^3}\right)^6$

58. $\left(\dfrac{x^7}{x^4}\right)^5$

59. $\dfrac{(x^5)^6}{(x^3)^4}$

60. $\dfrac{(x^7)^3}{(x^4)^5}$

61. $\dfrac{(x^{-2})^3(x^3)^{-2}}{x^{10}}$

62. $\dfrac{(x^{-4})^3(x^3)^{-4}}{x^{10}}$

63. $\dfrac{5a^8b^3}{20a^5b^{-4}}$

64. $\dfrac{7a^6b^{-2}}{21a^2b^{-5}}$

65. $\dfrac{(3x^{-2}y^8)^4}{(9x^4y^{-3})^2}$

66. $\dfrac{(6x^{-3}y^{-5})^2}{(3x^{-4}y^{-3})^4}$

67. $\left(\dfrac{8x^2y}{4x^4y^{-3}}\right)^4$

68. $\left(\dfrac{5x^4y^5}{10xy^{-2}}\right)^3$

69. $\left(\dfrac{x^{-5}y^2}{x^{-3}y^5}\right)^{-2}$

70. $\left(\dfrac{x^{-8}y^{-3}}{x^{-5}y^6}\right)^{-1}$

2.1 Problem Set

Write each expression as a perfect square.

71. $x^4y^2 = (\quad)^2$
72. $x^8y^6 = (\quad)^2$
73. $9a^2b^4 = (\quad)^2$
74. $225x^6y^{12} = (\quad)^2$

Write each expression as a perfect cube.

75. $8a^3 = (\quad)^3$
76. $27b^3 = (\quad)^3$
77. $64x^3y^{12} = (\quad)^3$
78. $216x^{15}y^{21} = (\quad)^3$

79. Let $x = 2$ in each of the following expressions and simplify.
  a. x^3x^2
  b. $(x^3)^2$
  c. x^5
  d. x^6

80. Let $x = -1$ in each of the following expressions and simplify.
  a. x^3x^4
  b. $(x^3)^4$
  c. x^7
  d. x^{12}

81. Let $x = 2$ in each of the following expressions and simplify.
  a. $\dfrac{x^5}{x^2}$
  b. x^3
  c. $\dfrac{x^2}{x^6}$
  d. x^{-4}

82. Let $x = -1$ in each of the following expressions and simplify.
  a. $\dfrac{x^{14}}{x^9}$
  b. x^5
  c. $\dfrac{x^{13}}{x^9}$
  d. x^4

83. Write each expression as a perfect square.
  a. $\dfrac{1}{49} = (\quad)^2$
  b. $\dfrac{1}{121} = (\quad)^2$
  c. $\dfrac{1}{4x^2} = (\quad)^2$
  d. $\dfrac{1}{64x^4} = (\quad)^2$

84. Write each expression as a perfect cube.
  a. $\dfrac{1}{125x^3} = (\quad)^3$
  b. $\dfrac{1}{64y^{12}} = (\quad)^3$
  c. $\dfrac{x^6}{216y^9} = (\quad)^3$
  d. $\dfrac{8a^9}{27b^{15}} = (\quad)^3$

Simplify.

85. $2 \cdot 2^{n-1}$
86. $3 \cdot 3^{n-1}$
87. $\dfrac{ar^6}{ar^3}$
88. $\dfrac{ar^7}{ar^4}$

You will see these problems later in the book. Multiply.

89. $8x^3 \cdot 10y^6$

90. $5y^2 \cdot 4x^2$

91. $8x^3 \cdot 9y^3$

92. $4y^3 \cdot 3x^2$

93. $3x \cdot 5y$

94. $3xy \cdot 5z$

95. $4x^6y^6 \cdot 3x$

96. $16x^4y^4 \cdot 3y$

97. $27a^6c^3 \cdot 2b^2c$

98. $8a^3b^3 \cdot 5a^2b$

Divide.

99. $\dfrac{10x^5}{5x^2}$

100. $\dfrac{-15x^4}{5x^2}$

101. $\dfrac{20x^3}{5x^2}$

102. $\dfrac{25x^7}{-5x^2}$

103. $\dfrac{8x^3y^5}{-2x^2y}$

104. $\dfrac{-16x^2y^2}{-2x^2y}$

105. $\dfrac{4x^4y^3}{-2x^2y}$

106. $\dfrac{10a^4b^2}{4a^2b^2}$

■ Applying the Concepts

107. Google Earth This Google Earth image is of the Luxor Hotel in Las Vegas, Nevada. The casino has a square base with sides of 525 feet. What is the area of the casino floor?

108. Google Earth This is a three dimensional model created by Google Earth of the Louvre Museum in Paris, France. The pyramid that dominates the Napoleon Courtyard has a square base with sides of 35.50 meters. What is the area of the base of the pyramid?

Getting Ready for the Next Section

Simplify.

109. $-4x + 9x$

110. $-6x - 2x$

111. $5x^2 + 3x^2$

112. $7x^2 + 3x^2$

113. $-8x^3 + 10x^3$

114. $4x^3 - 7x^3$

115. $2x + 3 - 2x - 8$

116. $9x - 4 - 9x - 10$

117. $-1(2x - 3)$

118. $-1(-3x + 1)$

119. $-3(-3x - 2)$

120. $-4(-5x + 3)$

121. $-500 + 27(100) - 0.1(100)^2$

122. $-500 + 27(170) - 0.1(170)^2$

Extending the Concepts

Assume all variable exponents represent positive integers and simplify each expression.

123. $x^{m+2} \cdot x^{-2m} \cdot x^{m-5}$

124. $x^{m-4} x^{m+9} x^{-2m}$

125. $(y^m)^2 (y^{-3})^m (y^{m+3})$

126. $(y^m)^{-4} (y^3)^m (y^{m+6})$

127. $\dfrac{x^{n+2}}{x^{n-3}}$

128. $\dfrac{x^{n-3}}{x^{n-7}}$

Left Blank Intentionally

2.2 Polynomials, Sums, and Differences

Objectives
A Give the degree of a polynomial.
B Add and subtract polynomials.
C Evaluate a polynomial for a given value of its variable.

The chart is from a company that duplicates DVDs. It shows the revenue and cost to duplicate a 30-minute DVD. From the chart you can see that 300 copies will bring in $900 in revenue, with a cost of $600. The profit is the difference between revenue and cost, or $300.

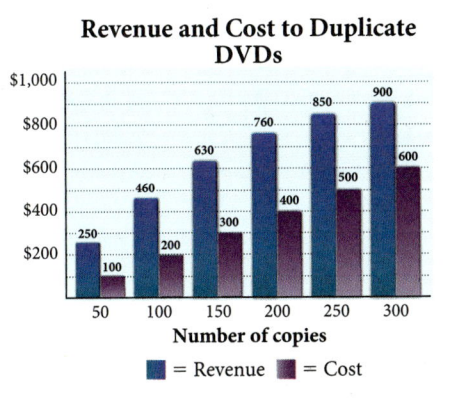

Revenue and Cost to Duplicate DVDs

(Bar chart showing revenue and cost for 50, 100, 150, 200, 250, 300 copies: Revenue = 250, 460, 630, 760, 850, 900; Cost = 100, 200, 300, 400, 500, 600)

■ = Revenue   ■ = Cost

The relationship between profit, revenue, and cost is one application of the polynomials we will study in this section. Let's begin with a definition that we will use to build polynomials.

Polynomials in General

> **Definition**
> A **term**, or **monomial**, is a constant or the product of a constant and one or more variables raised to whole-number exponents.

The following are monomials, or terms:

$$-16 \quad 3x^2y \quad -\frac{2}{5}a^3b^2c \quad xy^2z$$

The numerical part of each monomial is called the *numerical coefficient,* or just *coefficient* for short. For the preceding terms, the coefficients are -16, 3, $-\frac{2}{5}$, and 1. Notice that the coefficient for xy^2z is understood to be 1.

> **Definition**
> A **polynomial** is any finite sum of terms. Because subtraction can be written in terms of addition, finite differences are also included in this definition.

The following are polynomials:

$$2x^2 - 6x + 3 \quad -5x^2y + 2xy^2 \quad 4a - 5b + 6c + 7d$$

Polynomials can be classified further according to the number of terms present. If a polynomial consists of two terms, it is said to be a *binomial.* If it has three terms, it is called a *trinomial.* And, as stated, a polynomial with only one term is said to be a *monomial.*

PRACTICE PROBLEMS

Identify each expression as monomial, binomial, or trinomial, and give the degree of each.
1. $3x + 1$
2. $4x^2 + 2x + 5$
3. -17
4. $4x^5 - 7x^3$
5. $4x^3 - 5x^2 + 2x$

A Degree of a Polynomial

Definition
The **degree** of a polynomial with one variable is the highest power to which the variable is raised in any one term.

EXAMPLES

1. $6x^2 + 2x - 1$ — A trinomial of degree 2
2. $5x - 3$ — A binomial of degree 1
3. $7x^6 - 5x^3 + 2x - 4$ — A polynomial of degree 6
4. $-7x^4$ — A monomial of degree 4
5. 15 — A monomial of degree 0

Polynomials in one variable are usually written in decreasing powers of the variable. When this is the case, the coefficient of the first term is called the *leading coefficient*. In Example 1, the leading coefficient is 6. In Example 2, it is 5. The leading coefficient in Example 3 is 7.

Definition
Two or more terms that differ only in their numerical coefficients are called **similar**, or **like**, terms. Since similar terms differ only in their coefficients, they have identical variable parts.

B Addition and Subtraction of Polynomials

To add two polynomials, we simply apply the commutative and associative properties to group similar terms together and then use the distributive property as we have in the following example.

EXAMPLE 6

Add $5x^2 - 4x + 2$ and $3x^2 + 9x - 6$.

SOLUTION
$(5x^2 - 4x + 2) + (3x^2 + 9x - 6)$
$= (5x^2 + 3x^2) + (-4x + 9x) + (2 - 6)$ Commutative and associative properties
$= (5 + 3)x^2 + (-4 + 9)x + (2 - 6)$ Distributive property
$= 8x^2 + 5x + (-4)$
$= 8x^2 + 5x - 4$

6. Add $3x^2 + 2x - 5$ and $2x^2 - 7x + 3$.

Note In practice it is not necessary to show all the steps shown in Example 6. It is important to understand that addition of polynomials is equivalent to combining similar terms.

EXAMPLE 7

Find the sum of $-8x^3 + 7x^2 - 6x + 5$ and $10x^3 + 3x^2 - 2x - 6$.

SOLUTION We can add the two polynomials using the method of Example 6, or we can arrange similar terms in columns and add vertically. Using the column method, we have

$$\begin{array}{r} -8x^3 + 7x^2 - 6x + 5 \\ 10x^3 + 3x^2 - 2x - 6 \\ \hline 2x^3 + 10x^2 - 8x - 1 \end{array}$$

7. Find the sum of $x^3 + 7x^2 + 3x + 2$ and $-3x^3 - 2x^2 + 3x - 1$.

Answers
1. Binomial, 1
2. Trinomial, 2
3. Monomial, 0
4. Binomial, 5
5. Trinomial, 3
6. $5x^2 - 5x - 2$
7. $-2x^3 + 5x^2 + 6x + 1$

2.2 Polynomials, Sums, and Differences

To find the difference of two polynomials, we need to use the fact that the opposite of a sum is the sum of the opposites; that is,

$$-(a + b) = -a + (-b)$$

One way to remember this is to observe that $-(a + b)$ is equivalent to $-1(a + b) = (-1)a + (-1)b = -a + (-b)$.

If a negative sign directly precedes the parentheses surrounding a polynomial, we may remove the parentheses and the preceding negative sign by changing the sign of each term within the parentheses. For example:

$$-(3x + 4) = -3x + (-4) = -3x - 4$$
$$-(5x^2 - 6x + 9) = -5x^2 + 6x - 9$$
$$-(-x^2 + 7x - 3) = x^2 - 7x + 3$$

EXAMPLE 8
Subtract $(9x^2 - 3x + 5) - (4x^2 + 2x - 3)$.

SOLUTION We subtract by adding the opposite of each term in the polynomial that follows the subtraction sign:

$$(9x^2 - 3x + 5) - (4x^2 + 2x - 3)$$
$$= 9x^2 - 3x + 5 + (-4x^2) + (-2x) + 3 \quad \text{The opposite of a sum is the sum of the opposites}$$
$$= (9x^2 - 4x^2) + (-3x - 2x) + (5 + 3) \quad \text{Commutative and associative properties}$$
$$= 5x^2 - 5x + 8 \quad \text{Combine similar terms}$$

EXAMPLE 9
Subtract $4x^2 - 9x + 1$ from $-3x^2 + 5x - 2$.

SOLUTION Again, to subtract, we add the opposite:

$$(-3x^2 + 5x - 2) - (4x^2 - 9x + 1)$$
$$= -3x^2 + 5x - 2 - 4x^2 + 9x - 1$$
$$= (-3x^2 - 4x^2) + (5x + 9x) + (-2 - 1)$$
$$= -7x^2 + 14x - 3$$

EXAMPLE 10
Simplify $4x - 3[2 - (3x + 4)]$.

SOLUTION Removing the innermost parentheses first, we have

$$4x - 3[2 - (3x + 4)] = 4x - 3(2 - 3x - 4)$$
$$= 4x - 3(-3x - 2)$$
$$= 4x + 9x + 6$$
$$= 13x + 6$$

EXAMPLE 11
Simplify $(2x + 3) - [(3x + 1) - (x - 7)]$.

SOLUTION
$$(2x + 3) - [(3x + 1) - (x - 7)] = (2x + 3) - (3x + 1 - x + 7)$$
$$= (2x + 3) - (2x + 8)$$
$$= 2x + 3 - 2x - 8$$
$$= -5$$

C Evaluating Polynomials

In the example that follows we will find the value of a polynomial for a given value of the variable.

---

8. Subtract.
   $(4x^2 - 2x + 7) - (7x^2 - 3x + 1)$

9. Subtract $3x + 5$ from $7x - 4$.

10. Simplify $2x - 4[6 - (5x + 3)]$.

11. Simplify
    $(9x - 4) - [(2x + 5) - (x + 3)]$

Answers
8. $-3x^2 + x + 6$
9. $4x - 9$
10. $22x - 12$
11. $8x - 6$

Chapter 2 Exponents and Polynomials

12. Find the value of
$2x^3 - 3x^2 + 4x - 8$ when x is -2.

EXAMPLE 12 Find the value of $5x^3 - 3x^2 + 4x - 5$ when x is 2.

SOLUTION We begin by substituting 2 for x in the original polynomial:

When $x = 2$
the polynomial $5x^3 - 3x^2 + 4x - 5$
becomes $5 \cdot 2^3 - 3 \cdot 2^2 + 4 \cdot 2 - 5 = 5 \cdot 8 - 3 \cdot 4 + 4 \cdot 2 - 5$
$= 40 - 12 + 8 - 5$
$= 31$

Polynomials and Function Notation

Example 12 can be restated using function notation by calling the polynomial $P(x)$ and asking for $P(2)$. The solution would look like this:

If $P(x) = 5x^3 - 3x^2 + 4x - 5$
then $P(2) = 5 \cdot 2^3 - 3 \cdot 2^2 + 4 \cdot 2 - 5$
$= 31$

Our next example is stated in terms of function notation.

Three functions that occur very frequently in business and economics classes are profit, revenue, and cost functions. If a company manufactures and sells x items, then the revenue $R(x)$ is the total amount of money obtained by selling all x items. The cost $C(x)$ is the total amount of money it costs the company to manufacture the x items. The profit $P(x)$ obtained by selling all x items is the difference between the revenue and the cost and is given by the equation

$$P(x) = R(x) - C(x)$$

EXAMPLE 13 A company produces and sells copies of an accounting program for home computers. The total weekly cost (in dollars) to produce x copies of the program is $C(x) = 8x + 500$. Find its weekly profit if the total revenue obtained from selling all x programs is $R(x) = 35x - 0.1x^2$. How much profit will the company make if it produces and sells 100 programs a week? That is, find $P(100)$.

13. Use the information in Example 13 to find the profit associated with producing and selling 170 programs per week.

SOLUTION Using the equation $P(x) = R(x) - C(x)$ and the information given in the problem, we have

$P(x) = R(x) - C(x)$
$= 35x - 0.1x^2 - (8x + 500)$
$= 35x - 0.1x^2 - 8x - 500$
$= -500 + 27x - 0.1x^2$

If the company produces and sells 100 copies of the program, its weekly profit will be

$P(100) = -500 + 27(100) - 0.1(100)^2$
$= -500 + 27(100) - 0.1(10,000)$
$= -500 + 2,700 - 1,000$
$= 1,200$

The weekly profit is $1,200.

> ### Getting Ready for Class
>
> After reading through the preceding section, respond in your own words and in complete sentences.
>
> 1. Is $3x^2 + 2x - \dfrac{1}{x}$ a polynomial? Explain.
> 2. What are similar terms?
> 3. Explain in words how you subtract one polynomial from another.
> 4. What is revenue?

Answer
12. -44
13. $1,200

Problem Set 2.2

A Identify those of the following that are monomials, binomials, or trinomials. Give the degree of each, and name the leading coefficient. [Examples 1–5]

1. $5x^2 - 3x + 2$
2. $2x^2 + 4x - 1$
3. $3x - 5$
4. $5y + 3$
5. $8a^2 + 3a - 5$
6. $9a^2 - 8a - 4$
7. $4x^3 - 6x^2 + 5x - 3$
8. $9x^4 + 4x^3 - 2x^2 + x$
9. $-\dfrac{3}{4}$
10. -16
11. $4x - 5 + 6x^3$
12. $9x + 2 + 3x^3$

B Simplify each of the following by combining similar terms. [Examples 6–9]

13. $(4x + 2) + (3x - 1)$
14. $(8x - 5) + (-5x + 4)$
15. $2x^2 - 3x + 10x - 15$
16. $6x^2 - 4x - 15x + 10$
17. $12a^2 + 8ab - 15ab - 10b^2$
18. $28a^2 - 8ab + 7ab - 2b^2$
19. $(5x^2 - 6x + 1) - (4x^2 + 7x - 2)$
20. $(11x^2 - 8x) - (4x^2 - 2x - 7)$
21. $\left(\dfrac{1}{2}x^2 - \dfrac{1}{3}x - \dfrac{1}{6}\right) - \left(\dfrac{1}{4}x^2 + \dfrac{7}{12}x\right) + \left(\dfrac{1}{3}x - \dfrac{1}{12}\right)$
22. $\left(\dfrac{2}{3}x^2 - \dfrac{1}{2}x\right) - \left(\dfrac{1}{4}x^2 + \dfrac{1}{6}x\right) + \dfrac{1}{12} - \left(\dfrac{1}{2}x^2 + \dfrac{1}{4}\right)$

23. $(y^3 - 2y^2 - 3y + 4) - (2y^3 - y^2 + y - 3)$

24. $(8y^3 - 3y^2 + 7y + 2) - (-4y^3 + 6y^2 - 5y - 8)$

25. $(5x^3 - 4x^2) - (3x + 4) + (5x^2 - 7) - (3x^3 + 6)$

26. $(x^3 - x) - (x^2 + x) + (x^3 - 1) - (-3x + 2)$

27. $\left(\frac{4}{7}x^2 - \frac{1}{7}xy + \frac{1}{14}y^2\right) - \left(\frac{1}{2}x^2 - \frac{2}{7}xy - \frac{9}{14}y^2\right)$

28. $\left(\frac{1}{5}x^2 - \frac{1}{2}xy + \frac{1}{10}y^2\right) - \left(-\frac{3}{10}x^2 + \frac{2}{5}xy - \frac{1}{2}y^2\right)$

29. $(3a^3 + 2a^2b + ab^2 - b^3) - (6a^3 - 4a^2b + 6ab^2 - b^3)$

30. $(a^3 - 3a^2b + 3ab^2 - b^3) - (a^3 + 3a^2b + 3ab^2 + b^3)$

31. Subtract $2x^2 - 4x$ from $2x^2 - 7x$.

32. Subtract $-3x + 6$ from $-3x + 9$.

33. Find the sum of $x^2 - 6xy + y^2$ and $2x^2 - 6xy - y^2$.

34. Find the sum of $9x^3 - 6x^2 + 2$ and $3x^2 - 5x + 4$.

35. Subtract $-8x^5 - 4x^3 + 6$ from $9x^5 - 4x^3 - 6$.

36. Subtract $4x^4 - 3x^3 - 2x^2$ from $2x^4 + 3x^3 + 4x^2$.

37. Find the sum of $11a^2 + 3ab + 2b^2$, $9a^2 - 2ab + b^2$, and $-6a^2 - 3ab + 5b^2$.

38. Find the sum of $a^2 - ab - b^2$, $a^2 + ab - b^2$, and $a^2 + 2ab + b^2$.

B Simplify each of the following. Begin by working on the innermost parentheses. [Examples 10–11]

39. $-[2 - (4 - x)]$

40. $-[-3 - (x - 6)]$

41. $-5[-(x - 3) - (x + 2)]$

42. $-6[(2x - 5) - 3(8x - 2)]$

43. $4x - 5[3 - (x - 4)]$

44. $x - 7[3x - (2 - x)]$

45. $-(3x - 4y) - [(4x + 2y) - (3x + 7y)]$

46. $(8x - y) - [-(2x + y) - (-3x - 6y)]$

47. $4a - \{3a + 2[a - 5(a + 1) + 4]\}$

48. $6a - \{-2a - 6[2a + 3(a - 1) - 6]\}$

C [Examples 12–13]

49. Find the value of $2x^2 - 3x - 4$ when x is 2.

50. Find the value of $4x^2 + 3x - 2$ when x is -1.

51. Find the value of $\frac{3}{2}x^2 - \frac{3}{4}x + 1$
    a. when x is 12
    b. when x is -8

52. Find the value of $\frac{2}{5}x^2 - \frac{1}{10}x + 2$
    a. when x is 10
    b. when x is -10

53. Find the value of $x^3 - x^2 + x - 1$
    a. when x is 4
    b. when x is -2

54. Find the value of $x^3 + x^2 + x - 1$
    a. when x is 5
    b. when x is -2

55. Find the value of $11.5x - 0.05x^2$
    a. when x is 10
    b. when x is -10

56. Find the value of $11.5x - 0.01x^2$
    a. when x is 10
    b. when x is -10

57. Find the value of $600 + 1{,}000x - 100x^2$
    a. when x is -4
    b. when x is 4

58. Find the value of $500 + 800x - 100x^2$
    a. when x is -6
    b. when x is 8

Chapter 2 Exponents and Polynomials

■ Applying the Concepts

59. Education The chart shows the average income for people with different levels of education. In a high school's graduating class, x students plan to get their Bachelor's Degree and y students plan to go on and get their Master's Degree. The next year, twice as many students plan to get their Bachelor's Degree than the year before but the number of students that plan to go on and get their Master's Degree stays the same as the year before. Write an expression that describes the total income of each year and then find the total income for both years in terms of x and y.

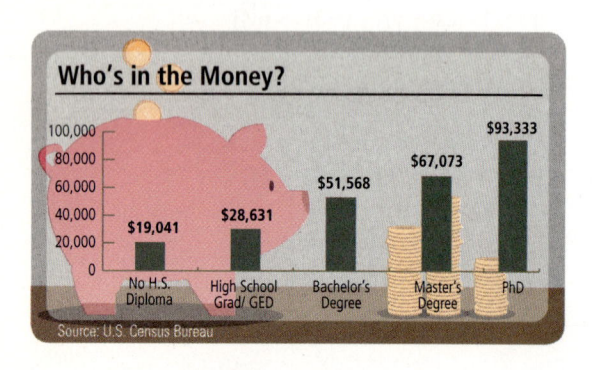

Who's in the Money?

- No H.S. Diploma: $19,041
- High School Grad/ GED: $28,631
- Bachelor's Degree: $51,568
- Master's Degree: $67,073
- PhD: $93,333

Source: U.S. Census Bureau

60. Classroom Energy The chart shows how much energy is wasted in the classroom by leaving appliances on. If twice as many ceiling fans were left on than printers and the DVD player was left on, write an expression that describes the energy used in this situation. Then find the value if 2 printers were left on.

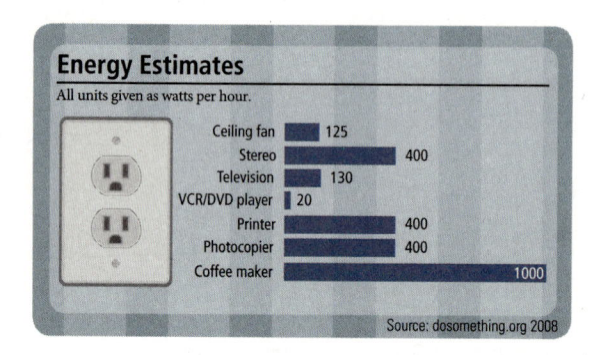

Energy Estimates
All units given as watts per hour.

- Ceiling fan: 125
- Stereo: 400
- Television: 130
- VCR/DVD player: 20
- Printer: 400
- Photocopier: 400
- Coffee maker: 1000

Source: dosomething.org 2008

■ Getting Ready for the Next Section

Simplify.

61. $2x^2 - 3x + 10x - 15$

62. $12a^2 + 8ab - 15ab - 10b^2$

63. $(6x^3 - 2x^2y + 8xy^2) + (-9x^2y + 3xy^2 - 12y^3)$

64. $(3x^3 - 15x^2 + 18x) + (2x^2 - 10x + 12)$

65. $4x^3(-3x)$

66. $5x^2(-4x)$

67. $4x^3(5x^2)$

68. $5x^2(3x^2)$

69. $(a^3)^2$

70. $(a^4)^2$

71. $11.5(130) - 0.05(130)^2$

72. $-0.05(130)^2 + 9.5(130) - 200$

Maintaining Your Skills

Simplify each expression.

73. $-1(5 - x)$

74. $-1(a - b)$

75. $-1(7 - x)$

76. $-1(6 - y)$

77. $5\left(x - \dfrac{1}{5}\right)$

78. $7\left(x + \dfrac{1}{7}\right)$

79. $x\left(1 - \dfrac{1}{x}\right)$

80. $a\left(1 + \dfrac{1}{a}\right)$

81. $12\left(\dfrac{1}{4}x + \dfrac{2}{3}y\right)$

82. $20\left(\dfrac{2}{5}x + \dfrac{1}{4}y\right)$

2.3 Multiplication of Polynomials

Objectives
A Multiply polynomials.

In the previous section we found the relationship between profit, revenue, and cost to be

$$P(x) = R(x) - C(x)$$

Revenue itself can be broken down further by another formula common in the business world. The revenue obtained from selling all x items is the product of the number of items sold and the price per item; that is,

Revenue = (number of items sold)(price of each item)

$$R = xp$$

Many times, x and p are polynomials, which means that the expression xp is the product of two polynomials. In this section we learn how to multiply polynomials, and in so doing, increase our understanding of the equations and formulas that describe business applications.

A Multiplying Polynomials

EXAMPLE 1 Find the product of $4x^3$ and $5x^2 - 3x + 1$.

SOLUTION To multiply, we apply the distributive property:

$$4x^3(5x^2 - 3x + 1)$$
$$= 4x^3(5x^2) + 4x^3(-3x) + 4x^3(1) \quad \text{Distributive property}$$
$$= 20x^5 - 12x^4 + 4x^3$$

Notice that we multiply coefficients and add exponents.

EXAMPLE 2 Multiply $2x - 3$ and $x + 5$.

SOLUTION Distributing the $2x - 3$ across the sum $x + 5$ gives us

$$(2x - 3)(x + 5)$$
$$= (2x - 3)x + (2x - 3)5 \quad \text{Distributive property}$$
$$= 2x(x) + (-3)x + 2x(5) + (-3)5 \quad \text{Distributive property}$$
$$= 2x^2 - 3x + 10x - 15$$
$$= 2x^2 + 7x - 15 \quad \text{Combine like terms}$$

Notice the third line in this example. It consists of all possible products of terms in the first binomial and those of the second binomial. We can generalize this into a rule for multiplying two polynomials.

> **Rule**
> To multiply two polynomials, multiply each term in the first polynomial by each term in the second polynomial.

Multiplying polynomials can be accomplished by a method that looks very similar to long multiplication with whole numbers.

PRACTICE PROBLEMS
1. Multiply $5x^2(3x^2 - 4x + 2)$.

2. Multiply $4x - 1$ and $x + 3$.

Answers
1. $15x^4 - 20x^3 + 10x^2$
2. $4x^2 + 11x - 3$

3. Multiply using the vertical method:
   $(3x + 2)(x^2 − 5x + 6)$

Note The vertical method of multiplying polynomials does not directly show the use of the distributive property. It is, however, very useful since it always gives the correct result and is easy to remember.

Note The FOIL method does not show the properties used in multiplying two binomials. It is simply a way of finding products of binomials quickly. Remember, the FOIL method applies only to products of two binomials. The vertical method applies to all products of polynomials with two or more terms.

Multiply using the FOIL method.
4. $(2a − 3b)(5a − b)$

Multiply using the FOIL method.
5. $(6 − 3t)(2 + 5t)$

Answers
3. $3x^3 − 13x^2 + 8x + 12$
4. $10a^2 − 17ab + 3b^2$
5. $12 + 24t − 15t^2$

Chapter 2 Exponents and Polynomials

EXAMPLE 3 Multiply $(2x − 3y)$ and $(3x^2 − xy + 4y^2)$ vertically.

SOLUTION
$$\begin{array}{r} 3x^2 - xy + 4y^2 \\ 2x - 3y \\ \hline 6x^3 - 2x^2y + 8xy^2 \\ - 9x^2y + 3xy^2 - 12y^3 \\ \hline 6x^3 - 11x^2y + 11xy^2 - 12y^3 \end{array}$$

Multiply $(3x^2 − xy + 4y^2)$ by $2x$
Multiply $(3x^2 − xy + 4y^2)$ by $−3y$
Add similar terms

Multiplying Binomials—The FOIL Method

Consider the product of $(2x − 5)$ and $(3x − 2)$. Distributing $(3x − 2)$ over $2x$ and $−5$, we have

$$(2x − 5)(3x − 2) = (2x)(3x − 2) + (−5)(3x − 2)$$
$$= (2x)(3x) + (2x)(−2) + (−5)(3x) + (−5)(−2)$$
$$= 6x^2 − 4x − 15x + 10$$
$$= 6x^2 − 19x + 10$$

Looking closely at the second and third lines, we notice the following:

1. $6x^2$ comes from multiplying the *first* terms in each binomial:

   $(2x − 5)(3x − 2)$ $2x(3x) = 6x^2$ *First* terms

2. $−4x$ comes from multiplying the *outside* terms in the product:

   $(2x − 5)(3x − 2)$ $2x(−2) = −4x$ *Outside* terms

3. $−15x$ comes from multiplying the *inside* terms in the product:

   $(2x − 5)(3x − 2)$ $−5(3x) = −15x$ *Inside* terms

4. 10 comes from multiplying the *last* two terms in the product:

   $(2x − 5)(3x − 2)$ $−5(−2) = 10$ *Last* terms

Once we know where the terms in the answer come from, we can reduce the number of steps used in finding the product:

$$(2x − 5)(3x − 2) = 6x^2 − 4x − 15x + 10 = 6x^2 − 19x + 10$$
$$\text{First}\ \ \text{Outside}\ \ \text{Inside}\ \ \text{Last}$$

EXAMPLE 4 Multiply $(4a − 5b)(3a + 2b)$ using the FOIL method.

SOLUTION
$$(4a − 5b)(3a + 2b) = 12a^2 + 8ab − 15ab − 10b^2$$
$$\text{F}\text{O}\text{I}\text{L}$$
$$= 12a^2 − 7ab − 10b^2$$

EXAMPLE 5 Multiply $(3 − 2t)(4 + 7t)$ using the FOIL method.

SOLUTION
$$(3 − 2t)(4 + 7t) = 12 + 21t − 8t − 14t^2$$
$$\text{F}\text{O}\text{I}\text{L}$$
$$= 12 + 13t − 14t^2$$

2.3 Multiplication of Polynomials

EXAMPLE 6 Multiply $\left(2x + \frac{1}{2}\right)\left(4x - \frac{1}{2}\right)$ using the FOIL method.

SOLUTION $\left(2x + \frac{1}{2}\right)\left(4x - \frac{1}{2}\right) = 8x^2 - x + 2x - \frac{1}{4} = 8x^2 + x - \frac{1}{4}$
FOIL

Multiply using the FOIL method.
6. $\left(3x + \frac{1}{4}\right)\left(4x - \frac{1}{3}\right)$

EXAMPLE 7 Multiply $(a^5 + 3)(a^5 - 7)$ using the FOIL method.

SOLUTION $(a^5 + 3)(a^5 - 7) = a^{10} - 7a^5 + 3a^5 - 21$
FOIL
$= a^{10} - 4a^5 - 21$

Multiply using the FOIL method.
7. $(a^4 - 2)(a^4 + 5)$

EXAMPLE 8 Multiply $(2x + 3)(5y - 4)$ using the FOIL method.

SOLUTION $(2x + 3)(5y - 4) = 10xy - 8x + 15y - 12$
FOIL

Multiply using the FOIL method.
8. $(7x - 2)(3y + 8)$

The Square of a Binomial

EXAMPLE 9 Find $(4x - 6)^2$.

SOLUTION Applying the definition of exponents and then the FOIL method, we have

$(4x - 6)^2 = (4x - 6)(4x - 6)$

$= 16x^2 - 24x - 24x + 36$
FOIL

$= 16x^2 - 48x + 36$

9. Expand and multiply $(3x - 2)^2$.

This example is the square of a binomial. This type of product occurs frequently enough in algebra that we have special formulas for it. Here are the formulas for binomial squares:

$(a + b)^2 = (a + b)(a + b) = a^2 + ab + ab + b^2 = a^2 + 2ab + b^2$
$(a - b)^2 = (a - b)(a - b) = a^2 - ab - ab + b^2 = a^2 - 2ab + b^2$

Observing the results in both cases, we have the following rule.

> **Rule**
> The square of a binomial is the sum of the square of the first term, twice the product of the two terms, and the square of the last term. Or:
>
> $(a + b)^2 = a^2 + 2ab + b^2$
> SquareTwice theSquare
> ofproduct$$of
> first$$of thelast
> termtwo termsterm
>
> $(a - b)^2 = a^2 - 2ab + b^2$

Answers
6. $12x^2 - \frac{1}{12}$
7. $a^8 + 3a^4 - 10$
8. $21xy + 56x - 6y - 16$
9. $9x^2 - 12x + 4$

Expand and simplify.

10. $(x - y)^2$
11. $(4t + 5)^2$
12. $(5x - 3y)^2$
13. $(6 - a^4)^2$

14. Multiply $(4x - 3)$ and $(4x + 3)$.

Multiply.

15. $(x + 2)(x - 2)$
16. $(5a + 7)(5a - 7)$
17. $(x^3 + 4)(x^3 - 4)$
18. $(x^4 - 5a)(x^4 + 5a)$

Answers
10. $x^2 - 2xy + y^2$
11. $16t^2 + 40t + 25$
12. $25x^2 - 30xy + 9y^2$
13. $36 - 12a^4 + a^8$
14. $16x^2 - 9$
15. $x^2 - 4$
16. $25a^2 - 49$
17. $x^6 - 16$
18. $x^8 - 25a^2$

EXAMPLES
Use the preceding formulas to expand each binomial square.

10. $(x + 7)^2 = x^2 + 2(x)(7) + 7^2 = x^2 + 14x + 49$
11. $(3t - 5)^2 = (3t)^2 - 2(3t)(5) + 5^2 = 9t^2 - 30t + 25$
12. $(4x + 2y)^2 = (4x)^2 + 2(4x)(2y) + (2y)^2 = 16x^2 + 16xy + 4y^2$
13. $(5 - a^3)^2 = 5^2 - 2(5)(a^3) + (a^3)^2 = 25 - 10a^3 + a^6$

Products Resulting in the Difference of Two Squares

Another frequently occurring kind of product is found when multiplying two binomials that differ only in the sign between their terms.

EXAMPLE 14
Multiply $(3x - 5)$ and $(3x + 5)$.

SOLUTION Applying the FOIL method, we have

$(3x - 5)(3x + 5) = 9x^2 + 15x - 15x - 25$ Two middle terms add to 0
          F   O   I   L
$= 9x^2 - 25$

The outside and inside products in Example 14 are opposites and therefore add to 0. Here it is in general:

$(a - b)(a + b) = a^2 + ab - ab - b^2$ Two middle terms add to 0
$= a^2 - b^2$

> **Rule**
> To multiply two binomials that differ only in the sign between their two terms, simply subtract the square of the second term from the square of the first term:
> $(a + b)(a - b) = a^2 - b^2$

The expression $a^2 - b^2$ is called the *difference of two squares*.

EXAMPLES
Find the following products.

15. $(x - 5)(x + 5) = x^2 - 25$
16. $(2a - 3)(2a + 3) = 4a^2 - 9$
17. $(x^2 + 4)(x^2 - 4) = x^4 - 16$
18. $(x^3 - 2a)(x^3 + 2a) = x^6 - 4a^2$

Getting Ready for Class

After reading through the preceding section, respond in your own words and in complete sentences.

1. Describe how the distributive property is used to multiply a monomial and a polynomial.
2. Describe how you would use the FOIL method to multiply two binomials.
3. Explain why $(x + 3)^2 \neq x^2 + 9$.
4. When will the product of two binomials result in a binomial?

Problem Set 2.3

A Multiply the following by applying the distributive property. [Examples 1–2]

1. $2x(6x^2 - 5x + 4)$
2. $-3x(5x^2 - 6x - 4)$
3. $-3a^2(a^3 - 6a^2 + 7)$
4. $4a^3(3a^2 - a + 1)$
5. $2a^2b(a^3 - ab + b^3)$
6. $5a^2b^2(8a^2 - 2ab + b^2)$

A Multiply the following vertically. [Example 3]

7. $(x - 5)(x + 3)$
8. $(x + 4)(x + 6)$
9. $(2x^2 - 3)(3x^2 - 5)$
10. $(3x^2 + 4)(2x^2 - 5)$
11. $(x + 3)(x^2 + 6x + 5)$
12. $(x - 2)(x^2 - 5x + 7)$
13. $(a - b)(a^2 + ab + b^2)$
14. $(a + b)(a^2 - ab + b^2)$
15. $(2x + y)(4x^2 - 2xy + y^2)$
16. $(x - 3y)(x^2 + 3xy + 9y^2)$
17. $(2a - 3b)(a^2 + ab + b^2)$
18. $(5a - 2b)(a^2 - ab - b^2)$

A Multiply the following using the FOIL method. [Examples 4–8]

19. $(x - 2)(x + 3)$
20. $(x + 2)(x - 3)$
21. $(2a + 3)(3a + 2)$
22. $(5a - 4)(2a + 1)$
23. $(5 - 3t)(4 + 2t)$
24. $(7 - t)(6 - 3t)$
25. $(x^3 + 3)(x^3 - 5)$
26. $(x^3 + 4)(x^3 - 7)$
27. $(y + 3)(y - 5)$
28. $(x - 9)(x + 4)$
29. $(4 - a)(3 + a)$
30. $(2x - 3)(x + 4)$

31. $(3x + 4)(2x + 1)$ **32.** $(5 - 2a)(3 - 3a)$ **33.** $(y - 4)(y - 6)$ **34.** $(4x - 1)(x - 3)$

35. $(5x - 6y)(4x + 3y)$ **36.** $(6x - 5y)(2x - 3y)$ **37.** $\left(3t + \dfrac{1}{3}\right)\left(6t - \dfrac{2}{3}\right)$ **38.** $\left(5t - \dfrac{1}{5}\right)\left(10t + \dfrac{3}{5}\right)$

A Find the following special products. [Examples 9–18]

39. $(5x + 2y)^2$ **40.** $(3x - 4y)^2$ **41.** $(5 - 3t^3)^2$

42. $(7 - 2t^4)^2$ **43.** $(2a + 3b)(2a - 3b)$ **44.** $(6a - 1)(6a + 1)$

45. $(3r^2 + 7s)(3r^2 - 7s)$ **46.** $(5r^2 - 2s)(5r^2 + 2s)$ **47.** $\left(y + \dfrac{3}{2}\right)^2$

48. $\left(y - \dfrac{7}{2}\right)^2$ **49.** $\left(a - \dfrac{1}{2}\right)^2$ **50.** $\left(a - \dfrac{5}{2}\right)^2$

51. $\left(x + \dfrac{1}{4}\right)^2$ **52.** $\left(x - \dfrac{3}{8}\right)^2$ **53.** $\left(t + \dfrac{1}{3}\right)^2$

54. $\left(t - \dfrac{2}{5}\right)^2$ **55.** $\left(\dfrac{1}{3}x - \dfrac{2}{5}\right)\left(\dfrac{1}{3}x + \dfrac{2}{5}\right)$ **56.** $\left(\dfrac{3}{4}x - \dfrac{1}{7}\right)\left(\dfrac{3}{4}x + \dfrac{1}{7}\right)$

Find the following products.

57. $(x - 2)^3$ **58.** $(4x + 1)^3$ **59.** $\left(x - \dfrac{1}{2}\right)^3$

60. $\left(x + \dfrac{1}{4}\right)^3$ **61.** $3(x - 1)(x - 2)(x - 3)$ **62.** $2(x + 1)(x + 2)(x + 3)$

63. $(b^2 + 8)(a^2 + 1)$ **64.** $(b^2 + 1)(a^4 - 5)$ **65.** $(x + 1)^2 + (x + 2)^2 + (x + 3)^2$

66. $(x - 1)^2 + (x - 2)^2 + (x - 3)^2$ **67.** $(2x + 3)^2 - (2x - 3)^2$ **68.** $(x - 3)^3 - (x + 3)^3$

Here are some problems you will see later in the book.

Simplify.

69. $(x + 3)^2 - 2(x + 3) - 8$

70. $(x - 2)^2 - 3(x - 2) - 10$

71. $(2a - 3)^2 - 9(2a - 3) + 20$

72. $(3a - 2)^2 + 2(3a - 2) - 3$

73. $2(4a + 2)^2 - 3(4a + 2) - 20$

74. $6(2a + 4)^2 - (2a + 4) - 2$

75. Let $a = 2$ and $b = 3$, and evaluate each of the following expressions.

$a^4 - b^4 \quad (a - b)^4 \quad (a^2 + b^2)(a + b)(a - b)$

76. Let $a = 2$ and $b = 3$, and evaluate each of the following expressions.

$a^3 + b^3 \quad (a + b)^3 \quad a^3 + 3a^2b + 3ab^2 + b^3$

A Applying the Concepts

Solar and Wind Energy The chart shows the cost to install either solar panels or a wind turbine. Use the chart to answer Problems 77 and 78.

77. A homeowner is buying a certain number of solar panel modules. He is going to get a discount on each module that is equal to 25 times the number of modules he buys. Write an equation that describes this situation, then simplify and find the cost if he buys 3 modules.

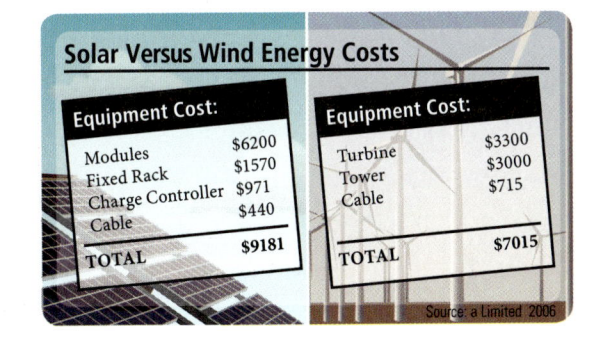

Solar Versus Wind Energy Costs

Equipment Cost:
Modules $6200
Fixed Rack $1570
Charge Controller $971
Cable $440
TOTAL $9181

Equipment Cost:
Turbine $3300
Tower $3000
Cable $715
TOTAL $7015

Source: a Limited 2006

78. A farmer is replacing several turbines in his field. He is going to get a discount on each turbine that is equal to 50 times the number of turbines he buys. Write an expression that describes this situation, then simplify and find the cost if he replaces 5 turbines.

79. Interest If you deposit $100 in an account with an interest rate r that is compounded annually, then the amount of money in that account at the end of 4 years is given by the formula $A = 100(1 + r)^4$. Expand the right side of this formula.

80. Interest If you deposit P dollars in an account with an annual interest rate r that is compounded twice a year, then at the end of a year the amount of money in that account is given by the formula

$$A = P\left(1 + \frac{r}{2}\right)^2$$

Expand the right side of this formula.

Getting Ready for the Next Section

81. $\dfrac{8a^3}{a}$

82. $\dfrac{-8a^2}{a}$

83. $\dfrac{-48a}{a}$

84. $\dfrac{-32a}{a}$

85. $\dfrac{16a^5b^4}{8a^2b^3}$

86. $\dfrac{12x^4y^5}{3x^3y^3}$

87. $\dfrac{-24a^5b^5}{8a^5b^3}$

88. $\dfrac{-15x^5y^3}{3x^3y^3}$

89. $\dfrac{x^3y^4}{-x^3}$

90. $\dfrac{x^2y^2}{-x^2}$

Extending the Concepts

91. Multiply $(x + y - 4)(x + y + 5)$ by first writing it like this:

$$[(x + y) - 4][(x + y) + 5]$$

and then applying the FOIL method.

92. Multiply $(x - 5 - y)(x - 5 + y)$ by first writing it like this:

$$[(x - 5) - y][(x - 5) + y]$$

and then applying the FOIL method.

Assume n is a positive integer and multiply.

93. $(x^n - 2)(x^n - 3)$

94. $(x^{2n} + 3)(x^{2n} - 3)$

95. $(2x^n + 3)(5x^n - 1)$

96. $(4x^n - 3)(7x^n + 2)$

97. $(x^n + 5)^2$

98. $(x^n - 2)^2$

99. $(x^n + 1)(x^{2n} - x^n + 1)$

100. $(x^{3n} - 3)(x^{6n} + 3x^{3n} + 9)$

The Greatest Common Factor

2.4

Objectives

A Factor by factoring out the greatest common factor.

B Factor by grouping

In general, factoring is the reverse of multiplication as the diagram here illustrates. Reading from left to right, we say the product of 3 and 7 is 21. Reading in the other direction, from right to left, we say 21 factors into 3 times 7. Or, 3 and 7 are factors of 21.

Multiplication

Factors : $3 \cdot 7 = 21$; Product

Factoring

A Greatest Common Factor

Definition
The **greatest common factor** for a polynomial is the largest monomial that divides (is a factor of) each term of the polynomial.

The greatest common factor for the polynomial $25x^5 + 20x^4 - 30x^3$ is $5x^3$ since it is the largest monomial that is a factor of each term. We can apply the distributive property and write

$$25x^5 + 20x^4 - 30x^3 = 5x^3(5x^2) + 5x^3(4x) - 5x^3(6)$$
$$= 5x^3(5x^2 + 4x - 6)$$

The last line is written in factored form.

Note The term largest monomial, as used here, refers to the monomial with the largest integer exponents whose coefficient has the greatest absolute value.

EXAMPLE 1
Factor the greatest common factor from
$$8a^3 - 8a^2 - 48a$$

SOLUTION The greatest common factor is $8a$. It is the largest monomial that divides each term of our polynomial. We can write each term in our polynomial as the product of $8a$ and another monomial. Then, we apply the distributive property to factor $8a$ from each term:

$$8a^3 - 8a^2 - 48a = 8a(a^2) - 8a(a) - 8a(6)$$
$$= 8a(a^2 - a - 6)$$

EXAMPLE 2
Factor the greatest common factor from
$$16a^5b^4 - 24a^2b^5 - 8a^3b^3$$

SOLUTION The largest monomial that divides each term is $8a^2b^3$. We write each term of the original polynomial in terms of $8a^2b^3$ and apply the distributive property to write the polynomial in factored form:

$$16a^5b^4 - 24a^2b^5 - 8a^3b^3 = 8a^2b^3(2a^3b) - 8a^2b^3(3b^2) - 8a^2b^3(a)$$
$$= 8a^2b^3(2a^3b - 3b^2 - a)$$

EXAMPLE 3
Factor the greatest common factor from
$$5x^2(a + b) - 6x(a + b) - 7(a + b)$$

SOLUTION The greatest common factor is $a + b$. Factoring it from each term, we have

$$5x^2(a + b) - 6x(a + b) - 7(a + b) = (a + b)(5x^2 - 6x - 7)$$

PRACTICE PROBLEMS

1. Factor the greatest common factor from $15a^7 - 25a^5 + 30a^3$.

2. Factor $12x^4y^5 - 9x^3y^4 - 15x^5y^3$.

3. Factor
$4(a + b)^4 - 6(a + b)^3 + 16(a + b)^2$

Answers
1. $5a^3(3a^4 - 5a^2 + 6)$
2. $3x^3y^3(4xy^2 - 3y - 5x^2)$
3. $2(a + b)^2[2(a + b)^2 - 3(a + b) + 8]$

B Factoring by Grouping

The polynomial $5x + 5y + x^2 + xy$ can be factored by noticing that the first two terms have a 5 in common, whereas the last two have an x in common. Applying the distributive property, we have

$$5x + 5y + x^2 + xy = 5(x + y) + x(x + y)$$

This last expression can be thought of as having two terms, $5(x + y)$ and $x(x + y)$, each of which has a common factor $(x + y)$. We apply the distributive property again to factor $(x + y)$ from each term:

$$5(x + y) + x(x + y) = (x + y)(5 + x)$$

4. Factor $ab^3 + b^3 + 6a + 6$.

EXAMPLE 4
Factor $a^2b^2 + b^2 + 8a^2 + 8$.

SOLUTION The first two terms have b^2 in common; the last two have 8 in common:

$$a^2b^2 + b^2 + 8a^2 + 8 = b^2(a^2 + 1) + 8(a^2 + 1)$$
$$= (a^2 + 1)(b^2 + 8)$$

5. Factor $15 - 3y^2 - 5x^2 + x^2y^2$.

EXAMPLE 5
Factor $15 - 5y^4 - 3x^3 + x^3y^4$.

SOLUTION Let's try factoring a 5 from the first two terms and an $-x^3$ from the last two terms:

$$15 - 5y^4 - 3x^3 + x^3y^4 = 5(3 - y^4) - x^3(3 - y^4)$$
$$= (3 - y^4)(5 - x^3)$$

6. Factor by grouping:
$x^3 + 5x^2 + 3x + 15$

EXAMPLE 6
Factor by grouping $x^3 + 2x^2 + 9x + 18$.

SOLUTION We begin by factoring x^2 from the first two terms and 9 from the second two terms:

$$x^3 + 2x^2 + 9x + 18 = x^2(x + 2) + 9(x + 2)$$
$$= (x + 2)(x^2 + 9)$$

Getting Ready for Class

After reading through the preceding section, respond in your own words and in complete sentences.

1. What is the greatest common factor for a polynomial?
2. After factoring a polynomial, how can you check your result?
3. When would you try to factor by grouping?
4. What is the relationship between multiplication and factoring?

Answers
4. $(a + 1)(b^3 + 6)$
5. $(5 - y^2)(3 - x^2)$
6. $(x + 5)(x^2 + 3)$

Problem Set 2.4

A Factor the greatest common factor from each of the following. (The answers in the back of the book all show greatest common factors whose coefficients are positive.) [Examples 1–3]

1. $10x^3 - 15x^2$
2. $12x^5 + 18x^7$
3. $9y^6 + 18y^3$

4. $24y^4 - 8y^2$
5. $9a^2b - 6ab^2$
6. $30a^3b^4 + 20a^4b^3$

7. $21xy^4 + 7x^2y^2$
8. $14x^6y^3 - 6x^2y^4$
9. $12x^3y^2 - 9xy^3$

10. $18x^2 - 12x^3$
11. $20xy^2 - 5xy$
12. $24a^2b + 20ab^3$

13. $18x^4y^3 - 14x^3y$
14. $21a^2b + 28ab^3$
15. $3a^2 - 21a + 30$

16. $3a^2 - 3a - 6$
17. $4x^3 - 16x^2 - 20x$
18. $2x^3 - 14x^2 + 20x$

19. $10x^4y^2 + 20x^3y^3 - 30x^2y^4$
20. $6x^4y^2 + 18x^3y^3 - 24x^2y^4$
21. $-x^2y + xy^2 - x^2y^2$

22. $-x^3y^2 - x^2y^3 - x^2y^2$
23. $4x^3y^2z - 8x^2y^2z^2 + 6xy^2z^3$
24. $7x^4y^3z^2 - 21x^2y^2z^2 - 14x^2y^3z^4$

25. $20a^2b^2c^2 - 30ab^2c + 25a^2bc^2$
26. $8a^3bc^5 - 48a^2b^4c + 16ab^3c^5$
27. $5x(a - 2b) - 3y(a - 2b)$

28. $3a(x - y) - 7b(x - y)$ **29.** $3x^2(x + y)^2 - 6y^2(x + y)^2$ **30.** $10x^3(2x - 3y) - 15x^2(2x - 3y)$

31. $2x^2(x + 5) + 7x(x + 5) + 6(x + 5)$ **32.** $2x^2(x + 2) + 13x(x + 2) + 15(x + 2)$

B Factor each of the following by grouping. [Examples 4–6]

33. $3xy + 3y + 2ax + 2a$ **34.** $5xy^2 + 5y^2 + 3ax + 3a$

35. $x^2y + x + 3xy + 3$ **36.** $x^3y^3 + 2x^3 + 5x^2y^3 + 10x^2$

37. $3xy^2 - 6y^2 + 4x - 8$ **38.** $8x^2y - 4x^2 + 6y - 3$

39. $x^2 - ax - bx + ab$ **40.** $ax - x^2 - bx + ab$

41. $ab + 5a - b - 5$ **42.** $x^2 - xy - ax + ay$

43. $a^4b^2 + a^4 - 5b^2 - 5$ **44.** $2a^2 - a^2b - bc^2 + 2c^2$

45. $x^3 + 3x^2 - 4x - 12$ **46.** $x^3 + 5x^2 - 4x - 20$

47. $x^3 + 2x^2 - 25x - 50$ **48.** $x^3 + 4x^2 - 9x - 36$

49. $2x^3 + 3x^2 - 8x - 12$

50. $3x^3 + 2x^2 - 27x - 18$

51. $4x^3 + 12x^2 - 9x - 27$

52. $9x^3 + 18x^2 - 4x - 8$

53. The greatest common factor of the binomial $3x - 9$ is 3. The greatest common factor of the binomial $6x - 2$ is 2. What is the greatest common factor of their product, $(3x - 9)(6x - 2)$, when it has been multiplied out?

54. The greatest common factors of the binomials $5x - 10$ and $2x + 4$ are 5 and 2, respectively. What is the greatest common factor of their product, $(5x - 10)(2x + 4)$, when it has been multiplied out?

Applying the Concepts

55. Investing If P dollars are placed in a savings account in which the rate of interest r is compounded yearly, then at the end of 1 year the amount of money in the account can be written as $P + Pr$. At the end of 2 years the amount of money in the account is

$$P + Pr + (P + Pr)r$$

Use factoring by grouping to show that this last expression can be written as $P(1 + r)^2$.

56. Investing At the end of 3 years, the amount of money in the savings account in Problem 55 will be

$$P(1 + r)^2 + P(1 + r)^2 r$$

Use factoring to show that this last expression can be written as $P(1 + r)^3$.

Getting Ready for the Next Section

Factor out the greatest common factor.

57. $3x^4 - 9x^3y - 18x^2y^2$

58. $5x^2 + 10x + 30$

59. $2x^2(x - 3) - 4x(x - 3) - 3(x - 3)$

60. $3x^2(x - 2) - 8x(x - 2) + 2(x - 2)$

Multiply.

61. $(x + 2)(3x − 1)$ **62.** $(x − 2)(3x + 1)$

63. $(x − 1)(3x − 2)$ **64.** $(x + 1)(3x + 2)$

65. $(x + 2)(x + 3)$ **66.** $(x − 2)(x − 3)$

67. $(2y + 5)(3y − 7)$ **68.** $(2y − 5)(3y + 7)$

69. $(4 − 3a)(5 − a)$ **70.** $(4 − 3a)(5 + a)$

Complete each table.

71.

Two Numbers a and b	Their Product ab	Their Sum $a + b$
1, −24		
−1, 24		
2, −12		
−2, 12		
3, −8		
−3, 8		
4, −6		
−4, 6		

72.

Two Numbers a and b	Their Product ab	Their Sum $a + b$
1, −54		
−1, 54		
2, −27		
−2, 27		
3, −18		
−3, 18		
6, −9		
−6, 9		

Special Factoring 2.5

Objectives
A Factor perfect square trinomials.
B Factor the difference of two squares.
C Factor the sum or difference of two cubes.

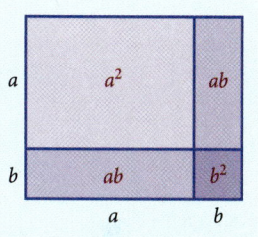

To find the area of the large square in the margin, we can square the length of its side, giving us $(a + b)^2$. However, we can add the areas of the four smaller figures to arrive at the same result.

Since the area of the large square is the same whether we find it by squaring a side or by adding the four smaller areas, we can write the following relationship:

$$(a + b)^2 = a^2 + 2ab + b^2$$

This is the formula for the square of a binomial. The figure gives us a geometric interpretation for one of the special multiplication formulas. We begin this section by looking at the special multiplication formulas from a factoring perspective.

A Perfect Square Trinomials

We previously listed some special products found in multiplying polynomials. Two of the formulas looked like this:

$$(a + b)^2 = a^2 + 2ab + b^2$$
$$(a - b)^2 = a^2 - 2ab + b^2$$

If we exchange the left and right sides of each formula, we have two special formulas for factoring:

$$a^2 + 2ab + b^2 = (a + b)^2$$
$$a^2 - 2ab + b^2 = (a - b)^2$$

The left side of each formula is called a *perfect square trinomial*. The right sides are binomial squares. Perfect square trinomials can always be factored using the usual methods for factoring trinomials. However, if we notice that the first and last terms of a trinomial are perfect squares, it is wise to see whether the trinomial factors as a binomial square before attempting to factor by the usual method.

EXAMPLE 1 Factor $x^2 - 6x + 9$.

SOLUTION Since the first and last terms are perfect squares, we attempt to factor according to the preceding formulas:

$$x^2 - 6x + 9 = (x - 3)^2$$

If we expand $(x - 3)^2$, we have $x^2 - 6x + 9$, indicating we have factored correctly.

EXAMPLES Factor each of the following perfect square trinomials.

SOLUTION
2. $16a^2 + 40ab + 25b^2 = (4a + 5b)^2$
3. $49 - 14t + t^2 = (7 - t)^2$
4. $9x^4 - 12x^2 + 4 = (3x^2 - 2)^2$
5. $(y + 3)^2 + 10(y + 3) + 25 = [(y + 3) + 5]^2 = (y + 8)^2$

EXAMPLE 6 Factor $8x^2 - 24xy + 18y^2$.

SOLUTION We begin by factoring the greatest common factor 2 from each term:

$$8x^2 - 24xy + 18y^2 = 2(4x^2 - 12xy + 9y^2)$$
$$= 2(2x - 3y)^2$$

PRACTICE PROBLEMS

1. Factor $x^2 - 10x + 25$.

Factor.
2. $9a^2 + 42ab + 49b^2$
3. $25 - 10t + t^2$
4. $16x^4 - 24x^2 + 9$
5. $(y + 2)^2 + 8(y + 2) + 16$

6. Factor $27x^2 - 36x + 12$.

Answers
1. $(x - 5)^2$
2. $(3a + 7b)^2$
3. $(5 - t)^2$
4. $(4x^2 - 3)^2$
5. $(y + 6)^2$
6. $3(3x - 2)^2$

2.5 Special Factoring

113

B The Difference of Two Squares

Recall the formula that results in the difference of two squares:
$(a + b)(a - b) = a^2 - b^2$. Writing this as a factoring formula, we have

$$a^2 - b^2 = (a + b)(a - b)$$

EXAMPLES Each of the following is the difference of two squares. Use the formula $a^2 - b^2 = (a + b)(a - b)$ to factor each one.

SOLUTION

7. $x^2 - 25 = x^2 - 5^2 = (x + 5)(x - 5)$
8. $49 - t^2 = 7^2 - t^2 = (7 + t)(7 - t)$
9. $81a^2 - 25b^2 = (9a)^2 - (5b)^2 = (9a + 5b)(9a - 5b)$
10. $4x^6 - 1 = (2x^3)^2 - 1^2 = (2x^3 + 1)(2x^3 - 1)$
11. $x^2 - \frac{4}{9} = x^2 - \left(\frac{2}{3}\right)^2 = \left(x + \frac{2}{3}\right)\left(x - \frac{2}{3}\right)$

As our next example shows, the difference of two fourth powers can be factored as the difference of two squares.

EXAMPLE 12 Factor $16x^4 - 81y^4$.

SOLUTION The first and last terms are perfect squares. We factor according to the preceding formula:

$$16x^4 - 81y^4 = (4x^2)^2 - (9y^2)^2$$
$$= (4x^2 + 9y^2)(4x^2 - 9y^2)$$

Notice that the second factor is also the difference of two squares. Factoring completely, we have

$$16x^4 - 81y^4 = (4x^2 + 9y^2)(2x + 3y)(2x - 3y)$$

Here is another example of the difference of two squares.

EXAMPLE 13 Factor $(x - 3)^2 - 25$.

SOLUTION This example has the form $a^2 - b^2$, where a is $x - 3$ and b is 5. We factor it according to the formula for the difference of two squares:

$(x - 3)^2 - 25 = (x - 3)^2 - 5^2$ Write 25 as 5^2
$\qquad\qquad\qquad = [(x - 3) + 5][(x - 3) - 5]$ Factor
$\qquad\qquad\qquad = (x + 2)(x - 8)$ Simplify

Notice in this example we could have expanded $(x - 3)^2$, subtracted 25, and then factored to obtain the same result:

$(x - 3)^2 - 25 = x^2 - 6x + 9 - 25$ Expand $(x - 3)^2$
$\qquad\qquad\qquad = x^2 - 6x - 16$ Simplify
$\qquad\qquad\qquad = (x - 8)(x + 2)$ Factor

Factor.

7. $x^2 - 16$
8. $64 - t^2$
9. $25x^2 - 36y^2$
10. $9x^6 - 1$
11. $x^2 - \frac{25}{64}$

12. Factor $x^4 - 81$.

Note The sum of two squares never factors into the product of two binomials; that is, if we were to attempt to factor $(4x^2 + 9y^2)$ in Example 12, we would be unable to find two binomials (or any other polynomials) whose product is $4x^2 + 9y^2$. The factors do not exist as polynomials.

13. Factor $(x - 4)^2 - 9$.

Answers
7. $(x + 4)(x - 4)$
8. $(8 + t)(8 - t)$
9. $(5x + 6y)(5x - 6y)$
10. $(3x^3 + 1)(3x^3 - 1)$
11. $\left(x + \frac{5}{8}\right)\left(x - \frac{5}{8}\right)$
12. $(x^2 + 9)(x + 3)(x - 3)$
13. $(x - 7)(x - 1)$

2.5 Special Factoring

EXAMPLE 14 Factor $x^2 - 10x + 25 - y^2$.

SOLUTION Notice the first three items form a perfect square trinomial; that is, $x^2 - 10x + 25 = (x - 5)^2$. If we replace the first three terms by $(x - 5)^2$, the expression that results has the form $a^2 - b^2$. We can factor as we did in Example 13:

$x^2 - 10x + 25 - y^2 = (x^2 - 10x + 25) - y^2$ Group first three terms together

$= (x - 5)^2 - y^2$ This has the form $a^2 - b^2$

$= [(x - 5) + y][(x - 5) - y]$ Factor according to the formula $a^2 - b^2 = (a + b)(a - b)$

$= (x - 5 + y)(x - 5 - y)$ Simplify

We could check this result by multiplying the two factors together. (You may want to do that to convince yourself that we have the correct result.)

EXAMPLE 15 Factor completely $x^3 + 2x^2 - 9x - 18$.

SOLUTION We use factoring by grouping to begin and then factor the difference of two squares:

$x^3 + 2x^2 - 9x - 18 = x^2(x + 2) - 9(x + 2)$

$= (x + 2)(x^2 - 9)$

$= (x + 2)(x + 3)(x - 3)$

C The Sum and Difference of Two Cubes

Here are the formulas for factoring the sum and difference of two cubes:

$a^3 + b^3 = (a + b)(a^2 - ab + b^2)$

$a^3 - b^3 = (a - b)(a^2 + ab + b^2)$

Since these formulas are unfamiliar, it is important that we verify them.

EXAMPLE 16 Verify the two formulas.

SOLUTION We verify the formulas by multiplying the right sides and comparing the results with the left sides:

$$\begin{array}{r} a^2 - ab + b^2 \\ \underline{a + b} \\ a^3 - a^2b + ab^2 \\ \underline{ a^2b - ab^2 + b^3} \\ a^3 + b^3 \end{array}$$

The first formula is correct.

$$\begin{array}{r} a^2 + ab + b^2 \\ \underline{a - b} \\ a^3 + a^2b + ab^2 \\ \underline{ -a^2b - ab^2 - b^3} \\ a^3 - b^3 \end{array}$$

The second formula is correct.

---

14. Factor $x^2 - 6x + 9 - y^2$.

15. Factor completely. $x^3 + 5x^2 - 4x - 20$

16. Multiply $(x - 3)(x^2 + 3x + 9)$.

Answers
14. $(x - 3 + y)(x - 3 - y)$
15. $(x + 5)(x + 2)(x - 2)$
16. $x^3 - 27$

Chapter 2 Exponents and Polynomials

Here are some examples using the formulas for factoring the sum and difference of two cubes.

EXAMPLE 17 Factor $64 + t^3$.

SOLUTION The first term is the cube of 4 and the second term is the cube of t. Therefore,
$$64 + t^3 = 4^3 + t^3$$
$$= (4 + t)(16 - 4t + t^2)$$

EXAMPLE 18 Factor $27x^3 + 125y^3$.

SOLUTION Writing both terms as perfect cubes, we have
$$27x^3 + 125y^3 = (3x)^3 + (5y)^3$$
$$= (3x + 5y)(9x^2 - 15xy + 25y^2)$$

EXAMPLE 19 Factor $a^3 - \frac{1}{8}$.

SOLUTION The first term is the cube of a, whereas the second term is the cube of $\frac{1}{2}$:
$$a^3 - \frac{1}{8} = a^3 - \left(\frac{1}{2}\right)^3$$
$$= \left(a - \frac{1}{2}\right)\left(a^2 + \frac{1}{2}a + \frac{1}{4}\right)$$

EXAMPLE 20 Factor $x^6 - y^6$.

SOLUTION We have a choice of how we want to write the two terms to begin. We can write the expression as the difference of two squares, $(x^3)^2 - (y^3)^2$, or as the difference of two cubes, $(x^2)^3 - (y^2)^3$. It is better to use the difference of two squares if we have a choice:
$$x^6 - y^6 = (x^3)^2 - (y^3)^2$$
$$= (x^3 - y^3)(x^3 + y^3)$$
$$= (x - y)(x^2 + xy + y^2)(x + y)(x^2 - xy + y^2)$$

Try this example again writing the first line as the difference of two cubes instead of the difference of two squares. It will become apparent why it is better to use the difference of two squares.

Getting Ready for Class

After reading through the preceding section, respond in your own words and in complete sentences.

1. In what cases can you factor a binomial?
2. What is a perfect square trinomial?
3. Is it possible to factor the sum of two squares?
4. Write the formula you use to factor the sum of two cubes.

Factor.
17. $27 + x^3$

18. $8x^3 + y^3$

19. $a^3 - \frac{1}{27}$

20. $x^6 - 1$

Answers
17. $(3 + x)(9 - 3x + x^2)$
18. $(2x + y)(4x^2 - 2xy + y^2)$
19. $\left(a - \frac{1}{3}\right)\left(a^2 + \frac{1}{3}a + \frac{1}{9}\right)$
20. $(x + 1)(x^2 - x + 1)(x - 1)(x^2 + x + 1)$

Problem Set 2.5

A Factor each perfect square trinomial. [Examples 1–6]

1. $x^2 - 6x + 9$
2. $x^2 + 10x + 25$
3. $a^2 - 12a + 36$
4. $x^2 + 8x + 16$

5. $y^2 - 18y + 81$
6. $a^2 + 24a + 144$
7. $x^2 - 14x + 49$
8. $36 - 12a + a^2$

9. $25 - 10t + t^2$
10. $64 + 16t + t^2$
11. $\frac{1}{9}x^2 + 2x + 9$
12. $\frac{1}{4}x^2 - 2x + 4$

13. $4y^4 - 12y^2 + 9$
14. $9y^4 + 12y^2 + 4$
15. $16a^2 + 40ab + 25b^2$
16. $25a^2 - 40ab + 16b^2$

17. $\frac{1}{25} + \frac{1}{10}t^2 + \frac{1}{16}t^4$
18. $\frac{1}{9} - \frac{1}{3}t^3 + \frac{1}{4}t^6$
19. $y^2 + 3y + \frac{9}{4}$
20. $y^2 - 7y + \frac{49}{4}$

21. $a^2 - a + \frac{1}{4}$
22. $a^2 - 5a + \frac{25}{4}$
23. $x^2 - \frac{1}{2}x + \frac{1}{16}$
24. $x^2 - \frac{3}{4}x + \frac{9}{64}$

25. $t^2 + \frac{2}{3}t + \frac{1}{9}$
26. $t^2 - \frac{4}{5}t + \frac{4}{25}$
27. $16x^2 - 48x + 36$
28. $36x^2 + 48x + 16$

29. $75a^3 + 30a^2 + 3a$
30. $45a^4 - 30a^3 + 5a^2$
31. $(x + 2)^2 + 6(x + 2) + 9$
32. $(x + 5)^2 + 4(x + 5) + 4$

B Factor each as the difference of two squares. Be sure to factor completely. [Examples 7–12]

33. $x^2 - 9$
34. $x^2 - 16$
35. $49x^2 - 64y^2$
36. $81x^2 - 49y^2$

37. $4a^2 - \frac{1}{4}$
38. $25a^2 - \frac{1}{25}$
39. $x^2 - \frac{9}{25}$
40. $x^2 - \frac{25}{36}$

41. $9x^2 - 16y^2$
42. $25x^2 - 49y^2$
43. $250 - 10t^2$
44. $640 - 10t^2$

B Factor each as the difference of two squares. Be sure to factor completely. [Examples 7–12]

45. $x^4 - 81$
46. $x^4 - 16$

47. $9x^6 - 1$
48. $25x^6 - 1$

49. $16a^4 - 81$
50. $81a^4 - 16b^4$

51. $\dfrac{1}{81} - \dfrac{y^4}{16}$
52. $\dfrac{1}{25} - \dfrac{y^4}{64}$

B Factor completely. [Examples 13–15]

53. $x^6 - y^6$
54. $x^6 - 1$
55. $2a^7 - 128a$

56. $128a^8 - 2a^2$
57. $(x - 2)^2 - 9$
58. $(x + 2)^2 - 9$

59. $(y + 4)^2 - 16$
60. $(y - 4)^2 - 16$
61. $x^2 - 10x + 25 - y^2$

62. $x^2 - 6x + 9 - y^2$
63. $a^2 + 8a + 16 - b^2$
64. $a^2 + 12a + 36 - b^2$

65. $x^2 + 2xy + y^2 - a^2$
66. $a^2 + 2ab + b^2 - y^2$
67. $x^3 + 3x^2 - 4x - 12$

68. $x^3 + 5x^2 - 4x - 20$

69. $x^3 + 2x^2 - 25x - 50$

70. $x^3 + 4x^2 - 9x - 36$

71. $2x^3 + 3x^2 - 8x - 12$

72. $3x^3 + 2x^2 - 27x - 18$

73. $4x^3 + 12x^2 - 9x - 27$

74. $9x^3 + 18x^2 - 4x - 8$

75. $(2x - 5)^2 - 100$

76. $(7a + 5)^2 - 64$

77. $(a - 3)^2 - (4b)^2$

78. $(2x - 5)^2 - (6y)^2$

79. $a^2 - 6a + 9 - 16b^2$

80. $x^2 - 10x + 25 - 9y^2$

81. $x^2(x + 4) - 6x(x + 4) + 9(x + 4)$

82. $x^2(x - 6) + 8x(x - 6) + 16(x - 6)$

C Factor each of the following as the sum or difference of two cubes. [Examples 16–20]

83. $x^3 - y^3$

84. $x^3 + y^3$

85. $a^3 + 8$

86. $a^3 - 8$

87. $27 + x^3$

88. $27 - x^3$

89. $y^3 - 1$

90. $y^3 + 1$

91. $10r^3 - 1{,}250$

92. $10r^3 + 1{,}250$

93. $64 + 27a^3$

94. $27 - 64a^3$

95. $8x^3 - 27y^3$

96. $27x^3 - 8y^3$

97. $t^3 + \dfrac{1}{27}$

98. $t^3 - \dfrac{1}{27}$

99. $27x^3 - \dfrac{1}{27}$

100. $8x^3 + \dfrac{1}{8}$

101. $64a^3 + 125b^3$

102. $125a^3 - 27b^3$

103. Find two values of b that will make $9x^2 + bx + 25$ a perfect square trinomial.

104. Find a value of c that will make $49x^2 - 42x + c$ a perfect square trinomial.

Getting Ready for the Next Section

Factor out the greatest common factor.

105. $y^3 + 25y$ **106.** $y^4 + 36y^2$ **107.** $2ab^5 + 8ab^4 + 2ab^3$ **108.** $3a^2b^3 + 6a^2b^2 - 3a^2b$

Factor by grouping.

109. $4x^2 - 6x + 2ax - 3a$ **110.** $6x^2 - 4x + 3ax - 2a$

Factor the difference of squares.

111. $x^2 - 4$ **112.** $x^2 - 9$

Factor the perfect square trinomial.

113. $x^2 - 6x + 9$ **114.** $x^2 - 10x + 25$

Factor.

115. $6a^2 - 11a + 4$ **116.** $6x^2 - x - 15$

Factor the sum or difference of cubes.

117. $x^3 + 8$ **118.** $x^3 - 27$

Extending the Concepts

Factor completely.

119. $a^2 - b^2 + 6b - 9$ **120.** $a^2 - b^2 - 18b - 81$ **121.** $(x-3)^2 - (y+5)^2$ **122.** $(a+7)^2 - (b-9)^2$

Find k such that each trinomial becomes a perfect square trinomial.

123. $kx^2 - 168xy + 49y^2$ **124.** $kx^2 + 110xy + 121y^2$ **125.** $49x^2 + kx + 81$ **126.** $64x^2 + kx + 169$

2.6 Factoring Trinomials

Objectives
A Factor trinomials in which the leading coefficient is 1.
B Factor trinomials in which the leading coefficient is a number other than 1.

A Factoring Trinomials with a Leading Coefficient of 1

Earlier in this chapter we multiplied binomials:

$$(x - 2)(x + 3) = x^2 + x - 6$$

$$(x + 5)(x + 2) = x^2 + 7x + 10$$

In each case the product of two binomials is a trinomial. The first term in the resulting trinomial is obtained by multiplying the first term in each binomial. The middle term comes from adding the product of the two inside terms with the product of the two outside terms. The last term is the product of the last term in each binomial.

In general,

$$(x + a)(x + b) = x^2 + ax + bx + ab$$
$$= x^2 + (a + b)x + ab$$

Writing this as a factoring problem, we have

$$x^2 + (a + b)x + ab = (x + a)(x + b)$$

To factor a trinomial with a leading coefficient of 1, we simply find the two numbers a and b whose sum is the coefficient of the middle term and whose product is the constant term.

EXAMPLE 1   Factor $x^2 + 2x - 15$.

SOLUTION Since the leading coefficient is 1, we need two integers whose product is -15 and whose sum is 2. The integers are 5 and -3.

$$x^2 + 2x - 15 = (x + 5)(x - 3)$$

In the preceding example we found factors of $x + 5$ and $x - 3$. These are the only two factors for $x^2 + 2x - 15$. There is no other pair of binomials $x + a$ and $x + b$ whose product is $x^2 + 2x - 15$.

EXAMPLE 2   Factor $x^2 - xy - 12y^2$.

SOLUTION We need two expressions whose product is $-12y^2$ and whose sum is $-y$. The expressions are $-4y$ and $3y$:

$$x^2 - xy - 12y^2 = (x - 4y)(x + 3y)$$

Checking this result gives

$$(x - 4y)(x + 3y) = x^2 + 3xy - 4xy - 12y^2$$
$$= x^2 - xy - 12y^2$$

EXAMPLE 3   Factor $x^2 - 8x + 6$.

SOLUTION Since there is no pair of integers whose product is 6 and whose sum is -8, the trinomial $x^2 - 8x + 6$ is not factorable. We say it is a *prime polynomial*.

PRACTICE PROBLEMS

1. Factor $x^2 - x - 12$.

2. Factor $x^2 + 2xy - 15y^2$.

3. Factor $x^2 + x + 1$.

Answers
1. $(x - 4)(x + 3)$
2. $(x + 5y)(x - 3y)$
3. Does not factor.

4. Factor $5x^2 + 25x + 30$.

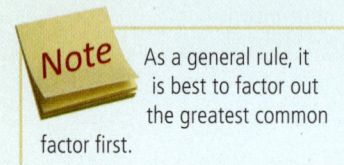

Note As a general rule, it is best to factor out the greatest common factor first.

5. Factor $3x^2 - x - 2$.

Answers
4. $5(x + 3)(x + 2)$
5. $(3x + 2)(x - 1)$

Chapter 2 Exponents and Polynomials

B Factoring When the Lead Coefficient is not 1

EXAMPLE 4 Factor $3x^4 - 15x^3y - 18x^2y^2$.

SOLUTION The leading coefficient is not 1. Each term is divisible by $3x^2$, however. Factoring this out to begin with we have

$$3x^4 - 15x^3y - 18x^2y^2 = 3x^2(x^2 - 5xy - 6y^2)$$

Factoring the resulting trinomial as in the previous examples gives

$$3x^2(x^2 - 5xy - 6y^2) = 3x^2(x - 6y)(x + y)$$

Factoring Other Trinomials by Trial and Error

We want to turn our attention now to trinomials with leading coefficients other than 1 and with no greatest common factor other than 1.

Suppose we want to factor $3x^2 - x - 2$. The factors will be a pair of binomials. The product of the first terms will be $3x^2$, and the product of the last terms will be -2. We can list all the possible factors along with their products as follows.

Possible Factors	First Term	Middle Term	Last Term
$(x + 2)(3x - 1)$	$3x^2$	$+5x$	-2
$(x - 2)(3x + 1)$	$3x^2$	$-5x$	-2
$(x + 1)(3x - 2)$	$3x^2$	$+x$	-2
$(x - 1)(3x + 2)$	$3x^2$	$-x$	-2

From the last line we see that the factors of $3x^2 - x - 2$ are $(x - 1)(3x + 2)$. That is,

$$3x^2 - x - 2 = (x - 1)(3x + 2)$$

To factor trinomials with leading coefficients other than 1, when the greatest common factor is 1, we must use trial and error or list all the possible factors. In either case the idea is this: look only at pairs of binomials whose products give the correct first and last terms, then look for the combination that will give the correct middle term.

EXAMPLE 5 Factor $2x^2 + 13xy + 15y^2$.

SOLUTION Listing all possible factors the product of whose first terms is $2x^2$ and the product of whose last terms is $15y^2$ yields

Possible Factors	Middle Term of Product
$(2x - 5y)(x - 3y)$	$-11xy$
$(2x - 3y)(x - 5y)$	$-13xy$
$(2x + 5y)(x + 3y)$	$+11xy$
$(2x + 3y)(x + 5y)$	$+13xy$
$(2x + 15y)(x + y)$	$+17xy$
$(2x - 15y)(x - y)$	$-17xy$
$(2x + y)(x + 15y)$	$+31xy$
$(2x - y)(x - 15y)$	$-31xy$

The fourth line has the correct middle term:

$$2x^2 + 13xy + 15y^2 = (2x + 3y)(x + 5y)$$

Actually, we did not need to check the first two pairs of possible factors in the preceding list. Because all the signs in the trinomial $2x^2 + 13xy + 15y^2$ are positive, the binomial factors must be of the form $(ax + b)(cx + d)$, where a, b, c, and d are all positive.

2.6 Factoring Trinomials

There are other ways to reduce the number of possible factors to consider. For example, if we were to factor the trinomial $2x^2 - 11x + 12$, we would not have to consider the pair of possible factors $(2x - 4)(x - 3)$. If the original trinomial has no greatest common factor other than 1, then neither of its binomial factors will either. The trinomial $2x^2 - 11x + 12$ has a greatest common factor of 1, but the possible factor $2x - 4$ has a greatest common factor of 2: $2x - 4 = 2(x - 2)$. Therefore, we do not need to consider $2x - 4$ as a possible factor.

EXAMPLE 6 Factor $12x^4 + 17x^2 + 6$.

SOLUTION This is a trinomial in x^2:
$$12x^4 + 17x^2 + 6 = (4x^2 + 3)(3x^2 + 2)$$

EXAMPLE 7 Factor $2x^2(x - 3) - 5x(x - 3) - 3(x - 3)$.

SOLUTION We begin by factoring out the greatest common factor $(x - 3)$. Then we factor the trinomial that remains.
$$2x^2(x - 3) - 5x(x - 3) - 3(x - 3) = (x - 3)(2x^2 - 5x - 3)$$
$$= (x - 3)(2x + 1)(x - 3)$$
$$= (x - 3)^2(2x + 1)$$

Another Method of Factoring Trinomials

As an alternative to the trial-and-error method of factoring trinomials, we present the following method. The new method does not require as much trial and error. To use this new method, we must rewrite our original trinomial in such a way that the factoring by grouping method can be applied.

Here are the steps we use to factor $ax^2 + bx + c$.

Step 1: Form the product ac.
Step 2: Find a pair of numbers whose product is ac and whose sum is b.
Step 3: Rewrite the polynomial to be factored so that the middle term bx is written as the sum of two terms whose coefficients are the two numbers found in step 2.
Step 4: Factor by grouping.

EXAMPLE 8 Factor $3x^2 - 10x - 8$ using these steps.

SOLUTION The trinomial $3x^2 - 10x - 8$ has the form $ax^2 + bx + c$, where $a = 3$, $b = -10$, and $c = -8$.

Step 1: The product ac is $3(-8) = -24$.
Step 2: We need to find two numbers whose product is -24 and whose sum is -10. Let's list all the pairs of numbers whose product is -24 to find the pair whose sum is -10.

Product	Sum
$1(-24) = -24$	$1 + (-24) = -23$
$-1(24) = -24$	$-1 + 24 = 23$
$2(-12) = -24$	$2 + (-12) = -10$
$-2(12) = -24$	$-2 + 12 = 10$
$3(-8) = -24$	$3 + (-8) = -5$
$-3(8) = -24$	$-3 + 8 = 5$
$4(-6) = -24$	$4 + (-6) = -2$
$-4(6) = -24$	$-4 + 6 = 2$

6. Factor $15x^4 + x^2 - 2$.

7. Factor $3x^2(x - 2) - 7x(x - 2) + 2(x - 2)$

8. Factor $8x^2 - 5x - 3$.

Answers
6. $(5x^2 + 2)(3x^2 - 1)$
7. $(x - 2)^2(3x - 1)$

Chapter 2 Exponents and Polynomials

As you can see, of all the pairs of numbers whose product is −24, only 2 and −12 have a sum of −10.

Step 3: We now rewrite our original trinomial so the middle term −10x is written as the sum of −12x and 2x:

$$3x^2 - 10x - 8 = 3x^2 - 12x + 2x - 8$$

Step 4: Factoring by grouping, we have

$$3x^2 - 12x + 2x - 8 = 3x(x - 4) + 2(x - 4)$$
$$= (x - 4)(3x + 2)$$

You can see that this method works by multiplying $x - 4$ and $3x + 2$ to get

$$3x^2 - 10x - 8$$

9. Factor $4x^2 + 11x + 6$.

EXAMPLE 9   Factor $9x^2 + 15x + 4$.

SOLUTION   In this case $a = 9$, $b = 15$, and $c = 4$. The product ac is $9 \cdot 4 = 36$. Listing all the pairs of numbers whose product is 36 with their corresponding sums, we have

Product	Sum
1(36) = 36	1 + 36 = 37
2(18) = 36	2 + 18 = 20
3(12) = 36	3 + 12 = 15
4(9) = 36	4 + 9 = 13
6(6) = 36	6 + 6 = 12

Notice we list only positive numbers since both the product and sum we are looking for are positive. The numbers 3 and 12 are the numbers we are looking for. Their product is 36, and their sum is 15. We now rewrite the original polynomial $9x^2 + 15x + 4$ with the middle term written as $3x + 12x$. We then factor by grouping:

$$9x^2 + 15x + 4 = 9x^2 + 3x + 12x + 4$$
$$= 3x(3x + 1) + 4(3x + 1)$$
$$= (3x + 1)(3x + 4)$$

The polynomial $9x^2 + 15x + 4$ factors into the product

$$(3x + 1)(3x + 4)$$

10. Factor $4x^2 - 4x - 15$.

EXAMPLE 10   Factor $8x^2 - 2x - 15$.

SOLUTION   The product ac is $8(-15) = -120$. There are many pairs of numbers whose product is −120. We are looking for the pair whose sum is also −2. The numbers are −12 and 10. Writing $-2x$ as $-12x + 10x$ and then factoring by grouping, we have

$$8x^2 - 2x - 15 = 8x^2 - 12x + 10x - 15$$
$$= 4x(2x - 3) + 5(2x - 3)$$
$$= (2x - 3)(4x + 5)$$

Getting Ready for Class

After reading through the preceding section, respond in your own words and in complete sentences.

Answers
8. $(x - 1)(8x + 3)$
9. $(x + 2)(4x + 3)$
10. $(2x - 5)(2x + 3)$

1. What is a prime polynomial?
2. When factoring polynomials, what should you look for first?
3. How can you check to see that you have factored a trinomial correctly?
4. Describe how to determine the binomial factors of $6x^2 + 5x - 25$.

Problem Set 2.6

A Factor the following trinomials. [Examples 1–3]

1. $x^2 + 7x + 12$
2. $x^2 + 7x + 10$
3. $x^2 + 3x + 2$
4. $x^2 + 7x + 6$

5. $a^2 + 10a + 21$
6. $a^2 - 7a + 12$
7. $x^2 - 7x + 10$
8. $x^2 - 3x + 2$

9. $y^2 - 10y + 21$
10. $y^2 - 7y + 6$
11. $x^2 - x - 12$
12. $x^2 - 4x - 5$

13. $y^2 + y - 12$
14. $y^2 + 3y - 18$
15. $x^2 + 5x - 14$
16. $x^2 - 5x - 24$

17. $r^2 - 8r - 9$
18. $r^2 - r - 2$
19. $x^2 - x - 30$
20. $x^2 + 8x + 12$

21. $a^2 + 15a + 56$
22. $a^2 - 9a + 20$
23. $y^2 - y - 42$
24. $y^2 + y - 42$

25. $x^2 + 13x + 42$
26. $x^2 - 13x + 42$

B Factor the following problems completely. First, factor out the greatest common factor, and then factor the remaining trinomial. [Example 4]

27. $2x^2 + 6x + 4$
28. $3x^2 - 6x - 9$
29. $3a^2 - 3a - 60$
30. $2a^2 - 18a + 28$

31. $100x^2 - 500x + 600$
32. $100x^2 - 900x + 2,000$
33. $100p^2 - 1,300p + 4,000$
34. $100p^2 - 1,200p + 3,200$

35. $x^4 - x^3 - 12x^2$
36. $x^4 - 11x^3 + 24x^2$
37. $2r^3 + 4r^2 - 30r$
38. $5r^3 + 45r^2 + 100r$

39. $2y^4 - 6y^3 - 8y^2$
40. $3r^3 - 3r^2 - 6r$
41. $x^5 + 4x^4 + 4x^3$
42. $x^5 + 13x^4 + 42x^3$

43. $3y^4 - 12y^3 - 15y^2$
44. $5y^4 - 10y^3 + 5y^2$
45. $4x^4 - 52x^3 + 144x^2$
46. $3x^3 - 3x^2 - 18x$

B Factor completely. Be sure to factor out the greatest common factor first if it is a number other than 1.

47. $2x^2 + x - 15$
48. $2x^2 - x - 15$
49. $2x^2 - 13x + 15$
50. $2x^2 + 13x + 15$

51. $2x^2 - 11x + 15$
52. $2x^2 + 11x + 15$
53. $2x^2 + 7x + 15$
54. $2x^2 + x + 15$

55. $2 + 7a + 6a^2$

56. $2 - 7a + 6a^2$

57. $60y^2 - 15y - 45$

58. $72y^2 + 60y - 72$

59. $6x^4 - x^3 - 2x^2$

60. $3x^4 + 2x^3 - 5x^2$

61. $40r^3 - 120r^2 + 90r$

62. $40r^3 + 200r^2 + 250r$

63. $4x^2 - 11xy - 3y^2$

64. $3x^2 + 19xy - 14y^2$

65. $10x^2 - 3xa - 18a^2$

66. $9x^2 + 9xa - 10a^2$

67. $18a^2 + 3ab - 28b^2$

68. $6a^2 - 7ab - 5b^2$

69. $600 + 800t - 800t^2$

70. $200 - 600t - 350t^2$

71. $9y^4 + 9y^3 - 10y^2$

72. $4y^5 + 7y^4 - 2y^3$

73. $24a^2 - 2a^3 - 12a^4$

74. $60a^2 + 65a^3 - 20a^4$

Maintaining Your Skills

75. $(2x - 3)(2x + 3)$

76. $(4 - 5x)(4 + 5x)$

77. $(2x - 3)^2$

78. $(4 - 5x)^2$

79. $(2x - 3)(4x^2 + 6x + 9)$

80. $(2x + 3)(4x^2 - 6x + 9)$

Getting Ready for the Next Section

For each problem below, place a number or expression inside the parentheses so that the resulting statement is true.

81. $\dfrac{25}{64} = (\quad)^2$

82. $\dfrac{4}{9} = (\quad)^2$

83. $x^6 = (\quad)^2$

84. $x^8 = (\quad)^2$

85. $16x^4 = (\quad)^2$

86. $81y^4 = (\quad)^2$

Write as a perfect cube.

87. $\dfrac{1}{8} = (\quad)^3$

88. $\dfrac{1}{27} = (\quad)^3$

89. $x^6 = (\quad)^3$

90. $x^{12} = (\quad)^3$

91. $27x^3 = (\quad)^3$

92. $125y^3 = (\quad)^3$

93. $8y^3 = (\quad)^3$

94. $1000x^3 = (\quad)^3$

Extending the Concepts

Factor completely.

95. $8x^6 + 26x^3y^2 + 15y^4$

96. $24x^4 + 6x^2y^3 - 45y^6$

97. $3x^2 + 295x - 500$

98. $3x^2 + 594x - 1{,}200$

99. $\dfrac{1}{8}x^2 + x + 2$

100. $\dfrac{1}{9}x^2 + x + 2$

101. $2x^2 + 1.5x + 0.25$

102. $6x^2 + 2x + 0.16$

2.7 A General Review of Factoring

Objectives
A Factor a variety of polynomials.

A Factoring Review

In this section we will review the different methods of factoring that we have presented in the previous sections of this chapter. This section is important because it will give you an opportunity to factor a variety of polynomials.

We begin this section by listing the steps that can be used to factor polynomials of any type.

Strategy To Factor a Polynomial

Step 1: If the polynomial has a greatest common factor other than 1, then factor out the greatest common factor.

Step 2: If the polynomial has two terms (it is a binomial), then see if it is the difference of two squares or the sum or difference of two cubes, and then factor accordingly. Remember, if it is the sum of two squares it will not factor.

Step 3: If the polynomial has three terms (a trinomial), then it is either a perfect square trinomial, which will factor into the square of a binomial, or it is not a perfect square trinomial, in which case we try to write it as the product of two binomials using the methods developed in this chapter.

Step 4: If the polynomial has more than three terms, then try to factor it by grouping.

Step 5: As a final check, see if any of the factors you have written can be factored further. If you have overlooked a common factor, you can catch it here.

Here are some examples illustrating how we use the steps in our list. There are no new factoring problems in this section. The problems here are all similar to the problems you have seen before. What is different is that they are not all of the same type.

EXAMPLE 1   Factor $2x^5 - 8x^3$.

SOLUTION First we check to see if the greatest common factor is other than 1. Since the greatest common factor is $2x^3$, we begin by factoring it out. Once we have done so, we notice that the binomial that remains is the difference of two squares, which we factor according to the formula $a^2 - b^2 = (a + b)(a - b)$.

$2x^5 - 8x^3 = 2x^3(x^2 - 4)$ Factor out the greatest common factor, $2x^3$
$ = 2x^3(x + 2)(x - 2)$ Factor the difference of two squares

EXAMPLE 2   Factor $3x^4 - 18x^3 + 27x^2$.

SOLUTION Step 1 is to factor out the greatest common factor $3x^2$. After we have done so, we notice that the trinomial that remains is a perfect square trinomial, which will factor as the square of a binomial.

$3x^4 - 18x^3 + 27x^2 = 3x^2(x^2 - 6x + 9)$ Factor out $3x^2$
$ = 3x^2(x - 3)^2$ $x^2 - 6x + 9$ is the square of $x - 3$

PRACTICE PROBLEMS

1. Factor $3x^8 - 27x^6$.

2. Factor $4x^4 + 40x^3 + 100x^2$.

Answers
1. $3x^6(x + 3)(x - 3)$
2. $4x^2(x + 5)^2$

3. Factor $y^4 + 36y^2$.

EXAMPLE 3 Factor $y^3 + 25y$.

SOLUTION We begin by factoring out the y that is common to both terms. The binomial that remains after we have done so is the sum of two squares, which does not factor, so after the first step, we are finished.

$$y^3 + 25y = y(y^2 + 25)$$

EXAMPLE 4 Factor $6a^2 - 11a + 4$.

4. Factor $6x^2 - x - 15$.

SOLUTION Here we have a trinomial that does not have a greatest common factor other than 1. Since it is not a perfect square trinomial, we factor it by trial and error. Without showing all the different possibilities, here is the answer.

$$6a^2 - 11a + 4 = (3a - 4)(2a - 1)$$

EXAMPLE 5 Factor $2x^4 + 16x$.

5. Factor $3x^5 - 81x^2$.

SOLUTION This binomial has a greatest common factor of $2x$. The binomial that remains after the $2x$ has been factored from each term is the sum of two cubes, which we factor according to the formula $a^3 + b^3 = (a + b)(a^2 - ab + b^2)$.

$$2x^4 + 16x = 2x(x^3 + 8) \quad \text{Factor } 2x \text{ from each term}$$
$$= 2x(x + 2)(x^2 - 2x + 4) \quad \text{The sum of two cubes}$$

EXAMPLE 6 Factor $2ab^5 + 8ab^4 + 2ab^3$.

6. Factor $3a^2b^3 + 6a^2b^2 - 3a^2b$.

SOLUTION The greatest common factor is $2ab^3$. We begin by factoring it from each term. After that we find that the trinomial that remains cannot be factored further.

$$2ab^5 + 8ab^4 + 2ab^3 = 2ab^3(b^2 + 4b + 1)$$

EXAMPLE 7 Factor $4x^2 - 6x + 2ax - 3a$.

7. Factor $x^2 - 10x + 25 - b^2$.
(*Hint:* Group the first three terms together.)

SOLUTION Our polynomial has four terms, so we factor by grouping.

$$4x^2 - 6x + 2ax - 3a = 2x(2x - 3) + a(2x - 3)$$
$$= (2x - 3)(2x + a)$$

Getting Ready for Class

After reading through the preceding section, respond in your own words and in complete sentences.

1. How do you know when you've factored completely?
2. If a polynomial has four terms, what method of factoring should you try?
3. What is the first step in factoring a polynomial?
4. What do we call a polynomial that does not factor?

Answers
3. $y^2(y^2 + 36)$
4. $(3x - 5)(2x + 3)$
5. $3x^2(x - 3)(x^2 + 3x + 9)$
6. $3a^2b(b^2 + 2b - 1)$
7. $(x - 5 + b)(x - 5 - b)$

Problem Set 2.7

A Factor each of the following polynomials completely. Once you are finished factoring, none of the factors you obtain should be factorable. Also, note that the even-numbered problems are not necessarily similar to the odd-numbered problems that precede them in this problem set. [Examples 1–7]

1. $x^2 - 81$
2. $x^2 - 18x + 81$
3. $x^2 + 2x - 15$

4. $15x^2 + 13x - 6$
5. $x^2(x + 2) + 6x(x + 2) + 9(x + 2)$
6. $12x^2 - 11x + 2$

7. $x^2y^2 + 2y^2 + x^2 + 2$
8. $21y^2 - 25y - 4$
9. $2a^3b + 6a^2b + 2ab$

10. $6a^2 - ab - 15b^2$
11. $x^2 + x + 1$
12. $x^2y + 3y + 2x^2 + 6$

13. $12a^2 - 75$
14. $18a^2 - 50$
15. $9x^2 - 12xy + 4y^2$

16. $x^3 - x^2$
17. $25 - 10t + t^2$
18. $t^2 + 4t + 4 - y^2$

19. $4x^3 + 16xy^2$
20. $16x^2 + 49y^2$
21. $2y^3 + 20y^2 + 50y$

22. $x^2 + 5bx - 2ax - 10ab$
23. $a^7 + 8a^4b^3$
24. $5a^2 - 45b^2$

25. $t^2 + 6t + 9 - x^2$
26. $36 + 12t + t^2$
27. $x^3 + 5x^2 - 9x - 45$

28. $x^3 + 5x^2 - 16x - 80$
29. $5a^2 + 10ab + 5b^2$
30. $3a^3b^2 + 15a^2b^2 + 3ab^2$

A Factor completely. [Examples 1–7]

31. $x^2 + 49$

32. $16 - x^4$

33. $3x^2 + 15xy + 18y^2$

34. $3x^2 + 27xy + 54y^2$

35. $9a^2 + 2a + \dfrac{1}{9}$

36. $18 - 2a^2$

37. $x^2(x - 3) - 14x(x - 3) + 49(x - 3)$

38. $x^2 + 3ax - 2bx - 6ab$

39. $x^2 - 64$

40. $9x^2 - 4$

41. $8 - 14x - 15x^2$

42. $5x^4 + 14x^2 - 3$

43. $49a^7 - 9a^5$

44. $a^6 - b^6$

45. $r^2 - \dfrac{1}{25}$

46. $27 - r^3$

47. $49x^2 + 9y^2$

48. $12x^4 - 62x^3 + 70x^2$

49. $100x^2 - 100x - 600$

50. $100x^2 - 100x - 1{,}200$

51. $25a^3 + 20a^2 + 3a$

52. $16a^5 - 54a^2$

53. $3x^4 - 14x^2 - 5$

54. $8 - 2x - 15x^2$

55. $24a^5b - 3a^2b$

56. $18a^4b^2 - 24a^3b^3 + 8a^2b^4$

57. $64 - r^3$

58. $r^2 - \dfrac{1}{9}$

59. $20x^4 - 45x^2$

60. $16x^3 + 16x^2 + 3x$

2.7 Problem Set   133

61. $400t^2 - 900$

62. $900 - 400t^2$

63. $16x^5 - 44x^4 + 30x^3$

64. $16x^2 + 16x - 1$

65. $y^6 - 1$

66. $25y^7 - 16y^5$

67. $50 - 2a^2$

68. $4a^2 + 2a + \frac{1}{4}$

69. $12x^4y^2 + 36x^3y^3 + 27x^2y^4$

70. $16x^3y^2 - 4xy^2$

71. $x^2 - 4x + 4 - y^2$

72. $x^2 - 12x + 36 - b^2$

73. $a^2 - \frac{4}{3}ab + \frac{4}{9}b^2$

74. $a^2 + \frac{3}{2}ab + \frac{9}{16}b^2$

75. $x^2 - \frac{4}{5}xy + \frac{4}{25}y^2$

76. $x^2 - \frac{8}{7}xy + \frac{16}{49}y^2$

77. $a^2 - \frac{5}{3}ab + \frac{25}{36}b^2$

78. $a^2 + \frac{5}{4}ab + \frac{25}{64}b^2$

79. $x^2 - \frac{8}{5}xy + \frac{16}{25}y^2$

80. $a^2 + \frac{3}{5}ab + \frac{9}{100}b^2$

81. $2x^2(x + 2) - 13x(x + 2) + 15(x + 2)$

82. $5x^2(x - 4) - 14x(x - 4) - 3(x - 4)$

83. $(x - 4)^3 + (x - 4)^4$

84. $(2x - 7)^5 + (2x - 7)^6$

85. $2y^3 - 54$

86. $81 + 3y^3$

87. $2a^3 - 128b^3$

88. $128a^3 + 2b^3$

89. $2x^3 + 432y^3$

90. $432x^3 - 2y^3$

Maintaining Your Skills

The following problems are taken from the book Algebra for the Practical Man, written by J. E. Thompson and published by D. Van Nostrand Company in 1931.

91. A man spent $112.80 for 108 geese and ducks, each goose costing 14 dimes and each duck 6 dimes. How many of each did he buy?

92. If 15 pounds of tea and 10 pounds of coffee together cost $15.50, while 25 pounds of tea and 13 pounds of coffee at the same prices cost $24.55, find the price per pound of each.

93. A number of oranges at the rate of three for $0.10 and apples at $0.15 a dozen cost, together, $6.80. Five times as many oranges and one-fourth as many apples at the same rates would have cost $25.45. How many of each were bought?

94. An estate is divided among three persons: A, B, and C. A's share is three times that of B and B's share is twice that of C. If A receives $9,000 more than C, how much does each receive?

Getting Ready for the Next Section

Simplify.

95. $x^2 + (x + 1)^2$

96. $x^2 + (x + 3)^2$

97. $\dfrac{16t^2 - 64t + 48}{16}$

98. $\dfrac{100p^2 - 1{,}300p + 4{,}000}{100}$

Factor each of the following.

99. $x^2 - 2x - 24$

100. $x^2 - x - 6$

101. $2x^3 - 5x^2 - 3x$

102. $3x^3 - 5x^2 - 2x$

103. $x^3 + 2x^2 - 9x - 18$

104. $x^3 + 5x^2 - 4x - 20$

Solve.

105. $x - 6 = 0$

106. $x + 4 = 0$

107. $2x + 1 = 0$

108. $3x + 1 = 0$

Solving Equations by Factoring

2.8

Objectives
A Solve equations by factoring.
B Apply the Blueprint for Problem Solving to solve application problems whose solutions involve quadratic equations.
C Solve problems that contain formulas that are quadratic.

In this section we will use our knowledge of factoring to solve equations. Most of the equations we will solve in this section are *quadratic equations*. Here is the definition of a quadratic equation.

> **Definition**
> Any equation that can be written in the form
> $$ax^2 + bx + c = 0$$
> where a, b, and c are constants and a is not 0 ($a \neq 0$) is called a **quadratic equation**. The form $ax^2 + bx + c = 0$ is called **standard form** for quadratic equations.

Each of the following is a quadratic equation:
$$2x^2 = 5x + 3 \qquad 5x^2 = 75 \qquad 4x^2 - 3x + 2 = 0$$

NOTATION For a quadratic equation written in standard form, the first term ax^2 is called the *quadratic term*; the second term bx is the *linear term*; and the last term c is called the *constant term*.

In the past we have noticed that the number 0 is a special number. There is another property of 0 that is the key to solving quadratic equations. It is called the *zero-factor property*.

Note The third equation is clearly a quadratic equation since it is in standard form. (Notice that a is 4, b is -3, and c is 2.) The first two equations are also quadratic because they could be put in the form $ax^2 + bx + c = 0$ by using the addition property of equality.

> **Zero-Factor Property**
> For all real numbers r and s,
> $$r \cdot s = 0 \quad \text{if and only if} \quad r = 0 \quad \text{or} \quad s = 0 \quad (\text{or both})$$

Note What the zero-factor property says in words is that we can't multiply and get 0 without multiplying by 0; that is, if we multiply two numbers and get 0, then one or both of the original two numbers we multiplied must have been 0.

A Solving Equations by Factoring

EXAMPLE 1 Solve $x^2 - 2x - 24 = 0$.

SOLUTION We begin by factoring the left side as $(x - 6)(x + 4)$ and get
$$(x - 6)(x + 4) = 0$$

Now both $(x - 6)$ and $(x + 4)$ represent real numbers. We notice that their product is 0. By the zero-factor property, one or both of them must be 0:
$$x - 6 = 0 \quad \text{or} \quad x + 4 = 0$$

We have used factoring and the zero-factor property to rewrite our original second-degree equation as two first-degree equations connected by the word *or*. Completing the solution, we solve the two first-degree equations:
$$x - 6 = 0 \quad \text{or} \quad x + 4 = 0$$
$$x = 6 \quad \text{or} \quad x = -4$$

We check our solutions in the original equation as follows:

Check $x = 6$ \qquad Check $x = -4$

$6^2 - 2(6) - 24 \stackrel{?}{=} 0 \qquad (-4)^2 - 2(-4) - 24 \stackrel{?}{=} 0$

$36 - 12 - 24 \stackrel{?}{=} 0 \qquad 16 + 8 - 24 \stackrel{?}{=} 0$

$0 = 0 \qquad\qquad\qquad 0 = 0$

PRACTICE PROBLEMS
1. Solve $x^2 - x - 6 = 0$.

Note We are placing a question mark over the equal sign because we don't know yet if the expression on the left will be equal to the expression on the right.

2.8 Solving Equations by Factoring

135

In both cases the result is a true statement, which means that both 6 and −4 are solutions to the original equation.

Although the next equation is not quadratic, the method we use is similar.

EXAMPLE 2 Solve $\frac{1}{3}x^3 = \frac{5}{6}x^2 + \frac{1}{2}x$.

SOLUTION We can simplify our work if we clear the equation of fractions. Multiplying both sides by the LCD, 6, we have

$$6 \cdot \frac{1}{3}x^3 = 6 \cdot \frac{5}{6}x^2 + 6 \cdot \frac{1}{2}x$$

$$2x^3 = 5x^2 + 3x$$

Next we add $-5x^2$ and $-3x$ to each side so that the right side will become 0.

$2x^3 - 5x^2 - 3x = 0$ Standard form

We factor the left side and then use the zero-factor property to set each factor to 0.

$x(2x^2 - 5x - 3) = 0$ Factor out the greatest common factor

$x(2x + 1)(x - 3) = 0$ Continue factoring

$x = 0$ or $2x + 1 = 0$ or $x - 3 = 0$ Zero-factor property

Solving each of the resulting equations, we have

$x = 0$ or $x = -\frac{1}{2}$ or $x = 3$

To generalize the preceding examples, here are the steps used in solving a quadratic equation by factoring.

Strategy To Solve an Equation by Factoring

Step 1: Write the equation in standard form.

Step 2: Factor the left side.

Step 3: Use the zero-factor property to set each factor equal to 0.

Step 4: Solve the resulting linear equations.

EXAMPLE 3 Solve $100x^2 = 300x$.

SOLUTION We begin by writing the equation in standard form and factoring:

$$100x^2 = 300x$$

$100x^2 - 300x = 0$ Standard form

$100x(x - 3) = 0$ Factor

Using the zero-factor property to set each factor to 0, we have

$100x = 0$ or $x - 3 = 0$

$x = 0$ or $x = 3$

The two solutions are 0 and 3.

---

2. Solve $\frac{1}{2}x^3 = \frac{5}{6}x^2 + \frac{1}{3}x$.

3. Solve $100x^2 = 500x$.

Answers
1. −2, 3
2. $0, -\frac{1}{3}, 2$
3. 0, 5

2.8 Solving Equations by Factoring

EXAMPLE 4 Solve $(x - 2)(x + 1) = 4$.

SOLUTION We begin by multiplying the two factors on the left side. (Notice that it would be incorrect to set each of the factors on the left side equal to 4. The fact that the product is 4 does not imply that either of the factors must be 4.)

$$(x - 2)(x + 1) = 4$$
$$x^2 - x - 2 = 4 \qquad \text{Multiply the left side}$$
$$x^2 - x - 6 = 0 \qquad \text{Standard form}$$
$$(x - 3)(x + 2) = 0 \qquad \text{Factor}$$
$$x - 3 = 0 \quad \text{or} \quad x + 2 = 0 \qquad \text{Zero-factor property}$$
$$x = 3 \quad \text{or} \quad x = -2$$

4. Solve $(x + 1)(x + 2) = 12$.

EXAMPLE 5 Solve for x: $x^3 + 2x^2 - 9x - 18 = 0$.

SOLUTION We start with factoring by grouping.

$$x^3 + 2x^2 - 9x - 18 = 0$$
$$x^2(x + 2) - 9(x + 2) = 0$$
$$(x + 2)(x^2 - 9) = 0$$
$$(x + 2)(x - 3)(x + 3) = 0 \qquad \text{The difference of two squares}$$
$$x + 2 = 0 \quad \text{or} \quad x - 3 = 0 \quad \text{or} \quad x + 3 = 0 \qquad \text{Set factors to 0}$$
$$x = -2 \quad \text{or} \quad x = 3 \quad \text{or} \quad x = -3$$

We have three solutions: -2, 3, and -3.

5. Solve $x^3 + 5x^2 - 4x - 20 = 0$.

B Applications

EXAMPLE 6 The sum of the squares of two consecutive integers is 25. Find the two integers.

SOLUTION We apply the Blueprint for Problem Solving to solve this application problem. Remember, step 1 in the blueprint is done mentally.

Step 1: *Read and list.*
Known items: Two consecutive integers. If we add their squares, the result is 25.
Unknown items: The two integers

Step 2: *Assign a variable and translate information.*
Let x = the first integer; then $x + 1$ = the next consecutive integer.

Step 3: *Reread and write an equation.*
Since the sum of the squares of the two integers is 25, the equation that describes the situation is

$$x^2 + (x + 1)^2 = 25$$

6. One integer is 3 more than another. The sum of their squares is 29. Find the two integers.

Answers
4. $-5, 2$
5. $-5, -2, 2$

Step 4: **Solve the equation.**

$$x^2 + (x+1)^2 = 25$$
$$x^2 + (x^2 + 2x + 1) = 25$$
$$2x^2 + 2x - 24 = 0$$
$$x^2 + x - 12 = 0$$
$$(x+4)(x-3) = 0$$
$$x = -4 \quad \text{or} \quad x = 3$$

Step 5: **Write the answer.**
If $x = -4$, then $x + 1 = -3$. If $x = 3$, then $x + 1 = 4$. The two integers are -4 and -3, or the two integers are 3 and 4.

Step 6: **Reread and check.**
The two integers in each pair are consecutive integers, and the sum of the squares of either pair is 25.

Another application of quadratic equations involves the Pythagorean theorem, an important theorem from geometry. The theorem gives the relationship between the sides of any right triangle (a triangle with a 90-degree angle). We state it here without proof.

Pythagorean Theorem
In any right triangle, the square of the longest side (hypotenuse) is equal to the sum of the squares of the other two sides (legs).

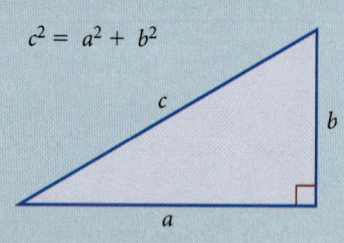

$$c^2 = a^2 + b^2$$

EXAMPLE 7 The lengths of the three sides of a right triangle are given by three consecutive integers. Find the lengths of the three sides.

SOLUTION **Step 1:** **Read and list.**
Known items: A right triangle. The three sides are three consecutive integers.
Unknown items: The three sides

Step 2: **Assign a variable and translate information.**
Let x = first integer (shortest side).
Then $x + 1$ = next consecutive integer
$x + 2$ = last consecutive integer (longest side)

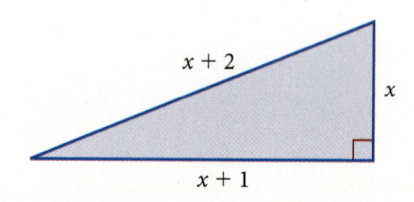

7. The longest side of a right triangle is 4 more than the shortest side. The third side is 2 more than the shortest side. Find the length of each side.

Answer
6. $-5, -2$, or $2, 5$

Step 3: **Reread and write an equation.**
By the Pythagorean theorem, we have

$$(x + 2)^2 = (x + 1)^2 + x^2$$

Step 4: **Solve the equation.**

$$x^2 + 4x + 4 = x^2 + 2x + 1 + x^2$$

$$x^2 - 2x - 3 = 0$$

$$(x - 3)(x + 1) = 0$$

$$x = 3 \quad \text{or} \quad x = -1$$

Step 5: **Write the answer.**
Since x is the length of a side in a triangle, it must be a positive number. Therefore, $x = -1$ cannot be used.
The shortest side is 3. The other two sides are 4 and 5.

Step 6: **Reread and check.**
The three sides are given by consecutive integers. The square of the longest side is equal to the sum of the squares of the two shorter sides.

EXAMPLE 8 Two boats leave from an island port at the same time. One travels due north at a speed of 12 miles per hour, and the other travels due west at a speed of 16 miles per hour. How long until the distance between the boats is 60 miles?

SOLUTION **Step 1:** **Read and list.**
Known items: The speed and direction of both boats. The distance between the boats.
Unknown items: The distance traveled by each boat, and the time.

Step 2: **Assign a variable and translate information.**
Let t = the time.
Then $12t$ = the distance traveled by boat going north
$16t$ = the distance traveled by boat going west
If we draw a diagram for the problem, we see that the distances traveled by the two boats form the legs of a right triangle. The hypotenuse of the triangle will be the distance between the boats, which is 60 miles.

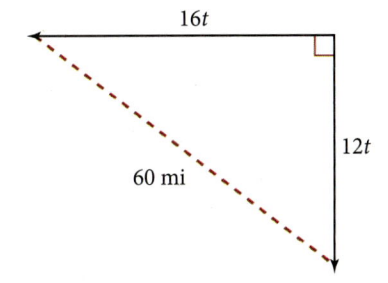

8. Two boats leave from an island port at the same time. One travels due east at a speed of 5 miles per hour, and the other travels due south at a speed of 12 miles per hour. How long until the distance between the boats is 52 miles?

Answer
7. 6, 8, 10

Step 3: *Reread and write an equation.*

By the Pythagorean theorem, we have

$$(16t)^2 + (12t)^2 = 60^2$$

Step 4: *Solve the equation.*

$$256t^2 + 144t^2 = 3600$$
$$400t^2 = 3600$$
$$400t^2 - 3600 = 0$$
$$t^2 - 9 = 0$$
$$(t + 3)(t - 3) = 0$$
$$t = -3 \quad \text{or} \quad t = 3$$

Step 5: *Write the answer.*

Because t is measuring time, it must be a positive number. Therefore, $t = -3$ cannot be used.

The two boats will be 60 miles apart after 3 hours.

Step 6: *Reread and check.*

The boat going north will travel $12 \cdot 3 = 36$ miles in 3 hours, and the boat going west wil travel $16 \cdot 3 = 48$ miles in three hours. The distance between them after 3 hours will be 60 miles ($48^2 + 36^2 = 60^2$).

C Formulas

Our next two examples involve formulas that are quadratic.

EXAMPLE 9   If an object is projected into the air with an initial vertical velocity of v feet/second, its height h, in feet, above the ground after t seconds will be given by

$$h = vt - 16t^2$$

Find t if $v = 64$ feet/second and $h = 48$ feet.

SOLUTION   Substituting $v = 64$ and $h = 48$ into the preceding formula, we have

$$48 = 64t - 16t^2$$

which is a quadratic equation. We write it in standard form and solve by factoring:

$$16t^2 - 64t + 48 = 0$$
$$t^2 - 4t + 3 = 0 \quad \text{Divide each side by 16}$$
$$(t - 1)(t - 3) = 0$$
$$t - 1 = 0 \quad \text{or} \quad t - 3 = 0$$
$$t = 1 \quad \text{or} \quad t = 3$$

Here is how we interpret our results: If an object is projected upward with an initial vertical velocity of 64 feet/second, it will be 48 feet above the ground after 1 second and after 3 seconds; that is, it passes 48 feet going up and also coming down.

9. Use the formula in Example 9 to find t if $v = 64$ feet/second and $h = 64$ feet.

Answers
8. 4 hours
9. 2 seconds

2.8 Solving Equations by Factoring

EXAMPLE 10 A manufacturer of small portable radios knows that the number of radios she can sell each week is related to the price of the radios by the equation $x = 1,300 - 100p$, where x is the number of radios and p is the price per radio. What price should she charge for each radio if she wants the weekly revenue to be $4,000?

SOLUTION The formula for total revenue is $R = xp$. Since we want R in terms of p, we substitute $1,300 - 100p$ for x in the equation $R = xp$:

If $\quad R = xp$

and $\quad x = 1,300 - 100p$

then $\quad R = (1,300 - 100p)p$

We want to find p when R is 4,000. Substituting 4,000 for R in the formula gives us

$$4,000 = (1,300 - 100p)p$$

$$4,000 = 1,300p - 100p^2$$

which is a quadratic equation. To write it in standard form, we add $100p^2$ and $-1,300p$ to each side, giving us

$$100p^2 - 1,300p + 4,000 = 0$$

$$p^2 - 13p + 40 = 0 \quad \text{Divide each side by 100}$$

$$(p - 5)(p - 8) = 0$$

$$p - 5 = 0 \quad \text{or} \quad p - 8 = 0$$

$$p = 5 \quad \text{or} \quad p = 8$$

If she sells the radios for $5 each or for $8 each she will have a weekly revenue of $4,000.

Getting Ready for Class

After reading through the preceding section, respond in your own words and in complete sentences.

1. Explain the Pythagorean theorem in words.
2. What is the first step in solving an equation by factoring?
3. Describe the zero-factor property in your own words.
4. Write an application problem for which the solution depends on solving the equation $x^2 + (x + 1)^2 = 313$.

---

10. Use the information in Example 10 to find the price that should be charged if the weekly revenue is to be $3,600.

Answer
10. $4 or $9

Left Blank Intentionally

Problem Set 2.8

A Solve each equation. [Examples 1–5]

1. $x^2 - 5x - 6 = 0$
2. $x^2 + 5x - 6 = 0$
3. $x^2 - 3x - 4 = 0$

4. $x^2 - 5x + 6 = 0$
5. $x^2 + 6x + 8 = 0$
6. $x^2 - 3x - 18 = 0$

7. $x^2 + 2x - 15 = 0$
8. $x^2 + 11x + 18 = 0$
9. $x^3 - 5x^2 + 6x = 0$

10. $x^3 + 5x^2 + 6x = 0$
11. $3y^2 + 11y - 4 = 0$
12. $3y^2 - y - 4 = 0$

13. $60x^2 - 130x + 60 = 0$
14. $90x^2 + 60x - 80 = 0$
15. $\frac{1}{10}t^2 - \frac{5}{2} = 0$

16. $\frac{2}{7}t^2 - \frac{7}{2} = 0$
17. $100x^4 = 400x^3 + 2{,}100x^2$
18. $100x^4 = -400x^3 + 2{,}100x^2$

19. $\frac{1}{5}y^2 - 2 = -\frac{3}{10}y$
20. $\frac{1}{2}y^2 + \frac{5}{3} = \frac{17}{6}y$
21. $9x^2 - 12x = 0$

22. $4x^2 + 4x = 0$
23. $0.02r + 0.01 = 0.15r^2$
24. $0.02r - 0.01 = -0.08r^2$

25. $9a^3 = 16a$
26. $16a^3 = 25a$
27. $-100x = 10x^2$

28. $800x = 100x^2$
29. $(x + 6)(x - 2) = -7$
30. $(x - 7)(x + 5) = -20$

31. $(y - 4)(y + 1) = -6$

32. $(y - 6)(y + 1) = -12$

33. $(x + 1)^2 = 3x + 7$

34. $(x + 2)^2 = 9x$

35. $(2r + 3)(2r - 1) = -(3r + 1)$

36. $(3r + 2)(r - 1) = -(7r - 7)$

37. $x^3 + 3x^2 - 4x - 12 = 0$

38. $x^3 + 5x^2 - 4x - 20 = 0$

39. $x^3 + 2x^2 - 25x - 50 = 0$

40. $x^3 + 4x^2 - 9x - 36 = 0$

41. $2x^3 + 3x^2 - 8x - 12 = 0$

42. $3x^3 + 2x^2 - 27x - 18 = 0$

43. $4x^3 + 12x^2 - 9x - 27 = 0$

44. $9x^3 + 18x^2 - 4x - 8 = 0$

A Problems 45–54 are problems you will see later in the book. Solve each equation. [Examples 1–5]

45. $3x^2 + x = 10$

46. $y^2 + y - 20 = 2y$

47. $12(x + 3) + 12(x - 3) = 3(x^2 - 9)$

48. $8(x + 2) + 8(x - 2) = 3(x^2 - 4)$

49. $(y + 3)^2 + y^2 = 9$

50. $(2y + 4)^2 + y^2 = 4$

51. $(x + 3)^2 + 1^2 = 2$

52. $(x - 3)^2 + (-1)^2 = 10$

53. $(x + 2)(x) = 2^3$

54. $(x + 3)(x) = 2^2$

55. Let $y = \left(x + \dfrac{3}{2}\right)^2$. Find all values for the variable x, for which $y = 0$.

56. Let $y = \left(x - \dfrac{5}{2}\right)^2$. Find all values for the variable x, for which $y = 0$.

57. Let $y = (x - 3)^2 - 25$. Find all values for the variable x, for which $y = 0$.

58. Let $y = 9x^3 + 18x^2 - 4x - 8$. Find all values for the variable x, for which $y = 0$.

Let $y = x^2 + 6x + 3$. Find all values for the variable x, for the following.

59. $y = -6$ **60.** $y = 19$ **61.** $y = 10$ **62.** $y = -2$

Let $y = x^2 - 5x$. Find all values for the variable x, for the following.

63. $y = 0$ **64.** $y = -6$ **65.** $y = 2x + 8$ **66.** $y = -2x + 10$

67. Solve each equation
   a. $9x - 25 = 0$
   b. $9x^2 - 25 = 0$
   c. $9x^2 - 25 = 56$
   d. $9x^2 - 25 = 30x - 50$

68. Solve each equation
   a. $5x - 6 = 0$
   b. $(5x - 6)^2 = 0$
   c. $25x^2 - 36 = 0$
   d. $25x^2 - 36 = 28$

A Applying the Concepts [Examples 6–8]

69. Distance Two cyclists leave from an intersection at the same time. One travels due north at a speed of 15 miles per hour, and the other travels due east at a speed of 20 miles per hour. How long until the distance between the two cyclists is 75 miles?

70. Distance Two airplanes leave from an airport at the same time. One travels due south at a speed of 480 miles per hour, and the other travels due west at a speed of 360 miles per hour. How long until the distance between the two airplanes is 2400 miles?

71. Consecutive Integers The square of the sum of two consecutive integers is 81. Find the two integers.

72. Consecutive Integers Find two consecutive even integers whose sum squared is 100.

73. Right Triangle A 25-foot ladder is leaning against a building. The base of the ladder is 7 feet from the side of the building. How high does the ladder reach along the side of the building?

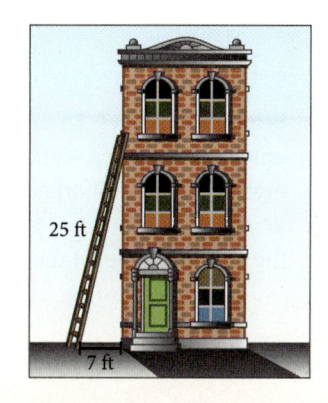

25 ft

7 ft

74. Right Triangle Noreen wants to place her 13-foot ramp against the side of her house so that the top of the ramp rests on a ledge that is 5 feet above the ground. How far will the base of the ramp be from the house?

75. **Right Triangle** The lengths of the three sides of a right triangle are given by three consecutive even integers. Find the lengths of the three sides.

76. **Right Triangle** The longest side of a right triangle is 3 less than twice the shortest side. The third side measures 12 inches. Find the length of the shortest side.

77. **Geometry** The length of a rectangle is 2 feet more than 3 times the width. If the area is 16 square feet, find the width and the length.

78. **Geometry** The length of a rectangle is 4 yards more than twice the width. If the area is 70 square yards, find the width and the length.

79. **Geometry** The base of a triangle is 2 inches more than 4 times the height. If the area is 36 square inches, find the base and the height.

80. **Geometry** The height of a triangle is 4 feet less than twice the base. If the area is 48 square feet, find the base and the height.

81. **Projectile Motion** If an object is thrown straight up into the air with an initial velocity of 32 feet per second, then its height above the ground at any time t is given by the formula $h = 32t - 16t^2$. Find the times at which the object is on the ground by letting $h = 0$ in the equation and solving for t.

82. **Projectile Motion** An object is projected into the air with an initial velocity of 64 feet per second. Its height at any time t is given by the formula $h = 64t - 16t^2$. Find the times at which the object is on the ground.

C The formula $h = vt - 16t^2$ gives the height h, in feet, of an object projected into the air with an initial vertical velocity v, in feet per second, after t seconds. [Examples 9–10]

83. **Projectile Motion** If an object is projected upward with an initial velocity of 48 feet per second, at what times will it reach a height of 32 feet above the ground?

84. **Projectile Motion** If an object is projected upward into the air with an initial velocity of 80 feet per second, at what times will it reach a height of 64 feet above the ground?

85. **Projectile Motion** An object is projected into the air with a vertical velocity of 24 feet per second. At what times will the object be on the ground? (It is on the ground when h is 0.)

86. **Projectile Motion** An object is projected into the air with a vertical velocity of 20 feet per second. At what times will the object be on the ground?

87. **Height of a Bullet** A bullet is fired into the air with an initial upward velocity of 80 feet per second from the top of a building 96 feet high. The equation that gives the height of the bullet at any time t is $h = 96 + 80t - 16t^2$. At what times will the bullet be 192 feet in the air?

88. **Height of an Arrow** An arrow is shot into the air with an upward velocity of 48 feet per second from a hill 32 feet high. The equation that gives the height of the arrow at any time t is $h = 32 + 48t - 16t^2$. Find the times at which the arrow will be 64 feet above the ground.

Chapter 2 Summary

Properties of Exponents [2.1]

If a and b represent real numbers and r and s represent integers, then

1. $a^r \cdot a^s = a^{r+s}$

2. $(a^r)^s = a^{r \cdot s}$

3. $(ab)^r = a^r \cdot b^r$

4. $a^{-r} = \dfrac{1}{a^r}$ $(a \neq 0)$

5. $\left(\dfrac{a}{b}\right)^r = \dfrac{a^r}{b^r}$ $(b \neq 0)$

6. $\dfrac{a^r}{a^s} = a^{r-s}$ $(a \neq 0)$

7. $a^1 = a$
$a^0 = 1$ $(a \neq 0)$

EXAMPLES

1. These expressions illustrate the properties of exponents.

   a. $x^2 \cdot x^3 = x^{2+3} = x^5$
   b. $(x^2)^3 = x^{2 \cdot 3} = x^6$
   c. $(3x)^2 = 3^2 \cdot x^2 = 9x^2$
   d. $2^{-3} = \dfrac{1}{2^3} = \dfrac{1}{8}$
   e. $\left(\dfrac{x}{5}\right)^2 = \dfrac{x^2}{5^2} = \dfrac{x^2}{25}$
   f. $\dfrac{x^7}{x^5} = x^{7-5} = x^2$
   g. $3^1 = 3$
      $3^0 = 1$

Addition of Polynomials [2.2]

To add two polynomials, simply combine the coefficients of similar terms.

2. $(3x^2 + 2x - 5) + (4x^2 - 7x + 2)$
   $= 7x^2 - 5x - 3$

Negative Signs Preceding Parentheses [2.2]

If there is a negative sign directly preceding the parentheses surrounding a polynomial, we may remove the parentheses and preceding negative sign by changing the sign of each term within the parentheses. (This procedure is actually just another application of the distributive property.)

3. $-(2x^2 - 8x - 9)$
   $= -2x^2 + 8x + 9$

Multiplication of Polynomials [2.3]

To multiply two polynomials, multiply each term in the first by each term in the second.

4. $(3x - 5)(x + 2)$
   $= 3x^2 + 6x - 5x - 10$
   $= 3x^2 + x - 10$

Special Products [2.3]

$(a + b)^2 = a^2 + 2ab + b^2$

$(a - b)^2 = a^2 - 2ab + b^2$

$(a + b)(a - b) = a^2 - b^2$

5. The following are examples of the three special products:

   $(x + 3)^2 = x^2 + 6x + 9$
   $(5 - x)^2 = 25 - 10x + x^2$
   $(x + 7)(x - 7) = x^2 - 49$

Business Applications [2.2, 2.3, 2.4, 2.8]

6. A company makes x items each week and sells them for p dollars each, according to the equation $p = 35 - 0.1x$. Then, the revenue is
$$R = x(35 - 0.1x) = 35x - 0.1x^2$$
If the total cost to make all x items is $C = 8x + 500$, then the profit gained by selling the x items is
$$P = 35x - 0.1x^2 - (8x + 500)$$
$$= -500 + 27x - 0.1x^2$$

If a company manufacturers and sells x items at p dollars per item, then the revenue R is given by the formula
$$R = xp$$
If the total cost to manufacture all x items is C, then the profit obtained from selling all x items is
$$P = R - C$$

Greatest Common Factor [2.4]

7. The greatest common factor of $10x^5 - 15x^4 + 30x^3$ is $5x^3$. Factoring it out of each term, we have
$5x^3(2x^2 - 3x + 6)$

The greatest common factor of a polynomial is the largest monomial (the monomial with the largest coefficient and highest exponent) that divides each term of the polynomial. The first step in factoring a polynomial is to factor the greatest common factor (if it is other than 1) out of each term.

Special Factoring [2.5]

8. Here are some binomials that have been factored this way.
$x^2 + 6x + 9 = (x + 3)^2$
$x^2 - 6x + 9 = (x - 3)^2$
$x^2 - 9 = (x + 3)(x - 3)$
$x^3 - 27 = (x - 3)(x^2 + 3x + 9)$
$x^3 + 27 = (x + 3)(x^2 - 3x + 9)$

$a^2 + 2ab + b^2 = (a + b)^2$	Perfect square trinomials
$a^2 - 2ab + b^2 = (a - b)^2$	
$a^2 - b^2 = (a - b)(a + b)$	Difference of two squares
$a^3 - b^3 = (a - b)(a^2 + ab + b^2)$	Difference of two cubes
$a^3 + b^3 = (a + b)(a^2 - ab + b^2)$	Sum of two cubes

Factoring Trinomials [2.6]

9. $x^2 + 5x + 6 = (x + 2)(x + 3)$
$x^2 - 5x + 6 = (x - 2)(x - 3)$
$x^2 + x - 6 = (x - 2)(x + 3)$
$x^2 - x - 6 = (x + 2)(x - 3)$

We factor a trinomial by writing it as the product of two binomials. (This refers to trinomials whose greatest common factor is 1.) Each factorable trinomial has a unique set of factors. Finding the factors is sometimes a matter of trial and error.

To Factor Polynomials in General [2.7]

Step 1: If the polynomial has a greatest common factor other than 1, then factor out the greatest common factor.

Step 2: If the polynomial has two terms (it is a binomial), then see if it is the difference of two squares, or the sum or difference of two cubes, and then factor accordingly. Remember, if it is the sum of two squares it will not factor.

Step 3: If the polynomial has three terms (a trinomial), then it is either a perfect square trinomial, which will factor into the square of a binomial, or it is not a perfect square trinomial, in which case you use one of the other methods.

Step 4: If the polynomial has more than three terms, then try to factor it by grouping.

Step 5: As a final check, see if any of the factors you have written can be factored further. If you have overlooked a common factor, you can catch it here.

10. Factor completely.

   a. $3x^3 - 6x^2 = 3x^2(x - 2)$

   b. $x^2 - 9 = (x + 3)(x - 3)$
   $x^3 - 8 = (x - 2)(x^2 + 2x + 4)$
   $x^3 + 27 = (x + 3)(x^2 - 3x + 9)$

   c. $x^2 - 6x + 9 = (x - 3)^2$
   $6x^2 - 7x - 5 = (2x + 1)(3x - 5)$

   d. $x^2 + ax + bx + ab$
   $= x(x + a) + b(x + a)$
   $= (x + a)(x + b)$

To Solve an Equation by Factoring [2.8]

Step 1: Write the equation in standard form.
Step 2: Factor the left side.
Step 3: Use the zero-factor property to set each factor equal to zero.
Step 4: Solve the resulting linear equations.

11. Solve $x^2 - 5x = -6$.
$x^2 - 5x + 6 = 0$
$(x - 3)(x - 2) = 0$
$x - 3 = 0$ or $x - 2 = 0$
$x = 3$ or $x = 2$

🚫 COMMON MISTAKES

When we subtract one polynomial from another, it is common to forget to add the opposite of each term in the second polynomial. For example

$$(6x - 5) - (3x + 4) = 6x - 5 - 3x + 4 \quad \text{Mistake}$$

This mistake occurs if the negative sign outside the second set of parentheses is not distributed over all terms inside the parentheses. To avoid this mistake, remember: the opposite of a sum is the sum of the opposites, or,

$$-(3x + 4) = -3x + (-4)$$

Left Blank Intentionally

Chapter 2 Review

Simplify each of the following. [2.1]

1. $x^3 \cdot x^7$
2. $(5x^3)^2$
3. $(2x^3y)^2(-2x^4y^2)^3$

Write with positive exponents, and then simplify. [2.1]

4. 2^{-3}
5. $\left(\dfrac{2}{3}\right)^{-2}$
6. $2^{-2} + 4^{-1}$

Simplify each expression. All answers should contain positive exponents only. (Assume all variables are nonnegative.) [2.1]

7. $\dfrac{a^{-4}}{a^5}$
8. $\dfrac{(4x^2)(-3x^3)^2}{(12x^{-2})^2}$
9. $\dfrac{x^n x^{3n}}{x^{4n-2}}$

Simplify by combining similar terms. [2.2]

10. $(6x^2 - 3x + 2) - (4x^2 + 2x - 5)$
11. $(x^3 - x) - (x^2 + x) + (x^3 - 3) - (x^2 + 1)$

12. Subtract $2x^2 - 3x + 1$ from $3x^2 - 5x - 2$.
13. Simplify $-3[2x - 4(3x + 1)]$.

14. Find the value of $2x^2 - 3x + 1$ when x is -2.

Multiply. [2.3]

15. $3x(4x^2 - 2x + 1)$
16. $2a^2b^3(a^2 + 2ab + b^2)$
17. $(6 - y)(3 - y)$

18. $(2x^2 - 1)(3x^2 + 4)$
19. $2t(t + 1)(t - 3)$
20. $(x + 3)(x^2 - 3x + 9)$

21. $(2x - 3)(4x^2 + 6x + 9)$
22. $(a^2 - 2)^2$
23. $(3x + 5)^2$

24. $(4x - 3y)^2$
25. $\left(x - \dfrac{1}{3}\right)\left(x + \dfrac{1}{3}\right)$
26. $(2a + b)(2a - b)$

27. $(x - 1)^3$
28. $(x^m + 2)(x^m - 2)$

Factor out the greatest common factor. [2.4]

29. $6x^4y - 9xy^4 + 18x^3y^3$
30. $4x^2(x + y)^2 - 8y^2(x + y)^2$

Factor by grouping. [2.4, 2.6]

31. $8x^2 + 10 - 4x^2y - 5y$
32. $x^3 + 8b^2 - x^3y^2 - 8y^2b^2$

Factor completely. [2.4, 2.5]

33. $x^2 - 5x + 6$

34. $2x^3 + 4x^2 - 30x$

35. $20a^2 - 41ab + 20b^2$

36. $6x^4 - 11x^3 - 10x^2$

37. $24x^2y - 6xy - 45y$

Factor completely. [2.6]

38. $x^4 - 16$

39. $3a^4 + 18a^2 + 27$

40. $a^3 - 8$

41. $5x^3 + 30x^2y + 45xy^2$

42. $3a^3b - 27ab^3$

43. $x^2 - 10x + 25 - y^2$

44. $36 - 25a^2$

45. $x^3 + 4x^2 - 9x - 36$

Solve each equation. [2.8]

46. $x^2 + 5x + 6 = 0$

47. $\frac{5}{6}y^2 = \frac{1}{4}y + \frac{1}{3}$

48. $9x^2 - 25 = 0$

49. $5x^2 = -10x$

50. $(x + 2)(x - 5) = 8$

51. $x^3 + 4x^2 - 9x - 36 = 0$

Solve each application. In each case be sure to show the equation used. [2.8]

52. Consecutive Numbers The product of two consecutive even integers is 80. Find the two integers.

53. Consecutive Numbers The sum of the squares of two consecutive integers is 41. Find the two integers.

Chapter 2 Cumulative Review

Simplify each of the following.

1. $(-5)^3$
2. $(-5)^{-3}$
3. $|-32 - 41| - 13$
4. $2^3 + 2(6^2 - 4^2)$
5. $96 \div 8 \cdot 4$
6. $55 \div 5 \cdot 11$
7. $\dfrac{3^3 + 5}{-2(5^2 - 3^2)}$
8. $\dfrac{2^3(17 - 2 \cdot 4)}{7^2 - 13}$

Let $P(x) = 11.5x - 0.01x^2$. Find the following.

9. $P(1)$
10. $P(-1)$

Let $Q(x) = \dfrac{3}{2}x^2 + \dfrac{3}{4}x - 5$. Find the following.

11. $Q(4)$
12. $Q(-4)$

Simplify each of the following expressions.

13. $6\left(x - \dfrac{1}{6}\right)$
14. $15\left(\dfrac{4}{3}x - \dfrac{3}{5}y\right)$
15. $(3x - 2) - (5x + 2)$
16. $3a - 2[4 - (a + 3)]$
17. $(3x^3 - 2) - (5x^2 + 2) + (5x^3 - 6) - (2x^2 + 4)$
18. $\left(\dfrac{5}{6}x^2 - \dfrac{2}{3}x\right) - \left(\dfrac{5}{3}x^2 + \dfrac{1}{2}x\right) + \left(\dfrac{5}{2}x^2 - \dfrac{1}{6}x\right) - \left(\dfrac{4}{3}x^2 + \dfrac{3}{2}x\right)$

Multiply or divide as indicated.

19. $(3x^2y)^4$
20. $(-3a^2b)(5a^3b^4)$
21. $\dfrac{32x^2y^6}{-16xy^2}$
22. $\dfrac{(-3a^2b)^4}{(a^3b^4)^2}$
23. $(3x - 2)(5x + 2)$
24. $5x(3x^2 + 2x + 7)$
25. $\left(x - \dfrac{2}{3}\right)^2$
26. $\left(t + \dfrac{2}{3}\right)^3$
27. $(4 \times 10^3)(7 \times 10^6)$
28. $\dfrac{8 \times 10^6}{2 \times 10^4}$

Factor each of the following expressions.

29. $x^2 - xy - ax + ay$
30. $6a^2 + a - 35$
31. $(x + 3)^2 - 25$
32. $\dfrac{1}{8} + t^3$

Solve the following equations.

33. $3x + 2 = 8$
34. $|2x - 4| + 7 = 13$
35. $\dfrac{3}{4}x - 7 = -34$
36. $-\dfrac{x}{2} = -\dfrac{3}{4} - \dfrac{3}{2}x$

37. $25a^3 = 36a$
38. $(x + 3)^2 = -2x - 7$

For problems 39 and 40, use the given value of x to evaluate $x^2 - 8x - 9$.

39. $x = -2$
40. $x = 2$

For problems 41 and 42, use the given value of x to evaluate $\dfrac{3x - 5}{2}$.

41. $x = -7$
42. $x = 7$

For problems 43 and 44, use the given values of a and b to evaluate $a^2 - 2ab + b^2$.

43. $a = 3, b = -2$
44. $a = -3, b = 2$

The illustration below shows the number of cattle (in millions) residing in various nations.

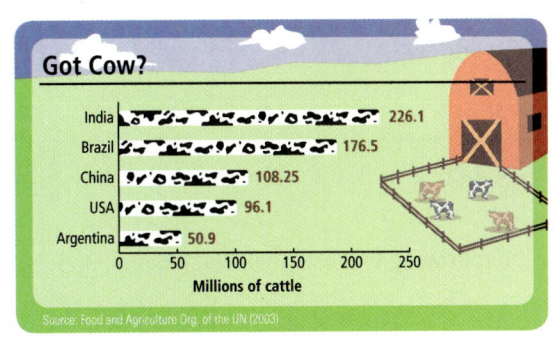

Got Cow?
- India: 226.1
- Brazil: 176.5
- China: 108.25
- USA: 96.1
- Argentina: 50.9

Source: Food and Agriculture Org. of the UN (2003)

Write in scientific notation the number of cows residing in the given country.

45. Brazil
46. USA

The illustration below shows the top five margins of victory for Nascar.

Close Calls in Nascar
- Craven/Busch, Darlington (2003): .002
- Earnhardt/Irvan, Talladega (1993): .005
- Harvick/Gordon, Atlanta (2001): .006
- Kahne/Kenseth, Rockinham (2004): .01
- Kenseth/Kahne, Atlanta (2000): .01

Source: NASCAR

Write in scientific notation the margin of victory for the given race.

47. Earnhardt/Irvan
48. Kahne/Kenseth

Chapter 2 Test

Simplify. All answers should contain positive exponents only. (Assume all variables are nonzero.) [2.1]

1. $\left(\dfrac{3}{4}\right)^{-2}$
2. 2^{-5}
3. $x^4 \cdot x^7 \cdot x^{-3}$
4. $\dfrac{a^{-5}}{a^{-7}}$
5. $(2x^2y)^3(2x^3y^4)^2$
6. $\dfrac{(2ab^3)^{-2}}{(a^{-4}b^3)^4}$

Let $P(x) = 15x - 0.01x^2$ and find the following.

7. $P(10)$
8. $P(-10)$

Simplify the following expressions. [2.2]

9. $\left(\dfrac{3}{4}x^3 - x^2 - \dfrac{3}{2}\right) - \left(\dfrac{1}{4}x^2 + 2x - \dfrac{1}{2}\right)$
10. $3 - 4[2x - 3(x + 6)]$

Multiply. [2.3]

11. $(3y - 7)(2y + 5)$
12. $(2x - 5)(x^2 + 4x - 3)$
13. $(8 - 3t^3)^2$
14. $(1 - 6y)(1 + 6y)$
15. $2x(x - 3)(2x + 5)$
16. $\left(5t^2 - \dfrac{1}{2}\right)\left(2t^2 + \dfrac{1}{5}\right)$

Factor the following expressions. [2.4, 2.5, 2.6, 2.7]

17. $x^2 + x - 12$
18. $12x^4 + 26x^2 - 10$
19. $16a^4 - 81y^4$
20. $7ax^2 - 14ay - b^2x^2 + 2b^2y$
21. $t^3 + \dfrac{1}{8}$
22. $4a^5b - 24a^4b^2 - 64a^3b^3$
23. $x^2 - 10x + 25 - b^2$
24. $81 - x^4$

Solve each equation [2.8]

25. $\dfrac{1}{5}x^2 = \dfrac{1}{3}x + \dfrac{2}{15}$
26. $100x^3 = 500x^2$
27. $(x + 1)(x + 2) = 12$
28. $x^3 + 2x^2 - 16x - 32 = 0$

Let $f(x) = x^2 - 5x + 6$. Find all values for the variable x for which $f(x) = g(x)$. [2.8]

29. $g(x) = 0$
30. $g(x) = x - 2$

31. Find the value of the variable x in the following figure. [2.8]

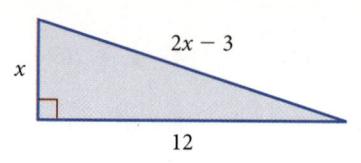

32. The area of the figure below is 35 square inches. Find the value of x. [2.8]

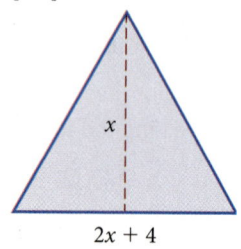

33. The area of the figure below is 48 square centimeters. Find the value of x. [2.8]

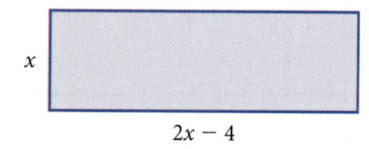

Profit, Revenue, and Cost A company making ceramic coffee cups finds that it can sell x cups per week at p dollars each, according to the formula $p = 25 - 0.2x$. If the total cost to produce and sell x coffee cups is $C = 2x + 100$, find

34. an equation for the revenue that gives the revenue in terms of x.
35. the profit equation.
36. the revenue brought in by selling 100 coffee cups.
37. the cost of producing 100 coffee cups.
38. the profit obtained by making and selling 100 coffee cups.

Chapter 2 Projects

EXPONENTS AND POLYNOMIALS

GROUP PROJECT

Discovering Pascal's Triangle

Number of People 3

Time Needed 20 minutes

Equipment Paper and pencils

Background The triangular array of numbers shown here is known as Pascal's triangle, after the French philosopher Blaise Pascal (1623–1662).

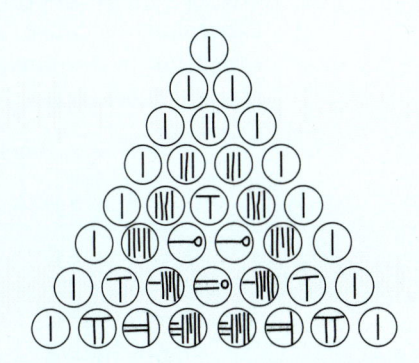

```
          1
        1   1
      1   2   1
    1   3   3   1
  1   4   6   4   1
1   5  10  10   5   1
```

Pascal's triangle in Japanese (1781)

Procedure Look at Pascal's triangle and discover how the numbers in each row of the triangle are obtained from the numbers in the row above it.

1. Once you have discovered how to extend the triangle, write the next two rows.

2. Pascal's triangle can be linked to the Fibonacci sequence by rewriting Pascal's triangle so that the 1s on the left side of the triangle line up under one another and the other columns are equally spaced to the right of the first column. Rewrite Pascal's triangle as indicated and then look along the diagonals of the new array until you discover how the Fibonacci sequence can be obtained from it.

3. The diagram above shows Pascal's triangle as written in Japanese in 1781. Use your knowledge of Pascal's triangle to translate the numbers written in Japanese into our number system. Then write down the Japanese numbers from 1 to 20.

RESEARCH PROJECT

Binomial Expansions

The title on the following diagram is *Binomial Expansions* because each line gives the expansion of the binomial $x + y$ raised to a whole-number power.

Binomial Expansions

$(x + y)^0 = 1$

$(x + y)^1 = x + y$

$(x + y)^2 = x^2 + 2xy + y^2$

$(x + y)^3 = x^3 + 3x^2y + 3xy^2 + y^3$

$(x + y)^4 =$

$(x + y)^5 =$

The fourth row in the diagram was completed by expanding $(x + y)^3$ using the methods developed in this chapter. Next, complete the diagram by expanding the binomials $(x + y)^4$ and $(x + y)^5$ using the multiplication procedures you have learned in this chapter. Finally, study the completed diagram until you see patterns that will allow you to continue the diagram one more row without using multiplication. (One pattern that you will see is Pascal's triangle, which we mentioned in the preceding group project.) When you are finished, write an essay in which you describe what you have done and the results you have obtained.

Rational Expressions and Rational Equations

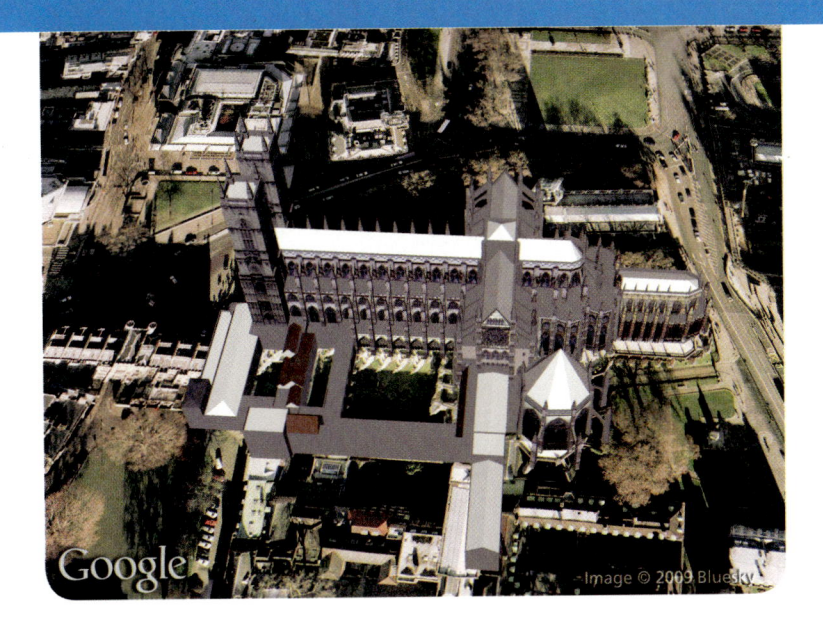

3

Chapter Outline

3.1 Reducing to Lowest Terms

3.2 Multiplication and Division of Rational Expressions

3.3 Addition and Subtraction of Rational Expressions

3.4 Equations Involving Rational Expressions

3.5 Simplified Form for Radicals

3.6 The Quadratic Formula

Introduction

The Google Earth image shows Westminster Abbey in England. Isaac Newton is one of the people credited with the invention of calculus. He died in London in 1727 and was buried in Westminster Abbey. The postage stamp below is one of many postage stamps depicting Newton's discovery of calculus.

Definition
The derivative of function $y = f(x)$ is given by the expression:
$$f'(a) = \lim_{x \to a} \frac{f(x) - f(a)}{x - a}$$

The definition next to the stamp is one of only two definitions that form the foundation of calculus. If the function in the definition is a polynomial function (the first functions encountered in calculus), then the expression $\frac{f(x) - f(a)}{x - a}$ is a rational expression. In this chapter we study rational expressions, and, along the way, we spend some time with this particular expression.

Chapter Pretest

Reduce to lowest terms. [3.1]

1. $\dfrac{x^2 + 2xy + y^2}{x + y}$

2. $\dfrac{4x^2 - 4x - 3}{2x^2 - x - 1}$

Multiply and divide as indicated. [3.2]

3. $\dfrac{a^2 - 9}{4a + 12} \cdot \dfrac{12(a + 2)^2}{a^2 - a - 6}$

4. $\dfrac{a^4 - 16}{2a^2 + a - 6} \div \dfrac{a^2 + 4}{8a - 12}$

5. $\dfrac{2x^2 + 3x + 1}{2x^2 + 6x + 18} \div \dfrac{2x^2 - 5x - 3}{x^3 - 27}$

Add and subtract as indicated. [3.3]

6. $\dfrac{3}{10} + \dfrac{6}{25}$

7. $\dfrac{1}{2} - \dfrac{5}{32} + \dfrac{3}{16}$

8. $\dfrac{a + 2}{a^2 - 25} - \dfrac{7}{a^2 - 25}$

9. $\dfrac{2}{x + 1} + \dfrac{4}{2x - 1}$

Write in simplified form. [3.5]

10. $\sqrt{27x^5y^3}$

11. $\sqrt[3]{128x^2y^7}$

12. $\sqrt{\dfrac{3}{5}}$

13. $\sqrt{\dfrac{8a^2b^7}{7c}}$

Solve each of the following equations. [3.4]

14. $5 + \dfrac{2}{x} = \dfrac{6}{5}$

15. $\dfrac{2x}{x + 4} - 3 = \dfrac{-8}{x + 4}$

16. $\dfrac{9}{y + 3} + \dfrac{y + 4}{7y} = \dfrac{1}{7}$

17. $1 - \dfrac{10}{x} = \dfrac{-21}{x^2}$

Use the quadratic formula to solve each equation. [3.6]

18. $(2x - 3)^2 = 36$

19. $(4x + 1)^2 = 12$

20. $y^2 + 8y - 16 = -25$

Getting Ready for Chapter 3

Reduce.

1. $\dfrac{6}{8}$

2. $\dfrac{10}{25}$

3. $\dfrac{200}{5}$

4. $\dfrac{240}{6}$

Multiply.

5. $250(3.14)$

6. $165(3.14)$

7. $-1(5 - x)$

8. $-1(a - b)$

9. $4x(2x - 4)$

Simplify.

10. $\dfrac{0 + 6}{0 - 3}$

11. $\dfrac{-6 + 6}{-6 - 3}$

12. $\dfrac{3 + 6}{3 - 3}$

13. $\dfrac{8x^3y^5}{-2x^2y}$

Use rules for exponents to simplify the following.

14. $x^8 \cdot x^{-3}$

15. $(y^2)^3$

16. $(3r^8s^{20})^5$

Find $\left(\dfrac{b}{2}\right)^2$ when

17. $b = 6$

18. $b = 4$

19. $b = 5$

Factor.

20. $x^2 - a^2$

21. $x^2 - 6x + 9$

22. $x^3 - x^2y$

23. $xy - y$

24. $2y^2 - 2$

25. $8y^3 - 27$

Chapter 3 Rational Expressions and Rational Equations

Reducing to Lowest Terms

3.1

Objectives
A Reduce rational expressions to lowest terms.
B Work with ratios.
C Find values for rational equations.

We will begin this section with the definition of a rational expression. We then will state the two basic properties associated with rational expressions, and go on to apply one of the properties to reduce rational expressions to lowest terms.

Recall that a *rational number* is any number that can be expressed as the ratio of two integers:

$$\text{Rational numbers} = \left\{ \frac{a}{b} \,\bigg|\, a \text{ and } b \text{ are integers}, b \neq 0 \right\}$$

A rational expression is defined similarly as any expression that can be written as the ratio of two polynomials:

$$\text{Rational expressions} = \left\{ \frac{P}{Q} \,\bigg|\, P \text{ and } Q \text{ are polynomials}, Q \neq 0 \right\}$$

Some examples of rational expressions are

$$\frac{2x-3}{x+5} \qquad \frac{x^2-5x-6}{x^2-1} \qquad \frac{a-b}{b-a}$$

Basic Properties

For rational expressions, multiplying the numerator and denominator by the same nonzero expression may change the form of the rational expression, but it will always produce an expression equivalent to the original one. The same is true when dividing the numerator and denominator by the same nonzero quantity.

Property

If P, Q, and K are polynomials with $Q \neq 0$ and $K \neq 0$, then

$$\frac{P}{Q} = \frac{PK}{QK} \quad \text{and} \quad \frac{P}{Q} = \frac{P/K}{Q/K}$$

Note These two statements are equivalent since division is defined as multiplication by the reciprocal. We choose to state them separately for clarity.

A Reducing to Lowest Terms

The fraction $\frac{6}{8}$ can be written in lowest terms as $\frac{3}{4}$. The process is shown here:

$$\frac{6}{8} = \frac{3 \cdot \overset{1}{\cancel{2}}}{4 \cdot \underset{1}{\cancel{2}}} = \frac{3}{4}$$

Reducing $\frac{6}{8}$ to $\frac{3}{4}$ involves dividing the numerator and denominator by 2, the factor they have in common. Before dividing out the common factor 2, we must notice that the common factor *is* 2! (This may be obvious since we are very familiar with the numbers 6 and 8 and therefore do not have to put much thought into finding what number divides both of them.)

We reduce rational expressions to lowest terms by first factoring the numerator and denominator and then dividing both numerator and denominator by any factors they have in common.

EXAMPLE 1 Reduce $\frac{x^2-9}{x-3}$ to lowest terms.

SOLUTION Factoring, we have

$$\frac{x^2-9}{x-3} = \frac{(x+3)(x-3)}{x-3}$$

PRACTICE PROBLEMS

1. Reduce to lowest terms.
   $$\frac{x^2-9}{x+3}$$

3.1 Reducing to Lowest Terms

159

Note The lines drawn through the $(x-3)$ in the numerator and denominator indicate that we have divided through by $(x-3)$. As the problems become more involved, these lines will help keep track of which factors have been divided out and which have not.

Reduce to lowest terms.
2. $\dfrac{y^2 - y - 6}{y^2 - 4}$

3. $\dfrac{3a^3 + 3}{6a^2 - 6a + 6}$

4. $\dfrac{x^2 + 4x + ax + 4a}{x^2 + ax + 4x + 4a}$

Answers
1. $x - 3$
2. $\dfrac{y-3}{y-2}$
3. $\dfrac{a+1}{2}$
4. 1

The numerator and denominator have the factor $x - 3$ in common. Dividing the numerator and denominator by $x - 3$, we have

$$\frac{(x+3)(x-3)}{x-3} = \frac{x+3}{1} = x+3$$

NOTE For the problem in Example 1, there is an implied restriction on the variable x: it cannot be 3. If x were 3, the expression $(x^2 - 9)/(x - 3)$ would become $0/0$, an expression that we cannot associate with a real number. For all problems involving rational expressions, we restrict the variable to only those values that result in a nonzero denominator. When we state the relationship

$$\frac{x^2 - 9}{x - 3} = x + 3$$

we are assuming that it is true for all values of x except $x = 3$.

Here are some other examples of reducing rational expressions to lowest terms.

EXAMPLE 2 Reduce $\dfrac{y^2 - 5y - 6}{y^2 - 1}$ to lowest terms.

SOLUTION Factoring, we have

$$\frac{y^2 - 5y - 6}{y^2 - 1} = \frac{(y-6)(y+1)}{(y-1)(y+1)} \quad \text{Factor numerator and denominator}$$

$$= \frac{y-6}{y-1} \quad \text{Divide out common factor } y + 1$$

EXAMPLE 3 Reduce $\dfrac{2a^3 - 16}{4a^2 - 12a + 8}$ to lowest terms.

SOLUTION Factoring, we have

$$\frac{2a^3 - 16}{4a^2 - 12a + 8} = \frac{2(a^3 - 8)}{4(a^2 - 3a + 2)}$$

$$= \frac{2(a-2)(a^2 + 2a + 4)}{4(a-2)(a-1)} \quad \text{Factor numerator and denominator}$$

$$= \frac{a^2 + 2a + 4}{2(a-1)} \quad \text{Divide out common factor } 2(a-2)$$

EXAMPLE 4 Reduce $\dfrac{x^2 - 3x + ax - 3a}{x^2 - ax - 3x + 3a}$ to lowest terms.

SOLUTION Factoring, we have

$$\frac{x^2 - 3x + ax - 3a}{x^2 - ax - 3x + 3a} = \frac{x(x-3) + a(x-3)}{x(x-a) - 3(x-a)}$$

$$= \frac{(x-3)(x+a)}{(x-a)(x-3)} \quad \text{Factor numerator and denominator}$$

$$= \frac{x+a}{x-a} \quad \text{Divide out common factor } x-3$$

The answer to Example 4 is $(x + a)/(x - a)$. The problem cannot be reduced further. It is a fairly common mistake to attempt to divide out an x or an a in this last expression. Remember, we can divide out only the factors common to the numerator and denominator of a rational expression. For the last expression in

Example 4, neither the numerator nor the denominator can be factored further; x is not a factor of the numerator or the denominator, and neither is a. The expression is in lowest terms.

The next example involves what we call a trick. The trick is to reverse the order of the terms in a difference by factoring -1 from each term. The next examples illustrate how this is done.

EXAMPLE 5 Reduce to lowest terms: $\dfrac{a-b}{b-a}$.

SOLUTION The relationship between $a - b$ and $b - a$ is that they are opposites. We can show this fact by factoring -1 from each term in the numerator:

$$\dfrac{a-b}{b-a} = \dfrac{-1(-a+b)}{b-a} \quad \text{Factor } -1 \text{ from each term in the numerator}$$

$$= \dfrac{-1(b-a)}{b-a} \quad \text{Reverse the order of the terms in the numerator}$$

$$= -1 \quad \text{Divide out common factor } b-a$$

5. Replace a with 7 and b with 4 and then simplify.
$$\dfrac{a-b}{b-a}$$

EXAMPLE 6 Reduce to lowest terms: $\dfrac{x^2-25}{5-x}$.

SOLUTION Begin by factoring the numerator:

$$\dfrac{x^2-25}{5-x} = \dfrac{(x-5)(x+5)}{5-x}$$

The factors $x - 5$ and $5 - x$ are similar but are not exactly the same. We can reverse the order of either by factoring -1 from it; that is: $5 - x = -1(-5 + x) = -1(x - 5)$.

$$\dfrac{(x-5)(x+5)}{5-x} = \dfrac{\cancel{(x-5)}(x+5)}{-1\cancel{(x-5)}}$$

$$= \dfrac{x+5}{-1}$$

$$= -(x+5)$$

6. Reduce to lowest terms.
$$\dfrac{7-x}{x^2-49}$$

B Ratios

You may recall from previous math classes that the ratio of a to b is the same as the fraction $\dfrac{a}{b}$. Here are two ratios that are used frequently in mathematics:

1. The number π is defined as the ratio of the circumference of a circle to the diameter of a circle; that is

$$\pi = \dfrac{C}{d}$$

Note Since the diameter of a circle is twice the radius, the formula for circumference is sometimes written $C = 2\pi r$. Neither formula should be confused with the formula for the area of a circle, which is $A = \pi r^2$.

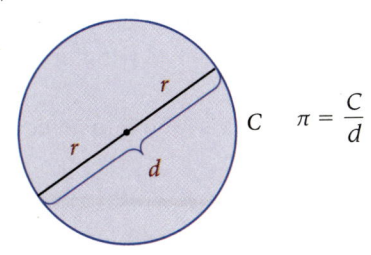

Multiplying both sides of this formula by d, we have the more common form $C = \pi d$.

Answers
5. -1
6. $\dfrac{-1}{x+7}$

Chapter 3 Rational Expressions and Rational Equations

2. The *average speed* of a moving object is defined to be the ratio of distance to time. If you drive your car for 5 hours and travel a distance of 200 miles, then your average rate of speed is

$$\text{Average speed} = \frac{200 \text{ miles}}{5 \text{ hours}} = 40 \text{ miles per hour}$$

The formula we use for the relationship between average speed r, distance d, and time t is

$$r = \frac{d}{t}$$

The formula is sometimes called the *rate equation*. Multiplying both sides by t, we have an equivalent form of the rate equation, $d = rt$.

C Rational Equations

Our next example involves both the formula for the circumference of a circle and the rate equation.

EXAMPLE 7 The first Ferris wheel was designed and built by George Ferris in 1893. The diameter of the wheel was 250 feet. It had 36 carriages, equally spaced around the wheel, each of which held a maximum of 40 people. One trip around the wheel took 20 minutes. Find the average speed of a rider on the first Ferris wheel. (Use 3.14 as an approximation for π.)

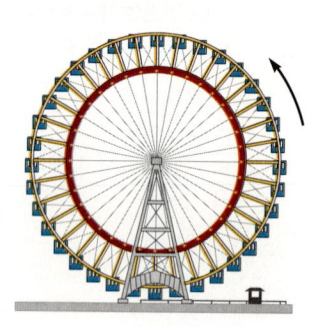

SOLUTION The distance traveled is the circumference of the wheel, which is

$$C = 250\pi = 250(3.14) = 785 \text{ feet}$$

To find the average speed, we divide the distance traveled by the amount of time it took to go once around the wheel.

$$r = \frac{d}{t} = \frac{785 \text{ feet}}{20 \text{ minutes}} = 39.3 \text{ feet per minute} \quad \text{(to the nearest tenth)}$$

Later in this chapter we will convert this ratio into an equivalent ratio that gives the speed of the rider in miles per hour.

The Ferris wheel in Example 7 has a circumference of 785 feet. In the next example we graph the relationship between the average speed of a person riding this wheel and the amount of time it takes the wheel to complete one revolution.

---

7. A Ferris wheel was built in St. Louis in 1986. It is named *Colossus*. The diameter of the wheel is 165 feet. It has 40 cars, each of which holds 6 passengers. A trip around the wheel takes 40 seconds. Find the average speed of a rider on *Colossus*. (Use 3.14 as an approximation for π, and round your answer to the nearest tenth.)

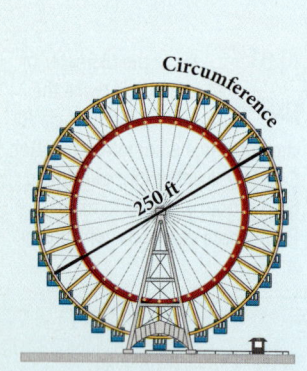

Answer
7. 13.0 feet per second

3.1 Reducing to Lowest Terms

EXAMPLE 8 A Ferris wheel has a circumference of 785 feet. If one complete revolution of the wheel takes 10 to 30 minutes, then the relationship between the average speed of a rider on the wheel and the amount of time it takes the wheel to complete one revolution is given by the equation

$$r = \frac{785}{t} \qquad 10 \leq t \leq 30$$

where $r(t)$ is the average speed (in feet per minute) and t is the amount of time (in minutes) it takes the wheel to complete one revolution. Graph the equation.

SOLUTION Since the variables r and t represent speed and time, both must be positive quantities. Therefore, the graph of this equation will lie in the first quadrant only. The following table displays the values of t and r found from the fnction, along with the graph of the equation (Figure 1). (Some of the numbers in te table have been rounded to the nearest tenth.)

Time to Complete One Revolution t	Speed (ft/min) r
10	78.5
15	52.3
20	39.3
25	31.4
30	26.2

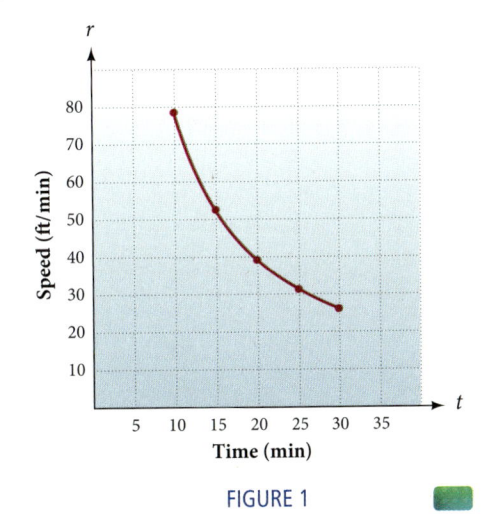

FIGURE 1

USING TECHNOLOGY

More About Example 8

If we use a graphing calculator to graph the equation in Example 8, it is not necessary to construct the table first. In fact, if we graph

$Y_1 = 785/X$ Window: X from 0 to 40, Y from 0 to 90

we can use the Trace and Zoom features together to produce the numbers in the table next to Figure 1. Graph the preceding equation, and zoom in on the point with x-coordinate 20 until you are convinced that the table values for x and y are correct.

Getting Ready for Class

After reading through the preceding section, respond in your own words and in complete sentences.

1. What is a rational expression?
2. Explain how to determine if a rational expression is in "lowest terms."
3. When is a rational expression undefined?
4. Explain the process we use to reduce a rational expression or a fraction to lowest terms.

---

8. Extend the table in Example 8 by finding r when $t = 35$. Then explain your result in words.

Answer
8. 22.4 to the nearest tenth. If one trip around the wheel takes 35 minutes, then the rider is moving at approximately 22.4 feet per minute.

Left Blank Intentionally

Problem Set 3.1

A 1. Simplify each expression. State any restrictions on the variable. [Examples 1–3]

a. $\dfrac{6+1}{36-1}$
b. $\dfrac{x+3}{x^2-9}$
c. $\dfrac{x^2-3x}{x^2-9}$
d. $\dfrac{x^3-27}{x^2-9}$

2. Simplify each expression. State any restrictions on the variable.

a. $\dfrac{64-80+25}{64-25}$
b. $\dfrac{x^2-10x+25}{x^2-25x}$
c. $\dfrac{x^2-26x+25}{x^2-25x}$
d. $\dfrac{x^2+5x+ax+5a}{x^2-25}$

A Reduce each rational expression to lowest terms. [Examples 1–4]

3. $\dfrac{x^2-16}{6x+24}$
4. $\dfrac{5x+25}{x^2-25}$
5. $\dfrac{12x-9y}{3x^2+3xy}$
6. $\dfrac{x^3-xy^2}{4x+4y}$

7. $\dfrac{a^4-81}{a-3}$
8. $\dfrac{a+4}{a^2-16}$
9. $\dfrac{a^2-4a-12}{a^2+8a+12}$
10. $\dfrac{a^2-7a+12}{a^2-9a+20}$

11. $\dfrac{20y^2-45}{10y^2-5y-15}$
12. $\dfrac{54y^2-6}{18y^2-60y+18}$
13. $\dfrac{a^3+b^3}{a^2-b^2}$
14. $\dfrac{a^2-b^2}{a^3-b^3}$

15. $\dfrac{x^2-9}{3x-9}$
16. $\dfrac{20a-28b}{4a^2-4ab}$
17. $\dfrac{x^4-16}{x-2}$
18. $\dfrac{12a-30}{6a^2-9a}$

19. $\dfrac{x^2-8x+15}{x^2+3x-18}$
20. $\dfrac{x^2-3x-28}{x^2-9x+14}$
21. $\dfrac{2x^2+3x-9}{x^2-x-12}$
22. $\dfrac{16x^2-52x+12}{12x^2-40x+12}$

Reduce to lowest terms.

23. $\dfrac{8x^4 - 8x}{4x^4 + 4x^3 + 4x^2}$

24. $\dfrac{6x^5 - 48x^2}{12x^3 + 24x^2 + 48x}$

25. $\dfrac{6x^2 + 7xy - 3y^2}{6x^2 + xy - y^2}$

26. $\dfrac{4x^2 - y^2}{4x^2 - 8xy - 5y^2}$

27. $\dfrac{ax + 2x + 3a + 6}{ay + 2y - 4a - 8}$

28. $\dfrac{ax - x - 5a + 5}{ax + x - 5a - 5}$

29. $\dfrac{x^2 + bx - 3x - 3b}{x^2 - 2bx - 3x + 6b}$

30. $\dfrac{x^2 - 3ax - 2x + 6a}{x^2 - 3ax + 2x - 6a}$

31. $\dfrac{x^3 + 3x^2 - 4x - 12}{x^2 + x - 6}$

32. $\dfrac{x^3 + 5x^2 - 4x - 20}{x^2 + 7x + 10}$

33. $\dfrac{4x^4 - 25}{6x^3 - 4x^2 + 15x - 10}$

34. $\dfrac{16x^4 - 49}{8x^3 - 12x^2 + 14x - 21}$

35. $\dfrac{x^3 - 8}{x^2 - 4}$

36. $\dfrac{y^2 - 9}{y^3 - 27}$

37. $\dfrac{64 + t^3}{16 - 4t + t^2}$

38. $\dfrac{25 + 5a + a^2}{125 - a^3}$

39. $\dfrac{8x^3 - 27}{4x^2 - 9}$

40. $\dfrac{25y^2 - 4}{125y^3 + 8}$

Refer to Examples 5 and 6 in this section, and reduce the following to lowest terms.

41. $\dfrac{x - 4}{4 - x}$

42. $\dfrac{6 - x}{x - 6}$

43. $\dfrac{y^2 - 36}{6 - y}$

44. $\dfrac{1 - y}{y^2 - 1}$

45. $\dfrac{1 - 9a^2}{9a^2 - 6a + 1}$

46. $\dfrac{1 - a^2}{a^2 - 2a + 1}$

Simplify each expression.

47. $\dfrac{(3x - 5) - (3a - 5)}{x - a}$

48. $\dfrac{(2x + 3) - (2a + 3)}{x - a}$

49. $\dfrac{(x^2 - 4) - (a^2 - 4)}{x - a}$

50. $\dfrac{(x^2 - 1) - (a^2 - 1)}{x - a}$

C Let $y_1 = \dfrac{x^2 - 4}{x - 2}$ and $y_2 = x + 2$, and evaluate the following expressions, if possible. Example 9

51. $y_1 = 0$ and $y_2 = 0$ **52.** $y_1 = 1$ and $y_2 = 1$ **53.** $y_1 = 2$ and $y_2 = 2$ **54.** $y_1 = 3$ and $y_2 = 3$

Let $y_1 = \dfrac{x^2 - 1}{x - 1}$ and $y_2 = x + 1$, and evaluate the following expressions, if possible.

55. $y_1 = 0$ and $y_2 = 0$ **56.** $y_1 = 1$ and $y_2 = 1$ **57.** $y_1 = 2$ and $y_2 = 2$ **58.** $y_1 = -1$ and $y_2 = -1$

[Example 10]

59. Graph the equation $y = \dfrac{x^2 - 4}{x - 2}$. Then explain how this graph is different from the graph of $y = x + 2$.

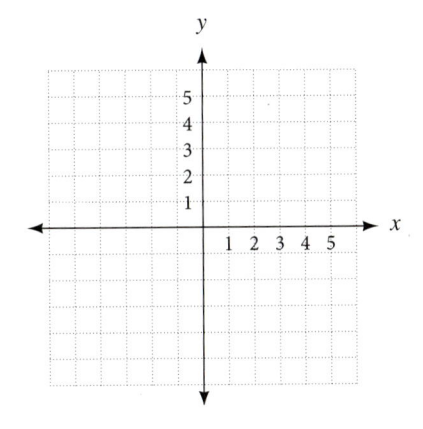

60. Graph the equation $y = \dfrac{x^2 - 1}{x - 1}$. Then explain how this graph is different from the graph of $y = x + 1$.

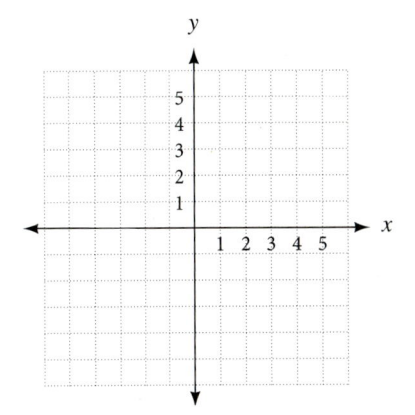

Applying the Concepts [Examples 7–8]

61. Diet The following rational equation shows the amount of weight lost by a person on weight-loss diet. The quantity W is the weight (in pounds) of the person after x weeks of dieting. Use the equation to fill in the table, rounding to the nearest pound.

$$W = \dfrac{80(2x + 15)}{x + 6}$$

Weeks x	Weight (pounds) W
0	
1	
4	
12	
24	

62. Drag Racing The following rational equation gives the speed V, in miles per hour, of a dragster at each second x during a quarter-mile race. Use the equation to fill in the table, rounding to the nearest tenth.

$$V = \dfrac{340x}{x + 3}$$

Time (sec) x	Speed (mi/hr) V
0	
1	
2	
3	
4	
5	
6	

Average Speed For Problems 77 and 78, use 3.14 as an approximation for π. Round answers to the nearest tenth.

63. A person riding a Ferris wheel with a diameter of 65 feet travels once around the wheel in 30 seconds. What is the average speed of the rider in feet per second?

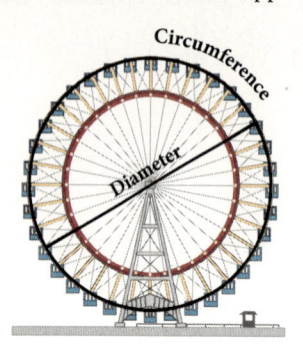

64. A person riding a Ferris wheel with a diameter of 102 feet travels once around the wheel in 3.5 minutes. What is the average speed of the rider in feet per minute?

Average Speed The abbreviation "rpm" stands for revolutions per minute. If a point on a circle rotates at 300 rpm, then it rotates through one complete revolution 300 times every minute. The length of time it takes to rotate once around the circle is $\frac{1}{300}$ minute. Use 3.14 as an approximation for π.

65. An audio CD, when placed in the CD player, rotates at 300 rpm (1 revolution takes $\frac{1}{300}$ minute). Find the average speed of a point 2 inches from the center of the CD. Then find the average speed of a point 1.5 inches from the center of the CD..

66. A power generating turbine rotates at 3,600 rpm. Find the average speed of a point 2 feet from the center of the turbine. Then find the average speed of a point 1.5 feet from the center.

■ Maintaining Your Skills

Subtract as indicated.

67. Subtract $x^2 + 2x + 1$ from $4x^2 - 5x + 5$.

68. Subtract $3x^2 - 5x + 2$ from $7x^2 + 6x + 4$.

69. Subtract $10x - 20$ from $10x - 11$.

70. Subtract $-6x - 18$ from $-6x + 5$.

71. Subtract $4x^3 - 8x^2$ from $4x^3$.

72. Subtract $2x^2 + 6x$ from $2x^2$.

■ Getting Ready for the Next Section

Divide.

73. $\dfrac{10x^5}{5x^2}$

74. $\dfrac{-15x^4}{5x^2}$

75. $\dfrac{4x^4y^3}{-2x^2y}$

76. $\dfrac{10a^4b^2}{4a^2b^2}$

77. $4,628 \div 25$

78. $7,546 \div 35$

Multiply.

79. $2x^2(2x - 4)$ **80.** $3x^2(x - 2)$ **81.** $(2x - 4)(2x^2 + 4x + 5)$ **82.** $(x - 2)(3x^2 + 6x + 15)$

Subtract.

83. $(2x^2 - 7x + 9) - (2x^2 - 4x)$ **84.** $(x^2 - 6xy - 7y^2) - (x^2 + xy)$

Factor.

85. $x^2 - a^2$ **86.** $x^2 - 1$ **87.** $x^2 - 6xy - 7y^2$ **88.** $2x^2 - 5xy + 3y^2$

■ Extending the Concepts

89. The graphs of two rational equations are given in Figures 3 and 4. Use the graphs to find the following.
   a. $y_1 = 2$ **b.** $y_1 = -1$ **c.** $y_1 = 0$ **d.** $y_2 = 3$ **e.** $y_2 = 6$ **f.** $y_2 = -1$

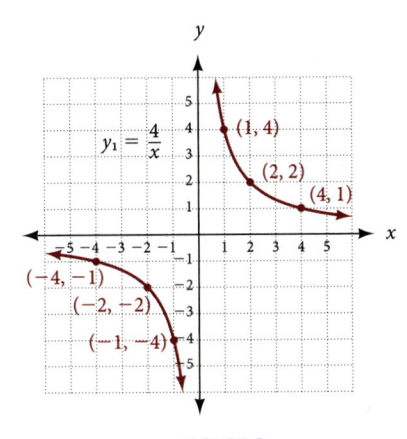

FIGURE 3

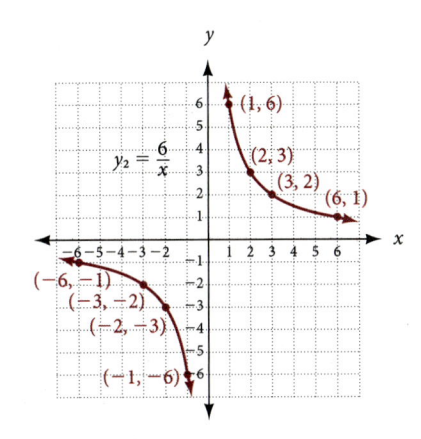

FIGURE 4

Left Blank Intentionally

Multiplication and Division of Rational Expressions

3.2

Objectives
A Multiply and divide rational expressions.

If you have ever taken a DVD to be duplicated, you know the amount you pay for the duplication service depends on the number of copies you have made: The more copies you have made, the lower the charge per copy. The following demand function gives the price (in dollars) per DVD $p(x)$ a company charges for making x copies of a 30-minute DVD. As you can see, it is a rational function.

$$p(x) = \frac{2(x+60)}{x+5}$$

The graph in Figure 1 shows this function from $x = 0$ to $x = 100$. As you can see, the more copies that are made, the lower the price per copy.

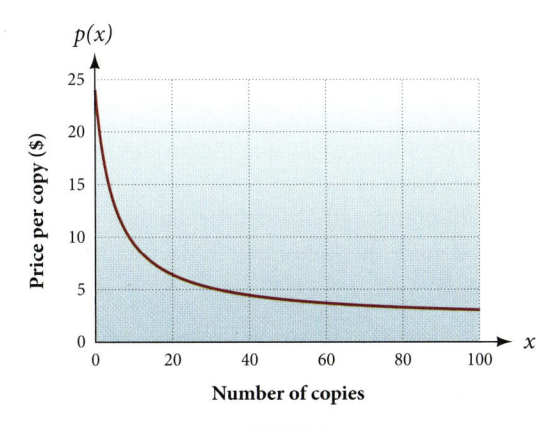

FIGURE 1

If we were interested in finding the revenue function for this situation, we would multiply the number of copies made x by the price per copy $p(x)$. This involves multiplication with a rational expression, which is one of the topics we cover in this section.

In the previous section we found the process of reducing rational expressions to lowest terms to be the same process used in reducing fractions to lowest terms. The similarity also holds for the process of multiplication or division of rational expressions.

Multiplication with fractions is the simplest of the four basic operations. To multiply two fractions we simply multiply numerators and multiply denominators; that is, if a, b, c, and d are real numbers, with $b \neq 0$ and $d \neq 0$, then

$$\frac{a}{b} \cdot \frac{c}{d} = \frac{ac}{bd}$$

A Multiplying and Dividing Rational Expressions

EXAMPLE 1 Multiply $\frac{6}{7} \cdot \frac{14}{18}$.

SOLUTION
$$\frac{6}{7} \cdot \frac{14}{18} = \frac{6(14)}{7(18)} \qquad \text{Multiply numerators and denominators}$$

$$= \frac{2 \cdot 3(2 \cdot 7)}{7(2 \cdot 3 \cdot 3)} \qquad \text{Factor}$$

$$= \frac{2}{3} \qquad \text{Divide out common factors}$$

PRACTICE PROBLEMS
1. Multiply $\frac{3}{4} \cdot \frac{12}{27}$.

Answer
1. $\frac{1}{3}$

Chapter 3 Rational Expressions and Rational Equations

Our next example is similar to some of the problems we worked in an earlier section. We multiply fractions whose numerators and denominators are monomials by multiplying numerators and multiplying denominators and then reducing to lowest terms. Here is how it looks.

2. Multiply $\dfrac{6x^4}{4y^9} \cdot \dfrac{12y^5}{3x^2}$.

EXAMPLE 2 Multiply $\dfrac{8x^3}{27y^8} \cdot \dfrac{9y^3}{12x^2}$.

SOLUTION We multiply numerators and denominators without actually carrying out the multiplication:

$$\dfrac{8x^3}{27y^8} \cdot \dfrac{9y^3}{12x^2} = \dfrac{8 \cdot 9x^3y^3}{27 \cdot 12x^2y^8} \quad \text{Multiply numerators} \atop \text{Multiply denominators}$$

$$= \dfrac{4 \cdot 2 \cdot 9x^3y^3}{9 \cdot 3 \cdot 4 \cdot 3x^2y^8} \quad \text{Factor coefficients}$$

$$= \dfrac{2x}{9y^5} \quad \text{Divide out common factors}$$

> **Note** Notice how we factor the coefficients just enough so that we can see the factors they have in common. If you want to show this step without showing the factoring, it would look like this:
>
> $$\dfrac{\overset{2}{\cancel{8}} \cdot 9x^3y^3}{\underset{3}{\cancel{27}} \cdot \underset{3}{\cancel{12}}x^2y^8}$$

The product of two rational expressions is the product of their numerators over the product of their denominators.

Once again, we should mention that the little slashes we have drawn through the factors are used to denote the factors we have divided out of the numerator and denominator.

3. Multiply $\dfrac{x+5}{x^2-25} \cdot \dfrac{x-5}{x^2-10x+25}$.

EXAMPLE 3 Multiply $\dfrac{x-3}{x^2-4} \cdot \dfrac{x+2}{x^2-6x+9}$.

SOLUTION We begin by multiplying numerators and denominators. We then factor all polynomials and divide out factors common to the numerator and denominator:

$$\dfrac{x-3}{x^2-4} \cdot \dfrac{x+2}{x^2-6x+9} = \dfrac{(x-3)(x+2)}{(x^2-4)(x^2-6x+9)} \quad \text{Multiply}$$

$$= \dfrac{\cancel{(x-3)}\cancel{(x+2)}}{\cancel{(x+2)}(x-2)\cancel{(x-3)}(x-3)} \quad \text{Factor}$$

$$= \dfrac{1}{(x-2)(x-3)} \quad \text{Divide out common factors}$$

The first two steps can be combined to save time. We can perform the multiplication and factoring steps together.

4. Multiply $\dfrac{3y^2-3y}{3y-12} \cdot \dfrac{y^2-2y-8}{y^2+3y+2}$.

EXAMPLE 4 Multiply $\dfrac{2y^2-4y}{2y^2-2} \cdot \dfrac{y^2-2y-3}{y^2-5y+6}$.

SOLUTION

$$\dfrac{2y^2-4y}{2y^2-2} \cdot \dfrac{y^2-2y-3}{y^2-5y+6} = \dfrac{\cancel{2}y\cancel{(y-2)}\cancel{(y-3)}\cancel{(y+1)}}{\cancel{2}\cancel{(y+1)}(y-1)\cancel{(y-3)}\cancel{(y-2)}}$$

$$= \dfrac{y}{y-1}$$

Notice in both of the preceding examples that we did not actually multiply the polynomials as we did in the chapter on exponents and polynomials. It would be senseless to do that since we would then have to factor each of the resulting products to reduce them to lowest terms.

Answers

2. $\dfrac{6x^2}{y^4}$
3. $\dfrac{1}{(x-5)^2}$
4. $\dfrac{y(y-1)}{y+1}$

3.2 Multiplication and Division of Rational Expressions

The quotient of two rational expressions is the product of the first and the reciprocal of the second; that is, we find the quotient of two rational expressions the same way we find the quotient of two fractions. Here is an example that reviews division with fractions.

EXAMPLE 5 Divide $\dfrac{6}{8} \div \dfrac{3}{5}$.

SOLUTION
$\dfrac{6}{8} \div \dfrac{3}{5} = \dfrac{6}{8} \cdot \dfrac{5}{3}$ Write division in terms of multiplication

$= \dfrac{6(5)}{8(3)}$ Multiply numerators and denominators

$= \dfrac{\cancel{2} \cdot \cancel{3}(5)}{\cancel{2} \cdot 2 \cdot 2(\cancel{3})}$ Factor

$= \dfrac{5}{4}$ Divide out common factors

To divide one rational expression by another, we use the definition of division to multiply by the reciprocal of the expression that follows the division symbol.

EXAMPLE 6 Divide $\dfrac{8x^3}{5y^2} \div \dfrac{4x^2}{10y^6}$.

SOLUTION First we rewrite the problem in terms of multiplication. Then we multiply.

$\dfrac{8x^3}{5y^2} \div \dfrac{4x^2}{10y^6} = \dfrac{8x^3}{5y^2} \cdot \dfrac{10y^6}{4x^2}$

$= \dfrac{\cancel{8}^2 \cdot \cancel{10}^2 x^3 y^6}{\cancel{4} \cdot \cancel{5} x^2 y^2}$

$= 4xy^4$

EXAMPLE 7 Divide $\dfrac{x^2 - y^2}{x^2 - 2xy + y^2} \div \dfrac{x^3 + y^3}{x^3 - x^2 y}$.

SOLUTION We begin by writing the problem as the product of the first and the reciprocal of the second and then proceed as in the previous two examples:

$\dfrac{x^2 - y^2}{x^2 - 2xy + y^2} \div \dfrac{x^3 + y^3}{x^3 - x^2 y}$ Multiply by the reciprocal of the divisor

$= \dfrac{x^2 - y^2}{x^2 - 2xy + y^2} \cdot \dfrac{x^3 - x^2 y}{x^3 + y^3}$

$= \dfrac{\cancel{(x-y)}\cancel{(x+y)}(x^2)\cancel{(x-y)}}{\cancel{(x-y)}\cancel{(x-y)}\cancel{(x+y)}(x^2 - xy + y^2)}$ Factor and multiply

$= \dfrac{x^2}{x^2 - xy + y^2}$ Divide out common factors

Here are some more examples of multiplication and division with rational expressions.

EXAMPLE 8 Perform the indicated operations.

$\dfrac{a^2 - 8a + 15}{a + 4} \cdot \dfrac{a + 2}{a^2 - 5a + 6} \div \dfrac{a^2 - 3a - 10}{a^2 + 2a - 8}$

---

5. Divide $\dfrac{5}{9} \div \dfrac{10}{27}$.

6. Divide $\dfrac{9x^4}{4y^3} \div \dfrac{3x^2}{8y^5}$.

7. Divide. $\dfrac{xy^2 - y^3}{x^2 - y^2} \div \dfrac{x^3 + y^3}{x^2 + 2xy + y^2}$.

8. Perform the indicated operations.

$\dfrac{a^2 + 3a - 4}{a - 4} \cdot \dfrac{a + 3}{a^2 - 4a + 3} \div \dfrac{a + 1}{a^2 - 2a - 3}$

Answers
5. $\dfrac{3}{2}$
6. $6x^2 y^2$
7. $\dfrac{y^2}{x^2 - xy + y^2}$

Chapter 3 Rational Expressions and Rational Equations

SOLUTION First we rewrite the division as multiplication by the reciprocal. Then we proceed as usual.

$$\frac{a^2 - 8a + 15}{a + 4} \cdot \frac{a + 2}{a^2 - 5a + 6} \div \frac{a^2 - 3a - 10}{a^2 + 2a - 8}$$ Change division to multiplication by the reciprocal

$$= \frac{(a^2 - 8a + 15)(a + 2)(a^2 + 2a - 8)}{(a + 4)(a^2 - 5a + 6)(a^2 - 3a - 10)}$$

$$= \frac{(a - 5)(a - 3)(a + 2)(a + 4)(a - 2)}{(a + 4)(a - 3)(a - 2)(a - 5)(a + 2)}$$ Factor

$$= 1$$ Divide out common factors

Our next example involves factoring by grouping. As you may have noticed, working the problems in this chapter gives you a very detailed review of factoring.

EXAMPLE 9 Multiply $\dfrac{xa + xb + ya + yb}{xa - xb - ya + yb} \cdot \dfrac{xa + xb - ya - yb}{xa - xb + ya - yb}$

SOLUTION We will factor each polynomial by grouping, which takes two steps.

$$\frac{xa + xb + ya + yb}{xa - xb - ya + yb} \cdot \frac{xa + xb - ya - yb}{xa - xb + ya - yb}$$

$$= \frac{x(a + b) + y(a + b)}{x(a - b) - y(a - b)} \cdot \frac{x(a + b) - y(a + b)}{x(a - b) + y(a - b)}$$ Factor by grouping

$$= \frac{(a + b)(x + y)(a + b)(x - y)}{(a - b)(x - y)(a - b)(x + y)}$$

$$= \frac{(a + b)^2}{(a - b)^2}$$

EXAMPLE 10 Multiply $(4x^2 - 36) \cdot \dfrac{12}{4x + 12}$.

SOLUTION We can think of $4x^2 - 36$ as having a denominator of 1. Thinking of it in this way allows us to proceed as we did in the previous examples.

$$(4x^2 - 36) \cdot \frac{12}{4x + 12}$$

$$= \frac{4x^2 - 36}{1} \cdot \frac{12}{4x + 12}$$ Write $4x^2 - 36$ with denominator 1

$$= \frac{4(x - 3)(x + 3) \cdot 12}{4(x + 3)}$$ Factor

$$= 12(x - 3)$$ Divide out common factors

Getting Ready for Class

After reading through the preceding section, respond in your own words and in complete sentences.

1. Summarize the steps used to multiply fractions.
2. What is the first step in multiplying two rational expressions?
3. Why is factoring important when multiplying and dividing rational expressions?
4. How is division with rational expressions different than multiplication of rational expressions?

---

9. Multiply.

$$\frac{xa + xb - ya - yb}{xa + 2x + ya + 2y} \cdot \frac{xa + 2x + ya + 2y}{xa + xb + ya + yb}$$

10. Multiply.

$$(5x^2 - 45) \cdot \frac{3}{5x - 15}$$

Answers
8. $\dfrac{(a + 4)(a + 3)}{a - 4}$
9. $\dfrac{x - y}{x + y}$
10. $3(x + 3)$

Problem Set 3.2

A Perform the indicated operations involving fractions. [Examples 1–2, 5–6]

1. $\dfrac{2}{9} \cdot \dfrac{3}{4}$
2. $\dfrac{5}{6} \cdot \dfrac{7}{8}$
3. $\dfrac{3}{4} \div \dfrac{1}{3}$
4. $\dfrac{3}{8} \div \dfrac{5}{4}$

5. $\dfrac{3}{7} \cdot \dfrac{14}{24} \div \dfrac{1}{2}$
6. $\dfrac{6}{5} \cdot \dfrac{10}{36} \div \dfrac{3}{4}$
7. $\dfrac{3}{4} \cdot \dfrac{16}{21} \div \dfrac{3}{14}$
8. $\dfrac{3}{5} \cdot \dfrac{10}{21} \div \dfrac{4}{3}$

9. $\dfrac{4}{9} \div \dfrac{5}{12} \cdot \dfrac{15}{8}$
10. $\dfrac{2}{3} \div \dfrac{7}{6} \cdot \dfrac{3}{4}$
11. $\dfrac{10x}{18y^2} \cdot \dfrac{9y^5}{2x^3}$
12. $\dfrac{12a^3}{7b} \cdot \dfrac{28b^2}{8a^4}$

13. $\dfrac{25x^4}{36y^3} \div \dfrac{10x}{9y}$
14. $\dfrac{16x^2y^3}{24xy^2} \cdot \dfrac{5x^3y}{10xy^2}$
15. $\dfrac{10x^2}{5y^2} \cdot \dfrac{15y^3}{2x^4}$
16. $\dfrac{8x^3}{7y^4} \cdot \dfrac{14y^6}{16x^2}$

17. $\dfrac{11a^2b}{5ab^2} \div \dfrac{22a^3b^2}{10ab^4}$
18. $\dfrac{8ab^3}{9a^2b} \div \dfrac{16a^2b^2}{18ab^3}$
19. $\dfrac{6x^2}{5y^3} \cdot \dfrac{11z^2}{2x^2} \div \dfrac{33z^5}{10y^8}$
20. $\dfrac{4x^3}{7y^2} \cdot \dfrac{6z^5}{5x^6} \div \dfrac{24z^2}{35x^6}$

A Perform the indicated operations. Be sure to write all answers in lowest terms. [Examples 3–4, 7]

21. $\dfrac{x^2 - 9}{x^2 - 4} \cdot \dfrac{x - 2}{x - 3}$
22. $\dfrac{x^2 - 16}{x^2 - 25} \cdot \dfrac{x - 5}{x - 4}$

23. $\dfrac{y^2 - 1}{y + 2} \cdot \dfrac{y^2 + 5y + 6}{y^2 + 2y - 3}$
24. $\dfrac{y - 1}{y^2 - y - 6} \cdot \dfrac{y^2 + 5y + 6}{y^2 - 1}$

25. $\dfrac{3x - 12}{x^2 - 4} \cdot \dfrac{x^2 + 6x + 8}{x - 4}$

26. $\dfrac{x^2 + 5x + 1}{4x - 4} \cdot \dfrac{x - 1}{x^2 + 5x + 1}$

A Perform the indicated operations. Be sure to write all answers in lowest terms. [Examples 8–10]

27. $\dfrac{5x + 2y}{25x^2 - 5xy - 6y^2} \cdot \dfrac{20x^2 - 7xy - 3y^2}{4x + y}$

28. $\dfrac{7x + 3y}{42x^2 - 17xy - 15y^2} \cdot \dfrac{12x^2 - 4xy - 5y^2}{2x + y}$

29. $\dfrac{a^2 - 5a + 6}{a^2 - 2a - 3} \div \dfrac{a - 5}{a^2 + 3a + 2}$

30. $\dfrac{a^2 + 7a + 12}{a - 5} \div \dfrac{a^2 + 9a + 18}{a^2 - 7a + 10}$

31. $\dfrac{4t^2 - 1}{6t^2 + t - 2} \div \dfrac{8t^3 + 1}{27t^3 + 8}$

32. $\dfrac{9t^2 - 1}{6t^2 + 7t - 3} \div \dfrac{27t^3 + 1}{8t^3 + 27}$

33. $\dfrac{2x^2 - 5x - 12}{4x^2 + 8x + 3} \div \dfrac{x^2 - 16}{2x^2 + 7x + 3}$

34. $\dfrac{x^2 - 2x + 1}{3x^2 + 7x - 20} \div \dfrac{x^2 + 3x - 4}{3x^2 - 2x - 5}$

35. $\dfrac{6a^2b + 2ab^2 - 20b^3}{4a^2b - 16b^3} \cdot \dfrac{10a^2 - 22ab + 4b^2}{27a^3 - 125b^3}$

36. $\dfrac{12a^2b - 3ab^2 - 42b^3}{9a^2 - 36b^2} \cdot \dfrac{6a^2 - 15ab + 6b^2}{8a^3b - b^4}$

37. $\dfrac{360x^3 - 490x}{36x^2 + 84x + 49} \cdot \dfrac{30x^2 + 83x + 56}{150x^3 + 65x^2 - 280x}$

38. $\dfrac{490x^2 - 640}{49x^2 - 112x + 64} \cdot \dfrac{28x^2 - 95x + 72}{56x^3 - 62x^2 - 144x}$

39. $\dfrac{x^5 - x^2}{5x^5 - 5x} \cdot \dfrac{10x^4 - 10x^2}{2x^4 + 2x^3 + 2x^2}$

40. $\dfrac{2x^4 - 16x}{3x^6 - 48x^2} \cdot \dfrac{6x^5 + 24x^3}{4x^4 + 8x^3 + 16x^2}$

3.2 Problem Set

41. $\dfrac{a^2 - 16b^2}{a^2 - 8ab + 16b^2} \cdot \dfrac{a^2 - 9ab + 20b^2}{a^2 - 7ab + 12b^2} \div \dfrac{a^2 - 25b^2}{a^2 - 6ab + 9b^2}$

42. $\dfrac{a^2 - 6ab + 9b^2}{a^2 - 4b^2} \cdot \dfrac{a^2 - 5ab + 6b^2}{(a - 3b)^2} \div \dfrac{a^2 - 9b^2}{a^2 - ab - 6b^2}$

43. $\dfrac{2y^2 - 7y - 15}{42y^2 - 29y - 5} \cdot \dfrac{12y^2 - 16y + 5}{7y^2 - 36y + 5} \div \dfrac{4y^2 - 9}{49y^2 - 1}$

44. $\dfrac{8y^2 + 18y - 5}{21y^2 - 16y + 3} \cdot \dfrac{35y^2 - 22y + 3}{6y^2 + 17y + 5} \div \dfrac{16y^2 - 1}{9y^2 - 1}$

45. $\dfrac{xy - 2x + 3y - 6}{xy + 2x - 4y - 8} \cdot \dfrac{xy + x - 4y - 4}{xy - x + 3y - 3}$

46. $\dfrac{ax + bx + 2a + 2b}{ax - 3a + bx - 3b} \cdot \dfrac{ax - bx - 3a + 3b}{ax - bx - 2a + 2b}$

47. $\dfrac{xy^2 - y^2 + 4xy - 4y}{xy - 3y + 4x - 12} \div \dfrac{xy^3 + 2xy^2 + y^3 + 2y^2}{xy^2 - 3y^2 + 2xy - 6y}$

48. $\dfrac{4xb - 8b + 12x - 24}{xb^2 + 3b^2 + 3xb + 9b} \div \dfrac{4xb - 8b - 8x + 16}{xb^2 + 3b^2 - 2xb - 6b}$

49. $\dfrac{2x^3 + 10x^2 - 8x - 40}{x^3 + 4x^2 - 9x - 36} \cdot \dfrac{x^2 + x - 12}{2x^2 + 14x + 20}$

50. $\dfrac{x^3 + 2x^2 - 9x - 18}{x^4 + 3x^3 - 4x^2 - 12x} \cdot \dfrac{x^3 + 5x^2 + 6x}{x^2 - x - 6}$

The next two problems are intended to give you practice reading, and paying attention to, the instructions that accompany the problems you are working. Working these problems is an excellent way to get ready for a test or a quiz.

51. Work each problem according to the instructions given.
    a. Simplify: $\dfrac{16 - 1}{64 - 1}$
    b. Reduce: $\dfrac{25x^2 - 9}{125x^3 - 27}$
    c. Multiply: $\dfrac{25x^2 - 9}{125x^3 - 27} \cdot \dfrac{5x - 3}{5x + 3}$
    d. Divide: $\dfrac{25x^2 - 9}{125x^3 - 27} \div \dfrac{5x - 3}{25x^2 + 15x + 9}$

52. Work each problem according to the instructions given.
    a. Simplify: $\dfrac{64 - 49}{64 + 112 + 49}$
    b. Reduce: $\dfrac{9x^2 - 49}{9x^2 + 42x + 49}$
    c. Multiply: $\dfrac{9x^2 - 49}{9x^2 + 42x + 49} \cdot \dfrac{3x + 7}{3x - 7}$
    d. Divide: $\dfrac{9x^2 - 49}{9x^2 + 42x + 49} \div \dfrac{3x + 7}{3x - 7}$

53. Let $y_1 = \dfrac{x^2 - x - 6}{x - 1}$ and $y_2 = \dfrac{x + 2}{x^2 - 4x + 3}$ find
    a. $y_1 \cdot y_2$
    b. $y_1 \div y_2$

54. Let $y_1 = \dfrac{x^2 - x - 12}{x^2 - 4x + 3}$ and $y_2 = \dfrac{x^2 - x - 12}{x^2 - 5x + 4}$ find
    a. $y_1 \cdot y_2$
    b. $y_1 \div y_2$

55. Let $y_1 = \dfrac{x^3 - 9x^2 - 3x + 27}{4x^2 - 12}$ and $y_2 = \dfrac{x^2 - 2x - 8}{x^2 - 81}$ find $y_1 \cdot y_2$.

56. Let $y_1 = \dfrac{x^2 - 7x + 12}{x^2 - 16}$ and $y_2 = \dfrac{x^2 - 4x + 3}{x^2 - 6x + 9}$ find $y_1 \cdot y_2$.

57. Let $y_1 = \dfrac{x^3 - 3x^2 - 4x + 12}{x + 2}$ and $y_2 = \dfrac{x^2 + 7x + 12}{x^2 - 5x + 6}$ find $y_1 \cdot y_2$.

58. Let $y_1 = 2x^2 - 9x + 9$ and $y_2 = \dfrac{2x + 8}{x^2 + x - 12}$ find $y_1 \cdot y_2$.

Use the method shown in Example 10 to find the following products.

59. $(3x - 6) \cdot \dfrac{x}{x - 2}$

60. $(4x + 8) \cdot \dfrac{x}{x + 2}$

61. $(x^2 - 25) \cdot \dfrac{2}{x - 5}$

62. $(x^2 - 49) \cdot \dfrac{5}{x + 7}$

63. $(x^2 - 3x + 2) \cdot \dfrac{3}{3x - 3}$

64. $(x^2 - 3x + 2) \cdot \dfrac{-1}{x - 2}$

65. $(y - 3)(y - 4)(y + 3) \cdot \dfrac{-1}{y^2 - 9}$

66. $(y + 1)(y + 4)(y - 1) \cdot \dfrac{3}{y^2 - 1}$

67. $a(a + 5)(a - 5) \cdot \dfrac{a + 1}{a^2 + 5a}$

68. $a(a + 3)(a - 3) \cdot \dfrac{a - 1}{a^2 - 3a}$

Divide.

69. $(x^2 - 2x - 8) \div \dfrac{x^2 - x - 6}{x - 4}$

70. $(2x^2 + 7x - 15) \div \dfrac{6x^2 + 21x - 45}{2x - 3}$

71. $(3 - x) \div \dfrac{x^2 - 9}{x - 1}$

72. $(-x^2 + 3x - 2) \div \dfrac{x^2 + 2x - 8}{x^2 + 3x - 4}$

73. $(xy - 2x - 3y + 6) \div \dfrac{x^2 - 2x - 3}{x^2 - 6x - 7}$

74. $(x^3 - x^2 - 9x + 9) \div \dfrac{x^3 - 3x^2 - 4x + 12}{6x - 12}$

Divide.

75. $\dfrac{x^2(x + 2) + 6x(x + 2) + 9(x + 2)}{x^2 - 2x - 8}$

76. $\dfrac{x^2(x - 3) - 14x(x - 3) + 49(x - 3)}{x^2 - 3x - 28}$

77. $\dfrac{2x^2(x + 3) - 5x(x + 3) + 3(x + 3)}{x^2 + x - 6}$

78. $\dfrac{3x^2(x + 1) - 4x(x + 1) + (x + 1)}{x^3 + 5x^2 - x - 5}$

Applying the Concepts

At the beginning of this section we introduced the demand equation shown here. Use it to work Problems 79–82.

$$y = \dfrac{2(x + 60)}{x + 5}$$

79. **Demand Equation** Use the demand equation to fill in the table. Then compare your results with the graph shown in Figure 1 of this section.

Number of Copies	Price per Copy ($)
1	
10	
20	
50	
100	

80. **Demand Equation** To find the revenue for selling 50 copies of a tape, we multiply the price per tape by 50. Find the revenue for selling 50 tapes.

81. **Revenue** Find the revenue for selling 100 DVDs.

82. **Revenue** Find the revenue equation.

83. **Area** The following box has a square top. The front face of the box has an area of $A = x^3 - 2x^2 - 2x - 3$. The height of the box is $h = x^2 + x + 1$. Find a formula for the area of the top square in terms of x.

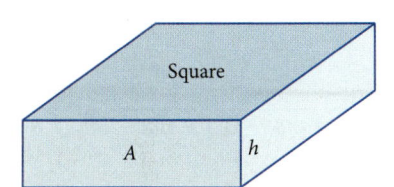

84. **Surface Area of a Cylinder** The surface area of the cylinder in the figure is defined as the area of its two circular bases and the lateral, or side, area. The surface area may be found by the formula

$$A = 2\pi r^2 + 2\pi rh$$

If the surface area is 6π, and $h = 2$, find r.

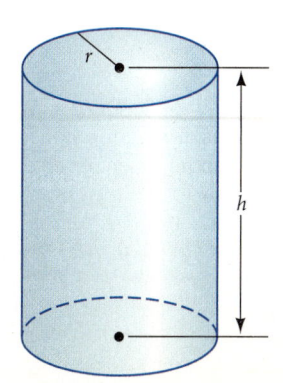

Maintaining Your Skills

Multiply.

85. $2x^2(5x^3 + 4x - 3)$

86. $3x^3(7x^2 - 4x - 8)$

87. $(3a - 1)(4a + 5)$

88. $(6a - 3)(2a + 1)$

89. $(3x + 7)(4y - 2)$

90. $(x + 2a)(2 - 3b)$

91. $(3 - t^2)^2$

92. $(2 - t^3)^2$

93. $3(x + 1)(x + 2)(x + 3)$

94. $4(x - 1)(x - 2)(x - 3)$

Getting Ready for the Next Section

Combine.

95. $\dfrac{4}{9} + \dfrac{2}{9}$

96. $\dfrac{3}{8} + \dfrac{1}{8}$

97. $\dfrac{3}{14} + \dfrac{7}{30}$

98. $\dfrac{3}{10} + \dfrac{11}{42}$

Multiply.

99. $-1(7 - x)$

100. $-1(3 - x)$

Factor.

101. $x^2 - 1$

102. $x^2 - 2x - 3$

103. $2x + 10$

104. $x^2 + 4x + 3$

105. $a^3 - b^3$

106. $8y^3 - 27$

Extending the concepts

107. $\dfrac{x^6 + y^6}{x^4 + 4x^2y^2 + 3y^4} \div \dfrac{x^4 + 3x^2y^2 + 2y^4}{x^4 + 5x^2y^2 + 6y^4}$

108. $\dfrac{x^2 + 9xy + 8y^2}{x^2 + 7xy - 8y^2} \div \dfrac{x^2 - y^2}{x^2 + 5xy - 6y^2}$

109. $\dfrac{a^2(2a + b) + 6a(2a + b) + 5(2a + b)}{3a^2(2a + b) - 2a(2a + b) + (2a + b)} \div \dfrac{a + 1}{a - 1}$

110. $\dfrac{2x^2(x - 3z) - 5x(x - 3z) + 2(x - 3z)}{4x^2(x - 3z) - 11x(x - 3z) + 6(x - 3z)} \div \dfrac{4x - 3}{4x + 1}$

111. $\dfrac{a^3 - a^2b}{ac - a} \div \left(\dfrac{a - b}{c - 1}\right)^2$

112. $\dfrac{p^3 + q^3}{q - p} \div \dfrac{(p + q)^2}{p^2 - q^2}$

Addition and Subtraction of Rational Expressions

3.3

Objectives

A Add and subtract rational expressions with the same denominator.

B Add and subtract rational expressions with different denominators.

This section is concerned with addition and subtraction of rational expressions. In the first part of this section we will look at addition of expressions that have the same denominator. In the second part of this section we will look at addition of expressions that have different denominators.

A Addition and Subtraction with the Same Denominator

To add two expressions that have the same denominator, we simply add numerators and put the sum over the common denominator. Since the process we use to add and subtract rational expressions is the same process used to add and subtract fractions, we will begin with an example involving fractions.

EXAMPLE 1 Add $\frac{4}{9} + \frac{2}{9}$.

SOLUTION We add fractions with the same denominator by using the distributive property. Here is a detailed look at the steps involved.

$$\frac{4}{9} + \frac{2}{9} = 4\left(\frac{1}{9}\right) + 2\left(\frac{1}{9}\right)$$

$$= (4 + 2)\left(\frac{1}{9}\right) \quad \text{Distributive property}$$

$$= 6\left(\frac{1}{9}\right)$$

$$= \frac{6}{9}$$

$$= \frac{2}{3} \quad \text{Divide numerator and denominator by common factor 3}$$

Note that the important thing about the fractions in this example is that they each have a denominator of 9. If they did not have the same denominator, we could not have written them as two terms with a factor of $\frac{1}{9}$ in common. Without the $\frac{1}{9}$ common to each term, we couldn't apply the distributive property. And without the distributive property, we would not have been able to add the two fractions.

In the examples that follow, we will not show all the steps we showed in Example 1. The steps are shown in Example 1 so that you will see why both fractions must have the same denominator before we can add them. In practice we simply add numerators and place the result over the common denominator.

We add and subtract rational expressions with the same denominator by combining numerators and writing the result over the common denominator. Then we reduce the result to lowest terms, if possible. Example 2 shows this process in detail. If you see the similarities between operations on rational numbers and operations on rational expressions, this chapter will look like an extension of rational numbers rather than a completely new set of topics.

EXAMPLE 2 Add $\frac{x}{x^2 - 1} + \frac{1}{x^2 - 1}$.

SOLUTION Since the denominators are the same, we simply add numerators:

$$\frac{x}{x^2 - 1} + \frac{1}{x^2 - 1} = \frac{x + 1}{x^2 - 1} \quad \text{Add numerators}$$

$$= \frac{\cancel{x + 1}}{(x - 1)\cancel{(x + 1)}} \quad \text{Factor denominator}$$

$$= \frac{1}{x - 1} \quad \text{Divide out common factor } x + 1$$

PRACTICE PROBLEMS

1. Add $\frac{3}{8} + \frac{1}{8}$.

2. Add $\frac{x}{x^2 - 9} + \frac{3}{x^2 - 9}$.

Answers
1. $\frac{1}{2}$
2. $\frac{1}{x - 3}$

3.3 Addition and Subtraction of Rational Expressions

3. Subtract $\dfrac{2x-7}{x-2} - \dfrac{x-5}{x-2}$.

Chapter 3 Rational Expressions and Rational Equations

Our next example involves subtraction of rational expressions. Pay careful attention to what happens to the signs of the terms in the numerator of the second expression when we subtract it from the first expression.

EXAMPLE 3 Subtract $\dfrac{2x-5}{x-2} - \dfrac{x-3}{x-2}$.

SOLUTION Since each expression has the same denominator, we simply subtract the numerator in the second expression from the numerator in the first expression and write the difference over the common denominator $x - 2$. We must be careful, however, that we subtract both terms in the second numerator. To ensure that we do, we will enclose that numerator in parentheses.

$$\dfrac{2x-5}{x-2} - \dfrac{x-3}{x-2} = \dfrac{2x - 5 - (x-3)}{x-2} \quad \text{Subtract numerators}$$

$$= \dfrac{2x - 5 - x + 3}{x-2} \quad \text{Remove parentheses}$$

$$= \dfrac{x-2}{x-2} \quad \text{Combine similar terms in the numerator}$$

$$= 1 \quad \text{Reduce (or divide)}$$

Note the +3 in the numerator of the second step. It is a very common mistake to write that as −3 by forgetting to subtract both terms in the numerator of the second expression. Whenever the expression we are subtracting has two or more terms in its numerator, we have to watch for this mistake.

Next we consider addition and subtraction of fractions and rational expressions that have different denominators.

B Addition and Subtraction with Different Denominators

Before we look at an example of addition of fractions with different denominators, we need to review the definition for the least common denominator.

> **Definition**
> The **least common denominator**, abbreviated LCD, for a set of denominators is the smallest expression that is divisible by each of the denominators.

The first step in combining two fractions is to find the LCD. Once we have the common denominator, we rewrite each fraction as an equivalent fraction with the common denominator. After that, we simply add or subtract as we did in our first three examples.

Example 4 is a review of the step-by-step procedure used to add two fractions with different denominators.

4. Add $\dfrac{3}{10} + \dfrac{11}{42}$.

EXAMPLE 4 Add $\dfrac{3}{14} + \dfrac{7}{30}$.

SOLUTION

Step 1: Find the LCD.
To do this, we first factor both denominators into prime factors.

Factor 14: $14 = 2 \cdot 7$

Factor 30: $30 = 2 \cdot 3 \cdot 5$

Answer
3. 1

3.3 Addition and Subtraction of Rational Expressions

Since the LCD must be divisible by 14, it must have factors of $2 \cdot 7$. It must also be divisible by 30 and, therefore, have factors of $2 \cdot 3 \cdot 5$. We do not need to repeat the 2 that appears in both the factors of 14 and those of 30. Therefore,
$$LCD = 2 \cdot 3 \cdot 5 \cdot 7 = 210$$

Step 2: Change to equivalent fractions.

Since we want each fraction to have a denominator of 210 and at the same time keep its original value, we multiply each by 1 in the appropriate form.

Change $\frac{3}{14}$ to a fraction with denominator 210:
$$\frac{3}{14} \cdot \frac{\mathbf{15}}{\mathbf{15}} = \frac{45}{210}$$

Change $\frac{7}{30}$ to a fraction with denominator 210:
$$\frac{7}{30} \cdot \frac{\mathbf{7}}{\mathbf{7}} = \frac{49}{210}$$

Step 3: Add numerators of equivalent fractions found in step 2:
$$\frac{45}{210} + \frac{49}{210} = \frac{94}{210}$$

Step 4: Reduce to lowest terms if necessary:
$$\frac{94}{210} = \frac{47}{105}$$

> **Note** When we multiply $\frac{3}{14}$ by $\frac{15}{15}$ we obtain a fraction with the same value as $\frac{3}{14}$ (because we multiplied by 1) but with the common denominator 210.

The main idea in adding fractions is to write each fraction again with the LCD for a denominator. In doing so, we must be sure not to change the value of either of the original fractions.

EXAMPLE 5 Add $\dfrac{-2}{x^2 - 2x - 3} + \dfrac{3}{x^2 - 9}$.

SOLUTION

Step 1: Factor each denominator and build the LCD from the factors:
$$\left.\begin{array}{l} x^2 - 2x - 3 = (x - 3)(x + 1) \\ x^2 - 9 = (x - 3)(x + 3) \end{array}\right\} LCD = (x - 3)(x + 3)(x + 1)$$

Step 2: Change each rational expression to an equivalent expression that has the LCD for a denominator:
$$\frac{-2}{x^2 - 2x - 3} = \frac{-2}{(x - 3)(x + 1)} \cdot \frac{\mathbf{(x + 3)}}{\mathbf{(x + 3)}} = \frac{-2x - 6}{(x - 3)(x + 3)(x + 1)}$$

$$\frac{3}{x^2 - 9} = \frac{3}{(x - 3)(x + 3)} \cdot \frac{\mathbf{(x + 1)}}{\mathbf{(x + 1)}} = \frac{3x + 3}{(x - 3)(x + 3)(x + 1)}$$

Step 3: Add numerators of the rational expressions found in step 2:
$$\frac{-2x - 6}{(x - 3)(x + 3)(x + 1)} + \frac{3x + 3}{(x - 3)(x + 3)(x + 1)} = \frac{x - 3}{(x - 3)(x + 3)(x + 1)}$$

Step 4: Reduce to lowest terms by dividing out the common factor $x - 3$:
$$= \frac{1}{(x + 3)(x + 1)}$$

5. Add $\dfrac{-3}{x^2 - 2x - 8} + \dfrac{4}{x^2 - 16}$.

Answers
4. $\dfrac{59}{105}$
5. $\dfrac{1}{(x + 4)(x + 2)}$

Chapter 3 Rational Expressions and Rational Equations

6. Add $\dfrac{x-4}{2x-6} + \dfrac{3}{x^2-9}$.

EXAMPLE 6 Subtract $\dfrac{x+4}{2x+10} - \dfrac{5}{x^2-25}$.

SOLUTION We begin by factoring each denominator:

$$\dfrac{x+4}{2x+10} - \dfrac{5}{x^2-25} = \dfrac{x+4}{2(x+5)} - \dfrac{5}{(x+5)(x-5)}$$

The LCD is $2(x+5)(x-5)$. Completing the solution we have

$$= \dfrac{x+4}{2(x+5)} \cdot \dfrac{(x-5)}{(x-5)} - \dfrac{5}{(x+5)(x-5)} \cdot \dfrac{2}{2}$$

$$= \dfrac{x^2-x-20}{2(x+5)(x-5)} - \dfrac{10}{2(x+5)(x-5)}$$

$$= \dfrac{x^2-x-30}{2(x+5)(x-5)}$$

To see if this expression will reduce, we factor the numerator into $(x-6)(x+5)$.

$$= \dfrac{(x-6)(x+5)}{2(x+5)(x-5)}$$

$$= \dfrac{x-6}{2(x-5)}$$

EXAMPLE 7 Subtract $\dfrac{2x-2}{x^2+4x+3} - \dfrac{x-1}{x^2+5x+6}$.

SOLUTION We factor each denominator and build the LCD from those factors:

$$\dfrac{2x-2}{x^2+4x+3} - \dfrac{x-1}{x^2+5x+6}$$

$$= \dfrac{2x-2}{(x+3)(x+1)} - \dfrac{x-1}{(x+3)(x+2)}$$

$$= \dfrac{2x-2}{(x+3)(x+1)} \cdot \dfrac{(x+2)}{(x+2)} - \dfrac{x-1}{(x+3)(x+2)} \cdot \dfrac{(x+1)}{(x+1)} \quad \text{Build the LCD}$$

$$= \dfrac{2x^2+2x-4}{(x+1)(x+2)(x+3)} - \dfrac{x^2-1}{(x+1)(x+2)(x+3)} \quad \text{Multiply out each numerator}$$

$$= \dfrac{(2x^2+2x-4) - (x^2-1)}{(x+1)(x+2)(x+3)}$$

$$= \dfrac{x^2+2x-3}{(x+1)(x+2)(x+3)} \quad \text{Subtract numerators}$$

$$= \dfrac{(x+3)(x-1)}{(x+1)(x+2)(x+3)} \quad \text{Factor numerator to see if we can reduce}$$

$$= \dfrac{x-1}{(x+1)(x+2)} \quad \text{Reduce}$$

7. Subtract.

$$\dfrac{2x-4}{x^2+5x+4} - \dfrac{x-4}{x^2+6x+8}$$

Answers

6. $\dfrac{x+2}{2(x+3)}$

7. $\dfrac{x-1}{(x+1)(x+2)}$

3.3 Addition and Subtraction of Rational Expressions

EXAMPLE 8 Add $\dfrac{x^2}{x-7} + \dfrac{6x+7}{7-x}$.

SOLUTION In the first section of this chapter we were able to reverse the terms in a factor such as $7 - x$ by factoring -1 from each term. In a problem like this, the same result can be obtained by multiplying the numerator and denominator by -1:

$$\frac{x^2}{x-7} + \frac{6x+7}{7-x} \cdot \frac{-1}{-1} = \frac{x^2}{x-7} + \frac{-6x-7}{x-7}$$

$$= \frac{x^2 - 6x - 7}{x-7} \qquad \text{Add numerators}$$

$$= \frac{(x-7)(x+1)}{(x-7)} \qquad \text{Factor numerator}$$

$$= x + 1 \qquad \text{Divide out } x - 7$$

For our next example we will look at a problem in which we combine a whole number and a rational expression.

EXAMPLE 9 Subtract $2 - \dfrac{9}{3x+1}$.

SOLUTION To subtract these two expressions, we think of 2 as a rational expression with a denominator of 1.

$$2 - \frac{9}{3x+1} = \frac{2}{1} - \frac{9}{3x+1}$$

The LCD is $3x + 1$. Multiplying the numerator and denominator of the first expression by $3x + 1$ gives us a rational expression equivalent to 2 but with a denominator of $3x + 1$.

$$\frac{2}{1} \cdot \frac{(3x+1)}{(3x+1)} - \frac{9}{3x+1} = \frac{6x + 2 - 9}{3x+1}$$

$$= \frac{6x - 7}{3x+1}$$

The numerator and denominator of this last expression do not have any factors in common other than 1, so the expression is in lowest terms.

EXAMPLE 10 Write an expression for the sum of a number and twice its reciprocal. Then, simplify that expression.

SOLUTION If x is the number, then its reciprocal is $\dfrac{1}{x}$. Twice its reciprocal is $\dfrac{2}{x}$. The sum of the number and twice its reciprocal is

$$x + \frac{2}{x}$$

To combine these two expressions, we think of the first term x as a rational expression with a denominator of 1. The least common denominator is x:

$$x + \frac{2}{x} = \frac{x}{1} + \frac{2}{x}$$

$$= \frac{x}{1} \cdot \frac{x}{x} + \frac{2}{x}$$

$$= \frac{x^2 + 2}{x}$$

---

8. Add $\dfrac{x^2}{x-4} + \dfrac{x+12}{4-x}$

9. Add $2 + \dfrac{25}{5x-1}$

10. One number is three times another. Write an expression for the sum of the reciprocals of the two numbers. Then simplify that expression.

Answers
8. $x + 3$
9. $\dfrac{10x + 23}{5x - 1}$
10. $\dfrac{4}{3x}$

Getting Ready for Class

After reading through the preceding section, respond in your own words and in complete sentences.

1. Briefly describe how you would add two rational expressions that have the same denominator.
2. Why is factoring important in finding a least common denominator?
3. What is the last step in adding or subtracting two rational expressions?
4. Explain how you would change the fraction $\frac{5}{x-3}$ to an equivalent fraction with denominator $x^2 - 9$.

Problem Set 3.3

B Combine the following fractions. [Example 4]

1. $\dfrac{3}{4} + \dfrac{1}{2}$
2. $\dfrac{5}{6} + \dfrac{1}{3}$
3. $\dfrac{2}{5} - \dfrac{1}{15}$
4. $\dfrac{5}{8} - \dfrac{1}{4}$

5. $\dfrac{5}{6} + \dfrac{7}{8}$
6. $\dfrac{3}{4} + \dfrac{2}{3}$
7. $\dfrac{9}{48} - \dfrac{3}{54}$
8. $\dfrac{6}{28} - \dfrac{5}{42}$

9. $\dfrac{3}{4} - \dfrac{1}{8} + \dfrac{2}{3}$
10. $\dfrac{1}{3} - \dfrac{5}{6} + \dfrac{5}{12}$

A Combine the following rational expressions. Reduce all answers to lowest terms. [Examples 2–3]

11. $\dfrac{x}{x+3} + \dfrac{3}{x+3}$
12. $\dfrac{5x}{5x+2} + \dfrac{2}{5x+2}$
13. $\dfrac{4}{y-4} - \dfrac{y}{y-4}$
14. $\dfrac{8}{y+8} + \dfrac{y}{y+8}$

15. $\dfrac{x}{x^2-y^2} - \dfrac{y}{x^2-y^2}$
16. $\dfrac{x}{x^2-y^2} + \dfrac{y}{x^2-y^2}$
17. $\dfrac{2x-3}{x-2} - \dfrac{x-1}{x-2}$
18. $\dfrac{2x-4}{x+2} - \dfrac{x-6}{x+2}$

19. $\dfrac{1}{a} + \dfrac{2}{a^2} - \dfrac{3}{a^3}$
20. $\dfrac{3}{a} + \dfrac{2}{a^2} - \dfrac{1}{a^3}$
21. $\dfrac{7x-2}{2x+1} - \dfrac{5x-3}{2x+1}$
22. $\dfrac{7x-1}{3x+2} - \dfrac{4x-3}{3x+2}$

23. Work each problem according to the instructions given.
   a. Multiply: $\dfrac{3}{8} \cdot \dfrac{1}{6}$
   b. Divide: $\dfrac{3}{8} \div \dfrac{1}{6}$
   c. Add: $\dfrac{3}{8} + \dfrac{1}{6}$
   d. Multiply: $\dfrac{x+3}{x-3} \cdot \dfrac{5x+15}{x^2-9}$
   e. Divide: $\dfrac{x+3}{x-3} \div \dfrac{5x+15}{x^2-9}$
   f. Subtract: $\dfrac{x+3}{x-3} - \dfrac{5x+15}{x^2-9}$

24. Work each problem according to the instructions given.
   a. Multiply: $\dfrac{16}{49} \cdot \dfrac{1}{28}$
   b. Divide: $\dfrac{16}{49} \div \dfrac{1}{28}$
   c. Add: $\dfrac{16}{49} - \dfrac{1}{28}$
   d. Multiply: $\dfrac{3x-2}{3x+2} \cdot \dfrac{15x+6}{9x^2-4}$
   e. Divide: $\dfrac{3x-2}{3x+2} \div \dfrac{15x+6}{9x^2-4}$
   f. Subtract: $\dfrac{3x-2}{3x+2} - \dfrac{15x+6}{9x^2-4}$

Chapter 3 Rational Expressions and Rational Equations

B Combine the following rational expressions. Reduce all answers to lowest terms. [Examples 5–10]

25. $\dfrac{2}{t^2} - \dfrac{3}{2t}$

26. $\dfrac{5}{3t} - \dfrac{4}{t^2}$

27. $\dfrac{3x+1}{2x-6} - \dfrac{x+2}{x-3}$

28. $\dfrac{x+1}{x-2} - \dfrac{4x+7}{5x-10}$

29. $\dfrac{x+1}{2x-2} - \dfrac{2}{x^2-1}$

30. $\dfrac{x+7}{2x+12} + \dfrac{6}{x^2-36}$

31. $\dfrac{1}{a-b} - \dfrac{3ab}{a^3-b^3}$

32. $\dfrac{1}{a+b} + \dfrac{3ab}{a^3+b^3}$

33. $\dfrac{1}{2y-3} - \dfrac{18y}{8y^3-27}$

34. $\dfrac{1}{3y-2} - \dfrac{18y}{27y^3-8}$

35. $\dfrac{x}{x^2-5x+6} - \dfrac{3}{3-x}$

36. $\dfrac{x}{x^2+4x+4} - \dfrac{2}{2+x}$

37. $\dfrac{2}{4t-5} + \dfrac{9}{8t^2-38t+35}$

38. $\dfrac{3}{2t-5} + \dfrac{21}{8t^2-14t-15}$

39. $\dfrac{1}{a^2-5a+6} + \dfrac{3}{a^2-a-2}$

40. $\dfrac{-3}{a^2+a-2} + \dfrac{5}{a^2-a-6}$

41. $\dfrac{1}{8x^3 - 1} - \dfrac{1}{4x^2 - 1}$

42. $\dfrac{1}{27x^3 - 1} - \dfrac{1}{9x^2 - 1}$

43. $\dfrac{4}{4x^2 - 9} - \dfrac{6}{8x^2 - 6x - 9}$

44. $\dfrac{9}{9x^2 + 6x - 8} - \dfrac{6}{9x^2 - 4}$

45. $\dfrac{4a}{a^2 + 6a + 5} - \dfrac{3a}{a^2 + 5a + 4}$

46. $\dfrac{3a}{a^2 + 7a + 10} - \dfrac{2a}{a^2 + 6a + 8}$

47. $\dfrac{2x - 1}{x^2 + x - 6} - \dfrac{x + 2}{x^2 + 5x + 6}$

48. $\dfrac{4x + 1}{x^2 + 5x + 4} - \dfrac{x + 3}{x^2 + 4x + 3}$

49. $\dfrac{2x - 8}{3x^2 + 8x + 4} + \dfrac{x + 3}{3x^2 + 5x + 2}$

50. $\dfrac{5x + 3}{2x^2 + 5x + 3} - \dfrac{3x + 9}{2x^2 + 7x + 6}$

51. $\dfrac{2}{x^2 + 5x + 6} - \dfrac{4}{x^2 + 4x + 3} + \dfrac{3}{x^2 + 3x + 2}$

52. $\dfrac{-5}{x^2 + 3x - 4} + \dfrac{5}{x^2 + 2x - 3} + \dfrac{1}{x^2 + 7x + 12}$

53. $\dfrac{2x + 8}{x^2 + 5x + 6} - \dfrac{x + 5}{x^2 + 4x + 3} - \dfrac{x - 1}{x^2 + 3x + 2}$

54. $\dfrac{2x + 11}{x^2 + 9x + 20} - \dfrac{x + 1}{x^2 + 7x + 12} - \dfrac{x + 6}{x^2 + 8x + 15}$

55. $2 + \dfrac{3}{2x + 1}$

56. $3 - \dfrac{2}{2x + 3}$

57. $5 + \dfrac{2}{4 - t}$

58. $7 + \dfrac{3}{5 - t}$

59. $x - \dfrac{4}{2x+3}$

60. $x - \dfrac{5}{3x+4} + 1$

61. $\dfrac{x}{x+2} + \dfrac{1}{2x+4} - \dfrac{3}{x^2+2x}$

62. $\dfrac{x}{x+3} + \dfrac{7}{3x+9} - \dfrac{2}{x^2+3x}$

63. $\dfrac{1}{x} + \dfrac{x}{2x+4} - \dfrac{2}{x^2+2x}$

64. $\dfrac{1}{x} + \dfrac{x}{3x+9} - \dfrac{3}{x^2+3x}$

65. Let $y_1 = \dfrac{2x-1}{4x-16}$ and $y_2 = \dfrac{x-3}{x-4}$; find $y_1 - y_2$.

66. Let $y_1 = \dfrac{2}{2x+4}$ and $y_2 = \dfrac{x}{x-4}$; find $y_1 + y_2$.

67. Let $y_1 = \dfrac{2}{x+4}$ and $y_2 = \dfrac{x-1}{x^2+3x-4}$; find $y_1 + y_2$.

68. Let $y_1 = \dfrac{5}{3t-2}$ and $y_2 = \dfrac{t-3}{3t^2+7t-6}$; find $y_1 - y_2$.

69. Let $y_1 = \dfrac{2x}{x^2-x-2}$ and $y_2 = \dfrac{5}{x^2+x-6}$; find $y_1 + y_2$.

70. Let $y_1 = \dfrac{7}{x^2-x-12}$ and $y_2 = \dfrac{5}{x^2+x-6}$; find $y_1 - y_2$.

71. Let $y_1 = \dfrac{x}{9x^2-4}$ and $y_2 = \dfrac{1}{3x^2-4x-4}$; find $y_1 + y_2$.

72. Let $y_1 = \dfrac{1}{16x^2-1}$ and $y_2 = \dfrac{1}{64x^3-1}$; find $y_1 - y_2$.

73. Let $y_1 = \dfrac{3x}{2x^2-x-1}$ and $y_2 = \dfrac{6}{2x^2-5x-3}$; find $y_1 + y_2$.

74. Let $y_1 = \dfrac{5x}{x^2-8x-9}$ and $y_2 = \dfrac{4x}{x^2-10x+9}$; find $y_1 - y_2$.

75. Let $y_1 = \dfrac{5x}{x^2-13x+36}$ and $y_2 = \dfrac{3x}{x^2-11x+28}$; find $y_1 - y_2$.

76. Let $y_1 = \dfrac{x+4}{x^2-6x-16}$ and $y_2 = \dfrac{x-6}{x^2-11x+24}$; find $y_1 - y_2$.

Applying the Concepts

77. Number Problem Write an expression for the sum of a number and 4 times its reciprocal. Then, simplify that expression.

78. Number Problem Write an expression for the sum of the reciprocals of two consecutive integers. Then, simplify that expression.

79. Optometry The formula

$$P = \frac{1}{a} + \frac{1}{b}$$

is used by optometrists to help determine how strong to make the lenses for a pair of eyeglasses. If a is 10 and b is 0.2, find the corresponding value of P.

80. Quadratic Formula Later in this chapter we will work with the quadratic formula. The derivation of the formula requires that you can add the fractions below. Add the fractions.

$$-\frac{c}{a} + \left(\frac{b}{2a}\right)^2$$

81. Elliptical Orbits Consider two objects, A and B, that move in the same direction along an elliptical path at constant but different velocities.

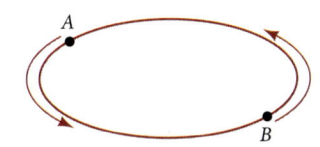

It can be shown that the time, T, it takes for the two objects to meet can be found from the formula

$$\frac{1}{T} = \frac{1}{t_A} - \frac{1}{t_B}$$

where t_A = time required for object A to orbit, and t_B = time required for object B to orbit.

a. If $t_A = 24$ months and $t_B = 30$ months, when will these two objects meet?

b. If $t_A = t_B$ what can one conclude?

82. Average Velocity If a car travels at a constant velocity, v_1, for 10 miles and then at a constant, but different velocity, v_2, for the next 10 miles, it can be shown that the cars average velocity, v_{avg}, over these 20 miles satisfies the equation

$$\frac{2}{v_{avg}} = \frac{1}{v_1} + \frac{1}{v_2}$$

Find the average velocity of a car that travels a constant 45 miles per hour for 10 miles and then increases to a constant 60 miles per hour for the next 10 miles. Round to the nearest tenth.

Maintaining Your Skills

Write each number in scientific notation.

83. 54,000   **84.** 768,000   **85.** 0.00034   **86.** 0.0359

Write each number in expanded form.

87. 6.44×10^3 **88.** 2.5×10^2 **89.** 6.44×10^{-3} **90.** 2.5×10^{-2}

Simplify each expression as much as possible. Write all answers in scientific notation.

91. $(3 \times 10^8)(4 \times 10^{-5})$ **92.** $\dfrac{8 \times 10^{-3}}{4 \times 10^{-6}}$

Getting Ready for the Next Section

Divide.

93. $\dfrac{3}{4} \div \dfrac{5}{8}$ **94.** $\dfrac{2}{3} \div \dfrac{5}{6}$

Multiply.

95. $x\left(1 + \dfrac{2}{x}\right)$ **96.** $3\left(x + \dfrac{1}{3}\right)$ **97.** $3x\left(\dfrac{1}{x} - \dfrac{1}{3}\right)$ **98.** $3x\left(\dfrac{1}{x} + \dfrac{1}{3}\right)$

Factor.

99. $x^2 - 4$ **100.** $x^2 - x - 6$

Extending the Concepts

Simplify.

101. $\left(1 - \dfrac{1}{x}\right)\left(1 - \dfrac{1}{x+1}\right)\left(1 - \dfrac{1}{x+2}\right)\left(1 - \dfrac{1}{x+3}\right)$ **102.** $\left(1 + \dfrac{1}{x}\right)\left(1 + \dfrac{1}{x+1}\right)\left(1 + \dfrac{1}{x+2}\right)\left(1 + \dfrac{1}{x+3}\right)$

103. $\left(\dfrac{a^2 - b^2}{u^2 - v^2}\right)\left(\dfrac{av - au}{b - a}\right) + \left(\dfrac{a^2 - av}{u + v}\right)\left(\dfrac{1}{a}\right)$ **104.** $\left(\dfrac{6r^2}{r^2 - 1}\right)\left(\dfrac{r + 1}{3}\right) - \dfrac{2r^2}{r - 1}$

105. $\dfrac{18x - 19}{4x^2 + 27x - 7} - \dfrac{12x - 41}{3x^2 + 17x - 28}$ **106.** $\dfrac{42 - 22y}{3y^2 - 13y - 10} - \dfrac{21 - 13y}{2y^2 - 9y - 5}$

107. $\left(\dfrac{1}{y^2 - 1} \div \dfrac{1}{y^2 + 1}\right)\left(\dfrac{y^3 + 1}{y^4 - 1}\right) + \dfrac{1}{(y + 1)^2(y - 1)}$ **108.** $\left(\dfrac{a^3 - 64}{a^2 - 16} \div \dfrac{a^2 - 4a + 16}{a^2 - 4} \div \dfrac{a^2 + 4a + 16}{a^3 + 64}\right) + 4 - a^2$

3.4 Equations Involving Rational Expressions

Objectives
A Solve equations containing rational expressions.
B Solve formulas containing rational expressions.

A Equations with Rational Expressions

The first step in solving an equation that contains one or more rational expressions is to find the LCD for all denominators in the equation. We then multiply both sides of the equation by the LCD to clear the equation of all fractions—that is, after we have multiplied through by the LCD, each term in the resulting equation will have a denominator of 1.

EXAMPLE 1 Solve $\frac{x}{2} - 3 = \frac{2}{3}$.

SOLUTION The LCD for 2 and 3 is 6. Multiplying both sides by 6, we have

$$6\left(\frac{x}{2} - 3\right) = 6\left(\frac{2}{3}\right)$$

$$6\left(\frac{x}{2}\right) - 6(3) = 6\left(\frac{2}{3}\right)$$

$$3x - 18 = 4$$

$$3x = 22$$

$$x = \frac{22}{3}$$

Multiplying both sides of an equation by the LCD clears the equation of fractions because the LCD has the property that all the denominators divide it evenly.

EXAMPLE 2 Solve $\frac{6}{a - 4} = \frac{3}{8}$.

SOLUTION The LCD for $a - 4$ and 8 is $8(a - 4)$. Multiplying both sides by this quantity yields

$$8(a - 4) \cdot \frac{6}{a - 4} = 8(a - 4) \cdot \frac{3}{8}$$

$$48 = (a - 4) \cdot 3$$

$$48 = 3a - 12$$

$$60 = 3a$$

$$20 = a$$

The solution set is {20}, which checks in the original equation.

When we multiply both sides of an equation by an expression containing the variable, we must be sure to check our solutions. The multiplication property of equality does not allow multiplication by 0. If the expression we multiply by contains the variable, then it has the possibility of being 0. In the last example we multiplied both sides by $8(a - 4)$. This gives a restriction $a \neq 4$ for any solution we come up with.

PRACTICE PROBLEMS

1. Solve $\frac{x}{3} + 1 = \frac{1}{2}$.

2. $\frac{2}{a + 5} = \frac{1}{3}$.

Answers
1. $-\frac{3}{2}$
2. 1

3. Solve $\dfrac{x}{x+1} - \dfrac{1}{2} = \dfrac{-1}{x+1}$.

Note In the process of solving the equation, we multiplied both sides by $3(x - 2)$, solved for x, and got $x = 2$ for our solution. But when x is 2, the quantity $3(x - 2) = 3(2 - 2) = 3(0) = 0$, which means we multiplied both sides of our equation by 0, which is not allowed under the multiplication property of equality.

EXAMPLE 3 Solve $\dfrac{x}{x-2} + \dfrac{2}{3} = \dfrac{2}{x-2}$.

SOLUTION The LCD is $3(x - 2)$. We are assuming $x \neq 2$ when we multiply both sides of the equation by $3(x - 2)$:

$$3(x-2) \cdot \left[\dfrac{x}{x-2} + \dfrac{2}{3}\right] = 3(x-2) \cdot \dfrac{2}{x-2}$$

$$3x + (x-2) \cdot 2 = 3 \cdot 2$$

$$3x + 2x - 4 = 6$$

$$5x - 4 = 6$$

$$5x = 10$$

$$x = 2$$

The only possible solution is $x = 2$. Checking this value back in the original equation gives

$$\dfrac{2}{2-2} + \dfrac{2}{3} \stackrel{?}{=} \dfrac{2}{2-2}$$

$$\dfrac{2}{0} + \dfrac{2}{3} \stackrel{?}{=} \dfrac{2}{0}$$

The first and last terms are undefined. The proposed solution, $x = 2$, does not check in the original equation. The solution set is the empty set. There is no solution to the original equation.

When the proposed solution to an equation is not actually a solution, it is called an *extraneous* solution. In the last example, $x = 2$ is an extraneous solution.

4. Solve.
$\dfrac{x}{x^2 - 9} - \dfrac{1}{x+3} = \dfrac{1}{4x - 12}$

EXAMPLE 4 Solve $\dfrac{5}{x^2 - 3x + 2} - \dfrac{1}{x-2} = \dfrac{1}{3x - 3}$.

SOLUTION Writing the equation again with the denominators in factored form, we have

$$\dfrac{5}{(x-2)(x-1)} - \dfrac{1}{x-2} = \dfrac{1}{3(x-1)}$$

The LCD is $3(x - 2)(x - 1)$. Multiplying through by the LCD, we have

$$3(x-2)(x-1) \cdot \dfrac{5}{(x-2)(x-1)} - 3(x-2)(x-1) \cdot \dfrac{1}{(x-2)}$$

$$= 3(x-2)(x-1) \cdot \dfrac{1}{3(x-1)}$$

$$3 \cdot 5 - 3(x-1) \cdot 1 = (x-2) \cdot 1$$

$$15 - 3x + 3 = x - 2$$

$$-3x + 18 = x - 2$$

$$-4x + 18 = -2$$

$$-4x = -20$$

$$x = 5$$

Note We can check the proposed solution in any of the equations obtained before multiplying through by the LCD. We cannot check the proposed solution in an equation obtained *after* multiplying through by the LCD since, if we have multiplied by 0, the resulting equations will not be equivalent to the original one.

Checking the proposed solution $x = 5$ in the original equation yields a true statement. Try it and see.

Answers
3. No solution
4. 9

3.4 Equations Involving Rational Expressions

EXAMPLE 5 Solve $3 + \dfrac{1}{x} = \dfrac{10}{x^2}$.

SOLUTION To clear the equation of denominators, we multiply both sides by x^2:

$$x^2\left(3 + \dfrac{1}{x}\right) = x^2\left(\dfrac{10}{x^2}\right)$$

$$3(x^2) + \left(\dfrac{1}{x}\right)(x^2) = \left(\dfrac{10}{x^2}\right)(x^2)$$

$$3x^2 + x = 10$$

Rewrite in standard form, and solve:

$$3x^2 + x - 10 = 0$$

$$(3x - 5)(x + 2) = 0$$

$$3x - 5 = 0 \quad \text{or} \quad x + 2 = 0$$

$$x = \dfrac{5}{3} \quad \text{or} \quad x = -2$$

The solution set is $\left\{-2, \dfrac{5}{3}\right\}$. Both solutions check in the original equation. Remember: We have to check *all solutions* any time we multiply both sides of the equation by an expression that contains the variable, just to be sure we haven't multiplied by 0.

5. Solve $1 - \dfrac{2}{x} = \dfrac{8}{x^2}$.

EXAMPLE 6 Solve $\dfrac{y - 4}{y^2 - 5y} = \dfrac{2}{y^2 - 25}$.

SOLUTION Factoring each denominator, we find the LCD is $y(y - 5)(y + 5)$. Multiplying each side of the equation by the LCD clears the equation of denominators and leads us to our possible solutions:

$$y(y - 5)(y + 5) \cdot \dfrac{y - 4}{y(y - 5)} = \dfrac{2}{(y - 5)(y + 5)} \cdot y(y - 5)(y + 5)$$

$$(y + 5)(y - 4) = 2y$$

$$y^2 + y - 20 = 2y \quad \text{Multiply out the left side}$$

$$y^2 - y - 20 = 0 \quad \text{Add } -2y \text{ to each side}$$

$$(y - 5)(y + 4) = 0$$

$$y - 5 = 0 \quad \text{or} \quad y + 4 = 0$$

$$y = 5 \quad \text{or} \quad y = -4$$

The two possible solutions are 5 and −4. If we substitute −4 for y in the original equation, we find that it leads to a true statement. It is, therefore, a solution. However, if we substitute 5 for y in the original equation, we find that both sides of the equation are undefined. The only solution to our original equation is $y = -4$. The other possible solution $y = 5$ is extraneous.

6. Solve $\dfrac{y + 1}{3(y + 4)} = \dfrac{8}{(y + 4)(y - 4)}$.

Answers
5. −2, 4
6. 7

7. Solve for y: $x = \dfrac{y+2}{y-1}$.

B Formulas

EXAMPLE 7 Solve for y: $x = \dfrac{y-4}{y-2}$.

SOLUTION To solve for y, we first multiply each side by $y - 2$ to obtain

$$x(y-2) = y - 4$$

$$xy - 2x = y - 4 \quad \text{Distributive property}$$

$$xy - y = 2x - 4 \quad \text{Collect all terms containing } y \text{ on the left side}$$

$$y(x-1) = 2x - 4 \quad \text{Factor } y \text{ from each term on the left side}$$

$$y = \dfrac{2x-4}{x-1} \quad \text{Divide each side by } x - 1$$

8. Solve the formula $\dfrac{1}{a} = \dfrac{1}{x} + \dfrac{1}{b}$ for x.

EXAMPLE 8 Solve the formula $\dfrac{1}{x} = \dfrac{1}{b} + \dfrac{1}{a}$ for x.

SOLUTION We begin by multiplying both sides by the least common denominator xab. As you can see from our previous examples, multiplying both sides of an equation by the LCD is equivalent to multiplying each term of both sides by the LCD:

$$xab \cdot \dfrac{1}{x} = \dfrac{1}{b} \cdot xab + \dfrac{1}{a} \cdot xab$$

$$ab = xa + xb$$

$$ab = (a+b)x \quad \text{Factor } x \text{ from the right side}$$

$$\dfrac{ab}{a+b} = x$$

We know we are finished because the variable we were solving for is alone on one side of the equation and does not appear on the other side.

Getting Ready for Class

After reading through the preceding section, respond in your own words and in complete sentences.

1. Explain how a least common denominator can be used to simplify an equation.
2. What is an extraneous solution?
3. Is it possible for an equation containing rational expressions to have no solutions?
4. What is the last step in solving an equation that contains rational expressions?

Answers
7. $y = \dfrac{x+2}{x-1}$
8. $x = \dfrac{ab}{b-a}$

Problem Set 3.4

A Solve each of the following equations. [Examples 1–3]

1. $\dfrac{x}{5} + 4 = \dfrac{5}{3}$
2. $\dfrac{x}{5} = \dfrac{x}{2} - 9$
3. $\dfrac{a}{3} + 2 = \dfrac{4}{5}$
4. $\dfrac{a}{4} + \dfrac{1}{2} = \dfrac{2}{3}$

5. $\dfrac{y}{2} + \dfrac{y}{4} + \dfrac{y}{6} = 3$
6. $\dfrac{y}{3} - \dfrac{y}{6} + \dfrac{y}{2} = 1$
7. $\dfrac{5}{2x} = \dfrac{1}{x} + \dfrac{3}{4}$
8. $\dfrac{1}{2a} = \dfrac{2}{a} - \dfrac{3}{8}$

9. $\dfrac{1}{x} = \dfrac{1}{3} - \dfrac{2}{3x}$
10. $\dfrac{5}{2x} = \dfrac{2}{x} - \dfrac{1}{12}$
11. $\dfrac{2x}{x-3} + 2 = \dfrac{2}{x-3}$
12. $\dfrac{2}{x+5} = \dfrac{2}{5} - \dfrac{x}{x+5}$

13. $1 - \dfrac{1}{x} = \dfrac{12}{x^2}$
14. $2 + \dfrac{5}{x} = \dfrac{3}{x^2}$
15. $y - \dfrac{4}{3y} = -\dfrac{1}{3}$
16. $\dfrac{y}{2} - \dfrac{4}{y} = -\dfrac{7}{2}$

Let $y_1 = \dfrac{1}{x-3}$ and $y_2 = \dfrac{1}{x+3}$, and find x if

17. $y_1 + y_2 = \dfrac{5}{8}$
18. $y_1 - y_2 = \dfrac{2}{9}$
19. $\dfrac{y_1}{y_2} = 5$

20. $\dfrac{y_2}{y_1} = 5$
21. $y_1 = y_2$
22. $y_1 = -y_2$

Let $y_1 = \dfrac{4}{x+2}$ and $y_2 = \dfrac{4}{x-2}$, and find x if

23. $y_1 + y_2 = \dfrac{24}{5}$

24. $y_1 - y_2 = -\dfrac{4}{3}$

25. $\dfrac{y_1}{y_2} = -5$

26. $\dfrac{y_2}{y_1} = -7$

27. $y_1 = y_2$

28. $y_1 = -y_2$

29. Solve each equation.

a. $6x - 2 = 0$

b. $\dfrac{6}{x} - 2 = 0$

c. $\dfrac{x}{6} - 2 = -\dfrac{1}{2}$

d. $\dfrac{6}{x} - 2 = -\dfrac{1}{2}$

e. $\dfrac{6}{x^2} + 6 = \dfrac{20}{x}$

30. Solve each equation.

a. $5x - 2 = 0$

b. $5 - \dfrac{2}{x} = 0$

c. $\dfrac{x}{2} - 5 = -\dfrac{3}{4}$

d. $\dfrac{2}{x} - 5 = -\dfrac{3}{4}$

e. $-\dfrac{3}{x} + \dfrac{2}{x^2} = 5$

31. Work each problem according to the instructions given.

a. Divide: $\dfrac{6}{x^2 - 2x - 8} \div \dfrac{x+3}{x+2}$

b. Add: $\dfrac{6}{x^2 - 2x - 8} + \dfrac{x+3}{x+2}$

c. Solve: $\dfrac{6}{x^2 - 2x - 8} + \dfrac{x+3}{x+2} = 2$

32. Work each problem according to the instructions given.

a. Divide: $\dfrac{-10}{x^2 - 25} \div \dfrac{x-4}{x-5}$

b. Add: $\dfrac{-10}{x^2 - 25} + \dfrac{x-4}{x-5}$

c. Solve: $\dfrac{-10}{x^2 - 25} + \dfrac{x-4}{x-5} = \dfrac{4}{5}$

A Solve each equation. [Examples 2–6]

33. $\dfrac{x+2}{x+1} = \dfrac{1}{x+1} + 2$

34. $\dfrac{x+6}{x+3} = \dfrac{3}{x+3} + 2$

35. $\dfrac{3}{a-2} = \dfrac{2}{a-3}$

36. $\dfrac{5}{a+1} = \dfrac{4}{a+2}$

37. $6 - \dfrac{5}{x^2} = \dfrac{7}{x}$

38. $10 - \dfrac{3}{x^2} = -\dfrac{1}{x}$

39. $\dfrac{1}{x-1} - \dfrac{1}{x+1} = \dfrac{3x}{x^2-1}$

40. $\dfrac{5}{x-1} + \dfrac{2}{x-1} = \dfrac{4}{x+1}$

41. $\dfrac{2}{x-3} + \dfrac{x}{x^2-9} = \dfrac{4}{x+3}$

42. $\dfrac{2}{x+5} + \dfrac{3}{x+4} = \dfrac{2x}{x^2+9x+20}$

43. $\dfrac{3}{2} - \dfrac{1}{x-4} = \dfrac{-2}{2x-8}$

44. $\dfrac{2}{x} - \dfrac{1}{x+1} = \dfrac{-2}{5x+5}$

45. $\dfrac{t-4}{t^2-3t} = \dfrac{-2}{t^2-9}$

46. $\dfrac{t+3}{t^2-2t} = \dfrac{10}{t^2-4}$

47. $\dfrac{3}{y-4} - \dfrac{2}{y+1} = \dfrac{5}{y^2-3y-4}$

48. $\dfrac{1}{y+2} - \dfrac{2}{y-3} = \dfrac{-2y}{y^2-y-6}$

49. $\dfrac{2}{1+a} = \dfrac{3}{1-a} + \dfrac{5}{a}$

50. $\dfrac{1}{a+3} - \dfrac{a}{a^2-9} = \dfrac{2}{3-a}$

51. $\dfrac{3}{2x-6} - \dfrac{x+1}{4x-12} = 4$

52. $\dfrac{2x-3}{5x+10} + \dfrac{3x-2}{4x+8} = 1$

53. $\dfrac{y+2}{y^2-y} - \dfrac{6}{y^2-1} = 0$

54. $\dfrac{y+3}{y^2-y} - \dfrac{8}{y^2-1} = 0$

55. $\dfrac{4}{2x-6} - \dfrac{12}{4x+12} = \dfrac{12}{x^2-9}$

56. $\dfrac{1}{x+2} + \dfrac{1}{x-2} = \dfrac{4}{x^2-4}$

57. $\dfrac{2}{y^2 - 7y + 12} - \dfrac{1}{y^2 - 9} = \dfrac{4}{y^2 - y - 12}$

58. $\dfrac{1}{y^2 + 5y + 4} + \dfrac{3}{y^2 - 1} = \dfrac{-1}{y^2 + 3y - 4}$

59. Solve the equation $6x^{-1} + 4 = 7$ by multiplying both sides by x. (Remember, $x^{-1} \cdot x = x^{-1} \cdot x^1 = x^0 = 1$.)

60. Solve the equation $3x^{-1} - 5 = 2x^{-1} - 3$ by multiplying both sides by x.

61. Solve the equation $1 + 5x^{-2} = 6x^{-1}$ by multiplying both sides by x^2.

62. Solve the equation $1 + 3x^{-2} = 4x^{-1}$ by multiplying both sides by x^2.

B [Examples 7–8]

63. Solve the formula $\dfrac{1}{x} = \dfrac{1}{b} - \dfrac{1}{a}$ for x.

64. Solve $\dfrac{1}{x} = \dfrac{1}{a} - \dfrac{1}{b}$ for x.

65. Solve for R in the formula $\dfrac{1}{R} = \dfrac{1}{R_1} + \dfrac{1}{R_2}$.

66. Solve for R in the formula $\dfrac{1}{R} = \dfrac{1}{R_1} + \dfrac{1}{R_2} + \dfrac{1}{R_3}$.

Solve for y.

67. $x = \dfrac{y - 3}{y - 1}$

68. $x = \dfrac{y - 2}{y - 3}$

69. $x = \dfrac{2y + 1}{3y + 1}$

70. $x = \dfrac{3y + 2}{5y + 1}$

● **Applying the Concepts**

71. **An Identity** An identity is an equation that is true for any value of the variable for which the expression is defined. Verify the following expression is an identity by simplifying the left side of the expression.

$$\dfrac{2}{x - y} - \dfrac{1}{y - x} = \dfrac{3}{x - y}$$

72. **Harmonic Mean** A number, h, is the harmonic mean of two numbers, n_1 and n_2, if $\dfrac{1}{h}$ is the mean (average) of $\dfrac{1}{n_1}$ and $\dfrac{1}{n_2}$.

a. Write an equation relating the harmonic mean, h, to two numbers, n_1 and n_2 then solve the equation for h.

b. Find the harmonic mean of 3 and 5.

73. Kayak Race In a kayak race, the participants must paddle a kayak 450 meters down a river and then return 450 meters up the river to the starting point (Figure 1). Susan has correctly deduced that the total time t (in seconds) depends on the speed c (in meters per second) of the water according to the following expression:

$$t = \frac{450}{v+c} + \frac{450}{v-c}$$

where v is the speed of the kayak relative to the water (the speed of the kayak in still water).
Fill in the following table.

Time t(sec)	Speed of Kayak Relative to the Water v(m/sec)	Current of the River c(m/sec)
240		1
300		2
	4	3
	3	1
540	3	
	3	3

FIGURE 1

74. Geometry From plane geometry and the principle of similar triangles, the relationship between y_1, y_2, and h shown in Figure 2 can be expressed as

$$\frac{1}{h} = \frac{1}{y_1} + \frac{1}{y_2}$$

Two poles are 12 feet high and 8 feet high. If a wire is attached to the top of each one and stretched to the bottom of the other, what is the height above the ground at which the two wires will meet?

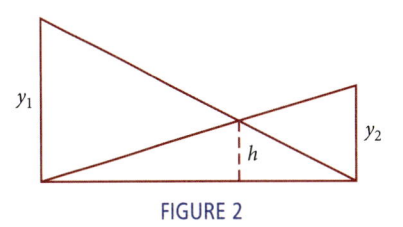

FIGURE 2

Maintaining Your Skills

75. Number Problem Twice the sum of a number and 3 is 16. Find the number.

76. Geometry The length of a rectangle is 3 less than twice the width. The perimeter is 42 meters. Find the length and width.

77. Consecutive Integers The sum of the squares of two consecutive integers is 61. Find the integers.

78. Consecutive Integers The square of the sum of two consecutive integers is 121. Find the two integers.

79. Geometry The lengths of the sides of a right triangle are given by three consecutive integers. Find the lengths of the three sides.

80. Geometry The longest side of a right triangle is 8 inches more than the shortest side. The other side is 7 inches more than the shortest side. Find the lengths of the three sides.

Getting Ready for the Next Section

Multiply.

81. $39.3 \cdot 60$

82. $1{,}100 \cdot 60 \cdot 60$

Divide. Round to the nearest tenth, if necessary.

83. $65{,}000 \div 5{,}280$

84. $3{,}960{,}000 \div 5{,}280$

Multiply.

85. $2x\left(\dfrac{1}{x} + \dfrac{1}{2x}\right)$

86. $3x\left(\dfrac{1}{x} + \dfrac{1}{3x}\right)$

Solve.

87. $12(x + 3) + 12(x - 3) = 3(x^2 - 9)$

88. $40 + 2x = 60 - 3x$

89. $\dfrac{1}{10} - \dfrac{1}{12} = \dfrac{1}{x}$

90. $\dfrac{1}{x} + \dfrac{1}{2x} = 2$

Extending the Concepts

Solve each equation.

91. $\dfrac{12}{x} + \dfrac{8}{x^2} - \dfrac{75}{x^3} - \dfrac{50}{x^4} = 0$

92. $\dfrac{45}{x} + \dfrac{18}{x^2} - \dfrac{80}{x^3} - \dfrac{32}{x^4} = 0$

93. $\dfrac{1}{x^3} - \dfrac{1}{3x^2} - \dfrac{1}{4x} + \dfrac{1}{12} = 0$

94. $\dfrac{1}{x^3} - \dfrac{1}{2x^2} - \dfrac{1}{9x} + \dfrac{1}{18} = 0$

95. Solve for x. $\dfrac{2}{x} + \dfrac{4}{x + a} = \dfrac{-6}{a - x}$

96. Solve for x. $\dfrac{1}{b - x} - \dfrac{1}{x} = \dfrac{-2}{b + x}$

97. Solve for v. $\dfrac{s - vt}{t^2} = -16$

98. Solve for r. $A = P\left(1 + \dfrac{r}{n}\right)$

99. Solve for f. $\dfrac{1}{p} = \dfrac{1}{f} + \dfrac{1}{g}$

100. Solve for p. $h = \dfrac{v^2}{2g} + \dfrac{p}{c}$

3.5 Simplified Form for Radicals

Objectives

A Write radical expressions in simplified form.

B Rationalize a denominator that contains only one term.

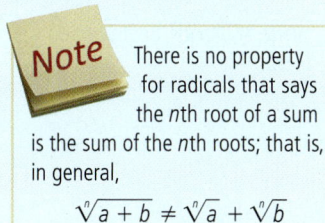

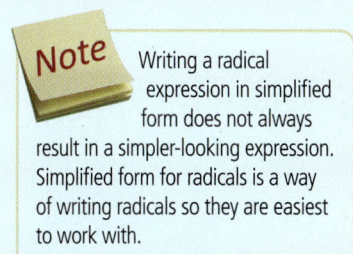

In this section we will define and use radical notation. There are a number of places in this book where square roots are used. The mathematics behind them is really very simple, and if you understand what it means to square a number, you will not have any trouble with square roots. The symbol $\sqrt{9}$ is used to represent the positive number we square to get 9. It must be 3 since 3 is positive and $3^2 = 9$. That is,

$$\sqrt{9} = 3$$

Here are some other square roots that you may see in this chapter.

$\sqrt{4} = 2$ $\qquad$ $\sqrt{16} = 4$

$\sqrt{25} = 5$ $\qquad$ $\sqrt{36} = 6$

We will begin our work in this section by stating two properties of radicals. Following this, we will give a definition for simplified form for radical expressions. The examples in this section show how we use the properties of radicals to write radical expressions in simplified form.

A Simplified Form

We will begin with the first two properties of radicals. We state them here without proof. For these two properties, we will assume a and b are nonnegative real numbers whenever n is an even number.

Property 1 for Radicals

$$\sqrt[n]{ab} = \sqrt[n]{a}\,\sqrt[n]{b}$$

In words: The nth root of a product is the product of the nth roots.

Property 2 for Radicals

$$\sqrt[n]{\frac{a}{b}} = \frac{\sqrt[n]{a}}{\sqrt[n]{b}} \qquad (b \neq 0)$$

In words: The nth root of a quotient is the quotient of the nth roots.

The two properties of radicals allow us to change the form of and simplify radical expressions without changing their value.

Simplified Form for Radical Expressions

A radical expression is in *simplified form* if

1. None of the factors of the radicand (the quantity under the radical sign) can be written as powers greater than or equal to the index—that is, no perfect squares can be factors of the quantity under a square root sign, no perfect cubes can be factors of what is under a cube root sign, and so forth;
2. There are no fractions under the radical sign; and
3. There are no radicals in the denominator.

Satisfying the first condition for simplified form actually amounts to taking as much out from under the radical sign as possible. The following examples illustrate the first condition for simplified form.

> **Note** There is no property for radicals that says the nth root of a sum is the sum of the nth roots; that is, in general,
> $$\sqrt[n]{a+b} \neq \sqrt[n]{a} + \sqrt[n]{b}$$

> **Note** Writing a radical expression in simplified form does not always result in a simpler-looking expression. Simplified form for radicals is a way of writing radicals so they are easiest to work with.

PRACTICE PROBLEMS

1. Write $\sqrt{18}$ in simplified form.

2. Write $\sqrt{50x^2y^3}$ in simplified form. Assume $x, y \geq 0$.

Note Unless we state otherwise, assume all variables throughout this section represent nonnegative numbers. Further, if a variable appears in a denominator, assume the variable cannot be 0.

3. Write $\sqrt[3]{54a^4b^3}$ in simplified form.

4. $\sqrt{75x^5y^8}$

5. $\sqrt[4]{48a^8b^5c^4}$

Answers
1. $3\sqrt{2}$
2. $5xy\sqrt{2y}$
3. $3ab\sqrt[3]{2a}$
4. $5x^2y^4\sqrt{3x}$
5. $2a^2bc\sqrt[4]{3b}$

Chapter 3 Rational Expressions and Rational Equations

EXAMPLE 1 Write $\sqrt{50}$ in simplified form.

SOLUTION The largest perfect square that divides 50 is 25. We write 50 as $25 \cdot 2$ and apply property 1 for radicals:

$\sqrt{50} = \sqrt{25 \cdot 2}$ $50 = 25 \cdot 2$
$\phantom{\sqrt{50}} = \sqrt{25}\sqrt{2}$ Property 1
$\phantom{\sqrt{50}} = 5\sqrt{2}$ $\sqrt{25} = 5$

We have taken as much as possible out from under the radical sign—in this case, factoring 25 from 50 and then writing $\sqrt{25}$ as 5.

EXAMPLE 2 Write in simplified form: $\sqrt{48x^4y^3}$, where $x, y \geq 0$.

SOLUTION The largest perfect square that is a factor of the radicand is $16x^4y^2$. Applying property 1 again, we have

$\sqrt{48x^4y^3} = \sqrt{16x^4y^2 \cdot 3y}$
$\phantom{\sqrt{48x^4y^3}} = \sqrt{16x^4y^2}\sqrt{3y}$
$\phantom{\sqrt{48x^4y^3}} = 4x^2y\sqrt{3y}$

EXAMPLE 3 Write $\sqrt[3]{40a^5b^4}$ in simplified form.

SOLUTION We now want to factor the largest perfect cube from the radicand. We write $40a^5b^4$ as $8a^3b^3 \cdot 5a^2b$ and proceed as we did in Examples 1 and 2.

$\sqrt[3]{40a^5b^4} = \sqrt[3]{8a^3b^3 \cdot 5a^2b}$
$\phantom{\sqrt[3]{40a^5b^4}} = \sqrt[3]{8a^3b^3}\sqrt[3]{5a^2b}$
$\phantom{\sqrt[3]{40a^5b^4}} = 2ab\sqrt[3]{5a^2b}$

Here are some further examples concerning the first condition for simplified form.

EXAMPLE 4 Write the expression in simplified form.

SOLUTION $\sqrt{12x^7y^6} = \sqrt{4x^6y^6 \cdot 3x}$
$\phantom{\sqrt{12x^7y^6}} = \sqrt{4x^6y^6}\sqrt{3x}$
$\phantom{\sqrt{12x^7y^6}} = 2x^3y^3\sqrt{3x}$

EXAMPLE 5 Write the expression in simplified form.

SOLUTION $\sqrt[3]{54a^6b^2c^4} = \sqrt[3]{27a^6c^3 \cdot 2b^2c}$
$\phantom{\sqrt[3]{54a^6b^2c^4}} = \sqrt[3]{27a^6c^3}\sqrt[3]{2b^2c}$
$\phantom{\sqrt[3]{54a^6b^2c^4}} = 3a^2c\sqrt[3]{2b^2c}$

3.5 Simplified Form for Radicals

B Rationalizing the Denominator

The second property of radicals is used to simplify a radical that contains a fraction.

EXAMPLE 6 Simplify $\sqrt{\frac{3}{4}}$.

SOLUTION Applying property 2 for radicals, we have

$$\sqrt{\frac{3}{4}} = \frac{\sqrt{3}}{\sqrt{4}} \quad \text{Property 2}$$

$$= \frac{\sqrt{3}}{2} \quad \sqrt{4} = 2$$

The last expression is in simplified form because it satisfies all three conditions for simplified form.

EXAMPLE 7 Write $\sqrt{\frac{5}{6}}$ in simplified form.

SOLUTION Proceeding as in Example 6, we have

$$\sqrt{\frac{5}{6}} = \frac{\sqrt{5}}{\sqrt{6}}$$

The resulting expression satisfies the second condition for simplified form since neither radical contains a fraction. It does, however, violate condition 3 since it has a radical in the denominator. Getting rid of the radical in the denominator is called *rationalizing the denominator* and is accomplished, in this case, by multiplying the numerator and denominator by $\sqrt{6}$:

$$\frac{\sqrt{5}}{\sqrt{6}} = \frac{\sqrt{5}}{\sqrt{6}} \cdot \frac{\sqrt{6}}{\sqrt{6}}$$

$$= \frac{\sqrt{30}}{\sqrt{6^2}}$$

$$= \frac{\sqrt{30}}{6}$$

EXAMPLE 8 Rationalize the denominator.

SOLUTION $\frac{4}{\sqrt{3}} = \frac{4}{\sqrt{3}} \cdot \frac{\sqrt{3}}{\sqrt{3}}$

$$= \frac{4\sqrt{3}}{\sqrt{3^2}}$$

$$= \frac{4\sqrt{3}}{3}$$

EXAMPLE 9 Rationalize the denominator.

SOLUTION $\frac{2\sqrt{3x}}{\sqrt{5y}} = \frac{2\sqrt{3x}}{\sqrt{5y}} \cdot \frac{\sqrt{5y}}{\sqrt{5y}}$

$$= \frac{2\sqrt{15xy}}{\sqrt{(5y)^2}}$$

$$= \frac{2\sqrt{15xy}}{5y}$$

---

6. Simplify $\sqrt{\frac{5}{9}}$.

7. Write $\sqrt{\frac{2}{3}}$ in simplified form.

Note The idea behind rationalizing the denominator is to produce a perfect square under the square root sign in the denominator. This is accomplished by multiplying both the numerator and denominator by the appropriate radical.

8. Rationalize the denominator.
$$\frac{5}{\sqrt{2}}$$

9. Rationalize the denominator.
$$\frac{3\sqrt{5x}}{\sqrt{2y}}$$

Answers

6. $\frac{\sqrt{5}}{3}$
7. $\frac{\sqrt{6}}{3}$
8. $\frac{5\sqrt{2}}{2}$
9. $\frac{3\sqrt{10xy}}{2y}$

10. Rationalize the denominator.

$$\frac{5}{\sqrt[3]{9}}$$

11. Simplify $\sqrt{\frac{48x^3y^4}{7z}}$.

Answers
10. $\frac{5\sqrt[3]{3}}{3}$
11. $\frac{4xy^2\sqrt{21xz}}{7z}$

Chapter 3 Rational Expressions and Rational Equations

When the denominator involves a cube root, we must multiply by a radical that will produce a perfect cube under the cube root sign in the denominator, as our next example illustrates.

EXAMPLE 10 Rationalize the denominator in $\frac{7}{\sqrt[3]{4}}$.

SOLUTION Since $4 = 2^2$, we can multiply both numerator and denominator by $\sqrt[3]{2}$ and obtain $\sqrt[3]{2^3}$ in the denominator.

$$\frac{7}{\sqrt[3]{4}} = \frac{7}{\sqrt[3]{2^2}}$$

$$= \frac{7}{\sqrt[3]{2^2}} \cdot \frac{\sqrt[3]{2}}{\sqrt[3]{2}}$$

$$= \frac{7\sqrt[3]{2}}{\sqrt[3]{2^3}}$$

$$= 7\sqrt[3]{2}$$

EXAMPLE 11 Simplify $\sqrt{\frac{12x^5y^3}{5z}}$.

SOLUTION We use property 2 to write the numerator and denominator as two separate radicals:

$$\sqrt{\frac{12x^5y^3}{5z}} = \frac{\sqrt{12x^5y^3}}{\sqrt{5z}}$$

Simplifying the numerator, we have

$$\frac{\sqrt{12x^5y^3}}{\sqrt{5z}} = \frac{\sqrt{4x^4y^2}\sqrt{3xy}}{\sqrt{5z}}$$

$$= \frac{2x^2y\sqrt{3xy}}{\sqrt{5z}}$$

To rationalize the denominator, we multiply the numerator and denominator by $\sqrt{5z}$:

$$\frac{2x^2y\sqrt{3xy}}{\sqrt{5z}} \cdot \frac{\sqrt{5z}}{\sqrt{5z}} = \frac{2x^2y\sqrt{15xyz}}{\sqrt{(5z)^2}}$$

$$= \frac{2x^2y\sqrt{15xyz}}{5z}$$

The Square Root of a Perfect Square

So far in this chapter we have assumed that all our variables are nonnegative when they appear under a square root symbol. There are times, however, when this is not the case.

Consider the following two statements:

$$\sqrt{3^2} = \sqrt{9} = 3 \quad \text{and} \quad \sqrt{(-3)^2} = \sqrt{9} = 3$$

Whether we operate on 3 or −3, the result is the same: Both expressions simplify to 3. The other operation we have worked with in the past that produces the same result is absolute value; that is,

$$|3| = 3 \quad \text{and} \quad |-3| = 3$$

3.5 Simplified Form for Radicals

This leads us to the next property of radicals.

> **Property 3 for Radicals**
> If a is a real number, then $\sqrt{a^2} = |a|$

The result of this discussion and property 3 is simply this:

If we know a is positive, then $\sqrt{a^2} = a$.

If we know a is negative, then $\sqrt{a^2} = |a|$.

If we don't know if a is positive or negative, then $\sqrt{a^2} = |a|$.

EXAMPLES Simplify each expression. Do *not* assume the variables represent positive numbers.

12. $\sqrt{9x^2} = 3|x|$
13. $\sqrt{x^3} = |x|\sqrt{x}$
14. $\sqrt{x^2 - 6x + 9} = \sqrt{(x-3)^2} = |x - 3|$
15. $\sqrt{x^3 - 5x^2} = \sqrt{x^2(x-5)} = |x|\sqrt{x-5}$

As you can see, we must use absolute value symbols when we take a square root of a perfect square, unless we know the base of the perfect square is a positive number. The same idea holds for higher even roots, but not for odd roots. With odd roots, no absolute value symbols are necessary.

EXAMPLES Simplify each expression.

16. $\sqrt[3]{(-2)^3} = \sqrt[3]{-8} = -2$
17. $\sqrt[3]{(-5)^3} = \sqrt[3]{-125} = -5$

We can extend this discussion to all roots as follows:

> **Extending Property 3 for Radicals**
> If a is a real number, then
> $\sqrt[n]{a^n} = |a|$ if n is even
> $\sqrt[n]{a^n} = a$ if n is odd

Getting Ready for Class

After reading through the preceding section, respond in your own words and in complete sentences.

1. Explain why this statement is false: "The square root of a sum is the sum of the square roots."
2. What is simplified form for an expression that contains a square root?
3. Why is it not necessarily true that $\sqrt{a^2} = a$?
4. What does it mean to rationalize the denominator in an expression?

Simplify each expression. Do *not* assume the variables represent nonnegative numbers.

12. $\sqrt{16x^2}$
13. $\sqrt{25x^3}$
14. $\sqrt{x^2 + 10x + 25}$
15. $\sqrt{2x^3 + 7x^2}$

Simplify each expression.

16. $\sqrt[3]{(-3)^3}$
17. $\sqrt[3]{(-1)^3}$

Answers
12. $4|x|$
13. $5|x|\sqrt{x}$
14. $|x + 5|$
15. $|x|\sqrt{2x + 7}$
16. -3
17. -1

Left Blank Intentionally

Problem Set 3.5

Use Property 1 for radicals to write each of the following expressions in simplified form. (Assume all variables are nonnegative through Problem 90.) [Examples 1–5]

1. $\sqrt{8}$
2. $\sqrt{32}$
3. $\sqrt{98}$
4. $\sqrt{75}$
5. $\sqrt{288}$

6. $\sqrt{128}$
7. $\sqrt{80}$
8. $\sqrt{200}$
9. $\sqrt{48}$
10. $\sqrt{27}$

11. $\sqrt{675}$
12. $\sqrt{972}$
13. $\sqrt[3]{54}$
14. $\sqrt[3]{24}$
15. $\sqrt[3]{128}$

16. $\sqrt[3]{162}$
17. $\sqrt[3]{432}$
18. $\sqrt[3]{1,536}$
19. $\sqrt[5]{64}$
20. $\sqrt[4]{48}$

21. $\sqrt{18x^3}$
22. $\sqrt{27x^5}$
23. $\sqrt{50x^5}$
24. $\sqrt{32y^7}$
25. $\sqrt{20a^3}$

26. $\sqrt{45x^2}$
27. $\sqrt{48y^6}$
28. $\sqrt{49x^5}$
29. $\sqrt[4]{32y^7}$
30. $\sqrt[5]{32y^7}$

31. $\sqrt[3]{40x^4y^7}$
32. $\sqrt[3]{128x^6y^2}$
33. $\sqrt{48a^2b^3c^4}$
34. $\sqrt{72a^4b^3c^2}$
35. $\sqrt[3]{48a^2b^3c^4}$

36. $\sqrt[3]{72a^4b^3c^2}$
37. $\sqrt[6]{64x^8y^{12}}$
38. $\sqrt[4]{32x^9y^{10}}$
39. $\sqrt[5]{243x^7y^{10}z^5}$
40. $\sqrt[5]{64x^8y^4z^{11}}$

Substitute the given number into the expression $\sqrt{b^2 - 4ac}$, and then simplify.

41. $a = 2, b = -6, c = 3$
42. $a = 6, b = 7, c = -5$
43. $a = 1, b = 2, c = 6$

44. $a = 2, b = 5, c = 3$
45. $a = \dfrac{1}{2}, b = -\dfrac{1}{2}, c = -\dfrac{5}{4}$
46. $a = \dfrac{7}{4}, b = -\dfrac{3}{4}, c = -2$

B Rationalize the denominator in each of the following expressions. [Examples 6–11]

47. $\dfrac{2}{\sqrt{3}}$
48. $\dfrac{3}{\sqrt{2}}$
49. $\dfrac{5}{\sqrt{6}}$
50. $\dfrac{7}{\sqrt{5}}$

51. $\sqrt{\dfrac{1}{2}}$
52. $\sqrt{\dfrac{1}{3}}$
53. $\sqrt{\dfrac{1}{5}}$
54. $\sqrt{\dfrac{1}{6}}$

55. $\dfrac{4}{\sqrt[3]{2}}$
56. $\dfrac{5}{\sqrt[3]{3}}$
57. $\dfrac{2}{\sqrt[3]{9}}$
58. $\dfrac{3}{\sqrt[3]{4}}$

59. $\sqrt[4]{\dfrac{3}{2x^2}}$
60. $\sqrt[4]{\dfrac{5}{3x^2}}$
61. $\sqrt[4]{\dfrac{8}{y}}$
62. $\sqrt[4]{\dfrac{27}{y}}$

63. $\sqrt[3]{\dfrac{4x}{3y}}$
64. $\sqrt[3]{\dfrac{7x}{6y}}$
65. $\sqrt[3]{\dfrac{2x}{9y}}$
66. $\sqrt[3]{\dfrac{5x}{4y}}$

67. $\sqrt[4]{\dfrac{1}{8x^3}}$
68. $\sqrt[4]{\dfrac{8}{9x^3}}$

Write each of the following in simplified form.

69. $\sqrt{\dfrac{27x^3}{5y}}$
70. $\sqrt{\dfrac{12x^5}{7y}}$
71. $\sqrt{\dfrac{75x^3y^2}{2z}}$
72. $\sqrt{\dfrac{50x^2y^3}{3z}}$

73. $\sqrt[3]{\dfrac{16a^4b^3}{9c}}$
74. $\sqrt[3]{\dfrac{54a^5b^4}{25c^2}}$
75. $\sqrt[3]{\dfrac{8x^3y^6}{9z}}$
76. $\sqrt[3]{\dfrac{27x^6y^3}{2z^2}}$

77. $\sqrt{\sqrt{x^2}}$ **78.** $\sqrt{\sqrt{2x^3}}$ **79.** $\sqrt[3]{\sqrt{xy}}$ **80.** $\sqrt{\sqrt{4x}}$

81. $\sqrt[3]{\sqrt[4]{a}}$ **82.** $\sqrt[6]{\sqrt[4]{x}}$ **83.** $\sqrt[3]{\sqrt[3]{6x^{10}}}$ **84.** $\sqrt[5]{\sqrt{x^{14}y^{11}z}}$

85. $\sqrt[4]{\sqrt[3]{a^{12}b^{24}c^{14}}}$ **86.** $\sqrt{\sqrt{4a^{17}}}$ **87.** $\sqrt[3]{\sqrt[5]{3a^{17}b^{16}c^{30}}}$ **88.** $\left(\sqrt{\sqrt[4]{x^4y^8z^9}}\right)^2$

89. $\left(\sqrt{\sqrt[3]{8ab^6}}\right)^2$ **90.** $\left(\sqrt[4]{\sqrt[3]{16x^8y^{12}z^3}}\right)^3$

Simplify each expression. Do *not* assume the variables represent positive numbers. [Examples 12–15]

91. $\sqrt{25x^2}$ **92.** $\sqrt{49x^2}$ **93.** $\sqrt{27x^3y^2}$ **94.** $\sqrt{40x^3y^2}$

95. $\sqrt{x^2 - 10x + 25}$ **96.** $\sqrt{x^2 - 16x + 64}$ **97.** $\sqrt{4x^2 + 12x + 9}$ **98.** $\sqrt{16x^2 + 40x + 25}$

99. $\sqrt{4a^4 + 16a^3 + 16a^2}$ **100.** $\sqrt{9a^4 + 18a^3 + 9a^2}$ **101.** $\sqrt{4x^3 - 8x^2}$ **102.** $\sqrt{18x^3 - 9x^2}$

103. Show that the statement $\sqrt{a + b} = \sqrt{a} + \sqrt{b}$ is not true by replacing *a* with 9 and *b* with 16 and simplifying both sides.

104. Find a pair of values for *a* and *b* that will make the statement $\sqrt{a + b} = \sqrt{a} + \sqrt{b}$ true.

Applying the Concepts

105. Diagonal Distance The distance d between opposite corners of a rectangular room with length l and width w is given by

$$d = \sqrt{l^2 + w^2}$$

How far is it between opposite corners of a living room that measures 10 by 15 feet?

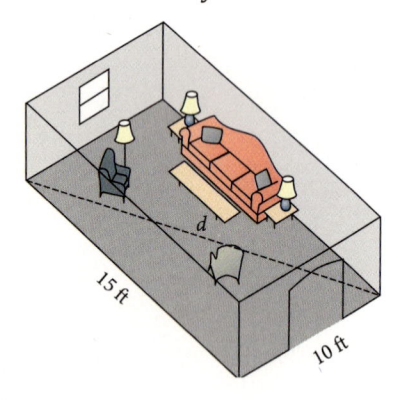

106. Radius of a Sphere The radius r of a sphere with volume V can be found by using the formula

$$r = \sqrt[3]{\frac{3V}{4\pi}}$$

Find the radius of a sphere with volume 9 cubic feet. Write your answer in simplified form. (Use $\frac{22}{7}$ for π.)

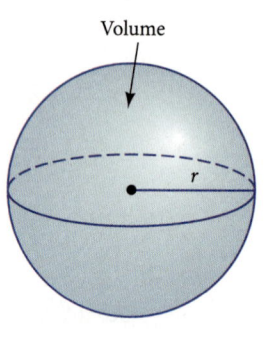

107. Diagonal of a Box The length of the diagonal of a rectangular box with length l, width w, and height h is given by $d = \sqrt{l^2 + w^2 + h^2}$.
  a. Find the length of the diagonal of a rectangular box that is 3 feet wide, 4 feet long, and 12 feet high.

  b. Find the length of the diagonal of a rectangular box that is 2 feet wide, 4 feet high, and 6 feet long.

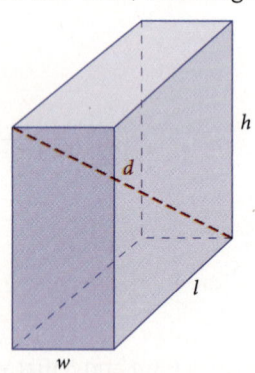

108. Distance to the Horizon If you are at a point k miles above the surface of the Earth, the distance you can see, in miles, is approximated by the equation $d = \sqrt{8000k + k^2}$.
  a. How far can you see from a point that is 1 mile above the surface of the Earth?

  b. How far can you see from a point that is 2 miles above the surface of the Earth?

  c. How far can you see from a point that is 3 miles above the surface of the Earth?

Getting Ready for the Next Section

Simplify the following.

109. $5x - 4x + 6x$

110. $12x + 8x - 7x$

111. $35xy^2 - 8xy^2$

112. $20a^2b + 33a^2b$

113. $\dfrac{1}{2}x + \dfrac{1}{3}x$

114. $\dfrac{2}{3}x + \dfrac{5}{8}x$

Write in simplified form for radicals.

115. $\sqrt{18}$

116. $\sqrt{8}$

117. $\sqrt{75xy^3}$

118. $\sqrt{12xy}$

119. $\sqrt[3]{8a^4b^2}$

120. $\sqrt[3]{27ab^2}$

Maintaining Your Skills

Perform the indicated operations.

121. $\dfrac{8xy^3}{9x^2y} \div \dfrac{16x^2y^2}{18xy^3}$

122. $\dfrac{25x^2}{5y^4} \cdot \dfrac{30y^3}{2x^5}$

123. $\dfrac{12a^2 - 4a - 5}{2a + 1} \cdot \dfrac{7a + 3}{42a^2 - 17a - 15}$

124. $\dfrac{20a^2 - 7a - 3}{4a + 1} \cdot \dfrac{25a^2 - 5a - 6}{5a + 2}$

125. $\dfrac{8x^3 + 27}{27x^3 + 1} \div \dfrac{6x^2 + 7x - 3}{9x^2 - 1}$

126. $\dfrac{27x^3 + 8}{8x^3 + 1} \div \dfrac{6x^2 + x - 2}{4x^2 - 1}$

Extending the Concepts

Factor each radicand into the product of prime factors. Then simplify each radical.

127. $\sqrt[3]{8,640}$

128. $\sqrt{8,640}$

129. $\sqrt[3]{10,584}$

130. $\sqrt{10,584}$

Assume a is a positive number, and rationalize each denominator.

131. $\dfrac{1}{\sqrt[10]{a^3}}$

132. $\dfrac{1}{\sqrt[12]{a^7}}$

133. $\dfrac{1}{\sqrt[20]{a^{11}}}$

134. $\dfrac{1}{\sqrt[15]{a^{13}}}$

135. Show that the two expressions $\sqrt{x^2 + 1}$ and $x + 1$ are not, in general, equal to each other by graphing $y = \sqrt{x^2 + 1}$ and $y = x + 1$ in the same viewing window.

136. Show that the two expressions $\sqrt{x^2 + 9}$ and $x + 3$ are not, in general, equal to each other by graphing $y = \sqrt{x^2 + 9}$ and $y = x + 3$ in the same viewing window.

137. Approximately how far apart are the graphs in Problem 135 when $x = 2$?

138. Approximately how far apart are the graphs in Problem 136 when $x = 2$?

139. For what value of x are the expressions $\sqrt{x^2 + 1}$ and $x + 1$ equal?

140. For what value of x are the expressions $\sqrt{x^2 + 9}$ and $x + 3$ equal?

The Quadratic Formula 3.6

We will begin this section by deriving the quadratic formula. The *quadratic formula* is a very useful tool in mathematics. It allows us to solve all types of quadratic equations.

Objectives

A Solve quadratic equations by the quadratic formula.

B Solve application problems using quadratic equations.

A The Quadratic Theorem

The Quadratic Theorem
For any quadratic equation in the form $ax^2 + bx + c = 0$, where $a \neq 0$, the two solutions are
$$x = \frac{-b + \sqrt{b^2 - 4ac}}{2a} \quad \text{and} \quad x = \frac{-b - \sqrt{b^2 - 4ac}}{2a}$$

PROOF We will derive the quadratic theorem by beginning with a quadratic equation in standard form.

$$ax^2 + bx + c = 0$$

$$ax^2 + bx = -c \quad \text{Add } -c \text{ to both sides}$$

$$x^2 + \frac{b}{a}x = -\frac{c}{a} \quad \text{Divide both sides by } a$$

Next, we add the square of $\frac{1}{2}$ of $\frac{b}{a}$ to both sides. $\left(\frac{1}{2} \text{ of } \frac{b}{a} \text{ is } \frac{b}{2a}\right)$

$$x^2 + \frac{b}{a}x + \left(\frac{b}{2a}\right)^2 = -\frac{c}{a} + \left(\frac{b}{2a}\right)^2$$

We now simplify the right side as a separate step. We square the second term and combine the two terms by writing each with the least common denominator $4a^2$.

$$-\frac{c}{a} + \left(\frac{b}{2a}\right)^2 = -\frac{c}{a} + \frac{b^2}{4a^2} = \frac{4a}{4a}\left(\frac{-c}{a}\right) + \frac{b^2}{4a^2} = \frac{-4ac + b^2}{4a^2}$$

It is convenient to write this last expression as

$$\frac{b^2 - 4ac}{4a^2}$$

Continuing with the proof, we have

$$x^2 + \frac{b}{a}x + \left(\frac{b}{2a}\right)^2 = \frac{b^2 - 4ac}{4a^2}$$

$$\left(x + \frac{b}{2a}\right)^2 = \frac{b^2 - 4ac}{4a^2} \quad \text{Write left side as a binomial square}$$

$$x + \frac{b}{2a} = \pm\frac{\sqrt{b^2 - 4ac}}{2a} \quad \text{Take the square root of both sides.}$$

$$x = -\frac{b}{2a} \pm \frac{\sqrt{b^2 - 4ac}}{2a} \quad \text{Add } -\frac{b}{2a} \text{ to both sides}$$

$$= \frac{-b \pm \sqrt{b^2 - 4ac}}{2a}$$

Note Although this text has not covered the completing the square method of solving quadratic equations, it is shown here as part of the proof.

Our proof is now complete. What we have is this: if our equation is in the form $ax^2 + bx + c = 0$ (standard form), where $a \neq 0$, the two solutions are always given by the formula

$$x = \frac{-b \pm \sqrt{b^2 - 4ac}}{2a}$$

PRACTICE PROBLEMS

1. Solve $6x^2 + 7x + 2 = 0$ using the quadratic formula.

Chapter 3 Rational Expressions and Rational Equations

This formula is known as the *quadratic formula*. If we substitute the coefficients a, b, and c of any quadratic equation in standard form into the formula, we need only perform some basic arithmetic to arrive at the solution set.

EXAMPLE 1 Use the quadratic formula to solve $6x^2 + 7x - 5 = 0$.

SOLUTION Using $a = 6$, $b = 7$, and $c = -5$ in the formula

$$x = \frac{-b \pm \sqrt{b^2 - 4ac}}{2a}$$

we have

$$x = \frac{-7 \pm \sqrt{49 - 4(6)(-5)}}{2(6)}$$

or

$$x = \frac{-7 \pm \sqrt{49 + 120}}{12}$$

$$= \frac{-7 \pm \sqrt{169}}{12}$$

$$= \frac{-7 \pm 13}{12}$$

We separate the last equation into the two statements

$$x = \frac{-7 + 13}{12} \quad \text{or} \quad x = \frac{-7 - 13}{12}$$

$$x = \frac{1}{2} \quad \text{or} \quad x = -\frac{5}{3}$$

The solution set is $\left\{\frac{1}{2}, -\frac{5}{3}\right\}$.

Whenever the solutions to a quadratic equation are rational numbers, as they are in Example 1, it means that the original equation was solvable by factoring. To illustrate, let's solve the equation from Example 1 again, but this time by factoring:

$6x^2 + 7x - 5 = 0$	Equation in standard form
$(3x + 5)(2x - 1) = 0$	Factor the left side
$3x + 5 = 0 \quad \text{or} \quad 2x - 1 = 0$	Set factors equal to 0
$x = -\frac{5}{3} \quad \text{or} \quad x = \frac{1}{2}$	

When an equation can be solved by factoring, then factoring is usually the faster method of solution. It is best to try to factor first, and then if you have trouble factoring, go to the quadratic formula. It always works.

2. Solve $\frac{x^2}{2} + x = \frac{1}{3}$.

EXAMPLE 2 Solve $\frac{x^2}{3} - x = -\frac{1}{2}$.

SOLUTION Multiplying through by 6 and writing the result in standard form, we have

$$2x^2 - 6x + 3 = 0$$

the left side of which is not factorable. Therefore, we use the quadratic formula with $a = 2$, $b = -6$, and $c = 3$. The two solutions are given by

$$x = \frac{-(-6) \pm \sqrt{36 - 4(2)(3)}}{2(2)}$$

$$= \frac{6 \pm \sqrt{12}}{4}$$

$$= \frac{6 \pm 2\sqrt{3}}{4} \qquad \sqrt{12} = \sqrt{4 \cdot 3} = \sqrt{4}\sqrt{3} = 2\sqrt{3}$$

Answer
1. $-\frac{1}{2}, -\frac{2}{3}$

We can reduce this last expression to lowest terms by factoring 2 from the numerator and denominator and then dividing the numerator and denominator by 2:

$$x = \frac{\cancel{2}(3 \pm \sqrt{3})}{\cancel{2} \cdot 2} = \frac{3 \pm \sqrt{3}}{2}$$

B Applications

EXAMPLE 3 If an object is thrown downward with an initial velocity of 20 feet per second, the distance $s(t)$, in feet, it travels in t seconds is given by the function $s(t) = 20t + 16t^2$. How long does it take the object to fall 40 feet?

SOLUTION We let $s(t) = 40$, and solve for t:

When $s(t) = 40$
the function $s(t) = 20t + 16t^2$
becomes $40 = 20t + 16t^2$
or $16t^2 + 20t - 40 = 0$
$4t^2 + 5t - 10 = 0$ Divide by 4

Using the quadratic formula, we have

$$t = \frac{-5 \pm \sqrt{25 - 4(4)(-10)}}{2(4)}$$

$$= \frac{-5 \pm \sqrt{185}}{8}$$

$$= \frac{-5 + \sqrt{185}}{8} \quad \text{or} \quad \frac{-5 - \sqrt{185}}{8}$$

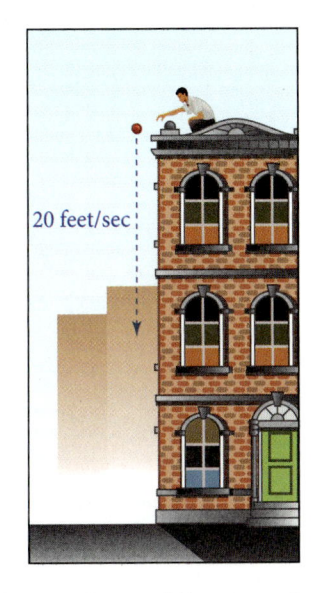

The second solution is impossible since it is a negative number and time t must be positive. It takes

$$t = \frac{-5 + \sqrt{185}}{8} \quad \text{or approximately} \quad \frac{-5 + 13.60}{8} \approx 1.08 \text{ seconds}$$

for the object to fall 40 feet.

Recall that the relationship between profit, revenue, and cost is given by the formula

$$P(x) = R(x) - C(x)$$

where $P(x)$ is the profit, $R(x)$ is the total revenue, and $C(x)$ is the total cost of producing and selling x items.

EXAMPLE 4 A company produces and sells copies of an accounting program for home computers. The total weekly cost (in dollars) to produce x copies of the program is $C(x) = 8x + 500$, and the weekly revenue for selling all x copies of the program is $R(x) = 35x - 0.1x^2$. How many programs must be sold each week for the weekly profit to be $1,200?

SOLUTION Substituting the given expressions for $R(x)$ and $C(x)$ in the equation $P(x) = R(x) - C(x)$, we have a polynomial in x that represents the weekly profit $P(x)$:

$$P(x) = R(x) - C(x)$$
$$= 35x - 0.1x^2 - (8x + 500)$$

---

3. An object thrown upward with an initial velocity of 32 feet per second rises and falls according to the equation

$$s = 32t - 16t^2$$

where s is the height of the object above the ground at any time t. At what times t will the object be 12 feet above the ground?

4. Use the information in Example 4 to find the number of programs the company must sell each week for its weekly profit to be $1,320.

Answers

2. $\dfrac{-3 \pm \sqrt{15}}{3}$

3. $\dfrac{3}{2}$ seconds, $\dfrac{1}{2}$ second

$$= 35x - 0.1x^2 - 8x - 500$$
$$= -500 + 27x - 0.1x^2$$

Setting this expression equal to 1,200, we have a quadratic equation to solve that gives us the number of programs x that need to be sold each week to bring in a profit of $1,200:

$$1,200 = -500 + 27x - 0.1x^2$$

We can write this equation in standard form by adding the opposite of each term on the right side of the equation to both sides of the equation. Doing so produces the following equation:

$$0.1x^2 - 27x + 1,700 = 0$$

Applying the quadratic formula to this equation with $a = 0.1$, $b = -27$, and $c = 1,700$, we have

$$x = \frac{27 \pm \sqrt{(-27)^2 - 4(0.1)(1,700)}}{2(0.1)}$$
$$= \frac{27 \pm \sqrt{729 - 680}}{0.2}$$
$$= \frac{27 \pm \sqrt{49}}{0.2}$$
$$= \frac{27 \pm 7}{0.2}$$

Writing this last expression as two separate expressions, we have our two solutions:

$$x = \frac{27 + 7}{0.2} \quad \text{or} \quad x = \frac{27 - 7}{0.2}$$
$$= \frac{34}{0.2} \qquad\qquad\qquad = \frac{20}{0.2}$$
$$= 170 \qquad\qquad\qquad\quad = 100$$

The weekly profit will be $1,200 if the company produces and sells 100 programs or 170 programs.

    What is interesting about the equation we solved in Example 4 is that it has rational solutions, meaning it could have been solved by factoring. But looking back at the equation, factoring does not seem like a reasonable method of solution because the coefficients are either very large or very small. So, there are times when using the quadratic formula is a faster method of solution, even though the equation you are solving is factorable.

Answer
4. 130 or 140 programs

USING TECHNOLOGY

Graphing Calculators:
More About Example 3

We can solve the problem discussed in Example 3 by graphing the function $Y_1 = 20X + 16X^2$ in a window with X from 0 to 2 (because X is taking the place of *t* and we know *t* is a positive quantity) and Y from 0 to 50 (because we are looking for X when Y_1 is 40). Graphing Y_1 gives a graph similar to the graph in Figure 1. Using the Zoom and Trace features at $Y_1 = 40$ gives us $X = 1.08$ to the nearest hundredth, matching the results we obtained by solving the original equation algebraically.

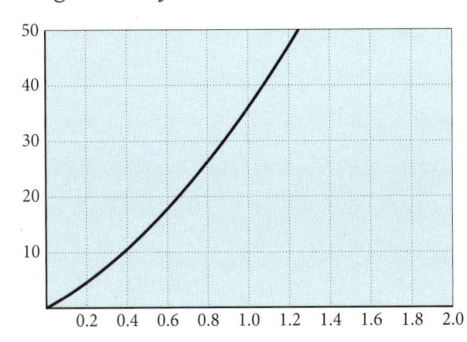

FIGURE 1

More About Example 4
To visualize the functions in Example 4, we set up our calculator this way:

$Y_1 = 35X - 0.1X^2$ Revenue function
$Y_2 = 8X + 500$ Cost function
$Y_3 = Y_1 - Y_2$ Profit function

Window: X from 0 to 350, Y from 0 to 3500

Graphing these functions produces graphs similar to the ones shown in Figure 2. The lower graph is the graph of the profit function. Using the Zoom and Trace features on the lower graph at $Y_3 = 1,200$ produces two corresponding values of X, 170 and 100, which match the results in Example 4.

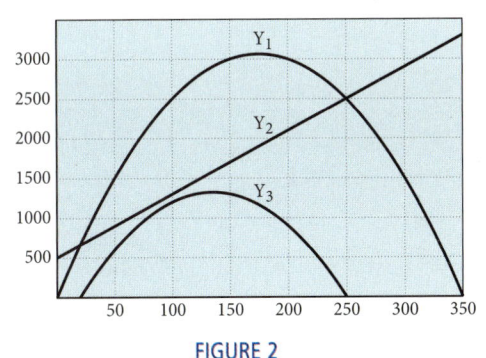

FIGURE 2

Getting Ready for Class

After reading through the preceding section, respond in your own words and in complete sentences.

1. What is the quadratic formula?
2. Under what circumstances should the quadratic formula be applied?
3. When would the quadratic formula result in rational solutions?
4. When will the quadratic formula result in only one solution?

Left Blank Intentionally

Problem Set 3.6

A Solve each equation in each problem using the quadratic formula. [Example 1]

1. a. $3x^2 + 4x - 2 = 0$

    b. $3x^2 - 4x - 2 = 0$

    c. $2x^2 + 4x - 3 = 0$

    d. $2x^2 + 6x - 3 = 0$

2. a. $3x^2 + 6x - 2 = 0$

    b. $3x^2 - 6x - 2 = 0$

    c. $3x^2 + 6x + 2 = 0$

    d. $2x^2 + 6x + 3 = 0$

3. a. $4x^2 + 2x - 4 = 0$

    b. $3x^2 - 8x + 2 = 0$

    c. $2x^2 - 5x - 1 = 0$

    d. $3x^2 - 2x - 6 = 0$

    e. $3x^2 + 2x - 3 = 0$

4. a. $4x^2 + 5x - 3 = 0$

    b. $4x^2 + 2x - 1 = 0$

    c. $2x^2 + 6x - 1 = 0$

    d. $2x^2 + 6x - 3 = 0$

    e. $x^2 + 8x + 3 = 0$

A Solve each equation. Use factoring or the quadratic formula, whichever is appropriate. (Try factoring first. If you have any difficulty factoring, then go right to the quadratic formula.) [Examples 1–2]

5. $\frac{1}{6}x^2 - \frac{1}{2}x + \frac{1}{3} = 0$

6. $\frac{1}{4}x^2 + \frac{1}{4}x - \frac{1}{2} = 0$

7. $\frac{x^2}{2} - 1 = \frac{2x}{3}$

8. $\frac{x^2}{2} - \frac{3}{2} = -\frac{2x}{3}$

9. $y^2 - 5y = 0$

10. $2y^2 + 10y = 0$

11. $30x^2 + 40x = 0$

12. $50x^2 - 20x = 0$

13. $\dfrac{2t^2}{3} - t = -\dfrac{1}{6}$

14. $\dfrac{t^2}{3} - \dfrac{t}{2} = \dfrac{1}{2}$

15. $0.01x^2 + 0.06x - 0.08 = 0$

16. $0.02x^2 - 0.04x - 0.05 = 0$

17. $2x - 3 = -2x^2$

18. $2x + 3 = 3x^2$

19. $100x^2 - 200x + 100 = 0$

20. $100x^2 - 600x + 900 = 0$

21. $\dfrac{1}{2}r^2 = \dfrac{1}{6}r + \dfrac{2}{3}$

22. $\dfrac{1}{4}r^2 = \dfrac{2}{5}r + \dfrac{1}{10}$

23. $(x - 3)(x - 5) = 1$

24. $(x - 3)(x + 1) = 6$

25. $(x + 3)^2 + (x - 8)(x - 1) = 16$

26. $(x - 2)^2 + (x + 2)(x + 1) = 9$

27. $\dfrac{x^2}{3} - \dfrac{5x}{6} = \dfrac{1}{2}$

28. $\dfrac{x^2}{6} - \dfrac{5}{6} = -\dfrac{x}{3}$

A Multiply both sides of each equation by its LCD. Then solve the resulting equation.

29. $\dfrac{1}{x + 1} - \dfrac{1}{x} = -\dfrac{1}{2}$

30. $\dfrac{1}{x + 1} + \dfrac{1}{x} = \dfrac{1}{3}$

31. $\dfrac{1}{y - 1} + \dfrac{1}{y + 1} = 1$

32. $\dfrac{2}{y + 2} + \dfrac{3}{y - 2} = 1$

33. $\dfrac{1}{x + 2} + \dfrac{1}{x + 3} = 1$

34. $\dfrac{1}{x + 3} + \dfrac{1}{x + 4} = 1$

35. $\dfrac{6}{r^2 - 1} - \dfrac{1}{2} = \dfrac{1}{r + 1}$

36. $2 + \dfrac{5}{r - 1} = \dfrac{12}{(r - 1)^2}$

A Each of the following equations has three solutions. Look for the greatest common factor, then use the quadratic formula to find all solutions.

37. $2x^3 + 2x^2 - 3x = 0$

38. $6x^3 - 4x^2 - 6x = 0$

39. $3y^4 = 6y^3 + 6y^2$

40. $4y^4 = 16y^3 + 20y^2$

41. $6t^5 + 4t^4 = 2t^3$

42. $8t^5 + 2t^4 = 10t^3$

43. Which two of the expressions below are equivalent?
  a. $\dfrac{6 + 2\sqrt{3}}{4}$
  b. $\dfrac{3 + \sqrt{3}}{2}$
  c. $6 + \dfrac{\sqrt{3}}{2}$

44. Which two of the expressions below are equivalent?
  a. $\dfrac{8 - 4\sqrt{2}}{4}$
  b. $2 - 4\sqrt{2}$
  c. $2 - \sqrt{2}$

45. Solve $3x^2 - 5x = 0$
  a. By factoring
  b. By the quadratic formula

46. Solve $3x^2 + 23x - 70 = 0$
  a. By factoring
  b. By the quadratic formula

47. Can the equation $x^2 - 4x - 7 = 0$ be solved by factoring? Solve the equation.

48. Can the equation $x^2 = 5$ be solved by factoring? Solve the equation.

49. Is $x = 2 + \sqrt{5}$ a solution to $x^2 - 4x = 1$?

50. Is $x = 2 - 2\sqrt{2}$ a solution to $(x - 2)^2 = 8$?

51. Let $y = x^2 - 2x - 3$. Find all values for the variable x, which produce the following values of y.
  a. $y = 0$
  b. $y = 11$
  c. $y = -2x + 1$
  d. $y = 2x + 1$

52. Let $y = x^2 + 16$. Find all values for the variable x, which produce the following values of y.
  a. $y = 24$
  b. $y = 20$
  c. $y = 8x$
  d. $y = -8x$

53. Let $y = \dfrac{10}{x^2}$. Find all values for the variable x, which produce the following values of y.
  a. $y = 3 + \dfrac{1}{x}$
  b. $y = 8 - \dfrac{17}{x^2}$
  c. $y = 0$
  d. $y = 10$

54. Let $h(x) = \dfrac{x + 2}{x}$. Find all values for the variable x, which produce the following values of y.
  a. $y = 2$
  b. $y = x + 2$
  c. $y = x - 2$
  d. $y = \dfrac{x - 2}{4}$

A Applying the Concepts [Examples 3–4]

55. Falling Object An object is thrown downward with an initial velocity of 5 feet per second. The relationship between the distance s it travels and time t is given by $s = 5t + 16t^2$. How long does it take the object to fall 74 feet?

56. Coin Toss A coin is tossed upward with an initial velocity of 32 feet per second from a height of 16 feet above the ground. The equation giving the object's height h at any time t is $h = 16 + 32t - 16t^2$. Does the object ever reach a height of 32 feet?

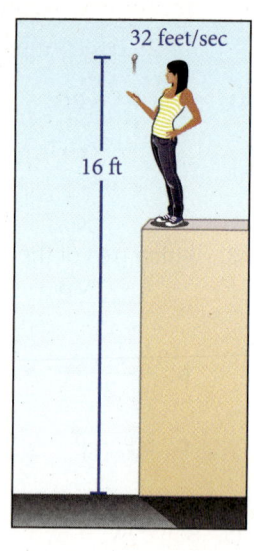

57. Profit The total cost (in dollars) for a company to manufacture and sell x items per week is $C = 60x + 300$, whereas the revenue brought in by selling all x items is $R = 100x - 0.5x^2$. How many items must be sold to obtain a weekly profit of $300?

58. Profit Suppose a company manufactures and sells x picture frames each month with a total cost of $C = 1{,}200 + 3.5x$ dollars. If the revenue obtained by selling x frames is $R = 9x - 0.002x^2$, find the number of frames it must sell each month if its monthly profit is to be $2,300.

59. Photograph Cropping The following figure shows a photographic image on a 10.5-centimeter by 8.2-centimeter background. The overall area of the background is to be reduced to 80% of its original area by cutting off (cropping) equal strips on all four sides. What is the width of the strip that is cut from each side?

60. Area of a Garden A garden measures 20.3 meters by 16.4 meters. To double the area of the garden, strips of equal width are added to all four sides.
  a. Draw a diagram that illustrates these conditions.
  b. What are the new overall dimensions of the garden?

61. Area and Perimeter A rectangle has a perimeter of 20 yards and an area of 15 square yards.
   a. Write two equations that state these facts in terms of the rectangle's length, l, and its width, w.
   b. Solve the two equations from part (a) to determine the actual length and width of the rectangle.
   c. Explain why two answers are possible to part (b).

62. Population Size Writing in 1829, former President James Madison made some predictions about the growth of the population of the United States. The populations he predicted fit the equation

$$y = 0.029x^2 - 1.39x + 42$$

where y is the population in millions of people x years from 1829.
   a. Use the equation to determine the approximate year President Madison would have predicted that the U.S. population would reach 100,000,000.
   b. If the U.S. population in 2006 was approximately 300 million, were President Madison's predictions accurate in the long term? Explain why or why not.

Maintaining Your Skills

Divide, using long division.

63. $\dfrac{8y^2 - 26y - 9}{2y - 7}$

64. $\dfrac{6y^2 + 7y - 18}{3y - 4}$

65. $\dfrac{x^3 + 9x^2 + 26x + 24}{x + 2}$

66. $\dfrac{x^3 + 6x^2 + 11x + 6}{x + 3}$

Getting Ready for the Next Section

Find the value of $b^2 - 4ac$ when

67. $a = 1, b = -3, c = -40$
68. $a = 2, b = 3, c = 4$
69. $a = 4, b = 12, c = 9$
70. $a = -3, b = 8, c = -1$

Solve.

71. $k^2 - 144 = 0$

72. $36 - 20k = 0$

Multiply.

73. $(x - 3)(x + 2)$
74. $(t - 5)(t + 5)$
75. $(x - 3)(x - 3)(x + 2)$
76. $(t - 5)(t + 5)(t - 3)$

Extending the Concepts

So far, all the equations we have solved have had coefficients that were rational numbers. Here are some equations that have irrational coefficients. Solve each equation.

77. $x^2 + \sqrt{3}x - 6 = 0$
78. $x^2 - \sqrt{5}x - 5 = 0$
79. $\sqrt{2}x^2 + 2x - \sqrt{2} = 0$

80. $\sqrt{7}x^2 + 2\sqrt{2}x - \sqrt{7} = 0$
81. $\sqrt{3}x^2 - 4\sqrt{2}x + \sqrt{3} = 0$
82. $\sqrt{5}x^2 - 3\sqrt{3}x + \sqrt{5} = 0$

Chapter 3 Summary

Rational Numbers and Expressions [3.1]

A *rational number* is any number that can be expressed as the ratio of two integers:

$$\text{Rational numbers} = \left\{ \frac{a}{b} \,\Big|\, a \text{ and } b \text{ are integers}, b \neq 0 \right\}$$

A *rational expression* is any quantity that can be expressed as the ratio of two polynomials:

$$\text{Rational expressions} = \left\{ \frac{P}{Q} \,\Big|\, P \text{ and } Q \text{ are polynomials}, Q \neq 0 \right\}$$

Properties of Rational Expressions [3.1]

If P, Q, and K are polynomials with $Q \neq 0$ and $K \neq 0$, then

$$\frac{P}{Q} = \frac{PK}{QK} \quad \text{and} \quad \frac{P}{Q} = \frac{P/K}{Q/K}$$

which is to say that multiplying or dividing the numerator and denominator of a rational expression by the same nonzero quantity always produces an equivalent rational expression.

Reducing to Lowest Terms [3.1]

To reduce a rational expression to lowest terms, we first factor the numerator and denominator and then divide the numerator and denominator by any factors they have in common.

Multiplication [3.2]

To multiply two rational numbers or rational expressions, multiply numerators and multiply denominators. In symbols,

$$\frac{P}{Q} \cdot \frac{R}{S} = \frac{PR}{QS} \quad (Q \neq 0 \text{ and } S \neq 0)$$

In practice, we don't really multiply, but rather, we factor and then divide out common factors.

Division [3.2]

To divide one rational expression by another, we use the definition of division to rewrite our division problem as an equivalent multiplication problem. To divide by a rational expression we multiply by its reciprocal. In symbols,

$$\frac{P}{Q} \div \frac{R}{S} = \frac{P}{Q} \cdot \frac{S}{R} = \frac{PS}{QR} \quad (Q \neq 0, S \neq 0, R \neq 0)$$

EXAMPLES

1. $\frac{3}{4}$ is a rational number. $\frac{x-3}{x^2-9}$ is a rational expression.

2. $\frac{x-3}{x^2-9} = \frac{\cancel{x-3}}{\cancel{(x-3)}(x+3)}$
$= \frac{1}{x+3}$

3. $\frac{x+1}{x^2-4} \cdot \frac{x+2}{3x+3}$
$= \frac{\cancel{(x+1)}\cancel{(x+2)}}{(x-2)\cancel{(x+2)}(3)\cancel{(x+1)}}$
$= \frac{1}{3(x-2)}$

4. $\frac{x^2-y^2}{x^3+y^3} \div \frac{x-y}{x^2-xy+y^2}$
$= \frac{x^2-y^2}{x^3+y^3} \cdot \frac{x^2-xy+y^2}{x-y}$
$= \frac{\cancel{(x+y)}\cancel{(x-y)}\cancel{(x^2-xy+y^2)}}{\cancel{(x+y)}\cancel{(x^2-xy+y^2)}\cancel{(x-y)}}$
$= 1$

Least Common Denominator [3.3]

5. The LCD for $\frac{2}{x-3}$ and $\frac{3}{5}$ is $5(x-3)$.

The *least common denominator*, LCD, for a set of denominators is the smallest quantity divisible by each of the denominators.

Addition and Subtraction [3.3]

6. $\frac{2}{x-3} + \frac{3}{5}$

$= \frac{2}{x-3} \cdot \frac{5}{5} + \frac{3}{5} \cdot \frac{x-3}{x-3}$

$= \frac{3x+1}{5(x-3)}$

If P, Q, and R represent polynomials, $R \neq 0$, then

$$\frac{P}{R} + \frac{Q}{R} = \frac{P+Q}{R} \quad \text{and} \quad \frac{P}{R} - \frac{Q}{R} = \frac{P-Q}{R}$$

When adding or subtracting rational expressions with different denominators, we must find the LCD for all denominators and change each rational expression to an equivalent expression that has the LCD.

Equations Involving Rational Expressions [3.4]

7. Solve $\frac{x}{2} + 3 = \frac{1}{3}$.

$6\left(\frac{x}{2}\right) + 6 \cdot 3 = 6 \cdot \frac{1}{3}$

$3x + 18 = 2$

$x = -\frac{16}{3}$

To solve an equation involving rational expressions, we first find the LCD for all denominators appearing on either side of the equation. We then multiply both sides by the LCD to clear the equation of all fractions and solve as usual.

Properties of Radicals [3.5]

8. $\sqrt{4 \cdot 5} = \sqrt{4}\sqrt{5} = 2\sqrt{5}$

$\sqrt{\frac{7}{9}} = \frac{\sqrt{7}}{\sqrt{9}} = \frac{\sqrt{7}}{3}$

If a and b are nonnegative real numbers whenever n is even, then

1. $\sqrt[n]{ab} = \sqrt[n]{a}\sqrt[n]{b}$
2. $\sqrt[n]{\frac{a}{b}} = \frac{\sqrt[n]{a}}{\sqrt[n]{b}}$ $(b \neq 0)$
3. $\sqrt[n]{a^n} = |a|$

Simplified Form for Radicals [3.5]

9. $\sqrt{\frac{4}{5}} = \frac{\sqrt{4}}{\sqrt{5}}$

$= \frac{2}{\sqrt{5}} \cdot \frac{\sqrt{5}}{\sqrt{5}}$

$= \frac{2\sqrt{5}}{5}$

A radical expression is said to be in *simplified form*

1. If there is no factor of the radicand that can be written as a power greater than or equal to the index;
2. If there are no fractions under the radical sign; and
3. If there are no radicals in the denominator.

The Quadratic Formula [3.6]

10. If $2x^2 + 3x - 4 = 0$, then

$x = \frac{-3 \pm \sqrt{9 - 4(2)(-4)}}{2(2)}$

$= \frac{-3 \pm \sqrt{41}}{4}$

For any quadratic equation in the form $ax^2 + bx + c = 0$, $a \neq 0$, the two solutions are

$$x = \frac{-b \pm \sqrt{b^2 - 4ac}}{2a}$$

This last equation is known as the *quadratic formula*.

Chapter 3 Review

Reduce to lowest terms. [3.1]

1. $\dfrac{125x^4yz^3}{35x^2y^4z^3}$

2. $\dfrac{a^3 - ab^2}{4a + 4b}$

3. $\dfrac{x^2 - 25}{x^2 + 10x + 25}$

4. $\dfrac{ax + x - 5a - 5}{ax - x - 5a + 5}$

Multiply and divide as indicated. [3.2]

5. $\dfrac{3}{4} \cdot \dfrac{12}{15} \div \dfrac{1}{3}$

6. $\dfrac{15x^2y}{8xy^2} \div \dfrac{10xy}{4x}$

7. $\dfrac{x^3 - 1}{x^4 - 1} \cdot \dfrac{x^2 - 1}{x^2 + x + 1}$

8. $\dfrac{a^2 + 5a + 6}{a + 1} \cdot \dfrac{a + 5}{a^2 + 2a - 3} \div \dfrac{a^2 + 7a + 10}{a^2 - 1}$

9. $\dfrac{ax + bx + 2a + 2b}{ax - 3a + bx - 3b} \div \dfrac{ax - bx - 2a + 2b}{ax - bx - 3a + 3b}$

10. $(4x^2 - 9) \cdot \dfrac{x + 3}{2x + 3}$

Add and subtract as indicated. [3.3]

11. $\dfrac{3}{5} - \dfrac{1}{10} + \dfrac{8}{15}$

12. $\dfrac{5}{x - 5} - \dfrac{x}{x - 5}$

13. $\dfrac{1}{x} + \dfrac{1}{x^2} + \dfrac{1}{x^3}$

14. $\dfrac{8}{y^2 - 16} - \dfrac{7}{y^2 - y - 12}$

15. $\dfrac{x - 2}{x^2 + 5x + 4} - \dfrac{x - 4}{2x^2 + 12x + 16}$

16. $3 + \dfrac{4}{5x - 2}$

Solve each equation. [3.4]

17. $\dfrac{3}{x - 1} = \dfrac{3}{5}$

18. $\dfrac{x + 1}{3} + \dfrac{x - 3}{4} = \dfrac{1}{6}$

19. $\dfrac{5}{y + 1} = \dfrac{4}{y + 2}$

20. $\dfrac{x + 6}{x + 3} - 2 = \dfrac{3}{x + 3}$

21. $\dfrac{4}{x^2 - x - 12} + \dfrac{1}{x^2 - 9} = \dfrac{2}{x^2 - 7x + 12}$

22. $\dfrac{a + 4}{a^2 + 5a} = \dfrac{-2}{a^2 - 25}$

Write each expression in simplified form for radicals. (Assume all variables represent nonnegative numbers.) [3.5]

23. $\sqrt{12}$

24. $\sqrt{50}$

25. $\sqrt[3]{16}$

26. $\sqrt{18x^2}$

27. $80a^3b^4c^2$

28. $\sqrt[4]{32a^4b^5c^6}$

Rationalize the denominator in each expression. [3.5]

29. $\dfrac{3}{\sqrt{2}}$

30. $\dfrac{6}{\sqrt[3]{2}}$

Write each expression in simplified form. (Assume all variables represent positive numbers.) [3.5]

31. $\sqrt{\dfrac{48x^3}{7y}}$

32. $\sqrt[3]{\dfrac{40x^2y^3}{3z}}$

Solve each equation. [3.6]

33. $\dfrac{1}{6}x^2 + \dfrac{1}{2}x - \dfrac{5}{3} = 0$

34. $8x^2 - 18x = 0$

35. $4t^2 - 8t - 19 = 0$

36. $100x^2 - 200x = 100$

37. $0.06a^2 + 0.05a = 0.04$

38. $9 - 6x = -x^2$

39. $(2x + 1)(x - 5) - (x + 3)(x - 2) = -17$

40. $2y^3 + 2y = 10y^2$

41. $5x^2 = -2x + 3$

42. $x^2 + 4x - 8 = 0$

43. $3 - \dfrac{2}{x} - \dfrac{1}{x^2} = 0$

44. $\dfrac{1}{x-3} + \dfrac{1}{x+2} = 1$

Chapter 3 Cumulative Review

Simplify.

1. $\left(-\dfrac{3}{2}\right)^3$

2. $\left(\dfrac{5}{6}\right)^{-2}$

3. $|-28 - 36| - 17$

4. $2^3 - 3(5^2 - 2^2)$

5. $64 \div 16 \cdot 4$

6. $125 \div 25 \cdot 5$

7. $\dfrac{3^4 - 6}{25(3^2 - 2^2)}$

8. $\dfrac{-2^3(19 - 5 \cdot 2)}{-7^2 + 13}$

9. $\left(\dfrac{x^{-5}y^4}{x^{-2}y^{-3}}\right)^{-1}$

10. $\left(\dfrac{a^{-1}b^{-2}}{a^3b^{-3}}\right)^{-2}$

11. $\sqrt[3]{32}$

12. $\sqrt{50x^3}$

13. $\dfrac{1}{\sqrt[3]{8^2}} + \dfrac{1}{\sqrt{25}}$

14. $\sqrt[3]{\left(\dfrac{8}{27}\right)^{-2}}$

15. Subtract $\dfrac{3}{4}$ from the product of -3 and $\dfrac{5}{12}$.

16. Write in symbols: The difference of $5a$ and $7b$ is greater than their sum.

Let $P(x) = 16x - 0.01x^2$ and find the following.

17. $P(10)$

18. $P(-10)$

Let $Q(x) = \dfrac{3}{5}x^2 + \dfrac{1}{2}x + 25$ and find the following.

19. $Q(10)$

20. $Q(-10)$

Simplify each of the following expressions.

21. $3y^3 - y = 5y^2$

22. $(x + 3)^2 - (x - 3)^2$

Subtract.

23. $\dfrac{6}{y^2 - 9} - \dfrac{5}{y^2 - y - 6}$

24. $\dfrac{y}{x^2 - y^2} - \dfrac{x}{x^2 - y^2}$

Multiply.

25. $\left(4t^2 + \dfrac{1}{3}\right)\left(3t^2 - \dfrac{1}{4}\right)$

26. $\dfrac{x^4 - 16}{x^3 - 8} \cdot \dfrac{x^2 + 2x + 4}{x^2 + 4}$

Divide.

27. $\dfrac{10x^{3n} - 15x^{4n}}{5x^n}$

28. $\dfrac{a^4 + a^3 - 1}{a + 2}$

Reduce to lowest terms. Write the answer with positive exponents only.

29. $\dfrac{x^{-5}}{x^{-8}}$

30. $\left(\dfrac{x^{-6}y^3}{x^{-3}y^{-4}}\right)^{-1}$

31. $\dfrac{x^3 + 2x^2 - 9x - 18}{x^2 - x - 6}$

32. $\dfrac{5x^2 - 26xy - 24y^2}{5x + 4y}$

Factor each of the following expressions.

33. $625a^4 - 16b^4$

34. $x^2 - 3x - 70$

35. $x^2 + 10x + 25 - y^2$

36. $y^3 + \dfrac{8}{27}x^3$

Solve the following equations.

37. $-\dfrac{3}{5}a + 3 = 15$

38. $7y - 6 = 2y + 9$

39. $\dfrac{2}{5}(15x - 2) - \dfrac{1}{5} = 5$

40. $\dfrac{2}{3}(9x - 2) + \dfrac{1}{3} = 4$

41. $\dfrac{3}{y - 2} = \dfrac{2}{y - 3}$

42. $2 - \dfrac{11}{x} = -\dfrac{12}{x^2}$

Rationalize the denominator.

43. $\dfrac{6}{\sqrt[4]{8}}$

44. $\sqrt[3]{\dfrac{8y}{4x}}$

Use the illustration to answer Questions 45 and 46.

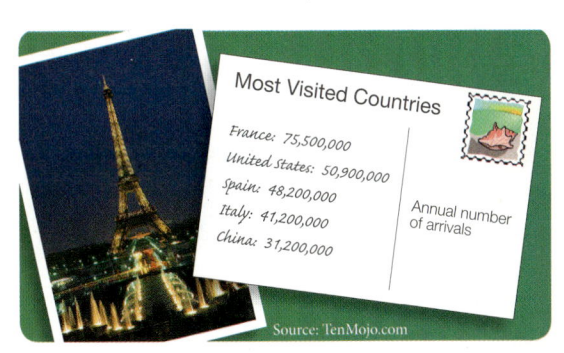

Most Visited Countries
France: 75,500,000
United States: 50,900,000
Spain: 48,200,000
Italy: 41,200,000
China: 31,200,000

Annual number of arrivals

Source: TenMojo.com

45. How many more arrivals were there in France than in Spain?

46. Find the difference in the number of people visting the United States and those visiting China.

Chapter 3 Test

Reduce to lowest terms. [3.1]

1. $\dfrac{x^2 - y^2}{x - y}$

2. $\dfrac{2x^2 - 5x + 3}{2x^2 - x - 3}$

3. **Average Speed** A person riding a Ferris wheel with a diameter of 75 feet travels once around the wheel in 40 seconds. What is the average speed of the rider in feet per second? Use 3.14 as an approximation for π and round your answer to the nearest tenth. [3.1]

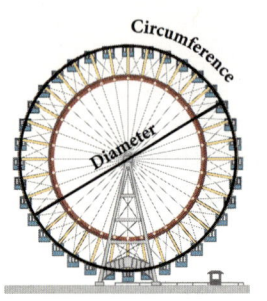

Multiply and divide as indicated. [3.2]

4. $\dfrac{a^2 - 16}{5a - 15} \cdot \dfrac{10(a - 3)^2}{a^2 - 7a + 12}$

5. $\dfrac{a^4 - 81}{a^2 + 9} \div \dfrac{a^2 - 8a + 15}{4a - 20}$

6. $\dfrac{x^3 - 8}{2x^2 - 9x + 10} \div \dfrac{x^2 + 2x + 4}{2x^2 + x - 15}$

Add and subtract as indicated. [3.3]

7. $\dfrac{4}{21} + \dfrac{6}{35}$

8. $\dfrac{3}{4} - \dfrac{1}{2} + \dfrac{5}{8}$

9. $\dfrac{a}{a^2 - 9} + \dfrac{3}{a^2 - 9}$

10. $\dfrac{1}{x} + \dfrac{2}{x - 3}$

11. $\dfrac{4x}{x^2 + 6x + 5} - \dfrac{3x}{x^2 + 5x + 4}$

12. $\dfrac{2x + 8}{x^2 + 4x + 3} - \dfrac{x + 4}{x^2 + 5x + 6}$

Solve each of the following equations. [3.4]

13. $\dfrac{1}{x} + 3 = \dfrac{4}{3}$

14. $\dfrac{x}{x - 3} + 3 = \dfrac{3}{x - 3}$

15. $\dfrac{y + 3}{2y} + \dfrac{5}{y - 1} = \dfrac{1}{2}$

16. $1 - \dfrac{1}{x} = \dfrac{6}{x^2}$

Write in simplified form. [3.5]

17. $\sqrt{125x^3y^5}$

18. $\sqrt[3]{40x^7y^8}$

19. $\sqrt{\dfrac{2}{3}}$

20. $\sqrt{\dfrac{12a^4b^3}{5c}}$

Solve each equation. [3.6]

21. $(2x + 4)^2 = 25$

22. $(2x - 6)^2 = 8$

23. $y^2 - 10y - 25 = -4$

24. $(y + 1)(y - 3) = 6$

25. $2x^2 - 3x - 1 = 0$

26. $\dfrac{1}{a + 2} + \dfrac{1}{3} = \dfrac{1}{a}$

27. **Projectile Motion** An object projected upward with an initial velocity of 32 feet per second will rise and fall according to the equation $s(t) = 32t - 16t^2$, where s is its distance above the ground at time t. At what times will the object be 12 feet above the ground? [3.6]

28. **Revenue** The total weekly cost for a company to make x ceramic coffee cups is given by the formula $C(x) = 2x + 100$. If the weekly revenue from selling all x cups is $R(x) = 25x - 0.2x^2$, how many cups must it sell a week to make a profit of $200 a week? [3.6]

Chapter 3 Projects

RATIONAL EXPRESSIONS AND RATIONAL EQUATIONS

GROUP PROJECT

Rational Expressions

Number of People 3

Time Needed 10–15 minutes

Equipment Pencil and paper

Procedure The four problems shown here all involve the same rational expressions. Often, students who have worked problems successfully on their homework have trouble when they take a test on rational expressions because the problems are mixed up and do not have similar instructions. Noticing similarities and differences between the types of problems involving rational expressions can help with this situation.

Problem 1: Add: $\dfrac{-2}{x^2 - 2x - 3} + \dfrac{3}{x^2 - 9}$

Problem 2: Divide: $\dfrac{-2}{x^2 - 2x - 3} \div \dfrac{3}{x^2 - 9}$

Problem 3: Solve: $\dfrac{-2}{x^2 - 2x - 3} + \dfrac{3}{x^2 - 9} = -1$

Problem 4: Simplify: $\dfrac{\dfrac{-2}{x^2 - 2x - 3}}{\dfrac{3}{x^2 - 9}}$

1. Which problems here do not require the use of a least common denominator?

2. Which two problems involve multiplying by the least common denominator?

3. Which of the problems will have an answer that is one or two numbers but no variables?

4. Work each of the four problems.

RESEARCH PROJECT

Ferris Wheel and *The Third Man*

Among the large Ferris wheels built around the turn of the twentieth century was one built in Vienna in 1897. It is the only one of those large wheels that is still in operation today. Known as the Riesenrad, it has a diameter of 197 feet and can carry a total of 800 people. A brochure that gives some statistics associated with the Riesenrad indicates that passengers riding it travel at 2 feet 6 inches per second. You can check the accuracy of this number by watching the movie *The Third Man.* In the movie, Orson Welles rides the Riesenrad through one complete revolution. Watch *The Third Man* so you can view the Riesenrad in operation. Use the pause button and the timer on your player to time how long it takes Orson Welles to ride once around the wheel. Then calculate his average speed during the ride. Use your results to either prove or disprove the claim that passengers travel at 2 feet 6 inches per second on the Riesenrad. When you have finished, write your procedures and results in essay form.

Functions

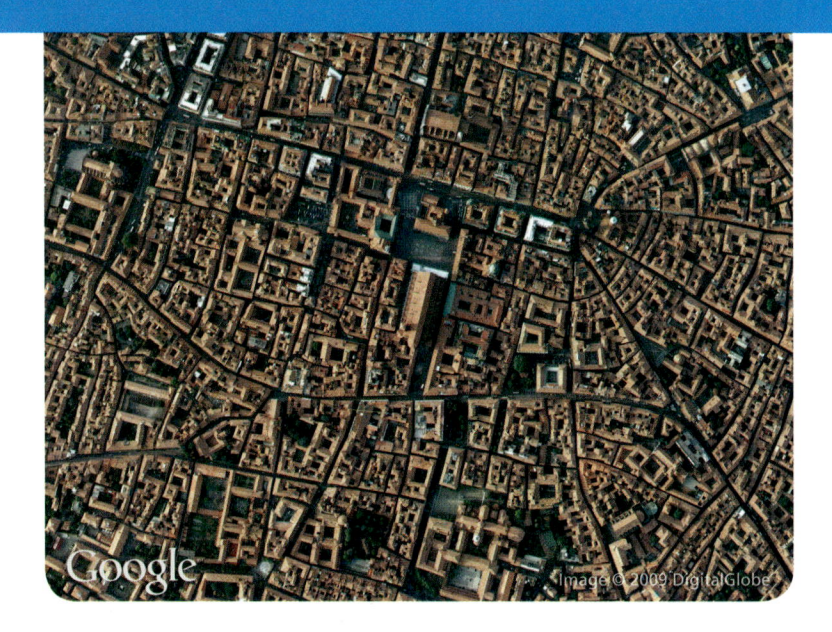

4

Chapter Outline

4.1 Paired Data and Graphing Ordered Pairs

4.2 Introduction to Functions

4.3 Function Notation

4.4 Algebra and Composition with Functions

Introduction

The photo from Google Earth is of the University of Bologna where Maria Agnesi (1718-1799) was the first female professor of mathematics. She is also recognized as the first woman to publish a book of mathematics. The title of the book is *Instituzioni analitiche ad uso della gioventù italiana*. A diagram similar to the one below appears in an 1801 English translation of her book.

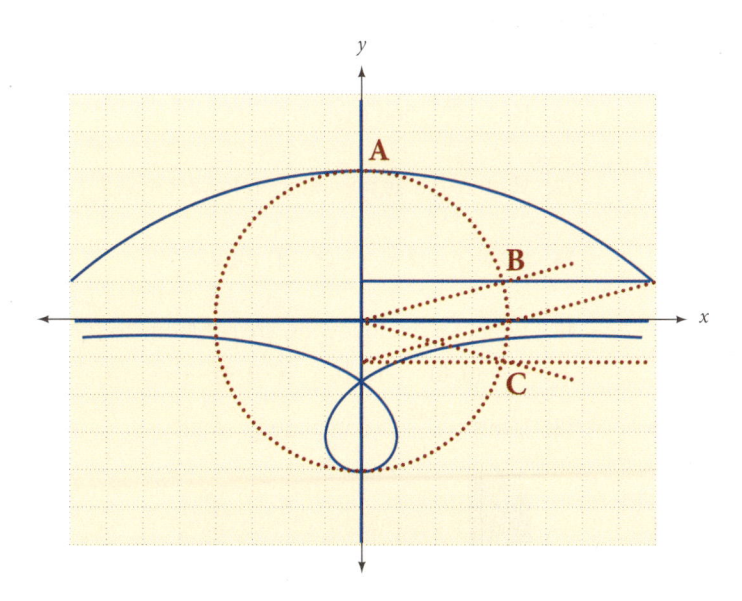

The diagram is drawn on a rectangular coordinate system. The rectangular coordinate system is the foundation upon which we build our study of graphing in this chapter.

Chapter Pretest

Graph the following ordered pairs. [4.1]

1. $(-1, 3)$
2. $(4, -2)$
3. $(-1, -3)$
4. $(0, 5)$

State whether each of the following graphs represent a function. [4.2]

5. 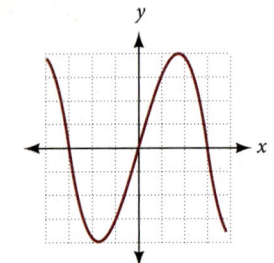
6. 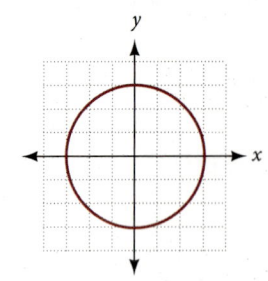
7. 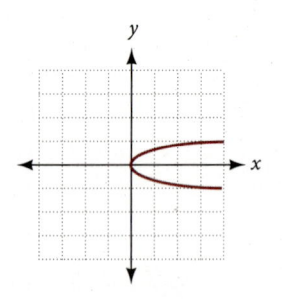
8.

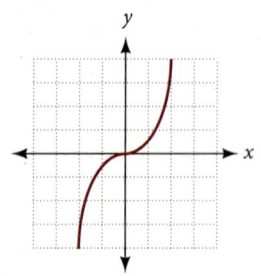

Let $f(x) = 3x - 1$, $g(x) = x^2 + 1$ and $h(x) = 2x^2 + x - 3$, and find the following. [4.3]

9. $f(-2)$
10. $h(1)$
11. $g(-4)$
12. $h(3)$

Let $f(x) = 2x + 1$, $g(x) = 3x$ and $h(x) = x^2 - 6x + 9$, and find the following. [4.4]

13. $f(1) + g(3)$
14. $g(0) - h(0)$
15. $h\left(f\left(\frac{1}{2}\right)\right)$
16. $f(h(g(0)))$

Getting Ready for Chapter 4

1. $16(4.5)^2$
2. $\dfrac{0.2(0.5)^2}{100}$
3. $-\dfrac{1}{3}(-3) + 2$
4. $4(-2) - 1$
5. $\dfrac{4 - (-2)}{3 - 1}$
6. $\dfrac{-3 - (-3)}{2 - (-1)}$
7. $\dfrac{4}{3}(3.14) \cdot 3^3$ (round to the nearest whole number)
8. $3(-2)^2 + 2(-2) - 1$
9. If $2x + 3y = 6$, find x when $y = 0$.
10. If $s = \dfrac{60}{t}$, find s when $t = 15$.
11. Write -0.06 as a fraction in lowest terms.
12. Write -0.6 as a fraction in lowest terms.
13. Solve $2x - 3y = 5$ for y.
14. Solve $y - 4 = -\dfrac{1}{2}(x + 1)$ for y.
15. Which of the following are solutions to $x + y = 5$?
    $(0, 0)$ $(5, 0)$ $(4, 3)$
16. Which of the following are solutions to $y = \dfrac{1}{2}x + 3$?
    $(0, 0)$ $(2, 0)$ $(-2, 4)$
17. Use $h = 64t - 16t^2$ to complete the table.

t	0	1	2	3	4
h					

18. Use $s = \dfrac{60}{t}$ to complete the table.

t	4	6	8	10
s				

19. If $y = \dfrac{K}{x}$, find K when $x = 3$ and $y = 45$.
20. If $y = Kxz^2$, find K when $x = 2$, $y = 81$, and $z = 9$.

Chapter 4 Functions

Paired Data and Graphing Ordered Pairs

4.1

Objectives

A Graph ordered pairs on a rectangular coordinate system.

B Graph linear equations by finding intercepts or by making a table.

C Graph horizontal and veritcal lines.

Table 1 and Figure 1 show the relationship between the speed of a racecar and the time elapsed since the start of the race. In Figure 1, the horizontal line that shows the elapsed time in seconds is called the *horizontal axis,* and the vertical line that shows the speed in miles per hour is called the *vertical axis.*

The data in Table 1 are called *paired data* because the information is organized so that each number in the first column is paired with a specific number in the second column. Each pair of numbers is associated with one of the solid bars in Figure 1. For example, the third bar in the bar chart is associated with the pair of numbers 3 seconds and 162.8 miles per hour. The first number, 3 seconds, is associated with the horizontal axis, and the second number, 162.8 miles per hour, is associated with the vertical axis.

Scatter Diagrams and Line Graphs

The information in Table 1 can be visualized with a *scatter diagram* and *line graph* as well. Figure 2 is a scatter diagram of the information in Table 1. We use dots instead of the bars shown in Figure 1 to show the speed of the racecar at each second during the race. Figure 3 is called a *line graph.* It is constructed by taking the dots in Figure 2 and connecting each one to the next with a straight line. Notice that we have labeled the axes in these two figures a little differently than we did with the bar chart by making the axes intersect at the number 0.

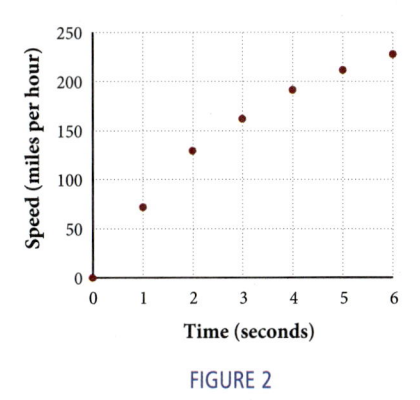

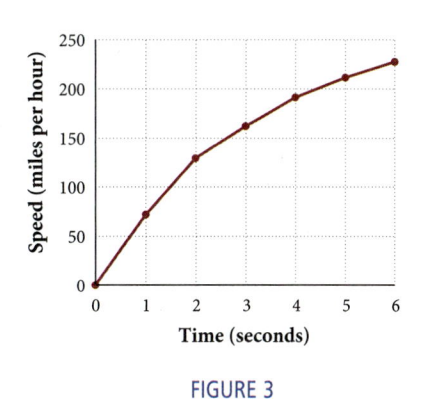

TABLE 1

SPEED OF A RACE CAR

Time in Seconds	Speed in Miles per Hour
0	0
1	72.7
2	129.9
3	162.8
4	192.2
5	212.4
6	228.1

FIGURE 2

FIGURE 3

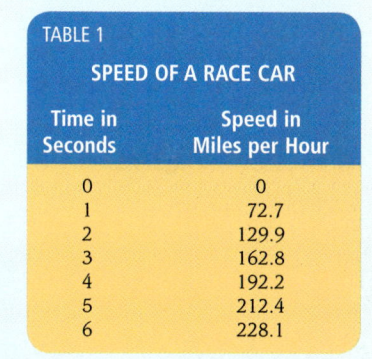

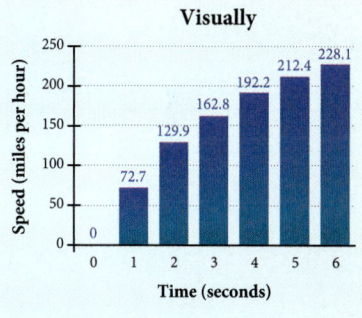

Visually

FIGURE 1

A Ordered Pairs

Paired data play an important role in equations that contain two variables. Working with these equations is easier if we standardize the terminology and notation associated with paired data. So here is a definition that will do just that.

Definition

A pair of numbers enclosed in parentheses and separated by a comma, such as (−2, 1), is called an ordered pair of numbers. The first number in the pair is called the **x-coordinate** of the ordered pair; the second number is called the **y-coordinate**. For the ordered pair (−2, 1), the **x-coordinate** is −2 and the **y-coordinate** is 1.

4.1 Paired Data and Graphing Ordered Pairs

237

Chapter 4 Functions

To standardize the way in which we display paired data visually, we use a rectangular coordinate system. A *rectangular coordinate system* is made by drawing two real number lines at right angles to each other. The two number lines, called *axes,* cross each other at 0. This point is called the *origin.* Positive directions are to the right and up. Negative directions are down and to the left. The rectangular coordinate system is shown in Figure 4.

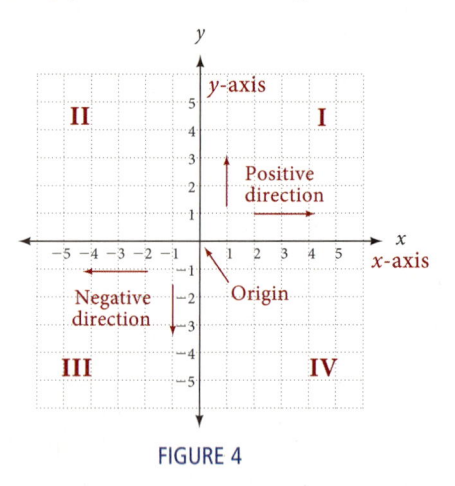

FIGURE 4

The horizontal number line is called the *x-axis* and the vertical number line is called the *y-axis.* The two number lines divide the coordinate system into four quadrants, which we number I through IV in a counterclockwise direction. Points on the axes are not considered as being in any quadrant.

To graph the ordered pair (a, b) on a rectangular system, we start at the origin and move a units right or left (right if a is positive, left if a is negative). Then we move b units up or down (up if b is positive and down if b is negative). The point where we end up is the graph of the ordered pair (a, b).

EXAMPLE 1 Plot (graph) the ordered pairs $(2, 5)$, $(-2, 5)$, $(-2, -5)$, and $(2, -5)$.

SOLUTION To graph the ordered pair $(2, 5)$, we start at the origin and move 2 units to the right, then 5 units up. We are now at the point whose coordinates are $(2, 5)$. We graph the other three ordered pairs in a similar manner (see Figure 5).

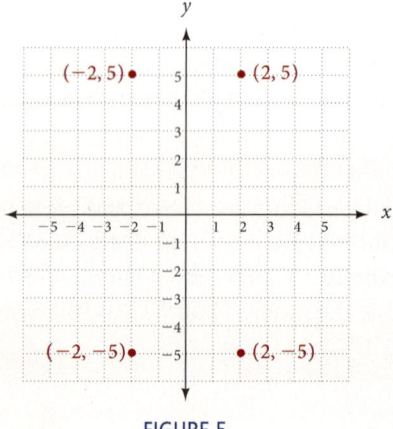

FIGURE 5

Note A rectangular coordinate system allows us to connect algebra and geometry by associating geometric shapes (the curves shown in the diagrams) with algebraic equations. The French philosopher and mathematician René Descartes (1596–1650) usually is credited with the invention of the rectangular coordinate system, which often is referred to as the Cartesian coordinate system in his honor. As a philosopher, Descartes is responsible for the statement, "I think, therefore, I am." Until Descartes invented his coordinate system in 1637, algebra and geometry were treated as separate subjects.

PRACTICE PROBLEMS

1. Plot the ordered pairs $(1, 3)$, $(-1, 3)$, $(-1, -3)$, and $(1, -3)$.

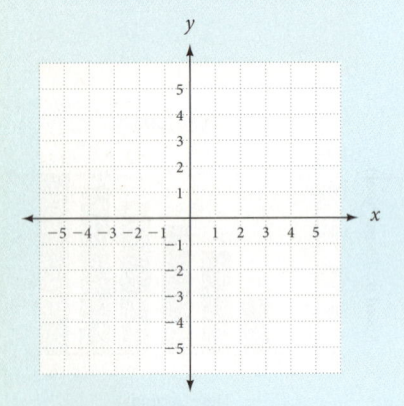

Note From Example 1 we see that any point in quadrant I has both its *x*- and *y*-coordinates positive $(+, +)$. Points in quadrant II have negative *x*-coordinates and positive *y*-coordinates $(-, +)$. In quadrant III both coordinates are negative $(-, -)$. In quadrant IV the form is $(+, -)$.

Answers
See Solutions to Selected Practice Problems for all answers in this section.

EXAMPLE 2 Graph the ordered pairs $(1, -3)$, $(0, 5)$, $(3, 0)$, $(0, -2)$, $(-1, 0)$, and $\left(\frac{1}{2}, 2\right)$.

SOLUTION

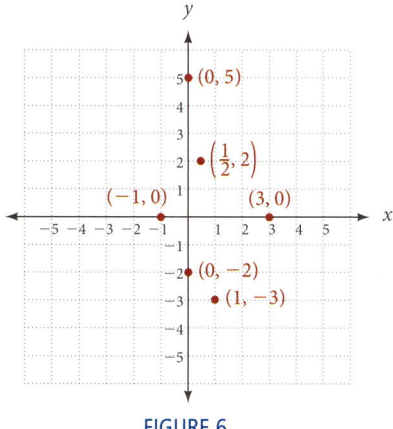

FIGURE 6

From Figure 6 we see that any point on the x-axis has a y-coordinate of 0 (it has no vertical displacement), and any point on the y-axis has an x-coordinate of 0 (no horizontal displacement).

B Linear Equations

We can plot a single point from an ordered pair, but to draw a line, we need two points or an equation in two variables.

> **Definition**
> Any equation that can be put in the form $ax + by = c$, where a, b, and c are real numbers and a and b are not both 0, is called a **linear equation in two variables**. The graph of any equation of this form is a straight line (that is why these equations are called "linear"). The form $ax + by = c$ is called **standard form.**

To graph a linear equation in two variables, we simply graph its solution set; that is, we draw a line through all the points whose coordinates satisfy the equation.

EXAMPLE 3 Graph the equation $y = -\frac{1}{3}x + 2$.

SOLUTION We need to find three ordered pairs that satisfy the equation. To do so, we can let x equal any numbers we choose and find corresponding values of y. But since every value of x we substitute into the equation is going to be multiplied by $-\frac{1}{3}$, let's use numbers for x that are divisible by 3, like -3, 0, and 3. That way, when we multiply them by $-\frac{1}{3}$, the result will be an integer.

2. Plot the ordered pairs $(4, -2)$, $\left(-\frac{1}{2}, 3\right)$, $(1, 0)$, $(0, -5)$, $(-6, 0)$, and $(0, 4)$.

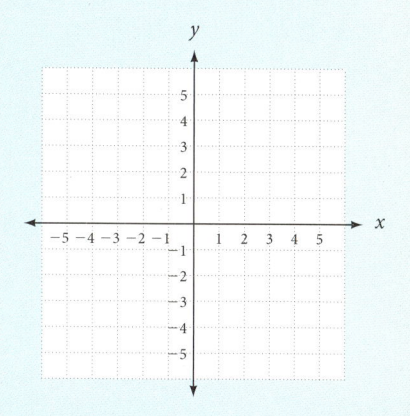

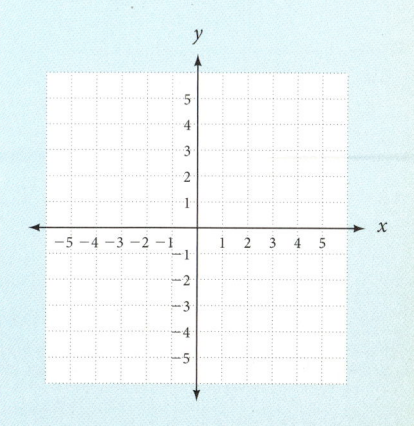

3. Graph $y = \frac{1}{2}x + 3$.

$$\text{Let } x = -3; \quad y = -\frac{1}{3}(-3) + 2$$
$$y = 1 + 2$$
$$y = 3$$

The ordered pair (−3, 3) is one solution.

$$\text{Let } x = 0; \quad y = -\frac{1}{3}(0) + 2$$
$$y = 0 + 2$$
$$y = 2$$

The ordered pair (0, 2) is a second solution.

$$\text{Let } x = 3; \quad y = -\frac{1}{3}(3) + 2$$
$$y = -1 + 2$$
$$y = 1$$

The ordered pair (3, 1) is a third solution.

In table form

x	y
−3	3
0	2
3	1

Plotting the ordered pairs (−3, 3), (0, 2), and (3, 1) and drawing a straight line through their graphs, we have the graph of the equation $y = -\frac{1}{3}x + 2$, as shown in Figure 7.

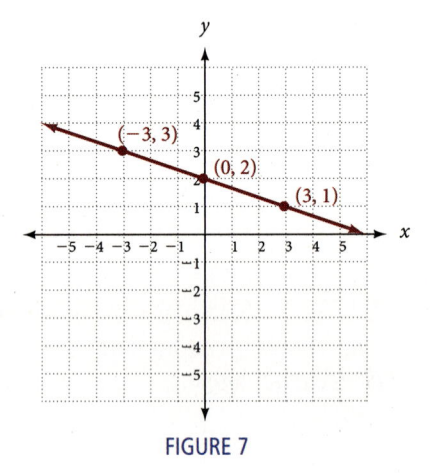

FIGURE 7

> **Note** It takes only two points to determine a straight line. We have included a third point for "insurance." If all three points do not line up in a straight line, we have made a mistake.

Example 3 illustrates again the connection between algebra and geometry that we mentioned earlier in this section. Descartes' rectangular coordinate system allows us to associate the equation $y = -\frac{1}{3}x + 2$ (an algebraic concept) with a specific straight line (a geometric concept). The study of the relationship between equations in algebra and their associated geometric figures is called *analytic geometry*.

Intercepts

Two important points on the graph of a straight line, if they exist, are the points where the graph crosses the axes.

> **Definition**
> The **x-intercept** of the graph of an equation is the x-coordinate of the point where the graph crosses the x-axis. The **y-intercept** is defined similarly.

4.1 Paired Data and Graphing Ordered Pairs

Because any point on the x-axis has a y-coordinate of 0, we can find the x-intercept by letting $y = 0$ and solving the equation for x. We find the y-intercept by letting $x = 0$ and solving for y.

EXAMPLE 4 Find the x- and y-intercepts for $2x + 3y = 6$; then graph the solution set.

SOLUTION To find the y-intercept we let $x = 0$.

When $x = 0$

we have $2(0) + 3y = 6$

$3y = 6$

$y = 2$

The y-intercept is 2, and the graph crosses the y-axis at the point (0, 2).

When $y = 0$

we have $2x + 3(0) = 6$

$2x = 6$

$x = 3$

The x-intercept is 3, so the graph crosses the x-axis at the point (3, 0). We use these results to graph the solution set for $2x + 3y = 6$. The graph is shown in Figure 8.

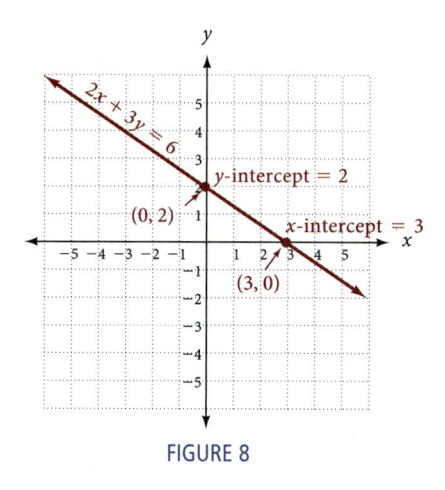

FIGURE 8

C Horizontal and Vertical Lines

EXAMPLE 5 Graph each of the following lines.

a. $y = \frac{1}{2}x$ b. $x = 3$ c. $y = -2$

SOLUTION

a. The line $y = \frac{1}{2}x$ passes through the origin because (0, 0) satisfies the equation. To sketch the graph we need at least one more point on the line. When x is 2, we obtain the point (2, 1), and when x is −4, we obtain the point (−4, −2). The graph of $y = \frac{1}{2}x$ is shown in Figure 9a.

---

4. Find the x- and y-intercepts for $2x - 3y = 6$, and then graph the solution set.

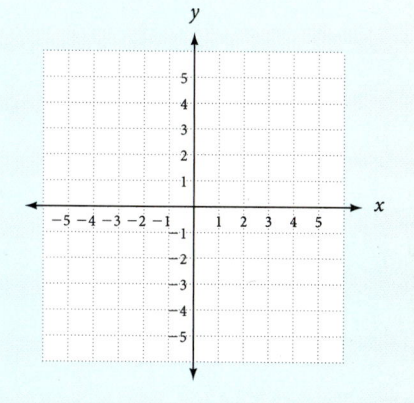

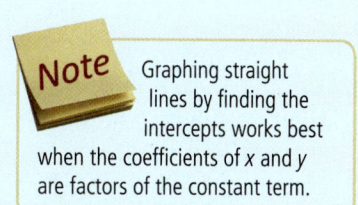

Note Graphing straight lines by finding the intercepts works best when the coefficients of x and y are factors of the constant term.

5. Graph the line $x = -1$ and the line $y = 4$.

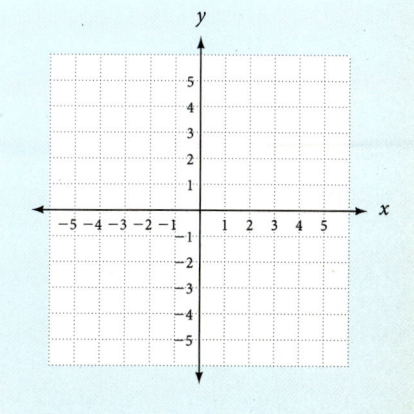

b. The line $x = 3$ is the set of all points whose x-coordinate is 3. The variable y does not appear in the equation, so the y-coordinate can be any number. Note that we can write our equation as a linear equation in two variables by writing it as $x + 0y = 3$. Because the product of 0 and y will always be 0, y can be any number. The graph of $x = 3$ is the vertical line shown in Figure 9b.

c. The line $y = -2$ is the set of all points whose y-coordinate is -2. The variable x does not appear in the equation, so the x-coordinate can be any number. Again, we can write our equation as a linear equation in two variables by writing it as $0x + y = -2$. Because the product of 0 and x will always be 0, x can be any number. The graph of $y = -2$ is the horizontal line shown in Figure 9c.

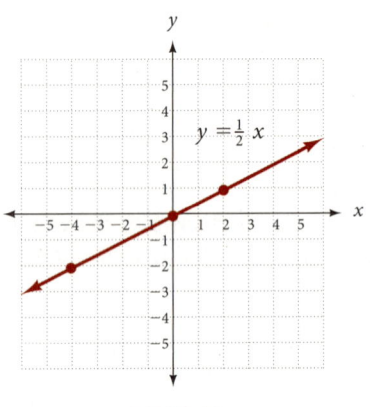

FIGURE 9A

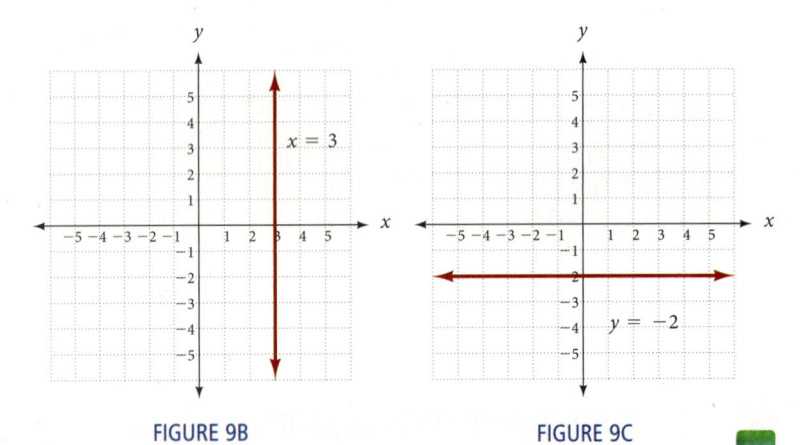

FIGURE 9B

FIGURE 9C

4.1 Paired Data and Graphing Ordered Pairs

FACTS FROM GEOMETRY: Special Equations and Their Graphs
For the equations below, m, a, and b are real numbers.

Through the Origin

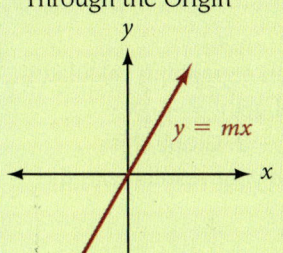

$y = mx$

Vertical Line

$x = a$

Horizontal Line

$y = b$

FIGURE 10A Any equation of the form $y = mx$ has a graph that passes through the origin.

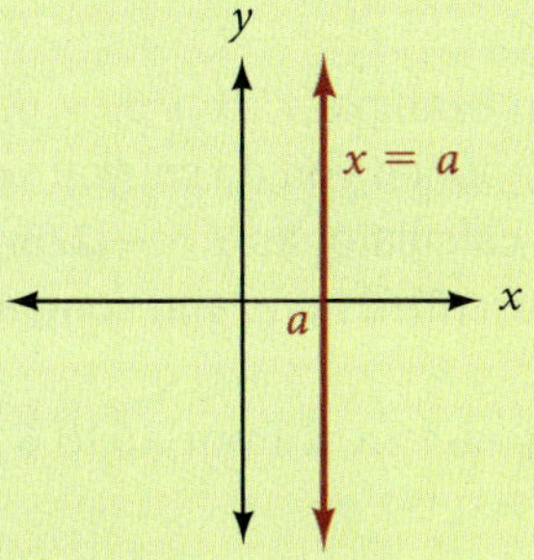

FIGURE 10B Any equation of the form $x = a$ has a vertical line for its graph.

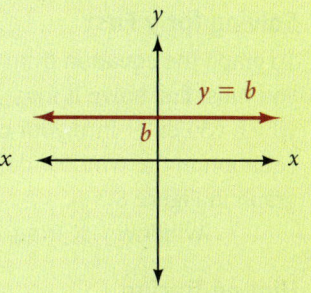

FIGURE 10C Any equation of the form $y = b$ has a horizontal line for its graph.

USING TECHNOLOGY

Graphing Calculators and Computer Graphing Programs

A variety of computer programs and graphing calculators are currently available to help us graph equations and then obtain information from those graphs much faster than we could with paper and pencil. We will not give instructions for all the available calculators. Most of the instructions we give are generic in form. You will have to use the manual that came with your calculator to find the specific instructions for your calculator.

Graphing with Trace and Zoom

All graphing calculators have the ability to graph a function and then trace over the points on the graph, giving their coordinates. Furthermore, all graphing calculators can zoom in and out on a graph that has been drawn. To graph a linear equation on a graphing calculator, we first set the graph window. Most calculators call the smallest value of x Xmin and the largest value of x Xmax. The counterpart values of y are Ymin and Ymax. We will use the notation

Window: X from -5 to 4, Y from -3 to 2

to stand for a window in which

Xmin = -5 Ymin = -3
Xmax = 4   Ymax = 2

Set your calculator with the following window:

Window: X from -10 to 10, Y from -10 to 10

Graph the equation $Y = -X + 8$. On the TI-82/83, you use the $\boxed{Y=}$ key to enter the equation; you enter a negative sign with the $\boxed{(-)}$ key, and a subtraction sign with the $\boxed{-}$ key. The graph will be similar to the one shown in Figure 11.

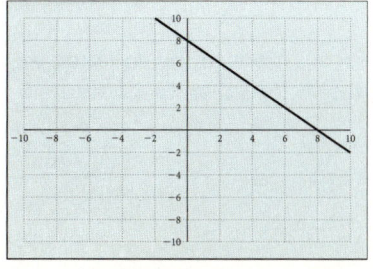

FIGURE 11

Use the Trace feature of your calculator to name three points on the graph. Next, use the Zoom feature of your calculator to zoom out so your window is twice as large.

Solving for y First

To graph the equation from Example 4, $2x + 3y = 6$, on a graphing calculator, you must first solve it for y. When you do so, you will get $y = -\frac{2}{3}x + 2$, which you enter into your calculator as $Y = -(2/3)X + 2$. Graph this equation in the window described here, and compare your results with the graph in Figure 8.

Window: X from -6 to 6, Y from -6 to 6

Hint on Tracing

If you are going to use the Trace feature and you want the x-coordinates to be exact numbers, set your window so that the range of X inputs is a multiple of the number of horizontal pixels on your calculator screen. On the TI-82/83, the screen has 94 pixels across. Here are a few convenient trace windows:

X from -4.7 to 4.7	To trace to the nearest tenth
X from -47 to 47	To trace to the nearest integer
X from 0 to 9.4	To trace to the nearest tenth
X from 0 to 94	To trace to the nearest integer
X from -94 to 94	To trace to the nearest even integer

STUDY SKILLS
Pay Attention to Instructions

Each of the following is a valid instruction with respect to the equation $y = 3x - 2$, and the result of applying the instructions will be different in each case:

Find x when y is 10.
Solve for x.
Graph the equation.
Find the intercepts.

There are many things to do with the equation $y = 3x - 2$. If you train yourself to pay attention to the instructions that accompany a problem as you work through the assigned problems, you will not find yourself confused about what to do with a problem when you see it on a test.

Getting Ready for Class

After reading through the preceding section, respond in your own words and in complete sentences.

1. Explain how you would construct a rectangular coordinate system from two real number lines.
2. Explain in words how you would graph the ordered pair (2, -3).
3. How can you tell if an ordered pair is a solution to the equation $y = 2x - 5$?
4. If you were looking for solutions to the equation $y = \frac{1}{3}x + 5$, why would it be easier to substitute 6 for x than to substitute 5 for x?

Problem Set 4.1

A Graph each of the following ordered pairs on a rectangular coordinate system. [Examples 1–2]

1. a. (1, 2)
   b. (−1, −2)
   c. (5, 0)
   d. (0, 2)
   e. (−5, −5)
   f. $\left(\frac{1}{2}, 2\right)$

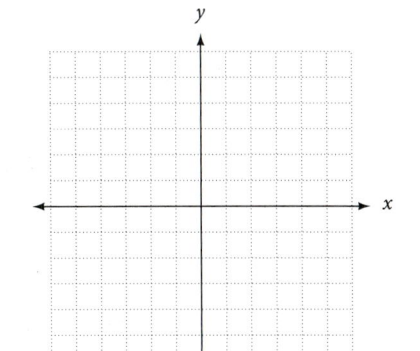

2. a. (−1, 2)
   b. (1, −2)
   c. (0, −3)
   d. (4, 0)
   e. (−4, −1)
   f. $\left(3, \frac{1}{4}\right)$

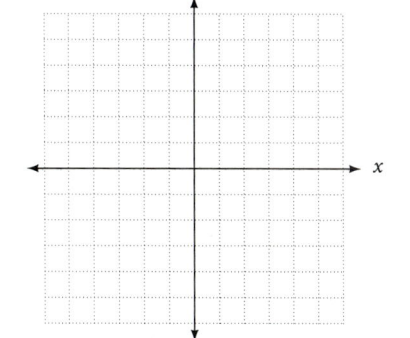

A Give the coordinates of each point.

3.

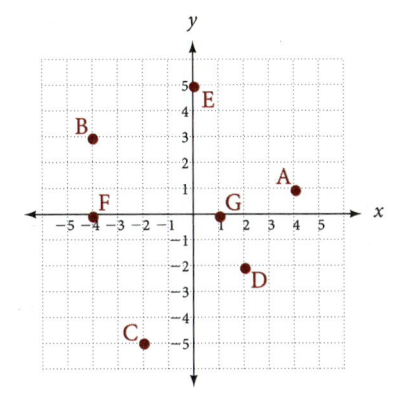

4.

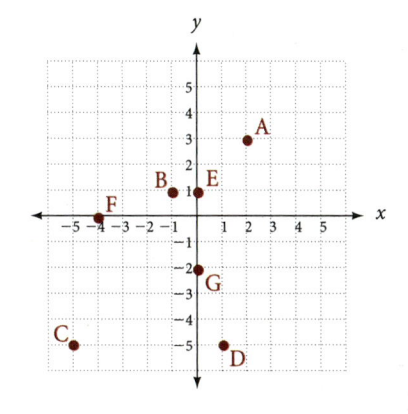

B Graph each of the following linear equations by first finding the intercepts. [Examples 3–4]

5. $2x - 3y = 6$

6. $y - 2x = 4$

7. $4x - 5y = 20$

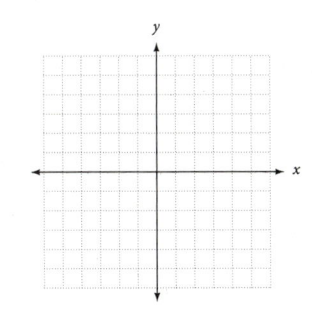

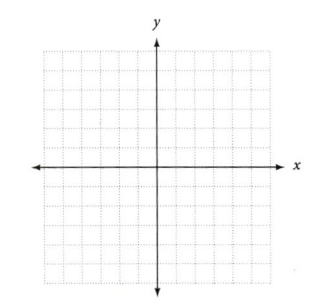

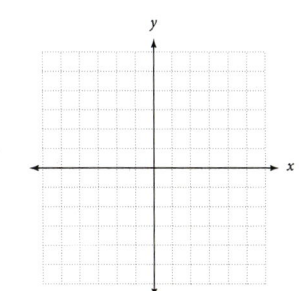

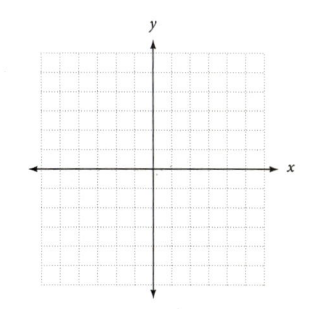

8. $5x - 3y - 15 = 0$

9. $y = 2x + 3$

10. $y = 3x - 2$

11. $-3x + 2y = 12$

12. $5x - 7y = -35$

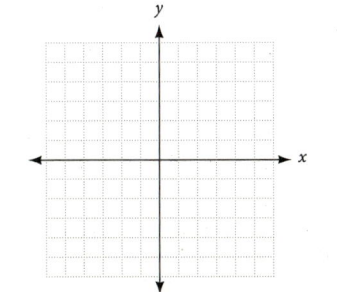

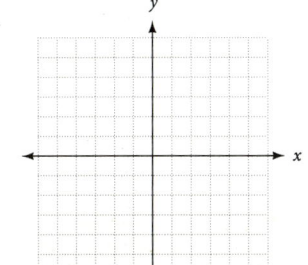

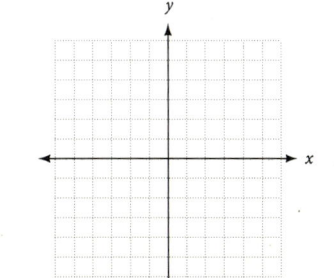

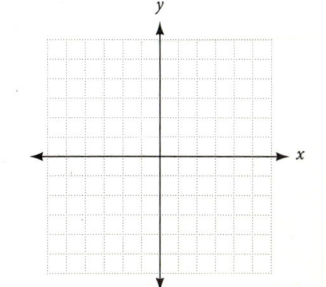

246 Chapter 4 Functions

13. $6x - 5y - 20 = 0$

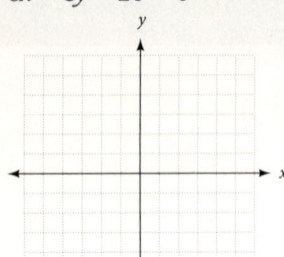

14. $-4x - 6y + 15 = 0$

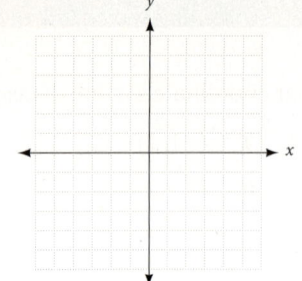

15. $y = 3x - 5$

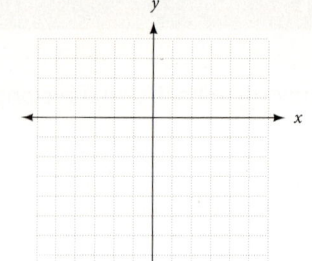

16. $y = -4x + 1$

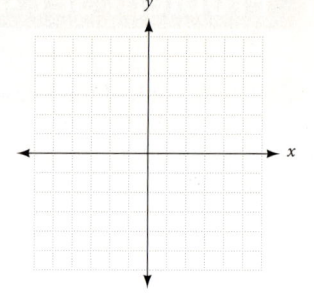

17. $\dfrac{x}{2} + \dfrac{y}{3} = 1$

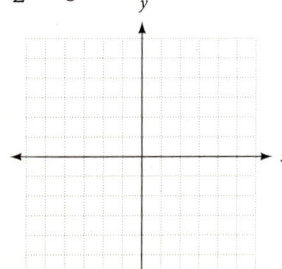

18. $\dfrac{x}{4} - \dfrac{y}{7} = 1$

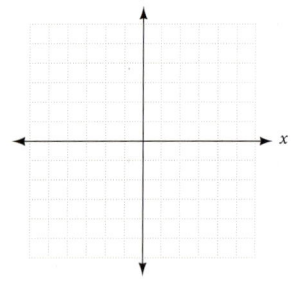

19. Which of the following tables could be produced from the equation $y = 2x - 6$?

a.
x	y
0	6
1	4
2	2
3	0

b.
x	y
0	-6
1	-4
2	-2
3	0

c.
x	y
0	-6
1	-5
2	-4
3	-3

20. Which of the following tables could be produced from the equation $3x - 5y = 15$?

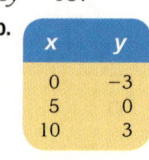

a.
x	y
0	5
-3	0
10	3

b.
x	y
0	-3
5	0
10	3

c.
x	y
0	-3
-5	0
10	-3

B Graph each of the following lines. [Example 5]

21. $y = \dfrac{1}{3}x$

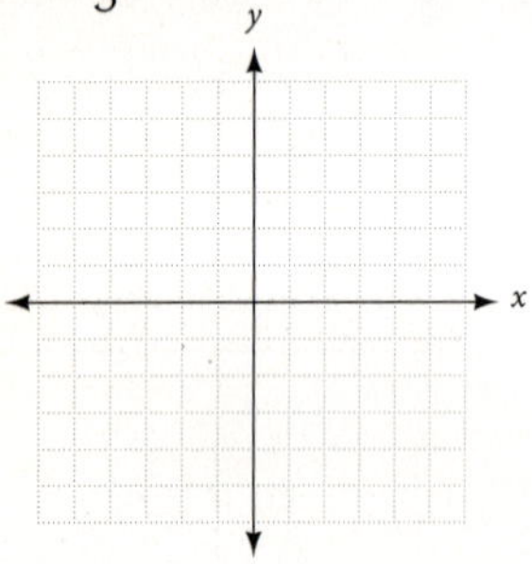

22. $y = \dfrac{1}{2}x$

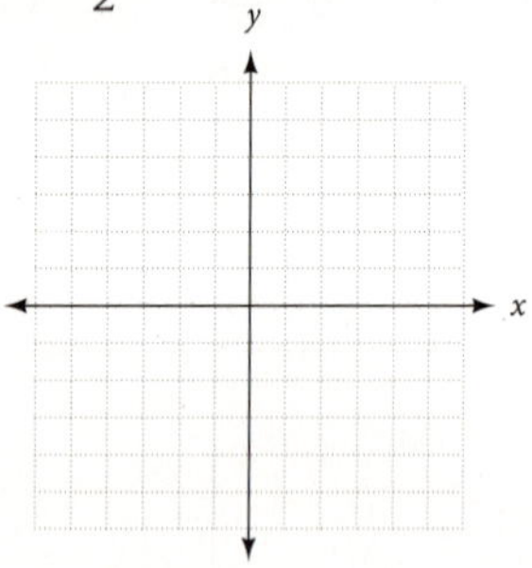

23. $-2x + y = -3$

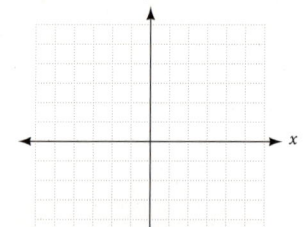

24. $-3x + y = -2$

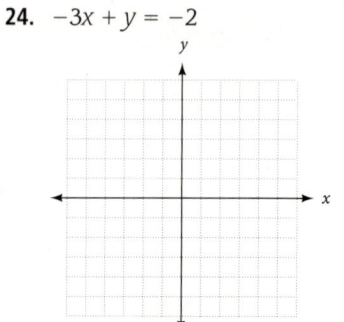

25. $y = -\dfrac{2}{3}x + 1$

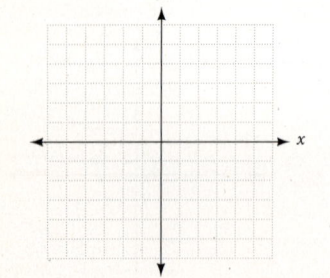

26. $y = -\dfrac{2}{3}x - 1$

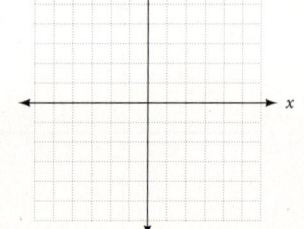

27. $\dfrac{x}{3} + \dfrac{y}{4} = 1$

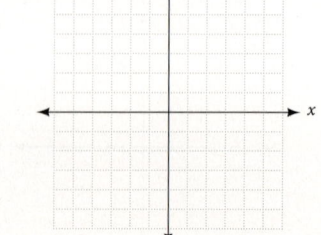

28. $\dfrac{x}{-2} + \dfrac{y}{3} = 1$

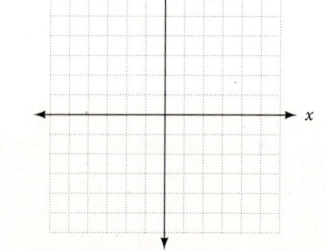

29. The graph shown here is the graph of which of the following equations:
   a. $3x - 2y = 6$
   b. $2x - 3y = 6$
   c. $2x + 3y = 6$

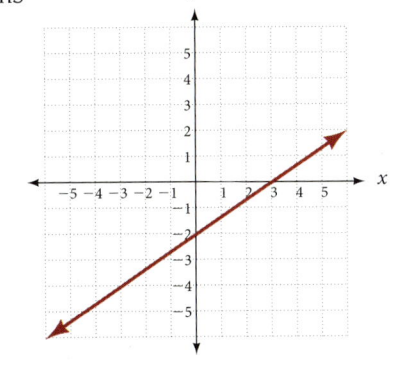

30. The graph shown here is the graph of which of the following equations:
   a. $3x - 2y = 8$
   b. $2x - 3y = 8$
   c. $2x + 3y = 8$

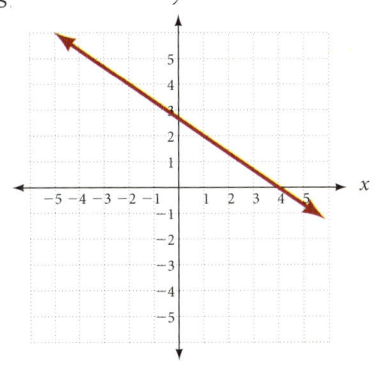

The next two problems are intended to give you practice reading, and paying attention to, the instructions that accompany the problems you are working. Working these problems is an excellent way to get ready for a test or a quiz.

31. Work each problem according to the instructions given:
   a. Solve: $4x + 12 = -16$
   b. Find x when y is 0: $4x + 12y = -16$
   c. Find y when x is 0: $4x + 12y = -16$
   d. Graph: $4x + 12y = -16$
   e. Solve for y: $4x + 12y = -16$

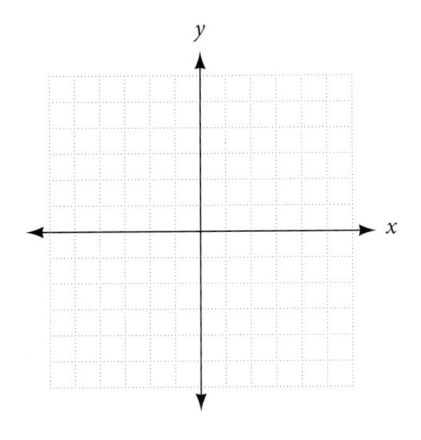

32. Work each problem according to the instructions given:
   a. Solve: $3x - 8 = -12$
   b. Find x when y is 0: $3x - 8y = -12$
   c. Find y when x is 0: $3x - 8y = -12$
   d. Graph: $3x - 8y = -12$
   e. Solve for y: $3x - 8y = -12$

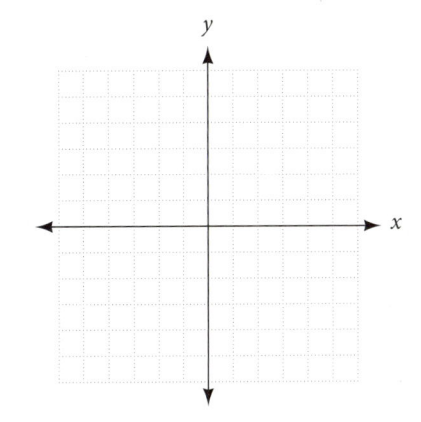

33. Graph each of the following lines.
   a. $y = 2x$ b. $x = -3$ c. $y = 2$

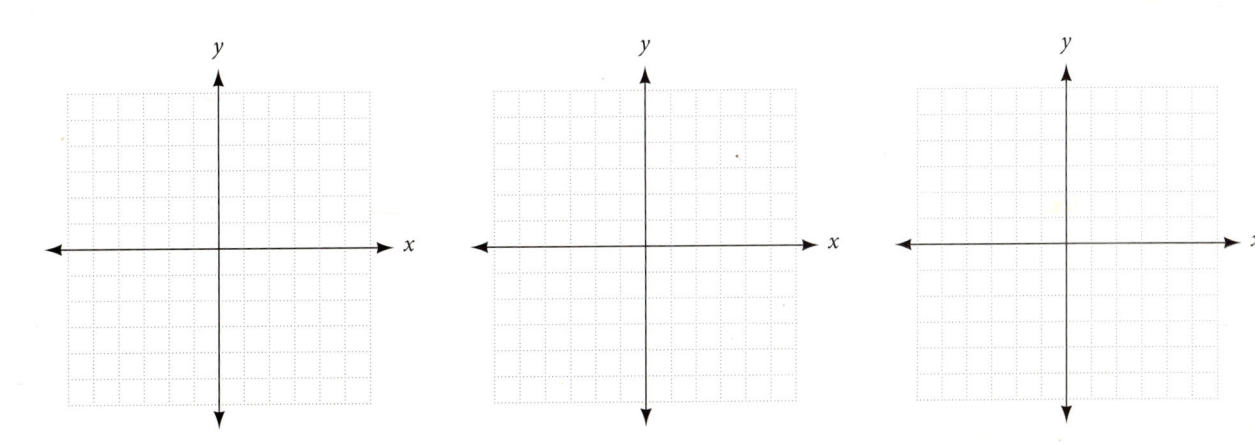

Chapter 4 Functions

34. Graph each of the following lines.
    a. $y = 3x$ b. $x = -2$ c. $y = 4$

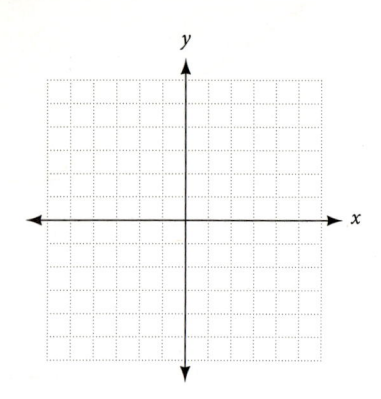

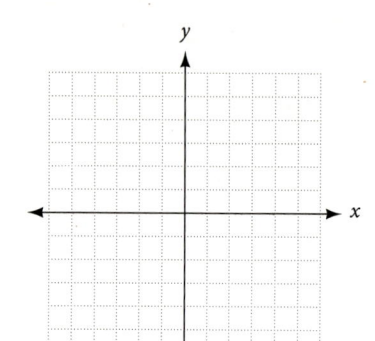

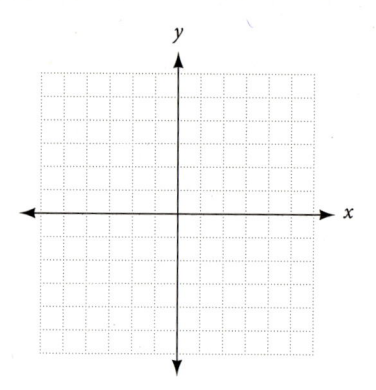

35. Graph each of the following lines.
    a. $y = -\frac{1}{2}x$ b. $x = 4$ c. $y = -3$

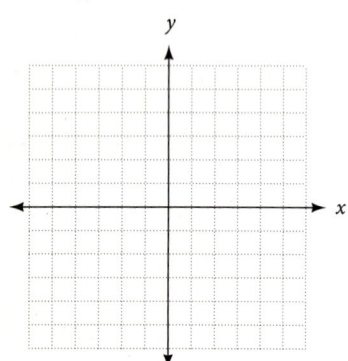

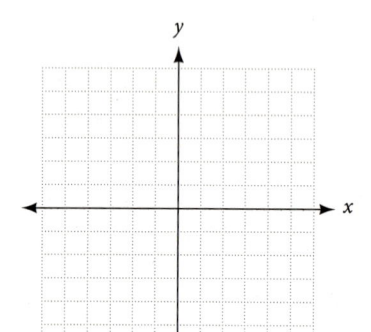

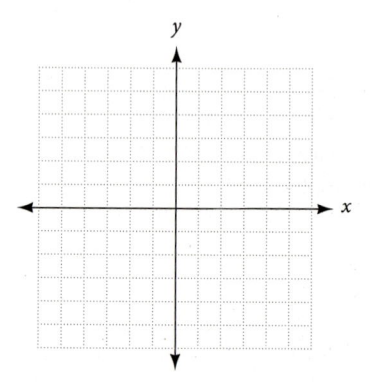

36. Graph each of the following lines.
    a. $y = -\frac{1}{3}x$ b. $x = 1$ c. $y = -5$

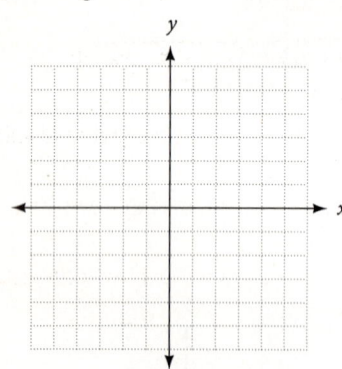

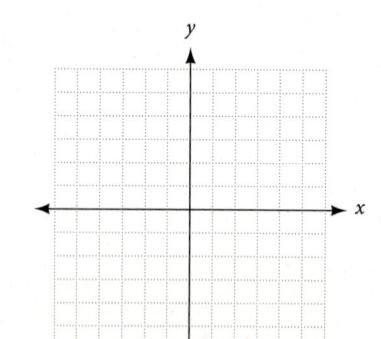

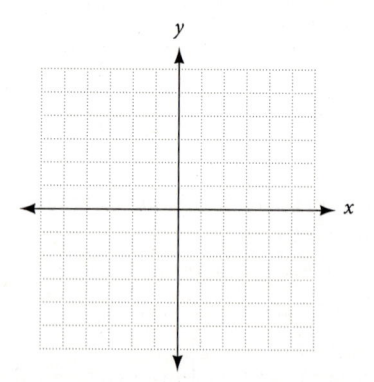

37. Graph the line $0.02x + 0.03y = 0.06$.

38. Graph the line $0.05x - 0.03y = 0.15$.

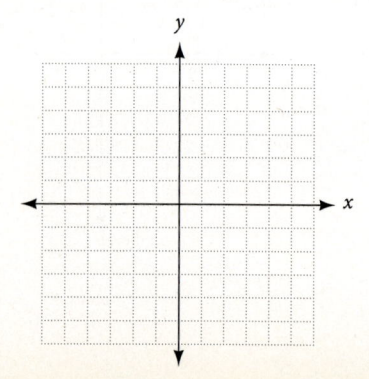

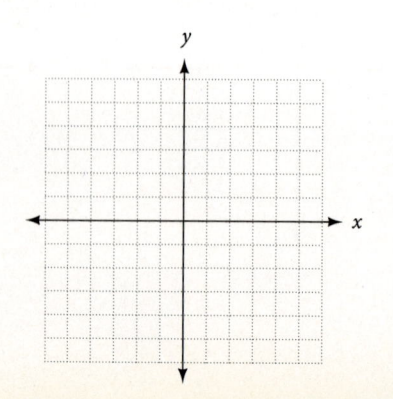

39. The ordered pairs that satisfy the equation $y = 3x$ all have the form $(x, 3x)$ because y is always 3 times x. Graph all ordered pairs of the form $(x, 3x)$.

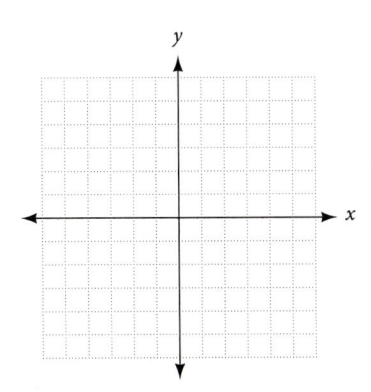

40. Graph all ordered pairs of the form $(x, -3x)$.

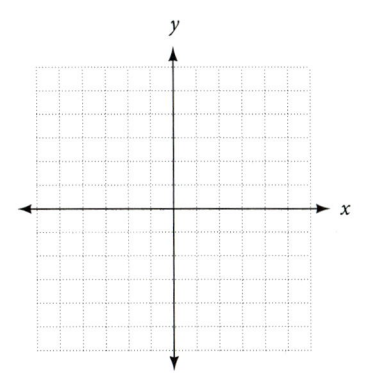

Applying the Concepts

41. Hourly Wages Jane takes a job at the local Marcy's department store. Her job pays $8.00 per hour. The graph shows how much Jane earns for working from 0 to 40 hours in a week.

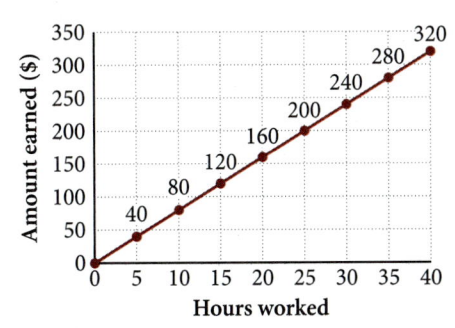

a. List three ordered pairs that lie on the line graph.

b. How much will she earn for working 40 hours?

c. If her check for one week is $240, how many hours did she work?

d. She works 35 hours one week, but her paycheck before deductions are subtracted out is for $260. Is this correct? Explain.

42. Hourly Wages Judy takes a job at Gigi's boutique. Her job pays $6.00 per hour plus $50 per week in commission. The graph shows how much Judy earns for working from 0 to 40 hours in a week.

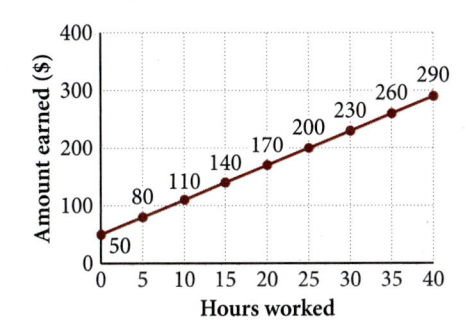

a. List three ordered pairs that lie on the line graph.

b. How much will she earn for working 40 hours?

c. If her check for one week is $230, how many hours did she work?

d. She works 35 hours one week, but her paycheck before deductions are subtracted out is for $260. Is this correct? Explain.

43. Non-Camera Phone Sales The table and bar chart shown here show the projected sales of non-camera phones. Use the information from the table and chart to construct a line graph.

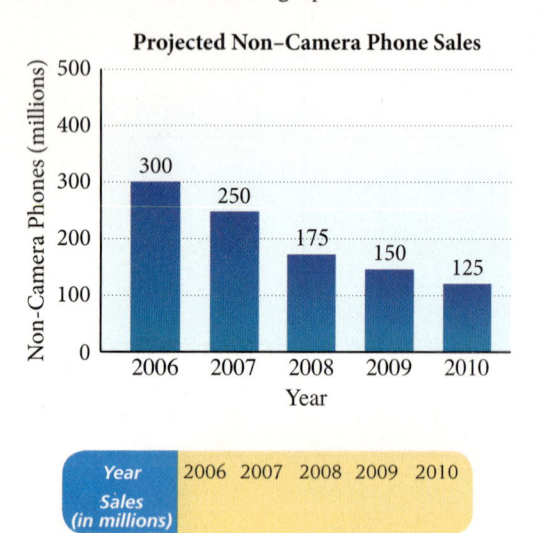

Projected Non–Camera Phone Sales

Year	2006	2007	2008	2009	2010
Sales (in millions)	300	250	175	150	125

44. Camera Phone Sales The table and bar chart shown here show the projected sales of camera phones from 2006 to 2010. Use the information from the table and chart to construct a line graph.

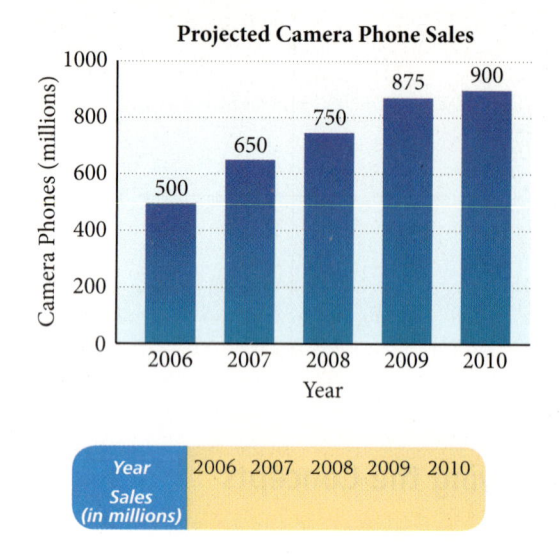

Projected Camera Phone Sales

Year	2006	2007	2008	2009	2010
Sales (in millions)	500	650	750	875	900

45. Kentucky Derby The graph gives the monetary bets placed at the Kentucky Derby for specific years. If x represents the year in question and y represents the total wagering for that year, write five ordered pairs that describe the information in the graph.

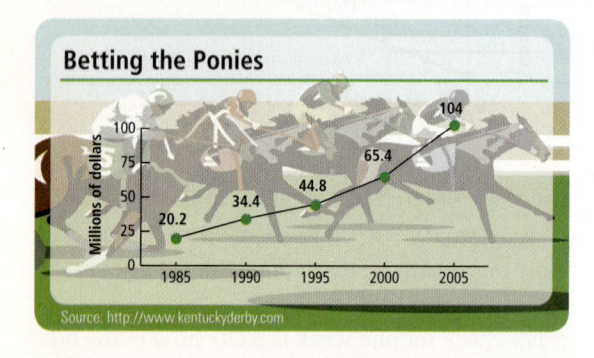

Betting the Ponies
- 1985: 20.2
- 1990: 34.4
- 1995: 44.8
- 2000: 65.4
- 2005: 104

Source: http://www.kentuckyderby.com

46. Age of New Mothers The graph here shows the increase in average age of first-time mothers in the U.S. since 1970. Estimate from the graph the average age of first-time mothers for the years 1970, 1980, 1990, and 2000 as a list of ordered pairs. (Round ages to the nearest whole number).

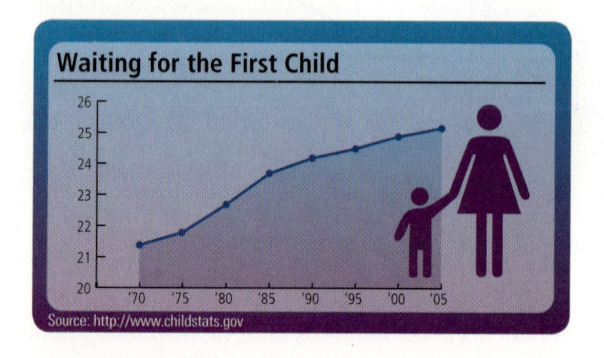

Waiting for the First Child

Source: http://www.childstats.gov

47. Mothers The graph used in the previous problem shows the average age of first-time mothers from 1970 to 2005. Use the chart to make a table of x and y values, where x is the year and y is the average age.

x (Year)	1970	1975	1980	1985	1990	1995	2000	2005
y (age)								

4.1 Problem Set

48. Camera Phones The chart shows the estimated number of camera phones and non-camera phones sold from 2004 to 2010. Use the chart to make a line graph of the total number of phones sold, and then list the ordered pairs.

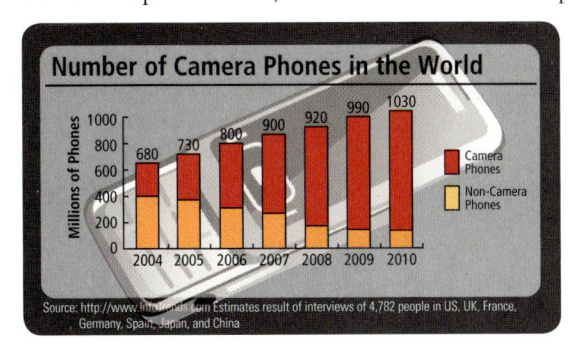

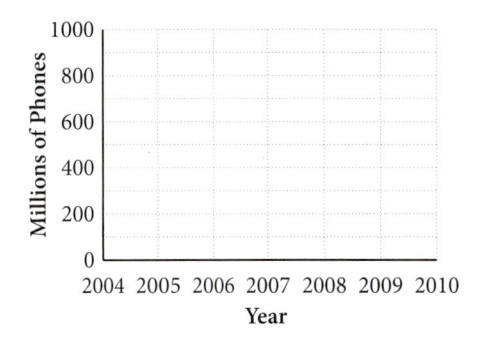

49. In the figure below, right triangle ABC has legs of length 5. Find the coordinates of points A and B.

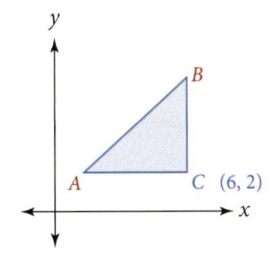

50. In the figure below, right triangle ABC has legs of length 5. Find the coordinates of points A and B.

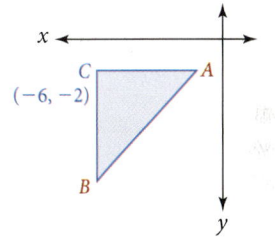

51. In the figure below, rectangle ABCD has a length of 5 and a width of 3. Find points A, B, and C.

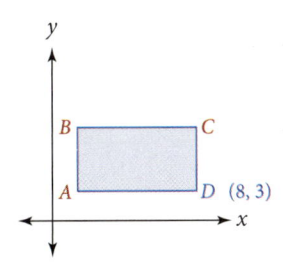

52. In the figure below, rectangle ABCD has a length of 5 and a width of 3. Find points A, B, and C.

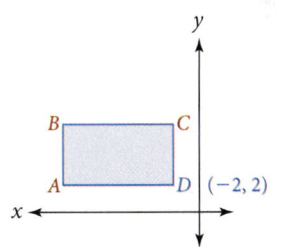

53. Solar Energy The graph shows the rise in solar thermal collectors from 1997 to 2006. Use the chart to answer the following questions.

  a. Does the graph contain the point (2000, 7,500)?
  b. Does the graph contain the point (2004, 15,000)?
  c. Does the graph contain the point (2005, 15,000)?

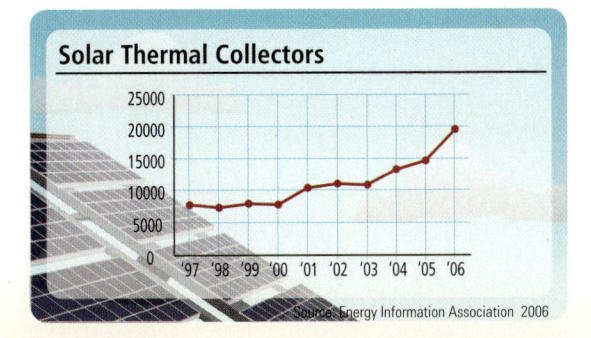

54. Health Care Costs The graph shows the projected rise in the cost of health care from 2002 to 2014. Using years as x and billions of dollars as y, write five ordered pairs that describe the information in the graph.

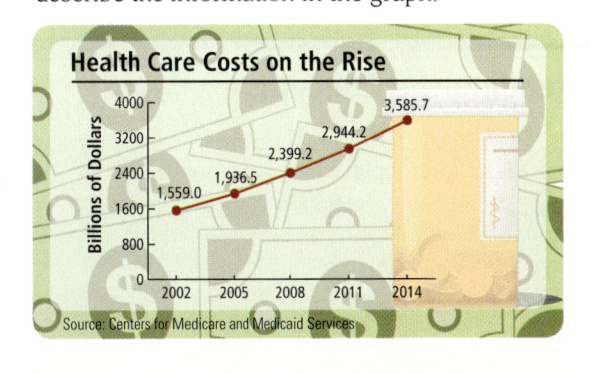

Maintaining Your Skills

The problems that follow review some of the more important skills you have learned in previous sections and chapters.

Solve each equation.

55. $5x - 4 = -3x + 12$

56. $\dfrac{1}{2} - \dfrac{y}{5} = -\dfrac{9}{10} + \dfrac{y}{2}$

57. $\dfrac{1}{2} - \dfrac{1}{8}(3t - 4) = -\dfrac{7}{8}t$

58. $3(5t - 1) - (3 - 2t) = 5t - 8$

59. $50 = \dfrac{K}{24}$

60. $3.5d = 16(3.5)^2$

61. $2(1) + y = 4$

62. $2(2y + 6) + 3y = 5$

63. $4\left(\dfrac{19}{15}\right) - 2y = 4$

64. $2x - 3\left(-\dfrac{5}{11}\right) = 4$

Getting Ready for the Next Section

65. Write -0.06 as a fraction with denominator 100.

66. Write -0.07 as a fraction with denominator 100.

67. If $y = 2x - 3$, find y when $x = 2$.

68. If $y = 2x - 3$, find x when $y = 5$.

Simplify.

69. $\dfrac{1 - (-3)}{-5 - (-2)}$

70. $\dfrac{-3 - 1}{-2 - (-5)}$

71. $\dfrac{-1 - 4}{3 - 3}$

72. $\dfrac{-3 - (-3)}{2 - (-1)}$

73. The product of $\dfrac{2}{3}$ and what number will result in

    a. 1? **b.** −1?

74. The product of 3 and what number will result in

    a. 1? **b.** −1?

Extending the Concepts

Find the x- and y-intercepts.

75. $ax + by = c$

76. $ax - by = c$

77. $\dfrac{x}{a} + \dfrac{y}{b} = 1$

78. $y = ax + b$

4.2 Introduction to Functions

Objectives
A Construct a table or a graph from a function rule.
B Identify the domain and range of a function or a relation.
C Determine whether a relation is also a function.

The ad shown in the margin appeared in the help wanted section of the local newspaper the day I was writing this section of the book. We can use the information in the ad to start an informal discussion of our next topic: functions.

An Informal Look at Functions

To begin with, suppose you have a job that pays $7.50 per hour and that you work anywhere from 0 to 40 hours per week. The amount of money you make in one week depends on the number of hours you work that week. In mathematics we say that your weekly earnings are a *function* of the number of hours you work. If we let the variable x represent hours and the variable y represent the money you make, then the relationship between x and y can be written as

$$y = 7.5x \quad \text{for} \quad 0 \leq x \leq 40$$

A Constructing Tables and Graphs

EXAMPLE 1 Construct a table and graph for the function

$$y = 7.5x \quad \text{for} \quad 0 \leq x \leq 40$$

SOLUTION Table 1 gives some of the paired data that satisfy the equation $y = 7.5x$. Figure 1 is the graph of the equation with the restriction $0 \leq x \leq 40$.

TABLE 1
WEEKLY WAGES

Hours Worked x	Rule $y = 7.5x$	Pay y
0	$y = 7.5(0)$	0
10	$y = 7.5(10)$	75
20	$y = 7.5(20)$	150
30	$y = 7.5(30)$	225
40	$y = 7.5(40)$	300

Ordered Pairs
(0, 0)
(10, 75)
(20, 150)
(30, 225)
(40, 300)

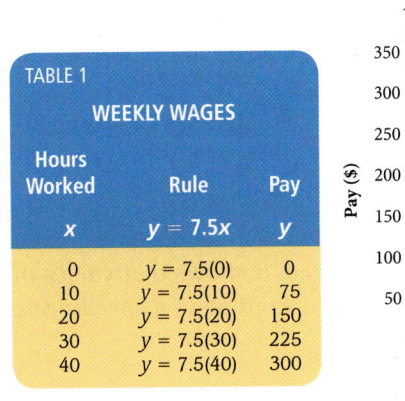

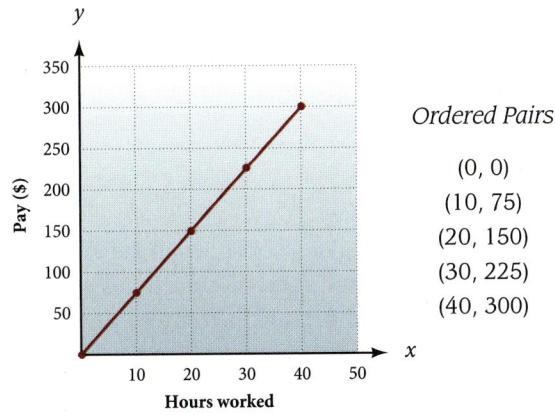

FIGURE 1 Weekly wages at $7.50 per hour

The equation $y = 7.5x$ with the restriction $0 \leq x \leq 40$, Table 1, and Figure 1 are three ways to describe the same relationship between the number of hours you work in 1 week and your gross pay for that week. In all three, we *input* values of x, and then use the function rule to *output* values of y.

HELP WANTED

YARD PERSON
Full time 40 hrs. with weekend work required. Cleaning & loading trucks. $7.50/hr.

Valid CDL with clean record & drug screen required.

Submit current MVR to KCI, 225 Suburban Rd., SLO 93405
(805) 555-3304.

PRACTICE PROBLEMS

1. Construct a table and graph for $y = 8x, \quad 0 \leq x \leq 40$

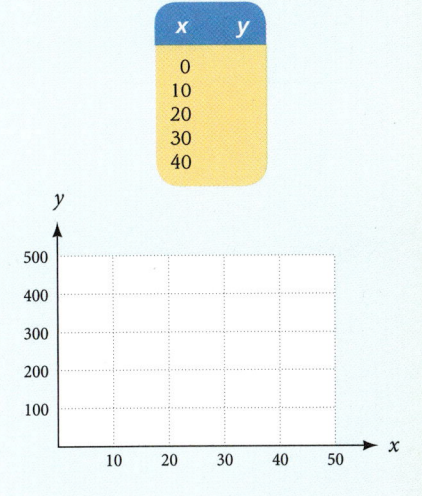

x	y
0	
10	
20	
30	
40	

Answer
1. See Solutions to Selected Practice Problems.

B Domain and Range of a Function

We began this discussion by saying that the number of hours worked during the week was from 0 to 40, so these are the values that x can assume. From the line graph in Figure 1, we see that the values of y range from 0 to 300. We call the complete set of values that x can assume the *domain* of the function. The values that are assigned to y are called the *range* of the function.

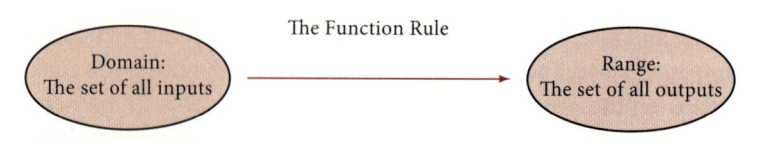

Domain: The set of all inputs — The Function Rule → Range: The set of all outputs

EXAMPLE 2 State the domain and range for the function

$$y = 7.5x, \quad 0 \leq x \leq 40$$

SOLUTION From the previous discussion we have

$$\text{Domain} = \{x \mid 0 \leq x \leq 40\}$$

$$\text{Range} = \{y \mid 0 \leq y \leq 300\}$$

Function Maps

Another way to visualize the relationship between x and y is with the diagram in Figure 2, which we call a *function map*:

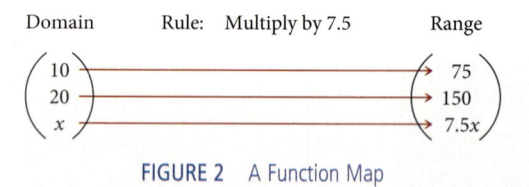

Domain   Rule: Multiply by 7.5   Range
10 → 75
20 → 150
x → 7.5x

FIGURE 2 A Function Map

Although Figure 2 does not show all the values that x and y can assume, it does give us a visual description of how x and y are related. It shows that values of y in the range come from values of x in the domain according to a specific rule (multiply by 7.5 each time).

A Formal Look at Functions

What is apparent from the preceding discussion is that we are working with paired data. The solutions to the equation $y = 7.5x$ are pairs of numbers; the points on the line graph in Figure 1 come from paired data; and the diagram in Figure 2 pairs numbers in the domain with numbers in the range. We are now ready for the formal definition of a function.

> **Definition**
> A **function** is a rule that pairs each element in one set, called the **domain**, with exactly one element from a second set, called the **range**.

In other words, a function is a rule for which each input is paired with exactly one output.

---

2. State the domain and range for
$y = 8x, \quad 0 \leq x \leq 40$

Answer
2. Domain = $\{x \mid 0 \leq x \leq 40\}$
   Range = $\{y \mid 0 \leq y \leq 320\}$

4.2 Introduction to Functions

Functions as Ordered Pairs

The function rule $y = 7.5x$ from Example 1 produces ordered pairs of numbers (x, y). The same thing happens with all functions: The function rule produces ordered pairs of numbers. We use this result to write an alternative definition for a function.

> **Alternative Definition**
> A **function** is a set of ordered pairs in which no two different ordered pairs have the same first coordinate. The set of all first coordinates is called the **domain** of the function. The set of all second coordinates is called the **range** of the function.

The restriction on first coordinates in the alternative definition keeps us from assigning a number in the domain to more than one number in the range.

C A Relation That Is Not a Function

You may be wondering if any sets of paired data fail to qualify as functions. The answer is yes, as the next example reveals.

EXAMPLE 3 Table 2 shows the prices of used Ford Mustangs that were listed in the local newspaper. The diagram in Figure 3 is called a *scatter diagram*. It gives a visual representation of the data in Table 2. Why is this data not a function?

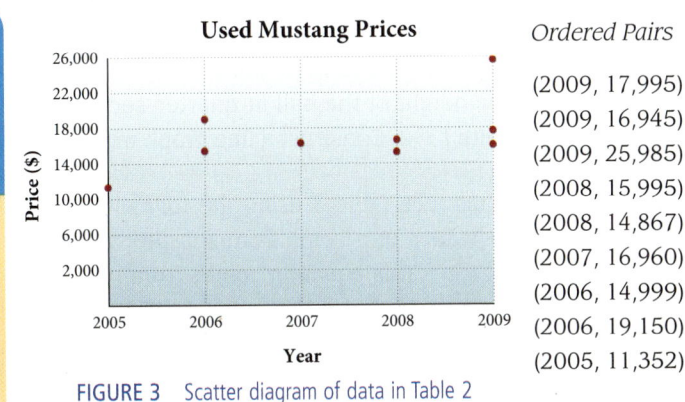

TABLE 2
USED MUSTANG PRICES

Year x	Price ($) y
2009	17,995
2009	16,945
2009	25,985
2008	15,995
2008	14,867
2007	16,960
2006	14,999
2006	19,150
2005	11,352

Ordered Pairs

(2009, 17,995)
(2009, 16,945)
(2009, 25,985)
(2008, 15,995)
(2008, 14,867)
(2007, 16,960)
(2006, 14,999)
(2006, 19,150)
(2005, 11,352)

FIGURE 3 Scatter diagram of data in Table 2

SOLUTION In Table 2, the year 2009 is paired with three different prices: $17,995, $16,945, and $25,985. That is enough to disqualify the data from belonging to a function. For a set of paired data to be considered a function, each number in the domain must be paired with exactly one number in the range.

Still, there is a relationship between the first coordinates and second coordinates in the used-car data in Example 3. It is not a function relationship, but it is a relationship. To classify all relationships specified by ordered pairs, whether they are functions or not, we include the following two definitions.

> **Definition**
> A **relation** is a rule that pairs each element in one set, called the **domain**, with one or more elements from a second set, called the **range**.

3. With respect to Table 2 and Figure 3, how many outputs are paired with each of the following inputs?
   a. 2008
   b. 2006

Answer
3. a. 2 b. 2

Chapter 4 Functions

> **Alternative Definition**
> A **relation** is a set of ordered pairs. The set of all first coordinates is the **domain** of the relation. The set of all second coordinates is the **range** of the relation.

Here are some facts that will help clarify the distinction between relations and functions.

1. Any rule that assigns numbers from one set to numbers in another set is a relation. If that rule makes the assignment so that no input has more than one output, then it is also a function.
2. Any set of ordered pairs is a relation. If none of the first coordinates of those ordered pairs is repeated, the set of ordered pairs is also a function.
3. Every function is a relation.
4. Not every relation is a function.

Graphing Relations and Functions

To give ourselves a wider perspective on functions and relations, we consider some equations whose graphs are not straight lines.

EXAMPLE 4 Kendra is tossing a softball into the air with an underhand motion. The distance of the ball above her hand at any time is given by the function

$$h = 32t - 16t^2 \quad \text{for} \quad 0 \leq t \leq 2$$

where h is the height of the ball in feet and t is the time in seconds. Construct a table that gives the height of the ball at quarter-second intervals, starting with $t = 0$ and ending with $t = 2$. Construct a line graph from the table.

SOLUTION We construct Table 3 using the following values of t: $0, \frac{1}{4}, \frac{1}{2}, \frac{3}{4}, 1, \frac{5}{4}, \frac{3}{2}, \frac{7}{4}, 2$. The values of h come from substituting these values of t into the equation $h = 32t - 16t^2$. (This equation comes from physics. If you take a physics class, you will learn how to derive this equation.) Then we construct the graph in Figure 4 from the table. The graph appears only in the first quadrant because neither t nor h can be negative.

TABLE 3

TOSSING A SOFTBALL INTO THE AIR

Time (sec) t	Function Rule $h = 32t - 16t^2$	Distance (ft) h
0	$h = 32(0) - 16(0)^2 = 0 - 0 = 0$	0
$\frac{1}{4}$	$h = 32\left(\frac{1}{4}\right) - 16\left(\frac{1}{4}\right)^2 = 8 - 1 = 7$	7
$\frac{1}{2}$	$h = 32\left(\frac{1}{2}\right) - 16\left(\frac{1}{2}\right)^2 = 16 - 4 = 12$	12
$\frac{3}{4}$	$h = 32\left(\frac{3}{4}\right) - 16\left(\frac{3}{4}\right)^2 = 24 - 9 = 15$	15
1	$h = 32(1) - 16(1)^2 = 32 - 16 = 16$	16
$\frac{5}{4}$	$h = 32\left(\frac{5}{4}\right) - 16\left(\frac{5}{4}\right)^2 = 40 - 25 = 15$	15
$\frac{3}{2}$	$h = 32\left(\frac{3}{2}\right) - 16\left(\frac{3}{2}\right)^2 = 48 - 36 = 12$	12
$\frac{7}{4}$	$h = 32\left(\frac{7}{4}\right) - 16\left(\frac{7}{4}\right)^2 = 56 - 49 = 7$	7
2	$h = 32(2) - 16(2)^2 = 64 - 64 = 0$	0

---

4. Kendra is tossing a softball into the air so that the distance h the ball is above her hand t seconds after she begins the toss is given by
$$h = 48t - 16t^2, \quad 0 \leq t \leq 3$$
Construct a table and line graph for this function.

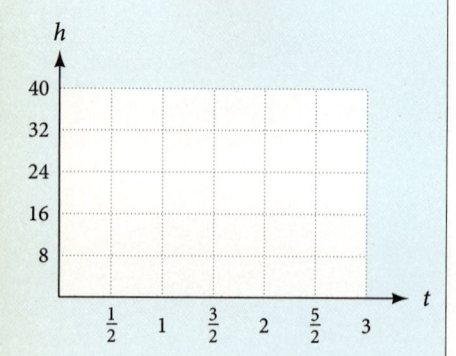

t	h
0	
$\frac{1}{2}$	
1	
$\frac{3}{2}$	
2	
$\frac{5}{2}$	
3	

4.2 Introduction to Functions

FIGURE 4

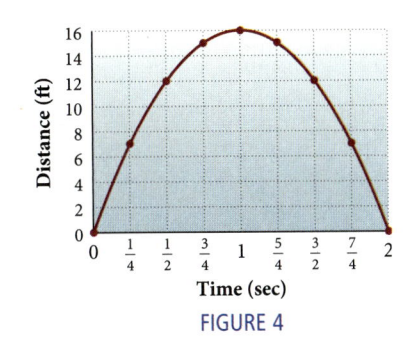

Here is a summary of what we know about functions as it applies to this example: We input values of *t* and output values of *h* according to the function rule

$$h = 32t - 16t^2 \quad \text{for} \quad 0 \leq t \leq 2$$

The domain is given by the inequality that follows the equation; it is

$$\text{Domain} = \{t \mid 0 \leq t \leq 2\}$$

The range is the set of all outputs that are possible by substituting the values of *t* from the domain into the equation. From our table and graph, it seems that the range is

$$\text{Range} = \{h \mid 0 \leq h \leq 16\}$$

USING TECHNOLOGY

More About Example 4

Most graphing calculators can easily produce the information in Table 3. Simply set Y_1 equal to $32X - 16X^2$. Then set up the table so it starts at 0 and increases by an increment of 0.25 each time. (On a TI-83/84, use the TBLSET key to set up the table.)

Table Setup

Table minimum = 0
Table increment = .25
Dependent variable: Auto
Independent variable: Auto

Y-Variables Setup

$Y_1 = 32X - 16X^2$

The table will look like this:

X	Y_1
0.00	0
0.25	7
0.50	12
0.75	15
1.00	16
1.25	15
1.50	12

Answer
4. See Solutions to Selected Practice Problems.

5. Use the equation

$$x = y^2 - 4$$

to fill in the following table. Then use the table to sketch the graph.

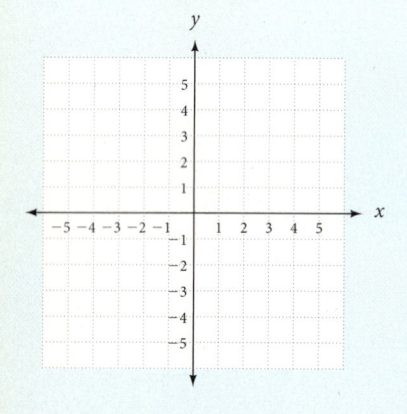

x	y
	−3
	−2
	−1
	0
	1
	2
	3

Chapter 4 Functions

EXAMPLE 5
Sketch the graph of $x = y^2$.

SOLUTION Without going into much detail, we graph the equation $x = y^2$ by finding a number of ordered pairs that satisfy the equation, plotting these points, and then drawing a smooth curve that connects them. A table of values for x and y that satisfy the equation follows, along with the graph of $x = y^2$ shown in Figure 5.

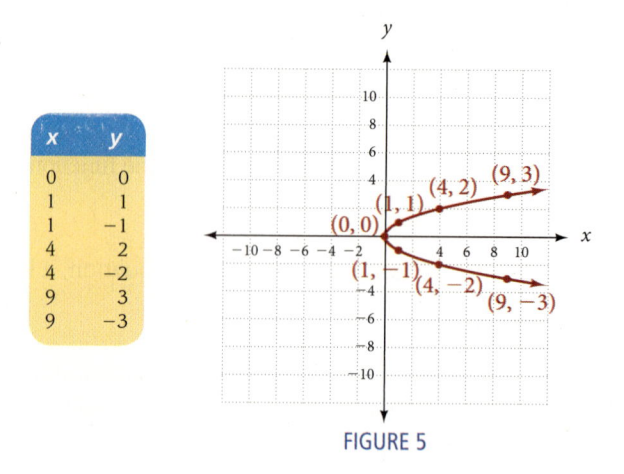

x	y
0	0
1	1
1	−1
4	2
4	−2
9	3
9	−3

FIGURE 5

As you can see from looking at the table and the graph in Figure 5, several ordered pairs whose graphs lie on the curve have repeated first coordinates. For instance, (1, 1) and (1, −1), (4, 2) and (4, −2), and (9, 3) and (9, −3). The graph is therefore not the graph of a function.

Vertical Line Test

Look back at the scatter diagram for used Mustang prices shown in Figure 3. Notice that some of the points on the diagram lie above and below each other along vertical lines. This is an indication that the data do not constitute a function. Two data points that lie on the same vertical line must have come from two ordered pairs with the same first coordinates.

Now, look at the graph shown in Figure 5. The reason this graph is the graph of a relation, but not of a function, is that some points on the graph have the same first coordinates—for example, the points (4, 2) and (4, −2). Furthermore, any time two points on a graph have the same first coordinates, those points must lie on a vertical line. [To convince yourself, connect the points (4, 2) and (4, −2) with a straight line. You will see that it must be a vertical line.] This allows us to write the following test that uses the graph to determine whether a relation is also a function.

> **Vertical Line Test**
> If a vertical line crosses the graph of a relation in more than one place, the relation cannot be a function. If no vertical line can be found that crosses a graph in more than one place, then the graph is the graph of a function.

If we look back to the graph of $h = 32t - 16t^2$ as shown in Figure 4, we see that no vertical line can be found that crosses this graph in more than one place. The graph shown in Figure 4 is therefore the graph of a function.

Answer
5. See Solutions to Selected Practice Problems.

4.2 Introduction to Functions

EXAMPLE 6 Graph $y = |x|$. Use the graph to determine whether we have the graph of a function. State the domain and range.

SOLUTION We let x take on values of $-4, -3, -2, -1, 0, 1, 2, 3,$ and 4. The corresponding values of y are shown in the table. The graph is shown in Figure 6.

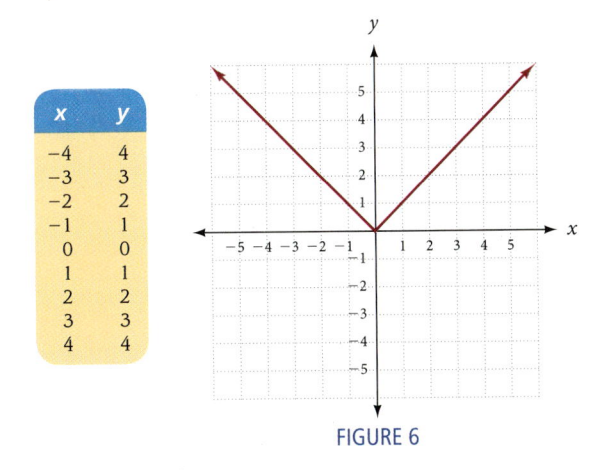

x	y
-4	4
-3	3
-2	2
-1	1
0	0
1	1
2	2
3	3
4	4

FIGURE 6

Because no vertical line can be found that crosses the graph in more than one place, $y = |x|$ is a function. The domain is all real numbers. The range is $\{y \mid y \geq 0\}$.

STUDY SKILLS
Be Focused, Not Distracted
I have students who begin their assignments by asking themselves, "Why am I taking this class?" or, "When am I ever going to use this stuff?" If you are asking yourself similar questions, you may be distracting yourself from doing the things that will produce the results you want in this course. Don't dwell on questions and evaluations of the class that can be used as excuses for not doing well. If you want to succeed in this course, focus your energy and efforts toward success.

Getting Ready for Class
After reading through the preceding section, respond in your own words and in complete sentences.

1. What is a function?
2. What is the vertical line test?
3. Is every line the graph of a function? Explain.
4. Which variable is usually associated with the domain of a function?

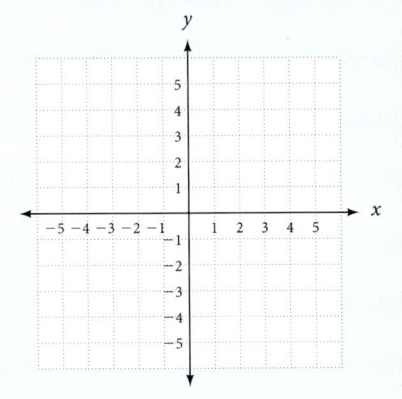

---

6. Graph $y = |x| - 3$.

Answer
6. See Solutions to Selected Practice Problems.

Left Blank Intentionally

Problem Set 4.2

B For each of the following relations, give the domain and range, and indicate which are also functions.

1. {(1, 3), (2, 5), (4, 1)}

2. {(3, 1), (5, 7), (2, 3)}

3. {(−1, 3), (1, 3), (2, −5)}

4. {(3, −4), (−1, 5), (3, 2)}

5. {(7, −1), (3, −1), (7, 4)}

6. {(5, −2), (3, −2), (5, −1)}

7. {(a, 3), (b, 4), (c, 3), (d, 5)}

8. {(a, 5), (b, 5), (c, 4), (d, 5)}

9. {(a, 1), (a, 2), (a, 3), (a, 4)}

10. {(a, 1), (b, 1), (c, 1), (d, 1)}

C State whether each of the following graphs represents a function. [Example 6]

11.

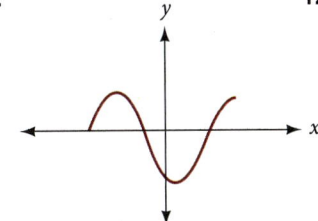

12.

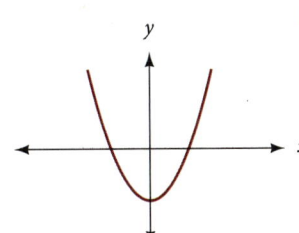

13.

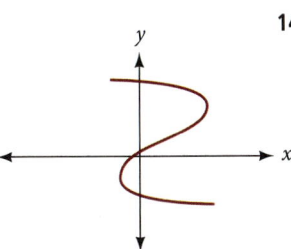

14.

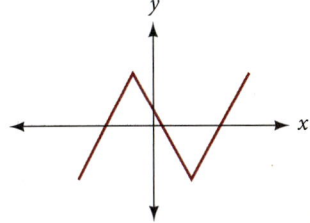

15.

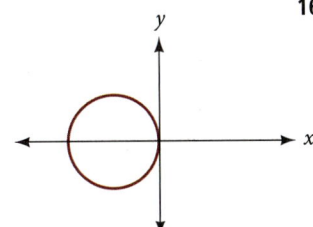

16.

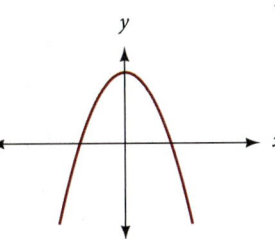

17.

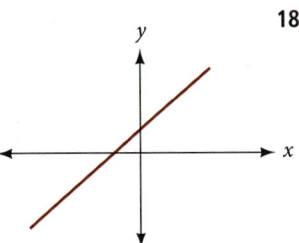

18.

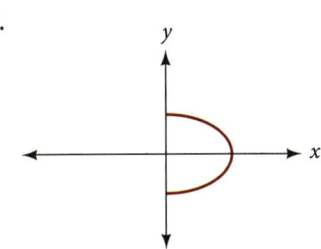

19.

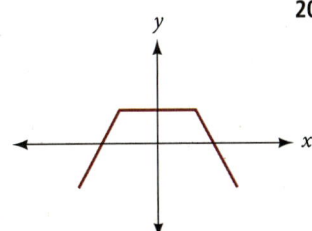

20.

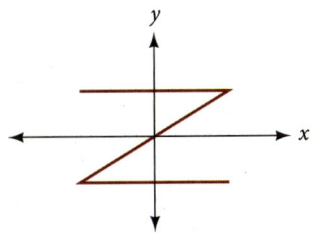

Chapter 4 Functions

Determine the domain and range of the following functions. Assume the entire function is shown.

21.

22.

23.

24.

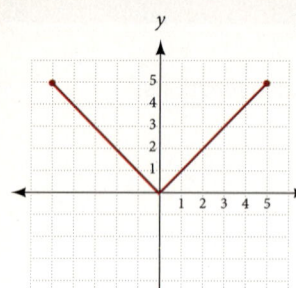

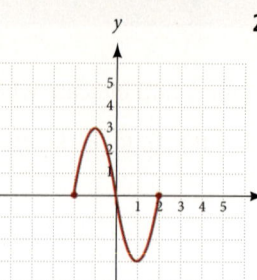

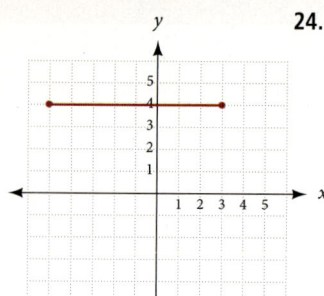

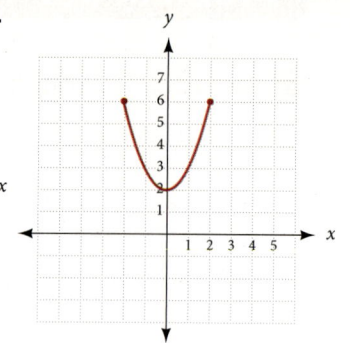

Graph each of the following relations. In each case, use the graph to find the domain and range, and indicate whether the graph is the graph of a function. [Example 5]

25. $y = x^2 - 1$

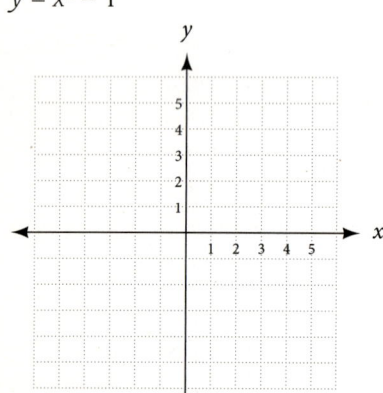

26. $y = x^2 + 1$

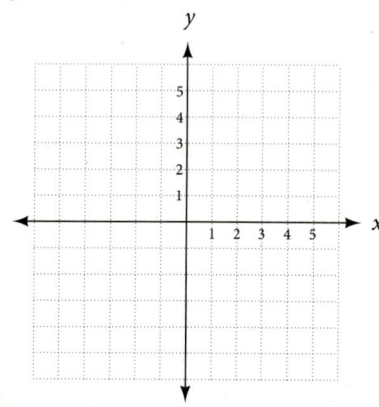

27. $y = x^2 + 4$

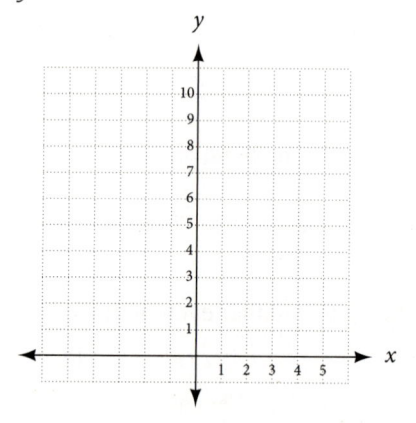

28. $y = x^2 - 9$

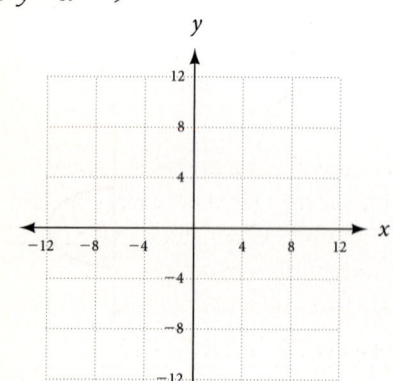

29. $x = y^2 - 1$

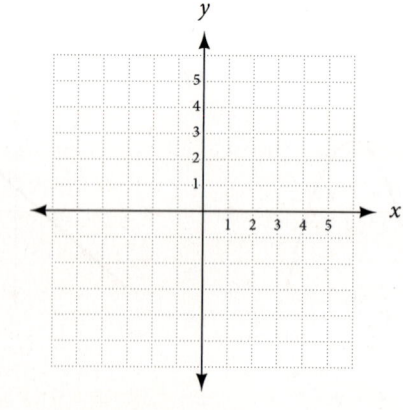

30. $x = y^2 + 1$

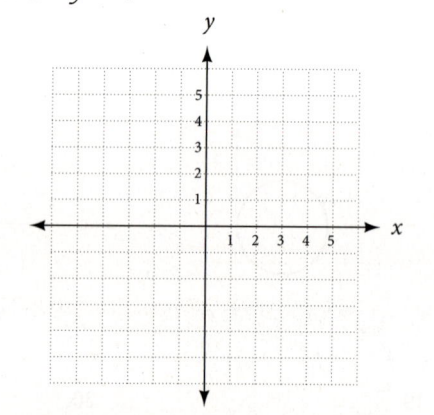

31. $x = y^2 + 4$

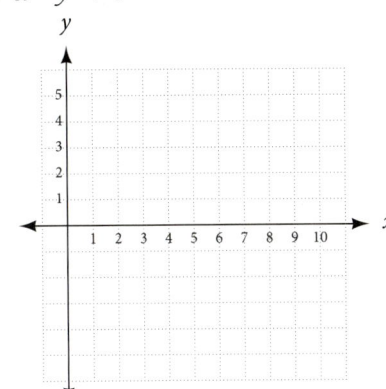

32. $x = y^2 - 9$

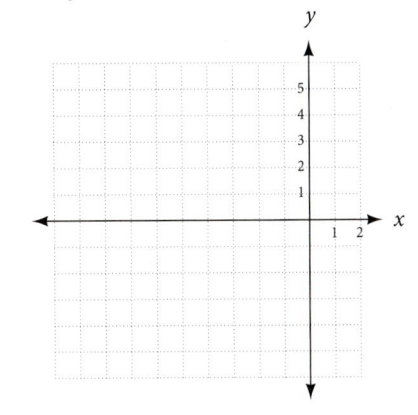

33. $y = |x - 2|$

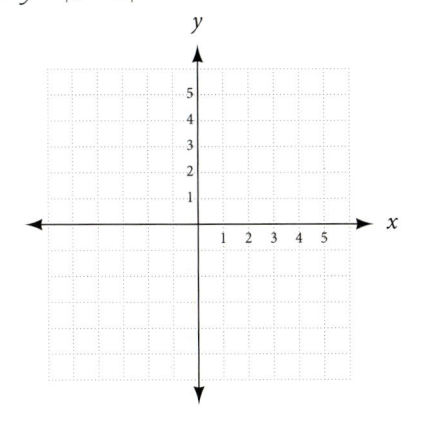

34. $y = |x + 2|$

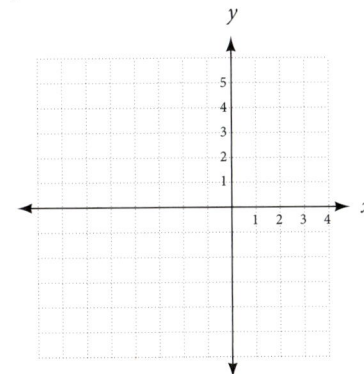

35. $y = |x| - 2$

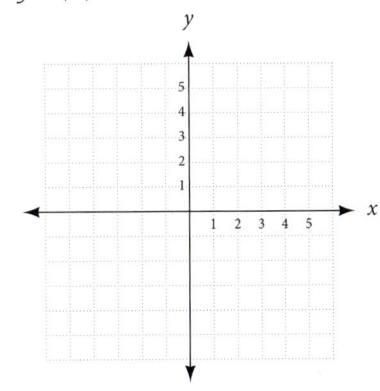

36. $y = |x| + 2$

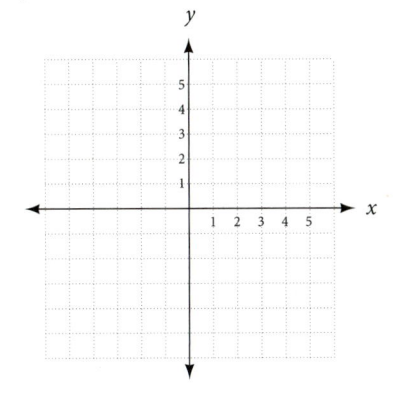

Applying the Concepts [Examples 3–4]

37. Camera Phones The chart shows the estimated number of camera phones and non-camera phones sold from 2004 to 2010. Using the chart, list all the values in the domain and range for the total phones sales.

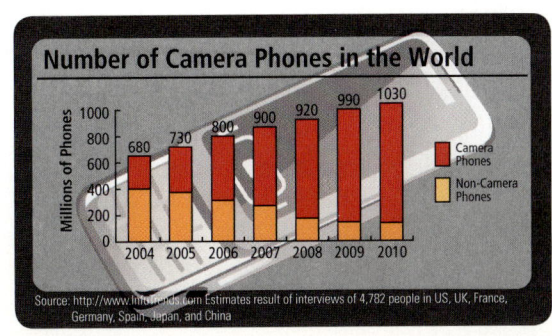

38. Light Bulbs The chart shows a comparison of power usage between incandescent and energy efficient light bulbs. Use the chart to state the domain and range of the function for an energy efficient bulb.

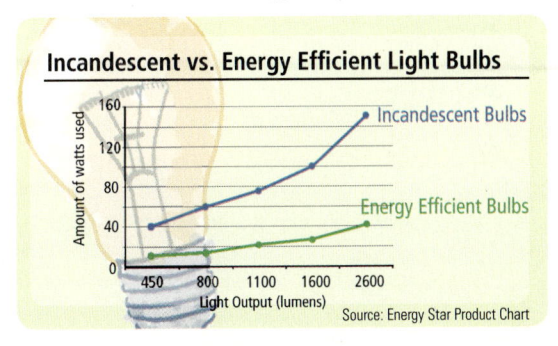

39. **Weekly Wages** Suppose you have a job that pays $8.50 per hour and you work anywhere from 10 to 40 hours per week.
    a. Write an equation, with a restriction on the variable x, that gives the amount of money, y, you will earn for working x hours in one week.

    b. Use the function rule to complete the table.

    c. Use the data from the table to graph the function.

    d. State the domain and range of this function.

    e. What is the minimum amount you can earn in a week with this job? What is the maximum amount?

TABLE 4
WEEKLY WAGES

Hours Worked	Function Rule	Gross Pay ($)
x		y
10		
20		
30		
40		

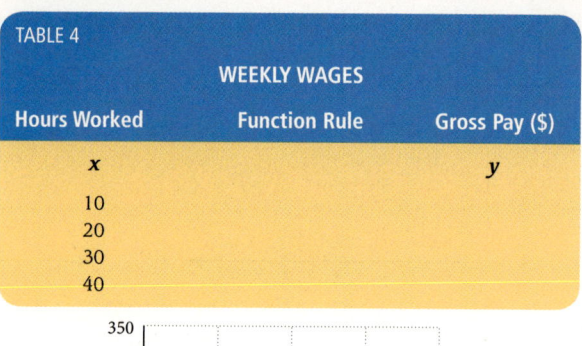

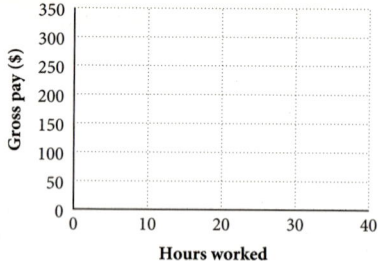

FIGURE 7 Template for line graph

40. **Weekly Wages** The ad shown here was in the local newspaper. Suppose you are hired for the job described in the ad.
    a. If x is the number of hours you work per week and y is your weekly gross pay, write the equation for y. (Be sure to include any restrictions on the variable x that are given in the ad.)

    b. Use the function rule to complete the table.

    c. Use the data from the table to graph the function.

    d. State the domain and range of this function.

    e. What is the minimum amount you can earn in a week with this job? What is the maximum amount?

TABLE 5
WEEKLY WAGES

Hours Worked	Function Rule	Gross Pay ($)
x		y
15		
20		
25		
30		

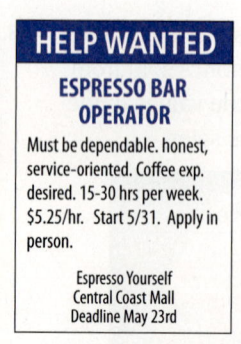

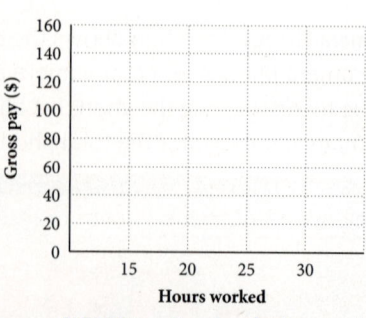

HELP WANTED
ESPRESSO BAR OPERATOR
Must be dependable, honest, service-oriented. Coffee exp. desired. 15-30 hrs per week. $5.25/hr. Start 5/31. Apply in person.

Espresso Yourself
Central Coast Mall
Deadline May 23rd

FIGURE 8 Template for line graph

4.2 Problem Set

41. Tossing a Coin Hali is tossing a quarter into the air with an underhand motion. The distance the quarter is above her hand at any time is given by the function

$$h = 16t - 16t^2 \quad \text{for} \quad 0 \le t \le 1$$

where h is the height of the quarter in feet, and t is the time in seconds.

a. Fill in the table.

b. State the domain and range of this function.

c. Use the data from the table to graph the function.

Time (sec) t	Function Rule $h = 16t - 16t^2$	Distance (ft) h
0		
0.1		
0.2		
0.3		
0.4		
0.5		
0.6		
0.7		
0.8		
0.9		
1		

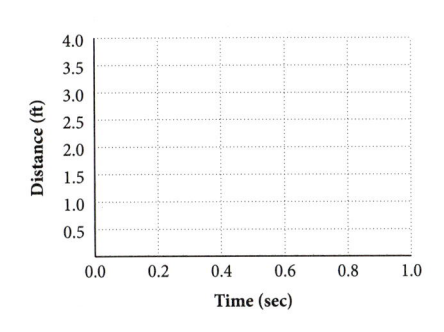

42. Intensity of Light The formula below gives the intensity of light that falls on a surface at various distances from a 100-watt light bulb:

$$I = \frac{120}{d^2} \quad \text{for} \quad d > 0$$

where I is the intensity of light (in lumens per square foot), and d is the distance (in feet) from the light bulb to the surface.

a. Fill in the table.

b. Use the data from the table to graph the function.

Distance (ft) d	Function Rule $I = \dfrac{120}{d^2}$	Intensity I
1		
2		
3		
4		
5		
6		

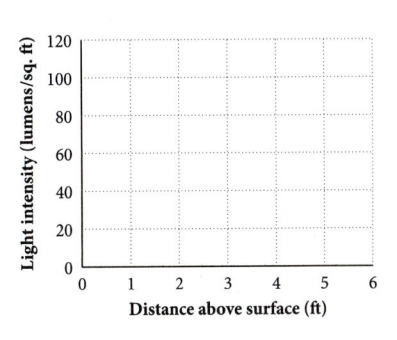

43. Area of a Circle The formula for the area A of a circle with radius r is given by $A = \pi r^2$. The formula shows that A is a function of r.
   a. Graph the function $A = \pi r^2$ for $0 \leq r \leq 3$. (On the graph, let the horizontal axis be the r-axis, and let the vertical axis be the A-axis.)

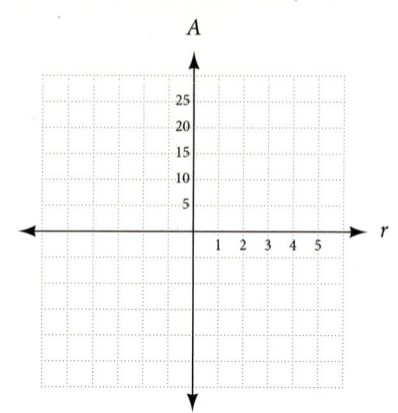

   b. State the domain and range of the function $A = \pi r^2$, $0 \leq r \leq 3$.

44. Area and Perimeter of a Rectangle A rectangle is 2 inches longer than it is wide. Let $x =$ the width, $P =$ the perimeter, and $A =$ the area of the rectangle.
   a. Write an equation that will give the perimeter P in terms of the width x of the rectangle. Are there any restrictions on the values that x can assume?

   b. Graph the relationship between P and x.

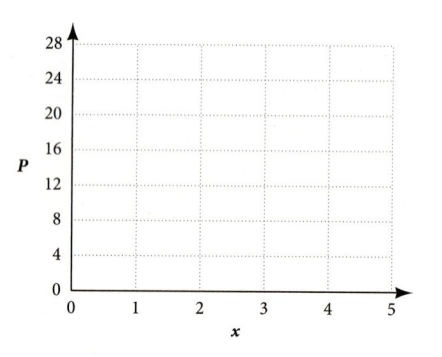

45. Tossing a Ball A ball is thrown straight up into the air from ground level. The relationship between the height h of the ball at any time t is illustrated by the following graph:

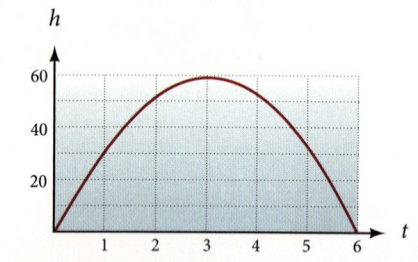

The horizontal axis represents time t, and the vertical axis represents height h.
   a. Is this graph the graph of a function?

   b. State the domain and range.

   c. At what time does the ball reach its maximum height?

   d. What is the maximum height of the ball?

   e. At what time does the ball hit the ground?

46. Company Profits The amount of profit a company earns is based on the number of items it sells. The relationship between the profit P and number of items it sells x, is illustrated by the following graph:

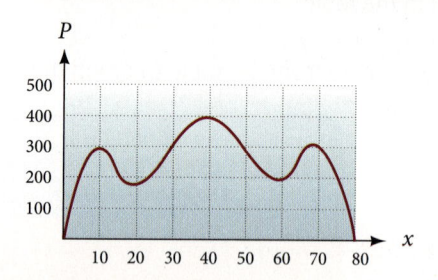

The horizontal axis represents items sold, x, and the vertical axis represents the profit, P.
   a. Is this graph the graph of a function?

   b. State the domain and range.

   c. How many items must the company sell to make their maximum profit?

   d. What is their maximum profit?

47. Profits Match each of the following statements to the appropriate graph indicated by labels I-IV.

 a. Sarah works 25 hours to earn $250
 b. Justin works 35 hours to earn $560
 c. Rosemary works 30 hours to earn $360
 d. Marcus works 40 hours to earn $320

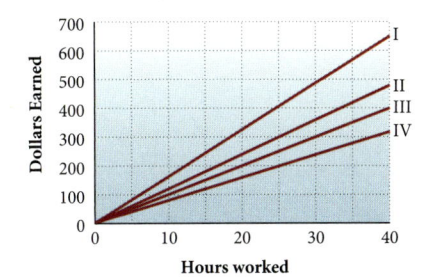

48. Find an equation for each of the functions shown in the graph. Show dollars earned, E as a function of hours worked, t. Then, indicate the domain and range of each function.

 a. Graph I: E =
    Domain = {t | }
    Range = {E | }

 b. Graph II: E =
    Domain = {t | }
    Range = {E | }

 c. Graph III: E =
    Domain = {t | }
    Range = {E | }

 d. Graph IV: E =
    Domain = {t | }
    Range = {E | }

■ Maintaining Your Skills

The problems that follow review some of the more important skills you have learned in previous sections and chapters.

For the equation $y = 3x - 2$:

49. Find y if x is 4.
50. Find y if x is 0.
51. Find y if x is −4.
52. Find y if x is −2.

For the equation $y = x^2 - 3$:

53. Find y if x is 2.
54. Find y if x is −2.
55. Find y if x is 0.
56. Find y if x is −4.

57. If $x - 2y = 4$ and $x = \frac{8}{5}$, find y.

58. If $5x - 10y = 15$, find y when x is 3.

59. Let $x = 0$ and $y = 0$ in $y = a(x - 80)^2 + 70$ and solve for a.

60. Find R if $p = 2.5$ and $R = (900 - 300p)p$.

Getting Ready for the Next Section

Simplify. Round to the nearest whole number if necessary.

61. 7.5(20)

62. 60 ÷ 7.5

63. 4(3.14)(9)

64. $\frac{4}{3}(3.14) \cdot 3^3$

65. 4(−2) − 1

66. $3(3)^2 + 2(3) - 1$

67. If $s = \frac{60}{t}$, find s when
   a. $t = 10$ **b.** $t = 8$

68. If $y = 3x^2 + 2x - 1$, find y when
   a. $x = 0$ **b.** $x = -2$

69. Find the value of $x^2 + 2$ for
   a. $x = 5$ **b.** $x = -2$

70. Find the value of $125 \cdot 2^t$ for
   a. $t = 0$ **b.** $t = 1$

Extending the Concepts

Graph each of the following relations. In each case, use the graph to find the domain and range, and indicate whether the graph is the graph of a function.

71. $y = 5 - |x|$

72. $y = |x| - 3$

73. $x = |y| + 3$

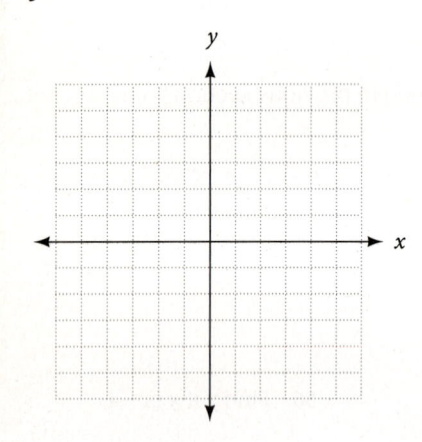

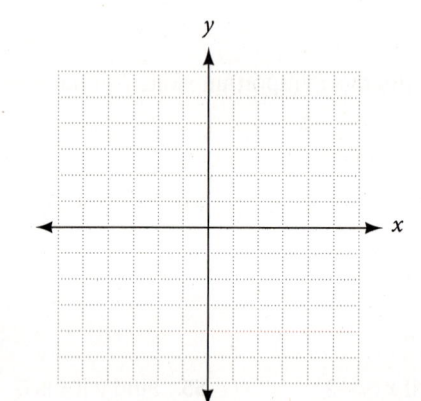

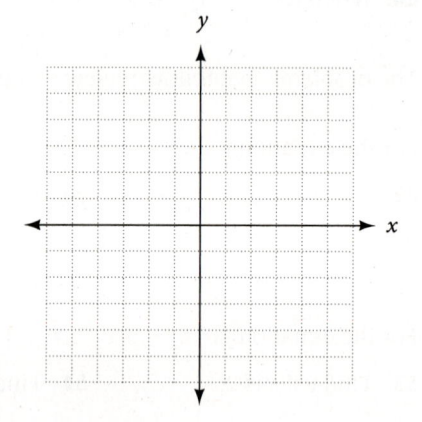

74. $x = 2 - |y|$

75. $|x| + |y| = 4$

76. $2|x| + |y| = 6$

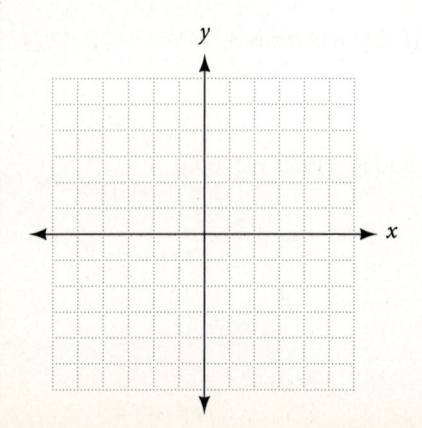

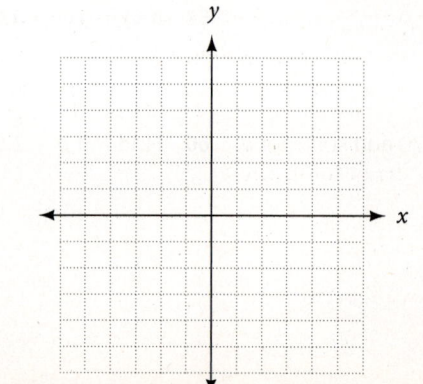

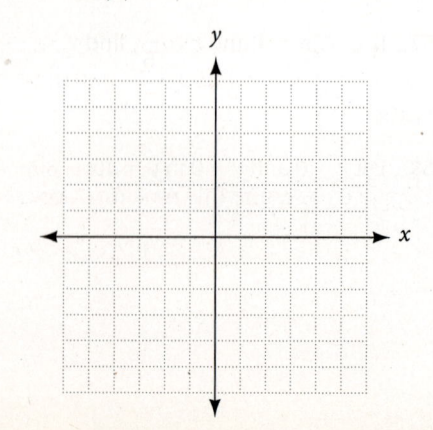

4.3 Function Notation

Objectives

A Use function notation to find the value of a function for a given value of the variable.

A Evaluate a Function at a Point

Let's return to the discussion that introduced us to functions. If a job pays $7.50 per hour for working from 0 to 40 hours a week, then the amount of money, y, earned in 1 week is a function of the number of hours worked, x. The exact relationship between x and y is written

$$y = 7.5x \quad \text{for} \quad 0 \leq x \leq 40$$

Because the amount of money earned, y, depends on the number of hours worked, x, we call y the *dependent variable* and x the *independent variable*. Furthermore, if we let f represent all the ordered pairs produced by the equation, then we can write

$$f = \{(x, y) \mid y = 7.5x \text{ and } 0 \leq x \leq 40\}$$

Once we have named a function with a letter, we can use an alternative notation to represent the dependent variable y. The alternative notation for y is $f(x)$. It is read "f of x" and can be used instead of the variable y when working with functions. The notation y and the notation $f(x)$ are equivalent—that is,

$$y = 7.5x \Leftrightarrow f(x) = 7.5x$$

When we use the notation $f(x)$ we are using *function notation*. The benefit of using function notation is that we can write more information with fewer symbols than we can by using just the variable y. For example, asking how much money a person will make for working 20 hours is simply a matter of asking for $f(20)$. Without function notation, we would have to say, "find the value of y that corresponds to a value of $x = 20$." To illustrate further, using the variable y, we can say, "y is 150 when x is 20." Using the notation $f(x)$, we simply say, "$f(20) = 150$." Each expression indicates that you will earn $150 for working 20 hours.

EXAMPLE 1
If $f(x) = 7.5x$, find $f(0)$, $f(10)$, and $f(20)$.

SOLUTION To find $f(0)$ we substitute 0 for x in the expression $7.5x$ and simplify. We find $f(10)$ and $f(20)$ in a similar manner—by substitution.

If $\qquad f(x) = 7.5x$

then $\qquad f(\mathbf{0}) = 7.5(\mathbf{0}) = 0$

$\qquad\qquad f(\mathbf{10}) = 7.5(\mathbf{10}) = 75$

$\qquad\qquad f(\mathbf{20}) = 7.5(\mathbf{20}) = 150$

If we changed the example in the discussion that opened this section so that the hourly wage was $6.50 per hour, we would have a new equation to work with:

$$y = 6.5x \quad \text{for} \quad 0 \leq x \leq 40$$

Suppose we name this new function with the letter g. Then

$$g = \{(x, y) \mid y = 6.5x \text{ and } 0 \leq x \leq 40\}$$

and

$$g(x) = 6.5x$$

PRACTICE PROBLEMS

1. If $f(x) = 8x$, find
   a. $f(0)$
   b. $f(5)$
   c. $f(10.5)$

Note Some students like to think of functions as machines. Values of x are put into the machine, which transforms them into values of $f(x)$, which then are output by the machine.

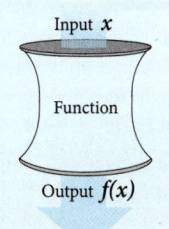

Input x

Function

Output $f(x)$

Answer
1. a. 0 b. 40 c. 84

Chapter 4 Functions

If we want to talk about both functions in the same discussion, having two different letters, f and g, makes it easy to distinguish between them. For example, because $f(x) = 7.5x$ and $g(x) = 6.5x$, asking how much money a person makes for working 20 hours is simply a matter of asking for $f(20)$ or $g(20)$, avoiding any confusion over which hourly wage we are talking about.

The diagrams shown in Figure 1 further illustrate the similarities and differences between the two functions we have been discussing.

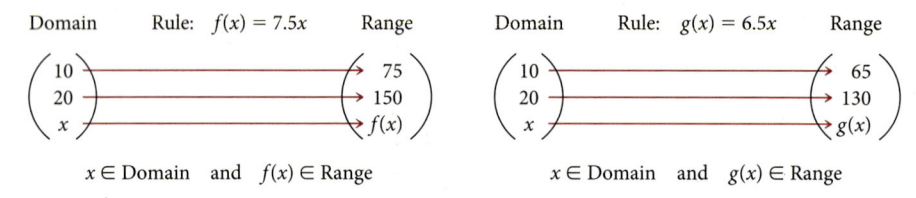

> **Note** The symbol $\in$ means "is a member of" a given set.

Domain   Rule: $f(x) = 7.5x$ Range
$\begin{pmatrix} 10 \\ 20 \\ x \end{pmatrix} \longrightarrow \begin{pmatrix} 75 \\ 150 \\ f(x) \end{pmatrix}$

$x \in$ Domain and $f(x) \in$ Range

Domain   Rule: $g(x) = 6.5x$ Range
$\begin{pmatrix} 10 \\ 20 \\ x \end{pmatrix} \longrightarrow \begin{pmatrix} 65 \\ 130 \\ g(x) \end{pmatrix}$

$x \in$ Domain and $g(x) \in$ Range

FIGURE 1 Function maps

Function Notation and Graphs

We can visualize the relationship between x and $f(x)$ or $g(x)$ on the graphs of the two functions. Figure 2 shows the graph of $f(x) = 7.5x$ along with two additional line segments. The horizontal line segment corresponds to $x = 20$, and the vertical line segment corresponds to $f(20)$. Figure 3 shows the graph of $g(x) = 6.5x$ along with the horizontal line segment that corresponds to $x = 20$, and the vertical line segment that corresponds to $g(20)$. (Note that the domain in each case is restricted to $0 \leq x \leq 40$.)

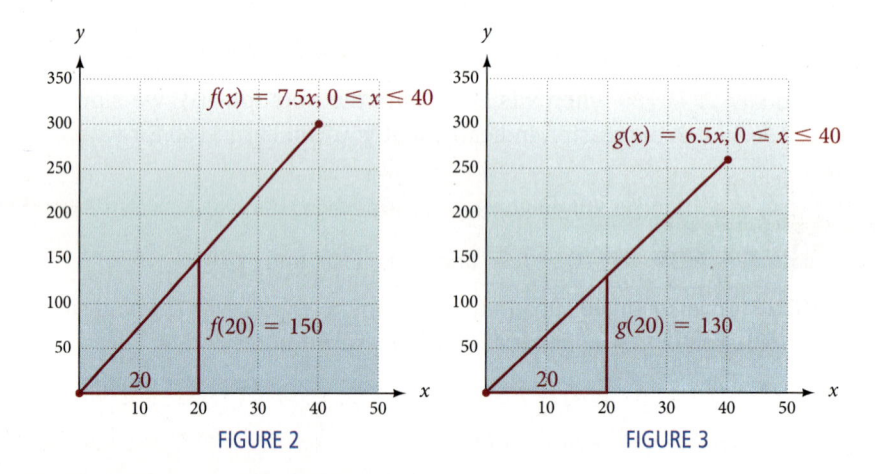

FIGURE 2 $f(x) = 7.5x, 0 \leq x \leq 40$, $f(20) = 150$

FIGURE 3 $g(x) = 6.5x, 0 \leq x \leq 40$, $g(20) = 130$

Using Function Notation

The remaining examples in this section show a variety of ways to use and interpret function notation.

EXAMPLE 2 If it takes Lorena t minutes to run a mile, then her average speed $s(t)$ in miles per hour is given by the formula

$$s(t) = \frac{60}{t} \quad \text{for} \quad t > 0$$

Find $s(10)$ and $s(8)$, and then explain what they mean.

SOLUTION To find $s(10)$, we substitute 10 for t in the equation and simplify:

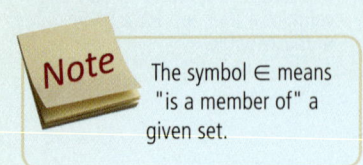

$$s(\mathbf{10}) = \frac{60}{\mathbf{10}} = 6$$

2. When Lorena runs a mile in t minutes, then her average speed in feet per seconds is given by $s(t) = \frac{88}{t}, t > 0$
   a. Find $s(8)$ and explain what it means.
   b. Find $s(11)$ and explain what it means.

In words: When Lorena runs a mile in 10 minutes, her average speed is 6 miles per hour.

We calculate $s(8)$ by substituting 8 for t in the equation. Doing so gives us

$$s(\mathbf{8}) = \frac{60}{\mathbf{8}} = 7.5$$

In words: Running a mile in 8 minutes is running at a rate of 7.5 miles per hour.

EXAMPLE 3 A painting is purchased as an investment for $125. If its value increases continuously so that it doubles every 5 years, then its value is given by the function

$$V(t) = 125 \cdot 2^{t/5} \quad \text{for} \quad t \geq 0$$

where t is the number of years since the painting was purchased, and $V(t)$ is its value (in dollars) at time t. Find $V(5)$ and $V(10)$, and explain what they mean.

SOLUTION The expression $V(5)$ is the value of the painting when $t = 5$ (5 years after it is purchased). We calculate $V(5)$ by substituting 5 for t in the equation $V(t) = 125 \cdot 2^{t/5}$. Here is our work:

$$V(\mathbf{5}) = 125 \cdot 2^{\mathbf{5}/5} = 125 \cdot 2^1 = 125 \cdot 2 = 250$$

In words: After 5 years, the painting is worth $250.

The expression $V(10)$ is the value of the painting after 10 years. To find this number, we substitute 10 for t in the equation:

$$V(\mathbf{10}) = 125 \cdot 2^{\mathbf{10}/5} = 125 \cdot 2^2 = 125 \cdot 4 = 500$$

In words: The value of the painting 10 years after it is purchased is $500.

EXAMPLE 4 A balloon has the shape of a sphere with a radius of 3 inches. Use the following formulas to find the volume and surface area of the balloon.

$$V(r) = \frac{4}{3}\pi r^3 \qquad S(r) = 4\pi r^2$$

SOLUTION As you can see, we have used function notation to write the two formulas for volume and surface area because each quantity is a function of the radius. To find these quantities when the radius is 3 inches, we evaluate $V(3)$ and $S(3)$:

$$V(\mathbf{3}) = \frac{4}{3}\pi \mathbf{3}^3 = \frac{4}{3}\pi 27 = 36\pi \text{ cubic inches, or 113 cubic inches} \quad \text{To the nearest whole number}$$

$$S(\mathbf{3}) = 4\pi \mathbf{3}^2 = 36\pi \text{ square inches, or 113 square inches} \quad \text{To the nearest whole number}$$

The fact that $V(3) = 36\pi$ means that the ordered pair $(3, 36\pi)$ belongs to the function V. Likewise, the fact that $S(3) = 36\pi$ tells us that the ordered pair $(3, 36\pi)$ is a member of function S.

We can generalize the discussion at the end of Example 4 this way:

$$(a, b) \in f \quad \text{if and only if} \quad f(a) = b$$

---

3. A medication has a half-life of 5 days. If the concentration of the medication in a patient's system is 80 ng/mL, and the patient stops taking it, then t days later the concentration will be

$$C(t) = 80\left(\frac{1}{2}\right)^{t/5}$$

Find each of the following, and explain what they mean.

a. $C(5)$

b. $C(10)$

4. The following formulas give the circumference and area of a circle with a radius of r. Use the formulas to find the circumference and area of a circular plate if the radius is 5 inches.

$$C(r) = 2\pi r$$
$$A(r) = \pi r^2$$

Answers
2. a. $s(8) = 11$; runs a mile in 8 minutes, average speed is 11 feet per second.
  b. $s(11) = 8$; runs a mile in 11 minutes, average speed is 8 feet per second.
3. a. $C(5) = 40$ ng/mL; after 5 days the concentration is 40 ng per mL.
  b. $C(10) = 20$ ng/mL; after 10 days the concentration is 20 ng per mL.
4. $C(5) = 10\pi \approx 31.4$ inches
  $A(5) = 25\pi \approx 78.5$ in^2

Chapter 4 Functions

> ### USING TECHNOLOGY
>
> **More About Example 4**
>
> If we look back at Example 4, we see that when the radius of a sphere is 3, the numerical values of the volume and surface area are equal. How unusual is this? Are there other values of r for which $V(r)$ and $S(r)$ are equal? We can answer this question by looking at the graphs of both V and S.
>
> To graph the function $V(r) = \frac{4}{3}\pi r^3$, set $Y_1 = 4\pi X^3/3$. To graph $S(r) = 4\pi r^2$, set $Y_2 = 4\pi X^2$. Graph the two functions in each of the following windows:
>
> Window 1: X from -4 to 4, Y from -2 to 10
>
> Window 2: X from 0 to 4, Y from 0 to 50
>
> Window 3: X from 0 to 4, Y from 0 to 150
>
> Then use the Trace and Zoom features of your calculator to locate the point in the first quadrant where the two graphs intersect. How do the coordinates of this point compare with the results in Example 4?

5. If $f(x) = 4x^2 - 3$, find
    a. $f(0)$
    b. $f(3)$
    c. $f(-2)$

EXAMPLE 5 If $f(x) = 3x^2 + 2x - 1$, find $f(0)$, $f(3)$, and $f(-2)$.

SOLUTION Because $f(x) = 3x^2 + 2x - 1$, we have

$f(\mathbf{0}) = 3(\mathbf{0})^2 + 2(\mathbf{0}) - 1 \quad = 0 + 0 - 1 = -1$

$f(\mathbf{3}) = 3(\mathbf{3})^2 + 2(\mathbf{3}) - 1 \quad = 27 + 6 - 1 = 32$

$f(-\mathbf{2}) = 3(-\mathbf{2})^2 + 2(-\mathbf{2}) - 1 = 12 - 4 - 1 = 7$

In Example 5, the function f is defined by the equation $f(x) = 3x^2 + 2x - 1$. We could just as easily have said $y = 3x^2 + 2x - 1$; that is, $y = f(x)$. Saying $f(-2) = 7$ is exactly the same as saying y is 7 when x is -2.

6. If $f(x) = 2x + 1$ and $g(x) = x^2 - 3$, find
    a. $f(5)$
    b. $g(5)$
    c. $f(-2)$
    d. $g(-2)$
    e. $f(a)$
    f. $g(a)$

EXAMPLE 6 If $f(x) = 4x - 1$ and $g(x) = x^2 + 2$, then

$f(\mathbf{5}) = 4(\mathbf{5}) - 1 = 19$ and $g(\mathbf{5}) = \mathbf{5}^2 + 2 = 27$

$f(-\mathbf{2}) = 4(-\mathbf{2}) - 1 = -9$ and $g(-\mathbf{2}) = (-\mathbf{2})^2 + 2 = 6$

$f(\mathbf{0}) = 4(\mathbf{0}) - 1 = -1$ and $g(\mathbf{0}) = \mathbf{0}^2 + 2 = 2$

$f(\mathbf{z}) = 4\mathbf{z} - 1$ and $g(\mathbf{z}) = \mathbf{z}^2 + 2$

$f(\mathbf{a}) = 4\mathbf{a} - 1$ and $g(\mathbf{a}) = \mathbf{a}^2 + 2$

> **STUDY SKILLS**
> **Continue to Set and Keep a Schedule**
> Sometimes I find that students who do well in Chapter 1 become overconfident. They will begin to put in less time with their homework. Don't make that mistake. Keep to the same schedule.

Answers
5. a. -3 b. 33 c. 13
6. a. 11 b. 22 c. -3 d. 1
    e. $2a + 1$ f. $a^2 - 3$

4.3 Function Notation

USING TECHNOLOGY

More About Example 6

Most graphing calculators can use tables to evaluate functions. To work Example 6 using a graphing calculator table, set Y_1 equal to $4X - 1$ and Y_2 equal to $X^2 + 2$. Then set the independent variable in the table to Ask instead of Auto. Go to your table and input 5, −2, and 0. Under Y_1 in the table, you will find $f(5)$, $f(-2)$, and $f(0)$. Under Y_2, you will find $g(5)$, $g(-2)$, and $g(0)$.

Table Setup

Table minimum = 0
Table increment = 1
Independent variable: Ask
Dependent variable: Ask

Y Variables Setup

$Y_1 = 4X - 1$
$Y_2 = X^2 + 2$

The table will look like this:

X	Y_1	Y_2
5	19	27
−2	−9	6
0	−1	2

Although the calculator asks us for a table increment, the increment doesn't matter since we are inputting the X-values ourselves.

EXAMPLE 7

If the function f is given by

$$f = \{(-2, 0), (3, -1), (2, 4), (7, 5)\}$$

then $f(-2) = 0$, $f(3) = -1$, $f(2) = 4$, and $f(7) = 5$.

EXAMPLE 8

If $f(x) = 2x^2$ and $g(x) = 3x - 1$, find

a. $f(g(2))$ **b.** $g(f(2))$

SOLUTION   The expression $f(g(2))$ is read "f of g of 2."

a. Since $g(2) = 3(2) - 1 = 5$,

$$f(g(2)) = f(5) = 2(5)^2 = 50$$

b. Since $f(2) = 2(2)^2 = 8$,

$$g(f(2)) = g(8) = 3(8) - 1 = 23$$

Getting Ready for Class

After reading through the preceding section, respond in your own words and in complete sentences.

1. Explain what you are calculating when you find $f(2)$ for a given function f.
2. If $s(t) = \dfrac{60}{t}$, how do you find $s(10)$?
3. If $f(2) = 3$ for a function f, what is the relationship between the numbers 2 and 3 and the graph of f?
4. If $f(6) = 0$ for a particular function f, then you can immediately graph one of the intercepts. Explain.

---

7. If $f = \{(-4, 1), (2, -3), (7, 9)\}$, find

   a. $f(-4)$
   b. $f(2)$
   c. $f(7)$

8. If $f(x) = 3x^2$ and $g(x) = 4x + 1$, find

   a. $f(g(2))$
   b. $g(f(2))$

Answers
7. **a.** 1   **b.** −3   **c.** 9
8. **a.** 243   **b.** 49

Left Blank Intentionally

Problem Set 4.3

A Let $f(x) = 2x - 5$ and $g(x) = x^2 + 3x + 4$. Evaluate the following. [Examples 1, 5–6]

1. $f(2)$
2. $f(3)$
3. $f(-3)$
4. $g(-2)$

5. $g(-1)$
6. $f(-4)$
7. $g(-3)$
8. $g(2)$

9. $g(4) + f(4)$
10. $f(2) - g(3)$
11. $f(3) - g(2)$
12. $g(-1) + f(-1)$

A Let $f(x) = 3x^2 - 4x + 1$ and $g(x) = 2x - 1$. Evaluate the following. [Examples 1, 5–6]

13. $f(0)$
14. $g(0)$
15. $g(-4)$
16. $f(1)$

17. $f(-1)$
18. $g(-1)$
19. $g(10)$
20. $f(10)$

21. $f(3)$
22. $g(3)$
23. $g\left(\dfrac{1}{2}\right)$
24. $g\left(\dfrac{1}{4}\right)$

25. $f(a)$
26. $g(b)$

A If $f = \{(1, 4), (-2, 0), (3, \frac{1}{2}), (\pi, 0)\}$ and $g = \{(1, 1), (-2, 2), (\frac{1}{2}, 0)\}$, find each of the following values of f and g. [Example 7]

27. $f(1)$ 28. $g(1)$ 29. $g\left(\frac{1}{2}\right)$

30. $f(3)$ 31. $g(-2)$ 32. $f(\pi)$

A Let $f(x) = 2x^2 - 8$ and $g(x) = \frac{1}{2}x + 1$. Evaluate each of the following. [Examples 1, 5–6, 8]

33. $f(0)$ 34. $g(0)$ 35. $g(-4)$

36. $f(1)$ 37. $f(a)$ 38. $g(z)$

39. $f(b)$ 40. $g(t)$ 41. $f(g(2))$

42. $g(f(2))$ 43. $g(f(-1))$ 44. $f(g(-2))$

45. $g(f(0))$ 46. $f(g(0))$

4.3 Problem Set

47. Graph the function $f(x) = \frac{1}{2}x + 2$. Then draw and label the line segments that represent $x = 4$ and $f(4)$.

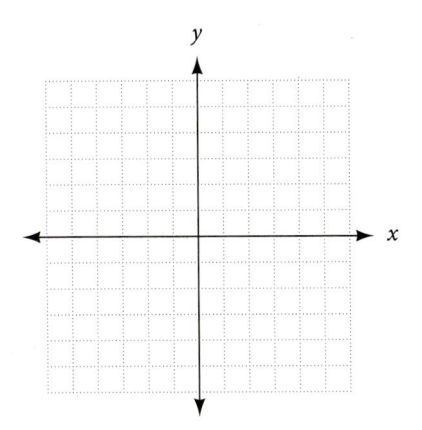

48. Graph the function $f(x) = -\frac{1}{2}x + 6$. Then draw and label the line segments that represent $x = 4$ and $f(4)$.

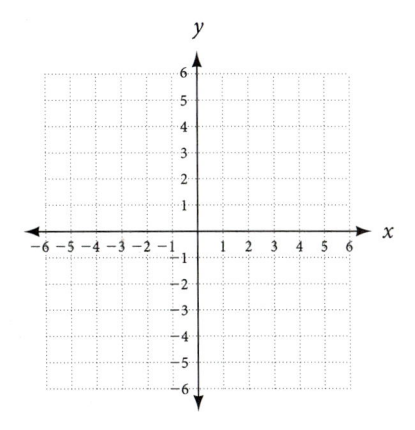

49. For the function $f(x) = \frac{1}{2}x + 2$, find the value of x for which $f(x) = x$.

50. For the function $f(x) = -\frac{1}{2}x + 6$, find the value of x for which $f(x) = x$.

51. Graph the function $f(x) = x^2$. Then draw and label the line segments that represent $x = 1$ and $f(1)$, $x = 2$ and $f(2)$, and, finally, $x = 3$ and $f(3)$.

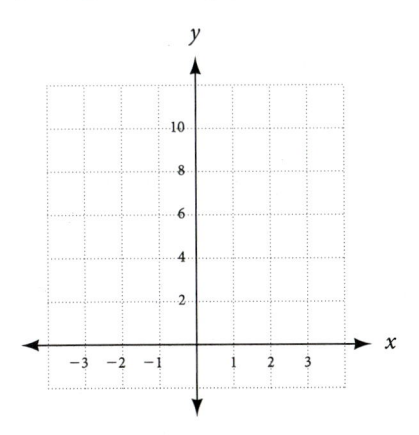

52. Graph the function $f(x) = x^2 - 2$. Then draw and label the line segments that represent $x = 2$ and $f(2)$, and the line segments corresponding to $x = 3$ and $f(3)$.

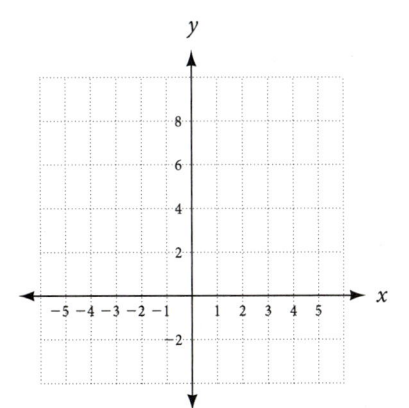

Applying the Concepts [Examples 2–4]

53. Investing in Art A painting is purchased as an investment for $150. If its value increases continuously so that it doubles every 3 years, then its value is given by the function

$$V(t) = 150 \cdot 2^{t/3} \quad \text{for} \quad t \geq 0$$

where t is the number of years since the painting was purchased, and $V(t)$ is its value (in dollars) at time t. Find $V(3)$ and $V(6)$, and then explain what they mean.

54. Average Speed If it takes Minke t minutes to run a mile, then her average speed $s(t)$, in miles per hour, is given by the formula

$$s(t) = \frac{60}{t} \quad \text{for} \quad t > 0$$

Find $s(4)$ and $s(5)$, and then explain what they mean.

55. Antidepressants The chart shows the sales of three different antidepressants from 2003 to 2005. Suppose the following:

f is a function of Zoloft sales with respect to time

g is a function of Effexor XR sales with respect to time

h is a function of Wellbutrin XL sales with respect to time

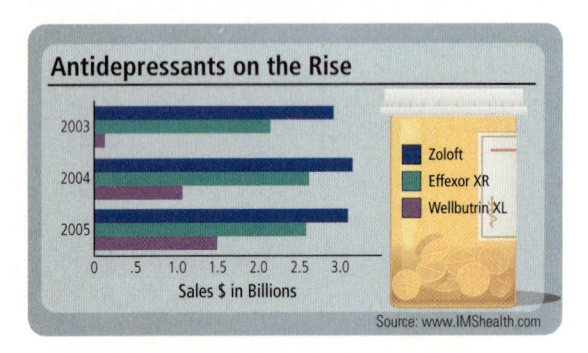

Use this information to determine if each if the following statements are true or false.

a. $g(2005) > g(2004)$
b. $h(2003) > h(2005)$
c. $f(2004) < f(2003)$

56. Light Bulbs The chart shows a comparison of power usage between incandescent and energy efficient light bulbs. Suppose that:

f is the function for an incandescent bulb

g is the function for an energy efficient bulb

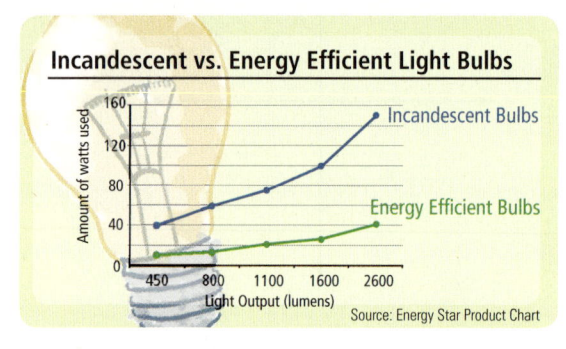

Use this information to evaluate the following.

a. $f(2600) - g(2600)$
b. $f(2600) - f(1100)$
c. $g(2600) - f(450)$

57. Antidepressant Sales Suppose x represents one of the years in the chart. Suppose further that we have three functions f, g, and h that do the following:

f pairs each year with the total sales of Zoloft in billions of dollars for that year.

g pairs each year with the total sales of Effexor in billions of dollars for that year.

h pairs each year with the total sales of Wellbutrin in billions of dollars for that year.

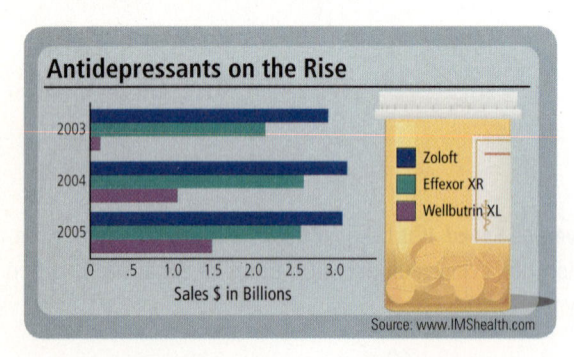

For each statement below, indicate whether the statement is true or false.

a. The domain of g is {2003, 2004, 2005}
b. $f(2003) < f(2004)$
c. $f(2004) > g(2004)$
d. $h(2005) > 1.5$
e. $h(2005) > h(2004) > h(2003)$

58. Mobile Phone Sales Suppose x represents one of the years in the chart. Suppose further that we have three functions f, g, and h that do the following:

f pairs each year with the number of camera phones sold that year.

g pairs each year with the number of non-camera phones sold that year.

h is such that $h(x) = f(x) + g(x)$.

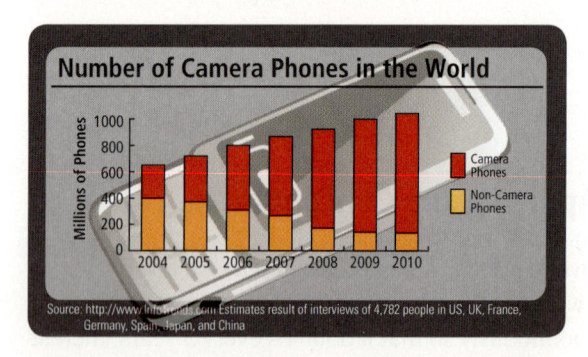

For each statement below, indicate whether the statement is true or false.

a. The domain of f is {2004, 2005, 2006, 2007, 2008, 2009, 2010}
b. $h(2005) = 741,000,000$
c. $f(2009) > g(2009)$
d. $f(2004) < f(2005)$
e. $h(2010) > h(2007) > h(2004)$

4.3 Problem Set

Straight-Line Depreciation Straight-line depreciation is an accounting method used to help spread the cost of new equipment over a number of years. It takes into account both the cost when new and the salvage value, which is the value of the equipment at the time it gets replaced.

59. **Value of a Copy Machine** The function $V(t) = -3,300t + 18,000$, where V is value and t is time in years, can be used to find the value of a large copy machine during the first 5 years of use.

    a. What is the value of the copier after 3 years and 9 months?

    b. What is the salvage value of this copier if it is replaced after 5 years?

    c. State the domain of this function.

    d. Sketch the graph of this function.

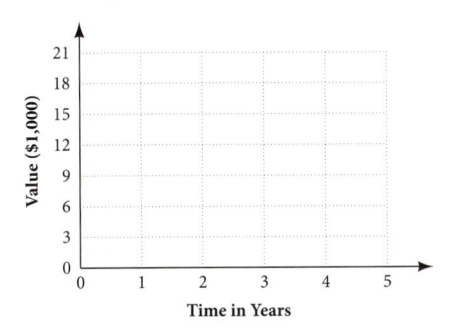

    e. What is the range of this function?

    f. After how many years will the copier be worth only $10,000?

60. **Value of a Forklift** The function $V = -16,500t + 125,000$, where V is value and t is time in years, can be used to find the value of an electric forklift during the first 6 years of use.

    a. What is the value of the forklift after 2 years and 3 months?

    b. What is the salvage value of this forklift if it is replaced after 6 years?

    c. State the domain of this function.

    d. Sketch the graph of this function.

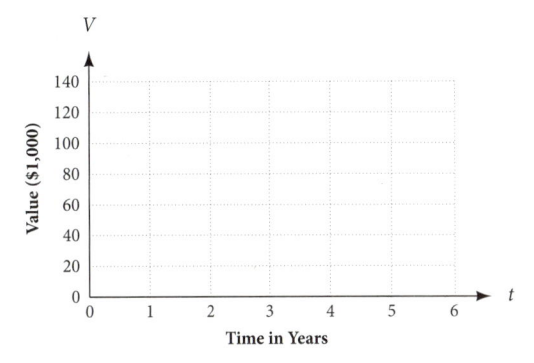

    e. What is the range of this function?

    f. After how many years will the forklift be worth only $45,000?

Getting Ready for the Next Section

Simplify.

61. $(35x - 0.1x^2) - (8x + 500)$

62. $70 + 0.6(M - 70)$

63. $(4x^2 + 3x + 2) + (2x^2 - 5x - 6)$

64. $(4x - 3) + (4x^2 - 7x + 3)$

65. $(4x^2 + 3x + 2) - (2x^2 - 5x - 6)$

66. $(x + 5)^2 - 2(x + 5)$

Multiply.

67. $0.6(M - 70)$

68. $x(35 - 0.1x)$

69. $(4x - 3)(x - 1)$

70. $(4x - 3)(4x^2 - 7x + 3)$

Extending the Concepts

The graphs of two functions are shown in Figures 4 and 5. Use the graphs to find the following.

71.
  a. $f(2)$
  b. $f(-4)$
  c. $g(0)$
  d. $g(3)$

72.
  a. $g(2)$
  b. $f(1)$
  c. $f(3)$
  d. $g(-3)$

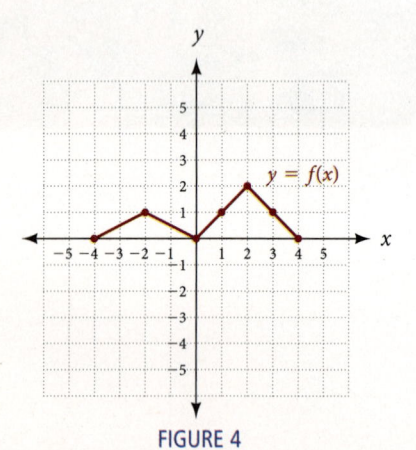

FIGURE 4

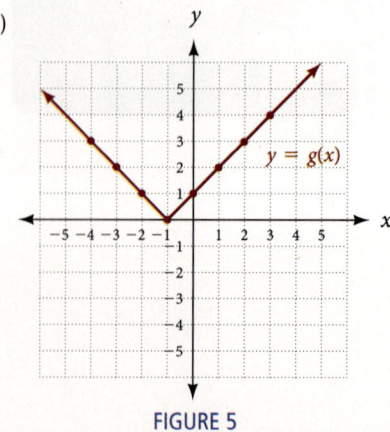

FIGURE 5

73. Step Function Figure 6 shows the graph of the step function C that was used to calculate the first-class postage on a large envelope weighing x ounces in 2010. Use the graph to answer questions (a) through (d).

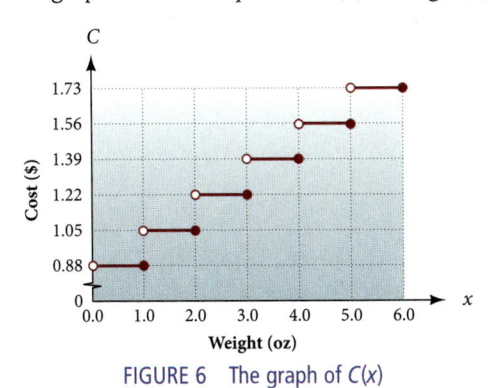

FIGURE 6 The graph of C(x)

a. Fill in the following table:

Weight (ounces)	0.6	1.0	1.1	2.5	3.6	4.8	5.0	5.3
Cost ($)								

b. If a large envelope costs $1.22 to mail, how much does it weigh? State your answer in words. State your answer as an inequality.

c. If the entire function is shown in Figure 6, state the domain.

d. State the range of the function shown in Figure 6.

74. Step Function A taxi ride in Boston at the time I am writing this problem is $1.50 for the first $\frac{1}{4}$ mile, and then $0.25 for each additional $\frac{1}{8}$ of a mile. The following graph shows how much you will pay for a taxi ride of 1 mile or less.

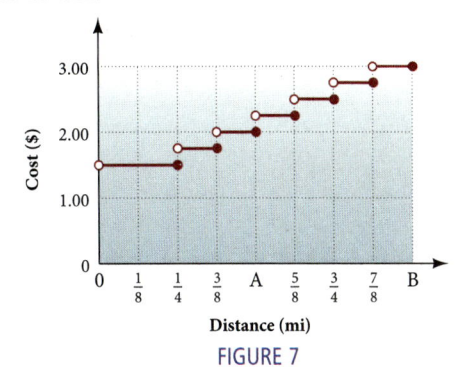

FIGURE 7

a. What is the most you will pay for this taxi ride?

b. How much does it cost to ride the taxi for $\frac{8}{10}$ of a mile?

c. Find the values of A and B on the horizontal axis.

d. If a taxi ride costs $2.50, what distance was the ride?

e. If the complete function is shown in Figure 7, find the domain and range of the function.

Left Blank Intentionally

4.4 Algebra and Composition with Functions

Objectives

A Find the sum, difference, product, and quotient of two functions

B For two functions f and g, find f(g(x)) and g(f(x)).

A company produces and sells copies of an accounting program for home computers. The price they charge for the program is related to the number of copies sold by the demand function

$$p(x) = 35 - 0.1x$$

We find the revenue for this business by multiplying the number of items sold by the price per item. When we do so, we are forming a new function by combining two existing functions; that is, if $n(x) = x$ is the number of items sold and $p(x) = 35 - 0.1x$ is the price per item, then revenue is

$$R(x) = n(x) \cdot p(x) = x(35 - 0.1x) = 35x - 0.1x^2$$

In this case, the revenue function is the product of two functions. When we combine functions in this manner, we are applying our rules for algebra to functions.

To carry this situation further, we know that the profit function is the difference between two functions. If the cost function for producing x copies of the accounting program is $C(x) = 8x + 500$, then the profit function is

$$P(x) = R(x) - C(x) = (35x - 0.1x^2) - (8x + 500) = -500 + 27x - 0.1x^2$$

The relationship between these last three functions is shown visually in Figure 1.

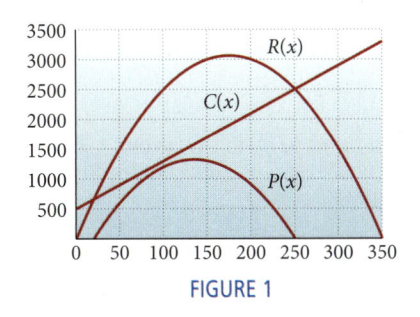

FIGURE 1

Again, when we combine functions in the manner shown, we are applying our rules for algebra to functions. To begin this section, we take a formal look at addition, subtraction, multiplication, and division with functions.

A Algebra with Functions

If we are given two functions f and g with a common domain, we can define four other functions as follows:

Definition

$(f + g)(x) = f(x) + g(x)$ The function $f + g$ is the sum of the functions f and g.

$(f - g)(x) = f(x) - g(x)$ The function $f - g$ is the difference of the functions f and g.

$(fg)(x) = f(x)g(x)$ The function fg is the product of the functions f and g.

$\left(\dfrac{f}{g}\right)(x) = \dfrac{f(x)}{g(x)}$ The function $\dfrac{f}{g}$ is the quotient of the functions f and g, where $g(x) \neq 0$.

PRACTICE PROBLEMS

1. If $f(x) = x^2 - 4$ and $g(x) = x + 2$, find formulas for
   a. $f + g$
   b. $f - g$

2. Let $f(x) = 3x + 2$, $g(x) = 3x^2 - 10x - 8$, and $h(x) = x - 4$. Find
   a. $f + g$
   b. fg
   c. $\dfrac{g}{f}$

Answers
1. a. $x^2 + x - 2$ b. $x^2 - x - 6$
2. a. $3x^2 - 7x - 6$
   b. $9x^3 - 24x^2 - 44x - 16$
   c. $h(x)$

Chapter 4 Functions

EXAMPLE 1 If $f(x) = 4x^2 + 3x$ and $g(x) = -5x - 6$, write the formula for the functions $f + g$ and $f - g$.

SOLUTION The function $f + g$ is defined by

$$(f + g)(x) = f(x) + g(x)$$
$$= (4x^2 + 3x) + (-5x - 6)$$
$$= 4x^2 - 2x - 6$$

The function $f - g$ is defined by

$$(f - g)(x) = f(x) - g(x)$$
$$= (4x^2 + 3x) - (-5x - 6)$$
$$= 4x^2 + 3x + 5x + 6$$
$$= 4x^2 + 8x + 6$$

EXAMPLE 2 Let $f(x) = 4x - 3$, and $g(x) = 4x^2$. Find $f + g$, fg, and $\dfrac{g}{f}$.

SOLUTION The function $f + g$, the sum of functions f and g, is defined by

$$(f + g)(x) = f(x) + g(x)$$
$$= (4x - 3) + 4x^2$$
$$= 4x^2 + 4x - 3$$

The product of the functions f and g, fg, is given by

$$(fg)(x) = f(x)g(x)$$
$$= (4x - 3)(4x^2)$$
$$= 16x^3 - 12x^2$$

The quotient of the functions g and f, $\dfrac{g}{f}$, is defined as

$$\left(\dfrac{g}{f}\right)(x) = \dfrac{g(x)}{f(x)}$$
$$= \dfrac{4x^2}{4x - 3}$$

B Composition of Functions

In addition to the four operations used to combine functions shown so far in this section, there is a fifth way to combine two functions to obtain a new function. It is called *composition of functions*. To illustrate the concept, recall the definition of training heart rate: Training heart rate, in beats per minute, is resting heart rate plus 60% of the difference between maximum heart rate and resting heart rate. If your resting heart rate is 70 beats per minute, then your training heart rate is a function of your maximum heart rate M:

$$T(M) = 70 + 0.6(M - 70) = 70 + 0.6M - 42 = 28 + 0.6M$$

But your maximum heart rate is found by subtracting your age in years from 220. So, if x represents your age in years, then your maximum heart rate is

$$M(x) = 220 - x$$

4.4 Algebra and Composition with Functions

Therefore, if your resting heart rate is 70 beats per minute and your age in years is x, then your training heart rate can be written as a function of x:

$$T(x) = 28 + 0.6(220 - x)$$

This last line is the composition of functions T and M. We input x into function M, which outputs $M(x)$. Then, we input $M(x)$ into function T, which outputs $T(M(x))$, which is the training heart rate as a function of age x. Here is a diagram of the situation, which is called a function map:

$$\text{Age} \quad\quad \text{Maximum heart rate} \quad\quad \text{Training heart rate}$$
$$x \xrightarrow{M} M(x) \xrightarrow{T} T(M(x))$$

FIGURE 2

Now let's generalize the preceding ideas into a formal development of composition of functions. To find the composition of two functions f and g, we first require that the range of g have numbers in common with the domain of f. Then the composition of f with g, denoted $f \circ g$, is defined this way:

$$(f \circ g)(x) = f(g(x))$$

To understand this new function, we begin with a number x, and we operate on it with g, giving us $g(x)$. Then we take $g(x)$ and operate on it with f, giving us $f(g(x))$. The only numbers we can use for the domain of the composition of f with g are numbers x in the domain of g, for which $g(x)$ is in the domain of f. The diagrams in Figure 3 illustrate the composition of f with g.

Function machines

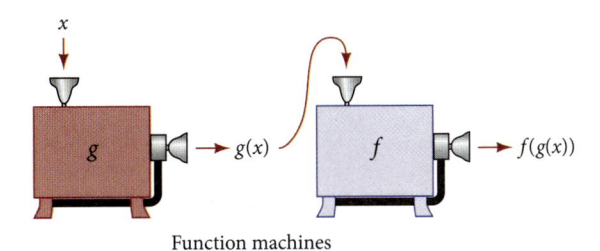

FIGURE 3

Composition of functions is not commutative. The composition of f with g, $f \circ g$, may therefore be different from the composition of g with f, $g \circ f$.

$$(g \circ f)(x) = g(f(x))$$

Again, the only numbers we can use for the domain of the composition of g with f are numbers in the domain of f, for which $f(x)$ is in the domain of g. The diagrams in Figure 4 illustrate the composition of g with f.

Function machines

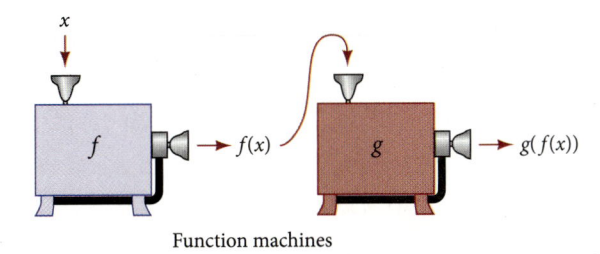

FIGURE 4

3. If $f(x) = x - 4$ and $g(x) = x^2 + 3x$, find
   a. $(f \circ g)(x)$
   b. $(g \circ f)(x)$

Chapter 4 Functions

EXAMPLE 3 If $f(x) = x + 5$ and $g(x) = 2x$, find $(f \circ g)(x)$.

SOLUTION The composition of f with g is

$$(f \circ g)(x) = f(g(x))$$
$$= f(2x)$$
$$= 2x + 5$$

The composition of g with f is

$$(g \circ f)(x) = g(f(x))$$
$$= g(x + 5)$$
$$= 2(x + 5)$$
$$= 2x + 10$$

Getting Ready for Class

After reading through the preceding section, respond in your own words and in complete sentences.

A. How are profit, revenue, and cost related?
B. How do you find maximum heart rate?
C. For functions f and g, how do you find the composition of f with g?
D. For functions f and g, how do you find the composition of g with f?

Answers
3. a. $x^2 + 3x - 4$ b. $x^2 - 5x + 4$

Problem Set 4.4

A Let $f(x) = 4x - 3$ and $g(x) = 2x + 5$. Write a formula for each of the following functions. [Example 1]

1. $f + g$
2. $f - g$
3. $g - f$
4. $g + f$

A If the functions f, g, and h are defined by $f(x) = 3x - 5$, $g(x) = x - 2$, and $h(x) = 3x^2$, write a formula for each of the following functions. [Examples 1–2]

5. $g + f$
6. $f + h$
7. $g + h$
8. $f - g$
9. $g - f$
10. $h - g$
11. fh
12. gh
13. $\dfrac{h}{f}$
14. $\dfrac{h}{g}$
15. $\dfrac{f}{h}$
16. $\dfrac{g}{h}$
17. $f + g + h$
18. $h - g + f$

A Let $f(x) = 2x + 1$, $g(x) = 4x + 2$, and $h(x) = 4x^2 + 4x + 1$, and find the following. [Examples 1–2]

19. $(f + g)(2)$

20. $(f - g)(-1)$

21. $(fg)(3)$

22. $\left(\dfrac{f}{g}\right)(-3)$

23. $\left(\dfrac{h}{g}\right)(1)$

24. $(hg)(1)$

25. $(fh)(0)$

26. $(h - g)(-4)$

27. $(f + g + h)(2)$

28. $(h - f + g)(0)$

29. $(h + fg)(3)$

30. $(h - fg)(5)$

B [Example 3]

31. Let $f(x) = x^2$ and $g(x) = x + 4$, and find
  a. $(f \circ g)(5)$

  b. $(g \circ f)(5)$

  c. $(f \circ g)(x)$

  d. $(g \circ f)(x)$

32. Let $f(x) = 3 - x$ and $g(x) = x^3 - 1$, and find
  a. $(f \circ g)(0)$

  b. $(g \circ f)(0)$

  c. $(f \circ g)(x)$

33. Let $f(x) = x^2 + 3x$ and $g(x) = 4x - 1$, and find
  a. $(f \circ g)(0)$

  b. $(g \circ f)(0)$

  c. $(g \circ f)(x)$

34. Let $f(x) = (x - 2)^2$ and $g(x) = x + 1$, and find
  a. $(f \circ g)(-1)$

  b. $(g \circ f)(-1)$

For each of the following pairs of functions f and g, show that $(f \circ g)(x) = (g \circ f)(x) = x$.

35. $f(x) = 5x - 4$ and $g(x) = \dfrac{x + 4}{5}$

36. $f(x) = \dfrac{x}{6} - 2$ and $g(x) = 6x + 12$

Use the graph to answer the following problems.

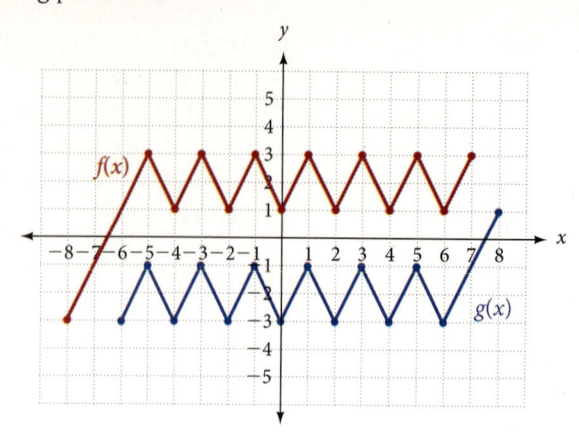

Evaluate.

37. $f(2) + 5$

38. $g(-2) - 5$

39. $f(-3) + g(-3)$

40. $f(5) - g(5)$

41. $(f \circ g)(0)$

42. $(g \circ f)(0)$

43. Find x if $f(x) = -3$

44. Find x if $g(x) = 1$

Use the graph to answer the following problems.

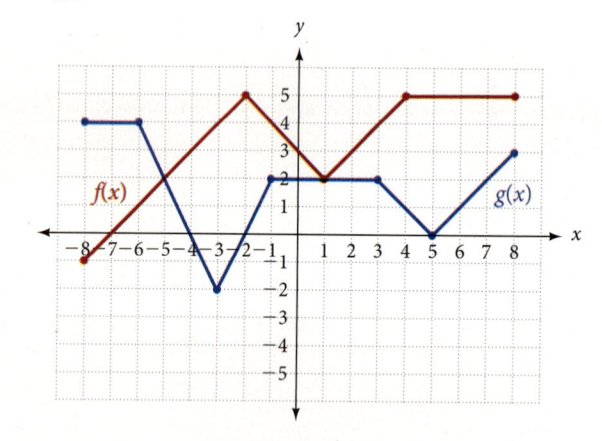

Evaluate.

45. $f(-3) + 2$

46. $g(3) - 3$

47. $f(2) + g(2)$

48. $f(-5) - g(-5)$

49. $(f \circ g)(0)$

50. $(g \circ f)(0)$

51. Find x if $f(x) = 1$

52. Find x if $g(x) = -2$

Applying the Concepts

53. Profit, Revenue, and Cost A company manufactures and sells prerecorded DVDs. Here are the equations they use in connection with their business.

Number of DVDs sold each day: $n(x) = x$

Selling price for each DVD: $p(x) = 11.5 - 0.05x$

Daily fixed costs: $f(x) = 200$

Daily variable costs: $v(x) = 2x$

Find the following functions.

a. Revenue = $R(x)$ = the product of the number of DVDs sold each day and the selling price of each DVD.

b. Cost = $C(x)$ = the sum of the fixed costs and the variable costs.

c. Profit = $P(x)$ = the difference between revenue and cost.

d. Average cost = $\overline{C}(x)$ = the quotient of cost and the number of DVDs sold each day.

54. Profit, Revenue, and Cost A company manufactures and sells CDs for home computers. Here are the equations they use in connection with their business.

Number of CDs sold each day: $n(x) = x$

Selling price for each CD: $p(x) = 3 - \frac{1}{300}x$

Daily fixed costs: $f(x) = 200$

Daily variable costs: $v(x) = 2x$

Find the following functions.

a. Revenue = $R(x)$ = the product of the number of CDs sold each day and the selling price of each CD.

b. Cost = $C(x)$ = the sum of the fixed costs and the variable costs.

c. Profit = $P(x)$ = the difference between revenue and cost.

d. Average cost = $\overline{C}(x)$ = the quotient of cost and the number of CDs sold each day.

55. Training Heart Rate Find the training heart rate function, $T(M)$, for a person with a resting heart rate of 62 beats per minute, then find the following.

a. Find the maximum heart rate function, $M(x)$, for a person x years of age.

b. What is the maximum heart rate for a 24-year-old person?

c. What is the training heart rate for a 24-year-old person with a resting heart rate of 62 beats per minute?

d. What is the training heart rate for a 36-year-old person with a resting heart rate of 62 beats per minute?

e. What is the training heart rate for a 48-year-old person with a resting heart rate of 62 beats per minute?

56. Training Heart Rate Find the training heart rate function, $T(M)$, for a person with a resting heart rate of 72 beats per minute, then find the following to the nearest whole number.

a. Find the maximum heart rate function, $M(x)$, for a person x years of age.

b. What is the maximum heart rate for a 20-year-old person?

c. What is the training heart rate for a 20-year-old person with a resting heart rate of 72 beats per minute?

d. What is the training heart rate for a 30-year-old person with a resting heart rate of 72 beats per minute?

e. What is the training heart rate for a 40-year-old person with a resting heart rate of 72 beats per minute?

Getting Ready for the Next Section

57. $16(3.5)^2$

58. $\dfrac{2{,}400}{100}$

59. $\dfrac{180}{45}$

60. $4(2)(4)^2$

61. $\dfrac{0.0005(200)}{(0.25)^2}$

62. $\dfrac{0.2(0.5)^2}{100}$

63. If $y = Kx$, find K if $x = 5$ and $y = 15$.

64. If $d = Kt^2$, find K if $t = 2$ and $d = 64$.

65. If $V = \dfrac{K}{P}$, find K if $P = 48$ and $V = 50$.

66. If $y = Kxz^2$, find K if $x = 5$, $z = 3$, and $y = 180$.

Chapter 4 Summary

Relations and Functions [4.2]

A *function* is a rule that pairs each element in one set, called the *domain*, with exactly one element from a second set, called the *range*.

A *relation* is any set of ordered pairs. The set of all first coordinates is called the *domain* of the relation, and the set of all second coordinates is the *range* of the relation. A function is a relation in which no two different ordered pairs have the same first coordinates.

EXAMPLES

1. The relation

$$\{(8, 1), (6, 1), (-3, 0)\}$$

is also a function because no ordered pairs have the same first coordinates. The domain is $\{8, 6, -3\}$ and the range is $\{1, 0\}$.

Vertical Line Test [4.2]

If a vertical line crosses the graph of a relation in more than one place, the relation cannot be a function. If no vertical line can be found that crosses the graph in more than one place, the relation must be a function.

2. The graph of $x = y^2$ shown in Figure 5 in Section 4.2 fails the vertical line test. It is not the graph of a function.

Function Notation [4.3]

The alternative notation for y is $f(x)$. It is read "f of x" and can be used instead of the variable y when working with functions. The notation y and the notation $f(x)$ are equivalent; that is, $y = f(x)$.

3. If $f(x) = 5x - 3$ then
$f(0) = 5(0) - 3$
$\quad = -3$
$f(1) = 5(1) - 3$
$\quad = 2$
$f(-2) = 5(-2) - 3$
$\quad = -13$
$f(a) = 5a - 3$

Algebra with Functions [4.4]

If f and g are any two functions with a common domain, then:

$(f + g)(x) = f(x) + g(x)$ The function $f + g$ is the sum of the functions f and g.

$(f - g)(x) = f(x) - g(x)$ The function $f - g$ is the difference of the functions f and g.

$(fg)(x) = f(x)g(x)$ The function fg is the product of the functions f and g.

$\dfrac{f}{g}(x) = \dfrac{f(x)}{g(x)}$ The function $\dfrac{f}{g}$ is the quotient of the functions f and g, where $g(x) \neq 0$.

Chapter 4 Functions

> **⊘ COMMON MISTAKES**
>
> 1. When graphing ordered pairs, the most common mistake is to associate the first coordinate with the *y*-axis and the second with the *x*-axis. If you make this mistake you would graph (3, 1) by going up 3 and to the right 1, which is just the reverse of what you should do. Remember, the first coordinate is always associated with the horizontal axis, and the second coordinate is always associated with the vertical axis.

Chapter 4 Review

Graph the following ordered pairs. [4.1]

1. (0, 3)
2. (−2, 1)
3. (4, −2)
4. $\left(-\frac{1}{2}, 3\right)$

Graph the points (2, 3) and (−1, −3), and draw a straight line that passes through them. Then, answer the following questions. [4.1]

5. Does the graph of (1, 1) lie on the line?
6. Does the graph of (−2, 1) lie on the line?
7. Does the graph of (0, −1) lie on the line?
8. Does the graph of (−1, 2) lie on the line?

Give the domain and range, and indicate which are also functions. [4.2]

9. {(0, 3), (1, 2), (−1, 2)}
10. {(−1, 3), (−1, 4), (0, 2)}

11.

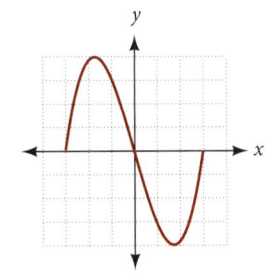

12.

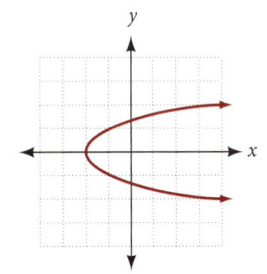

State the domain and range of each relation, and then indicate which relations are also functions. [4.2]

13. {(2, 4), (3, 3), (4, 2)}
14. {(6, 3), (−4, 3), (−2, 0)}

If f = {(2, −1), (−3, 0), $\left(4, \frac{1}{2}\right)$, (π, 2)} and g = {(2, 2), (−1, 4), (0, 0)}, find the following. [4.3, 4.4]

15. $f(-3)$
16. $f(2) + g(2)$

Let $f(x) = 2x^2 - 4x + 1$ and $g(x) = 3x + 2$, and evaluate each of the following. [4.3, 4.4]

17. $f(0)$
18. $g(a)$
19. $f(g(0))$
20. $f(g(1))$

Chapter 4 Review   295

Left Blank Intentionally

Chapter 4 Cumulative Review

Simplify each of the following.

1. -5^2
2. $-|-6|$
3. $5^2 + 7^2$
4. $(5 + 7)^2$
5. $48 \div 8 \cdot 6$
6. $75 \div 15 \cdot 5$
7. $30 - 15 \div 3 + 2$
8. $30 - 15 \div (3 + 2)$

Find the value of each expression when x is 4.

9. $x^2 - 10x + 25$
10. $x^2 - 25$

Reduce the following fractions to lowest terms.

11. $\dfrac{336}{432}$
12. $\dfrac{721}{927}$

Subtract.

13. $\dfrac{3}{15} - \dfrac{2}{20}$
14. $\dfrac{6}{28} - \dfrac{5}{42}$

15. Add $-\dfrac{3}{4}$ to the product of -3 and $\dfrac{5}{12}$.

16. Subtract $\dfrac{2}{5}$ from the product of -3 and $\dfrac{6}{15}$.

Simplify each of the following expressions.

17. $8\left(\dfrac{3}{4}x - \dfrac{1}{2}y\right)$
18. $12\left(\dfrac{5}{6}x + \dfrac{3}{4}y\right)$

Solve the following equations.

19. $\dfrac{2}{3}a - 4 = 6$
20. $\dfrac{3}{4}(8x - 3) + \dfrac{3}{4} = 2$

21. $-4 + 3(3x + 2) = 7$
22. $3x - (x + 2) = -6$

23. $|2x - 3| - 7 = 1$
24. $|2y + 3| = 2y - 7$

Solve the following expressions for y and simplify.

25. $y - 5 = -3(x + 2)$
26. $y + 3 = \dfrac{1}{2}(3x - 4)$

If $f(x) = x^2 - 3x$, $g(x) = x - 1$ and $h(x) = 3x - 1$, find the following.

27. $h(0)$
28. $f(-1) + g(4)$
29. $(g \circ f)(-2)$
30. $(h - g)(x)$

31. Specify the domain and range for the relation $\{(-1, 3), (2, -1), (3, 3)\}$. Is the relation also a function?

Chapter 4 Test

Give the coordinates of each numbered point. [4.1]

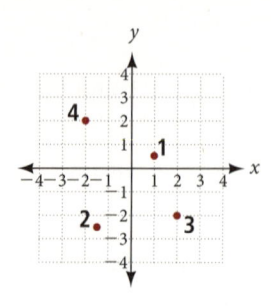

Do the following coordinates lie on the graph below? [4.1]

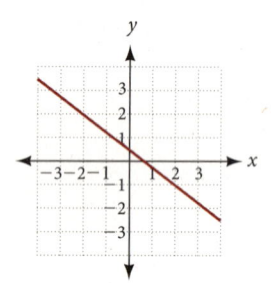

5. $(-2, 2)$ 6. $\left(-\frac{1}{2}, \frac{1}{2}\right)$ 7. $(-1, 2)$ 8. $(2, -1)$

State the domain and range for the following relations, and indicate which relations are also functions. [4.2]

9. $\{(-2, 0), (-3, 0), (-2, 1)\}$ 10. $y = x^2 - 9$

11. $\{(0, 0), (1, 3), (2, 5)\}$ 12. $y = 3x - 5$

Determine the domain and range of the following functions. Assume the *entire* function is shown. [4.2]

13.

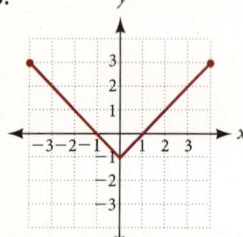

14.

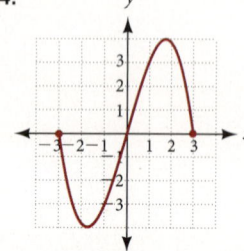

Let $f(x) = x - 2$, $g(x) = 3x + 4$ and $h(x) = 3x^2 - 2x - 8$, and find the following. [4.3, 4.4]

15. $f(3) + g(2)$ 16. $h(0) + g(0)$

17. $f(g(2))$ 18. $g(f(2))$

Use the graph to answer Questions 19–22. Evaluate. [4.3, 4.4]

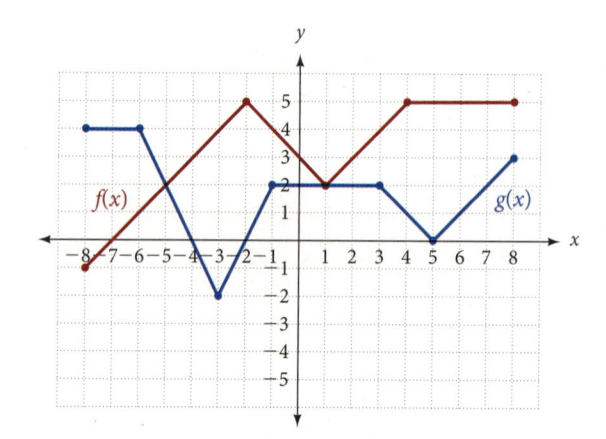

19. $f(-6) + g(-6)$ 20. $(f - g)(1)$

21. $(f \circ g)(5)$ 22. $(g \circ f)(5)$

Chapter 4 Projects

FUNCTIONS

GROUP PROJECT

Light Intensity

Number of People 2–3

Time Needed 15 minutes

Equipment Paper and pencil

Background I found the following diagram while shopping for some track lighting for my home. I was impressed by the diagram because it displays a lot of useful information in a very efficient manner. As the diagram indicates, the amount of light that falls on a surface depends on how far above the surface the light is placed and how much the light spreads out on the surface. Assume that this light illuminates a circle on a flat surface, and work the following problems.

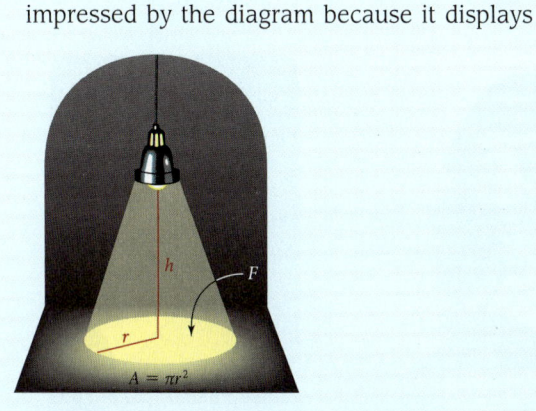

Procedure a. Fill in each table.

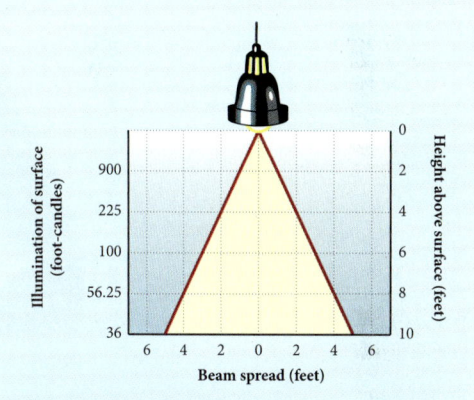

Height Above Surface (ft)	Illumination (foot-candles)
2	
4	
6	
8	
10	

Distance Above Surface (ft)	Area of Illuminated Region (ft²)
2	
4	
6	
8	
10	

b. Use the templates in Figures 1 and 2 to construct line graphs from the data in the tables.

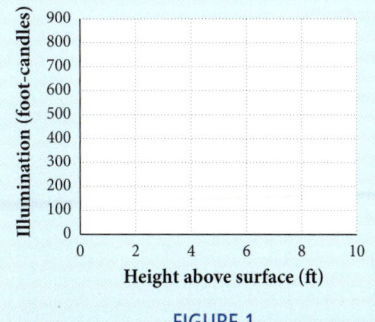

FIGURE 1

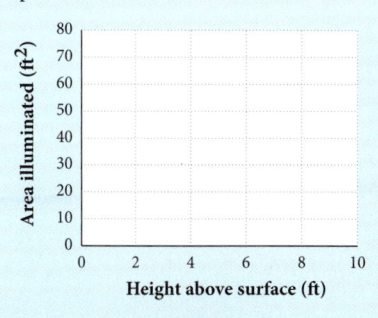

FIGURE 2

c. Let F represent the number of foot-candles that fall on the surface, h the distance the light source is above the surface, and A the area of the illuminated region. Write an equation that shows the relationship between A and h, then write another equation that gives the relationship between F and h.

RESEARCH PROJECT

Descartes and Pascal

René Descartes, the inventor of the rectangular coordinate system, is the person who made the statement, "I think, therefore, I am." Blaise Pascal, another French philosopher, is responsible for the statement, "The heart has its reasons which reason does not know." Although Pascal and Descartes were contemporaries, the philosophies of the two men differed greatly. Research the philosophy of both Descartes and Pascal, and then write an essay that gives the main points of each man's philosophy. In the essay, show how the quotations given here fit in with the philosophy of the man responsible for the quotation.

René Descartes, 1596–1650

Blaise Pascal, 1623–1662

Geometry

5

Chapter Outline
5.1 Angles
5.2 Parallel Lines
5.3 Classification of Triangles
5.4 Congruent Triangles
5.5 Parallelograms
5.6 Similar Figures and Proportions
5.7 The Pythagorean Theorem
5.8 Right Triangle Trigonometry
5.9 Solving Right Triangles
5.10 Area and Perimeter

Introduction

The Google Earth image here shows the Nile River in Africa. The Nile is the longest river in the world, measuring 4,160 miles and stretching across ten different countries. Rivers across the world serve as important means of transportation, particularly in less developed countries, like those in Africa.

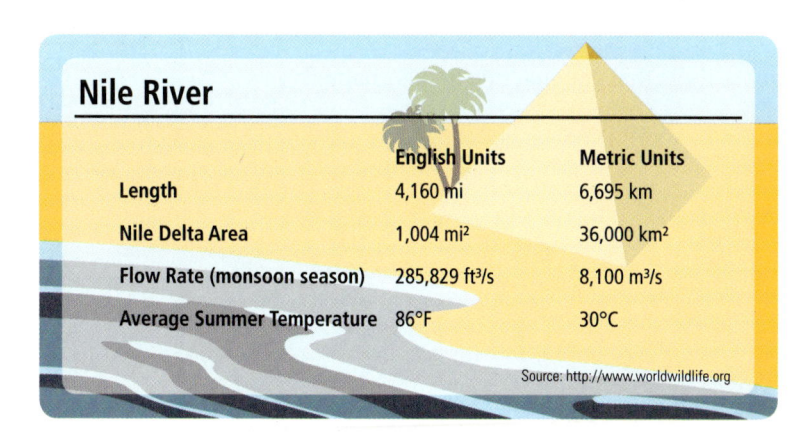

Nile River

	English Units	Metric Units
Length	4,160 mi	6,695 km
Nile Delta Area	1,004 mi²	36,000 km²
Flow Rate (monsoon season)	285,829 ft³/s	8,100 m³/s
Average Summer Temperature	86°F	30°C

Source: http://www.worldwildlife.org

301

Symbols Used in this Chapter

SYMBOL	MEANING
△	Triangle
∠	Angle
▱	Parallelogram
≅	Congruent
~	Similar
√	Positive square root

EXAMPLES	
△ABC ≅ △DEF	Triangle *ABC* is congruent to triangle *DEF*.
△RST ~ △XYZ	Triangle *RST* is similar to triangle *XYZ*.
m∠A = 90°	The measure of angle *A* is 90 degrees.

A Reminder about Square Roots

There are a number of places in this book where square roots are used. The mathematics behind them is really very simple, and if you understand what it means to square a number, you will not have any trouble with square roots. The symbol $\sqrt{9}$ is used to represent the positive number we square to get 9. It must be 3 since 3 is positive and $3^2 = 9$. That is,

$$\sqrt{9} = 3$$

Here are some other square roots that you may see in this chapter.

$$\sqrt{4} = 2 \qquad \sqrt{16} = 4$$

$$\sqrt{25} = 5 \qquad \sqrt{36} = 6$$

Angles 5.1

Plane Geometry

Learning Geometry is based on understanding the basic terms; point, line, and plane. From these three terms, we will build our understanding of Geometry. While these basic terms are undefined, we use the following descriptions to help better understand them.

Objectives
- **A** Understand degree measure.
- **B** Solve problems involving complementary and supplementary angles.
- **C** Use congruence of vertical angles.

Point
A point is the simplest way to describe location. A point has no size or dimensions, so it has no length, area or volume. We will use capital letters to represent points.

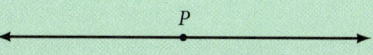

• P

Line
A line is a one-dimensional quantity with no thickness that extends indefinitely. The one dimension of a line is length. We will use two points or a lowercase letter to represent a line.

l

A B

Plane
A plane is a two-dimensional quantity with no thickness that extendes indefinitely. The two dimensions of a plane are length and width. Planes are described by three points not on the same line.

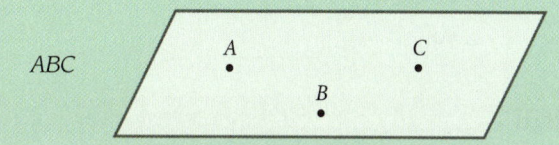

ABC

Three or more points on the same line are called *collinear*, while points on the same plane are called *coplanar*.

Here are some simple definitions of geometric terms based on the above descriptions of points, lines, and planes:

Definition
A **line segment** is a section of line that is bound by two points at its ends, called *endpoints*. A line segment includes all the points between two endpoints.

Definition
An **endpoint** is a point at the end of a line segment or the point at the start of a ray.

5.1 Angles 303

Chapter 5 Geometry

> **Definition**
> A **ray** is a line segment that extends indefinately in one direction, and starts with a point, or endpoint.

> **Definition**
> The **midpoint** of a line segment is the point that divides the segment into two segments of equal length.

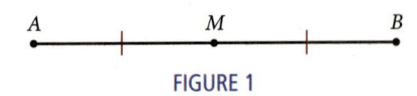

FIGURE 1

If M is the midpoint of line segment AB, then $\overline{AM} = \overline{MB}$.

EXAMPLE 1 If B is the midpoint of $\overline{AC}$, find x.

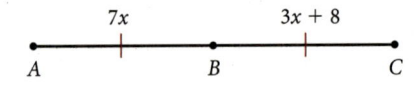

SOLUTION Since B is the midpoint of $\overline{AC}$,

$$\overline{AB} = \overline{BC}$$
$$7x = 3x + 8$$
$$4x = 8$$
$$x = 2$$

Therefore, $\overline{AB} = 7x$, and $\overline{BC} = 3x + 8$

$\phantom{Therefore, \overline{AB}} = 7(2) = 3(2) + 8$

$\phantom{Therefore, \overline{AB}} = 14 = 14$

Angles in General

An angle is formed by two rays with the same end point. The common end point is called the *vertex* of the angle, and the rays are called the *sides* of the angle.

In Figure 2 the vertex of angle θ (theta) is labeled O, and A and B are points on each side of θ. Angle θ can also be denoted by AOB, where the letter associated with the vertex is written between the letters associated with the points on each side.

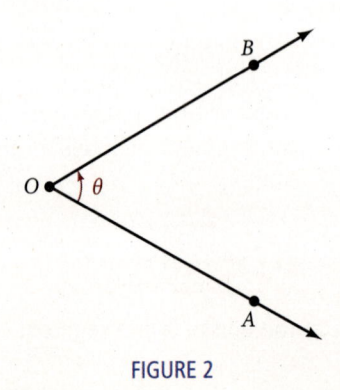

FIGURE 2

PRACTICE PROBLEMS

1. If B is the midpoint of $\overline{AC}$, find x.

Answer
1. $x = 2$

Angle Bisector

An *angle bisector* is a ray that divides an angle into two angles with equal measures.

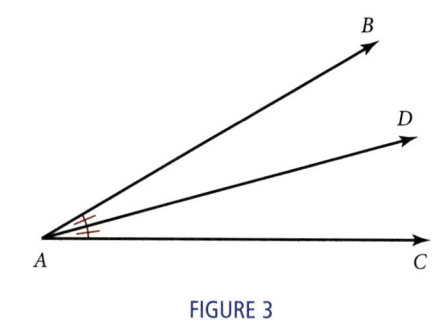

FIGURE 3

If $\overline{AD}$ is an angle bisector, then the measure of $\angle BAD$ = the measure of $\angle DAC$.

EXAMPLE 2 $\overline{BC}$ bisects $\angle ABD$. Solve for x.

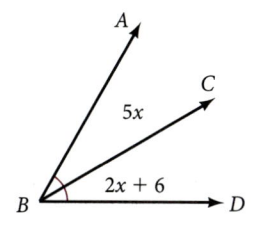

SOLUTION Since $\overline{BC}$ bisects $\angle ABD$,

$m\angle ABC = m\angle CBD$

$5x = 2x + 6$

$3x = 6$

$x = 2$

Therefore, $m\angle ABC = 5x$, and $m\angle CBD = 2x + 6$

$ = 5(2) = 2(2) + 6$

$ = 10 = 10$

A Degree Measure

One way to measure the size of an angle is with degree measure. The angle formed by rotating a ray through one complete revolution has a measure of 360 degrees, written 360° (Figure 4).

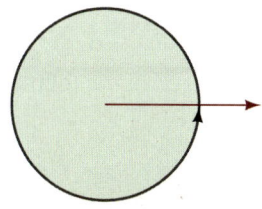

One complete revolution = 360°

FIGURE 4

2. $\overline{BC}$ bisects $\angle ABD$. Solve for x.

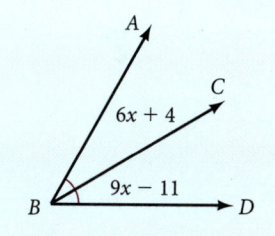

Answer
2. $x = 5$

Chapter 5 Geometry

One degree (1°), then, is $\frac{1}{360}$ of a full rotation. Likewise, 180° is one-half of a full rotation, and 90° is half of that (or one quarter of a rotation). Angles that measure 90° are called *right angles*, while angles that measure 180° are called *straight angles*. Angles that measure between 0° and 90° are called *acute angles*, while angles that measure between 90° and 180° are called *obtuse angles*.

FIGURE 5

> Note The little square by the vertex of the right angle in Figure 5 is used to indicate that the angle is a right angle. You will see this symbol often in the book.

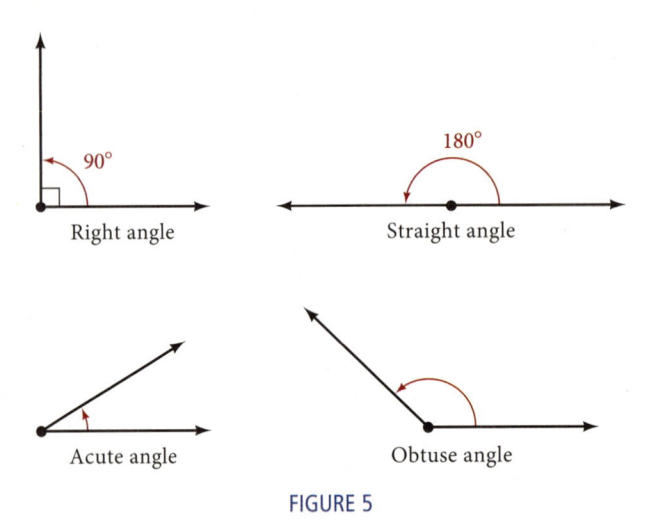

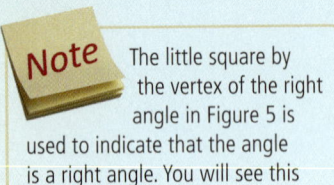

B Complementary and Supplementary Angles

We can apply our knowledge of algebra to help solve some simple geometry problems. Before we do, however, we need to review some of the vocabulary associated with angles.

> **Definition**
> In geometry, two angles that add to 90° are called **complementary angles**. In a similar manner, two angles that add to 180° are called **supplementary angles**. The following diagrams illustrate the relationships between angles that are complementary and between angles that are supplementary.

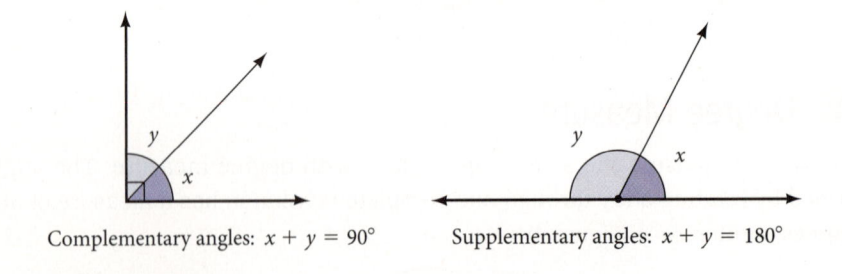

Complementary angles: $x + y = 90°$

Supplementary angles: $x + y = 180°$

EXAMPLE 3   Find x in each of the following diagrams.

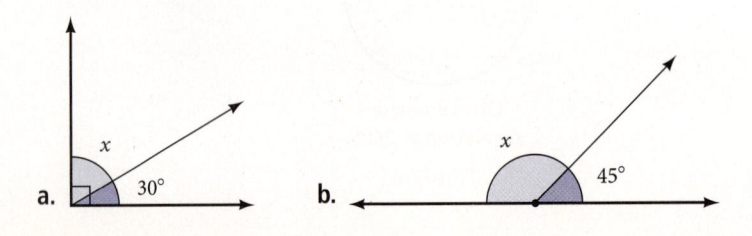

a. 30°

b. 45°

3. Find x in each of the following diagrams.

a. 45°

b. 60°

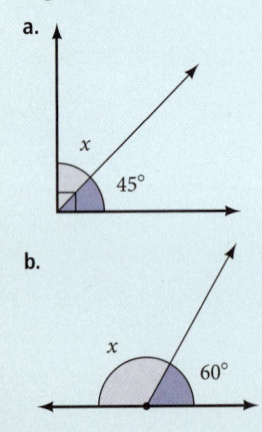

Answer
3. a. 45°   b. 120°

SOLUTION We use subtraction to find each angle.

a. Because the two angles are complementary, we can find x by subtracting 30° from 90°:

$$x = 90° - 30° = 60°$$

We say 30° and 60° are complementary angles. The complement of 30° is 60°.

b. The two angles in the diagram are supplementary. To find x, we subtract 45° from 180°:

$$x = 180° - 45° = 135°$$

We say 45° and 135° are supplementary angles. The supplement of 45° is 135°.

We defined complementary angles as angles that add to 90°. If x and y are complementary angles, then

$$x + y = 90°$$

If we solve this formula for y, we obtain a formula equivalent to our original formula:

$$y = 90° - x$$

Because y is the complement of x, we can generalize by saying that the complement of angle x is the angle $90° - x$. By a similar reasoning process, we can say that the supplement of angle x is the angle $180° - x$. To summarize, if x is an angle, then

The complement of x is $90° - x$, and

The supplement of x is $180° - x$

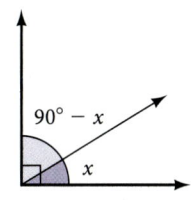

Complementary angles

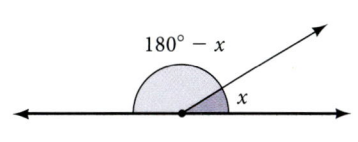

Supplementary angles

EXAMPLE 4 Give the complement and supplement of each angle.
- **a.** 40°
- **b.** 110°
- **c.** θ

SOLUTION
- **a.** The complement of 40° is $90° - 40° = 50°$.
  The supplement of 40° is $180° - 40° = 140°$.
- **b.** The complement of 110° is $90° - 110° = -20°$.
  The supplement of 110° is $180° - 110° = 70°$.
- **c.** The complement of θ is $90° - \theta$ since $\theta + (90° - \theta) = 90°$.
  The supplement of θ is $180° - \theta$ since $\theta + (180° - \theta) = 180°$.

4. Give the complement and supplement of each angle.
   - **a.** 70°
   - **b.** 135°
   - **c.** $3t$

Answer
4. **a.** 20°; 110° **b.** −45°; 45°
   c. $90° - 3t$; $180° - 3t$

5. Find x in each of the following diagrams.

a.

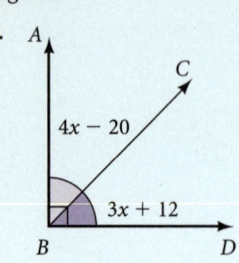

$4x - 20$
$3x + 12$

b.

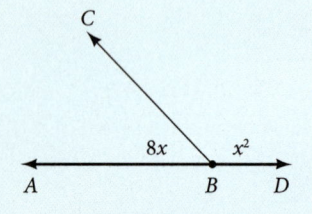

$8x$ x^2

Chapter 5 Geometry

EXAMPLE 5 Find x in each of the following diagrams.

a.

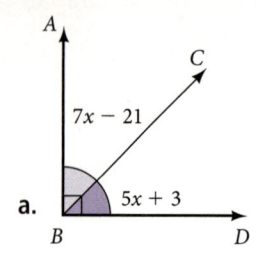

$7x - 21$
$5x + 3$

b.

$3x$ x^2

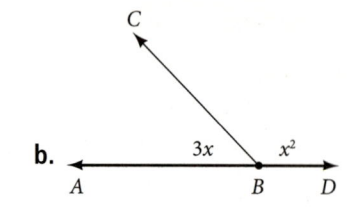

SOLUTION

a. Since $\angle ABD$ is a right angle, $\angle ABC$ and $\angle CBD$ are complementary, meaning their sum is 90°. We have

$$(7x - 21) + (5x + 3) = 90$$
$$12x - 18 = 90$$
$$12x = 108$$
$$x = 9$$

b. Since $\angle ABC$ and $\angle CBD$ are supplementary their sum is 180°. We have

$$3x + x^2 = 180$$
$$x^2 + 3x - 180 = 0 \qquad \text{Standard form for a quadratic equation}$$
$$(x + 15)(x - 12) = 0 \qquad \text{Factor } x^2 + 3x - 180$$
$$x + 15 = 0, \text{ or } x - 12 = 0$$
$$x = -15, \text{ or } x = 12$$

However, the measure of an angle cannot be a negative number, so $x = 12$ is the only answer.

C Vertical Angles

When two lines in the same plane intersect, two pairs of vertical angles are formed. Figure 6 shows lines ℓ_1 and ℓ_2 intersecting to form $\angle 1$, $\angle 2$, $\angle 3$, and $\angle 4$.

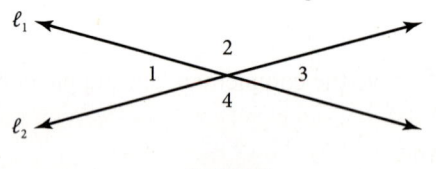

FIGURE 6

$\angle 1$ and $\angle 3$ are vertical

$\angle 2$ and $\angle 4$ are vertical

As you can see from the figure, vertical angles are on opposite sides of the vertex.

Definition
Two angles are **vertical angles** if and only if they are non-adjacent and formed by intersecting lines.

Answer
5. a. $x = 14$ b. $x = 10$

5.1 Angles

> **Definition**
> Two angles are **congruent** if, when superimposed on top of each other, all of their points coincide.

From this definition we have the following Theorem:

> **Theorems for Vertical Angles**
> If two angles are vertical then they are congruent and have equal measure.

EXAMPLE 6 Use the properties of vertical angles to find the value of x and the measure of $\angle 1$ and $\angle 2$ given in Figure 7.

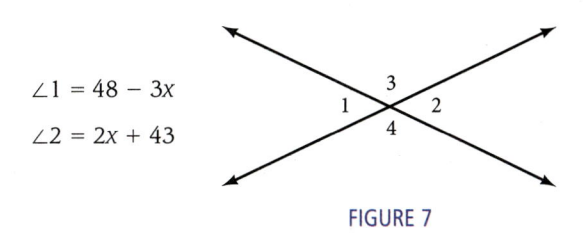

$\angle 1 = 48 - 3x$

$\angle 2 = 2x + 43$

FIGURE 7

SOLUTION Since vertical angles have equal measure, we set the two expressions equal to each other and solve the resulting linear equation for x. We have:

$$\angle 1 = \angle 2$$
$$48 - 3x = 2x + 43$$
$$-5x = -5$$
$$x = 1$$

Now we can use $x = 1$ to find the measure of the angles;

$$\angle 1 = 48 - 3(1)$$
$$= 48 - 3$$
$$= 45°$$

Even though we know that these angles are equal, we can check our work by making sure the expression for $\angle 2$ also simplifies to 45°;

$$\angle 2 = 2(1) + 43$$
$$= 2 + 43$$
$$= 45°$$

6. Rework Example 4 for the following angle measures.

$$\angle 1 = 58 - 4x$$
$$\angle 2 = 5x + 40$$

Answer
6. $\angle 1 = 50°$, $\angle 2 = 50°$

Left Blank Intentionally

Problem Set 5.1

B is the midpoint of $\overline{AC}$. Find x.

1. $A \quad 6(x-3) \quad B \quad x+2 \quad C$

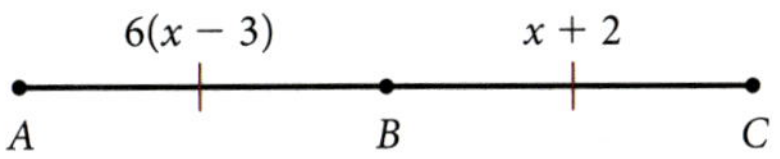

2. $A \quad 5(x+2) \quad B \quad x+6 \quad C$

3. $A \quad 8(x-2) \quad B \quad 2(x+7) \quad C$

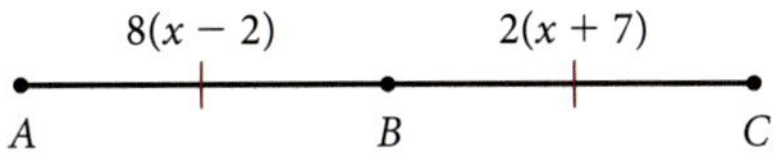

4. $A \quad 6(x-3) \quad B \quad 3(x+4) \quad C$

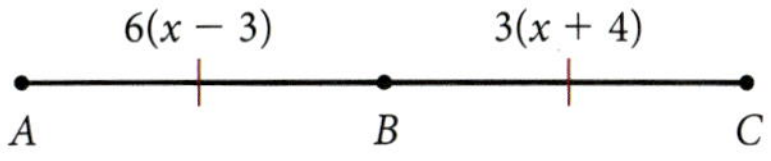

M is the midpoint of $\overline{AC}$. Find x.

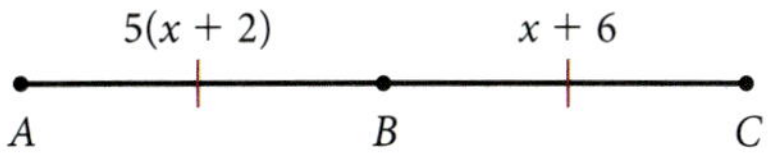

5. $A \quad x^2 \quad M \quad 6x-9 \quad C$

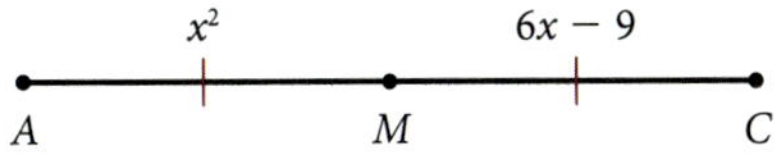

6. $A \quad x^2-3x \quad M \quad 10 \quad C$

7. $A \quad x^2-x \quad M \quad 4x-6 \quad C$

8. $A \quad x(x-4) \quad M \quad 6(x-4) \quad C$

9. $A \quad x^2+2x \quad M \quad 15 \quad C$

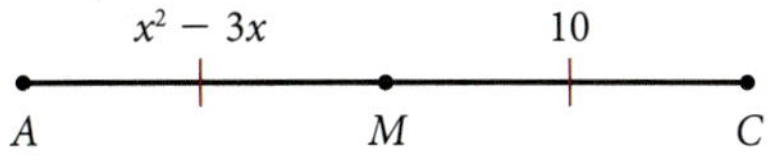

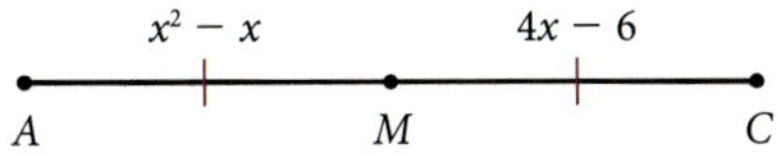

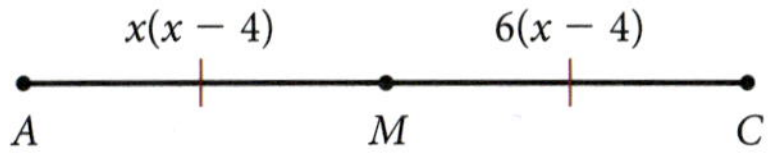

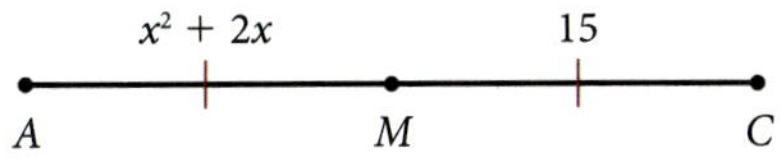

10. $A \quad 2x^2-5x \quad M \quad 3 \quad C$

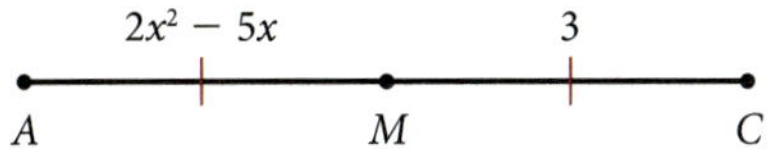

For Problems 11–18, $\overline{BC}$ bisects $\angle ABD$. Solve for x.

11. $7x+3$, $4x+9$

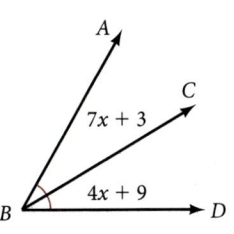

12. $6x+5$, $7x+1$

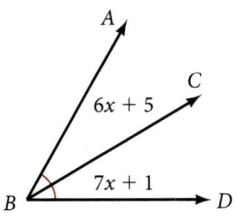

13. $5x+13$, $7x-9$

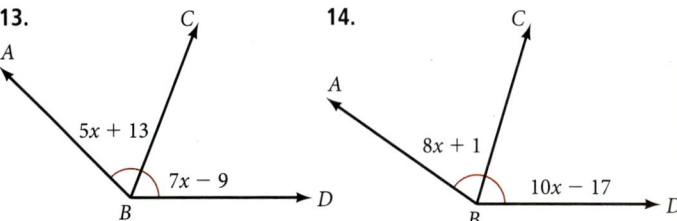

14. $8x+1$, $10x-17$

15. $3x+6$, $5x$

16. $11x$, $4x+21$

17. $6x+15$, $9x$

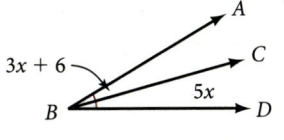

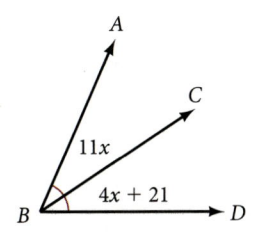

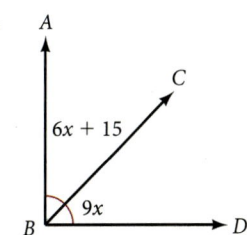

18. $2x$, $5x-33$

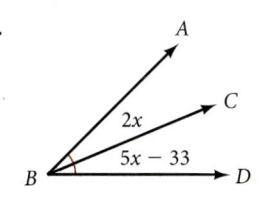

Chapter 5 Geometry

Solve for *x* in problems 19–34.

19. $2x + 3$, $5x + 3$

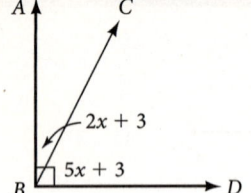

20. $9x + 11$, $4x - 12$

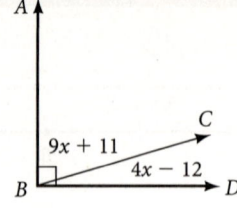

21. $5x - 10$, $25x - 20$

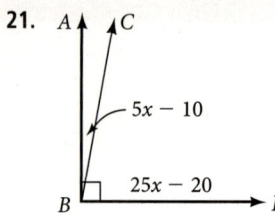

22. $5x - 4$, $6x + 6$

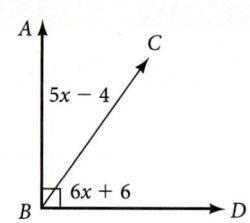

23. x^2, $3x + 2$

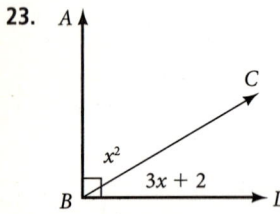

24. $2x^2 - 4$, $2x + 10$

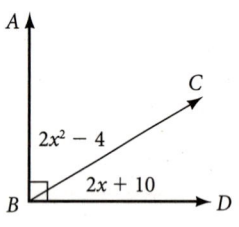

25. $6x - 1$, x^2

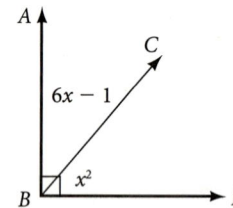

26. $x^2 + x$, $5x - 1$

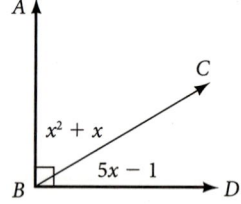

27. $10x + 2$, $7x - 9$

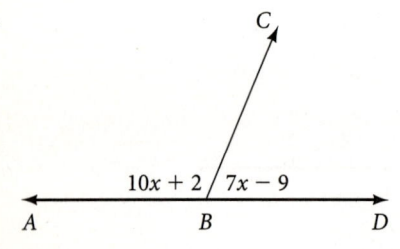

28. $10x + 5$, $12x - 1$

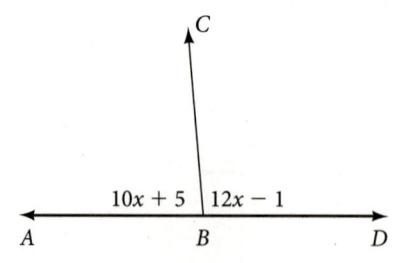

29. $13x - 5$, $2x + 5$

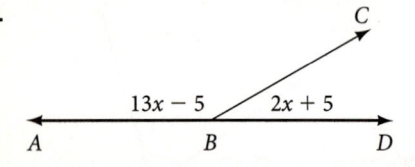

30. $15x + 15$, $15(x + 1)$

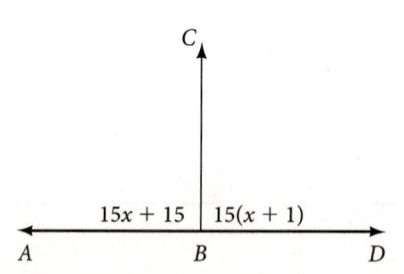

31. $5x + 4$, x^2

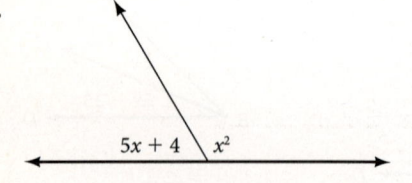

32. $7x$, $x^2 + x$

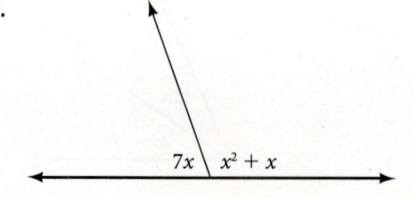

33. $(x-3)^2$ $(x+3)^2$

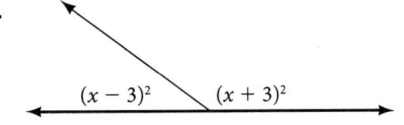

34. $4x$ $x^2 - x$

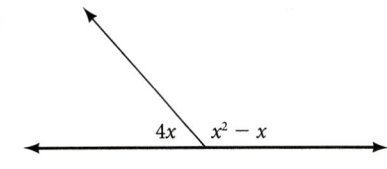

Use the properties of vertical angles to solve for *x* in problems 35–42.

35. $4x + 3$ $6x - 11$

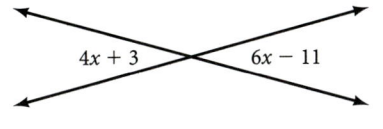

36. $5x + 2$ $7x - 20$

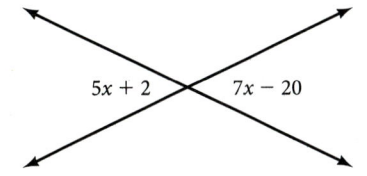

37. $21x + 2$ $19x + 14$

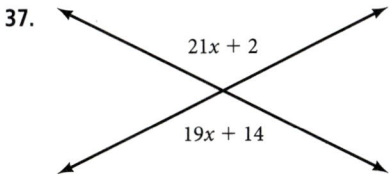

38. $11x - 35$ $7x + 25$

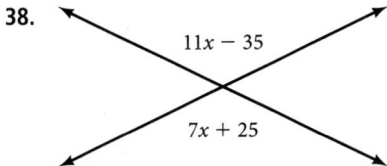

39. $2x^2$ $x^2 + 2x + 15$

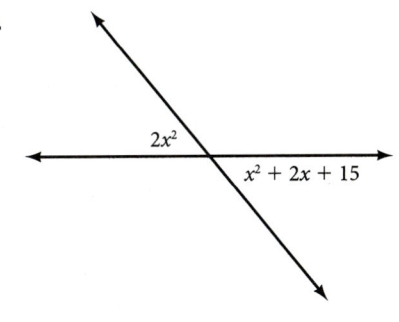

40. $x^2 - 2x$ $6x$

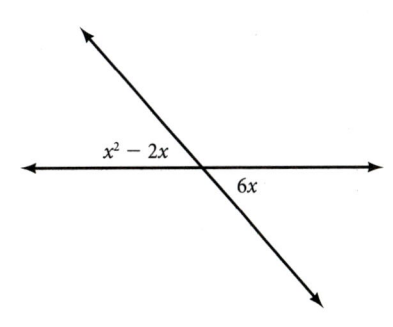

41. $x^2 + x - 10$ $10x + 12$

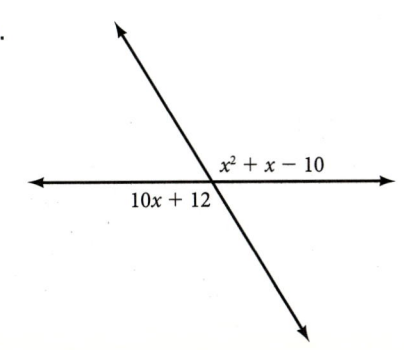

42. $4x^2$ $3x^2 + 5x + 6$

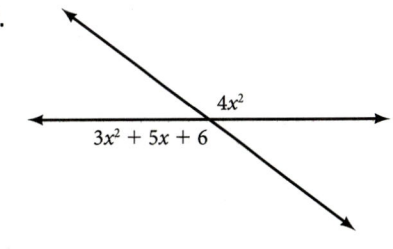

Use the properties of vertical and supplementary angles to solve for x, y, and z in problems 43–50.

43. $4x + 5$, x, y, z

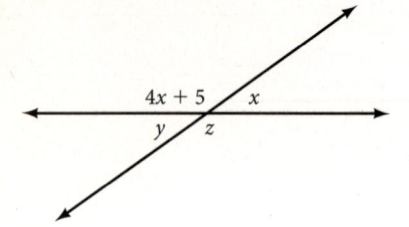

44. $5x - 12$, x, y, z

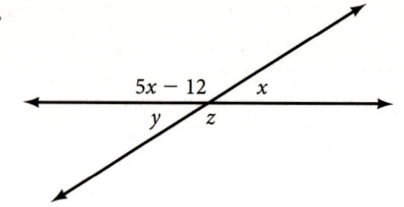

45. $6x - 16$, x, y, z

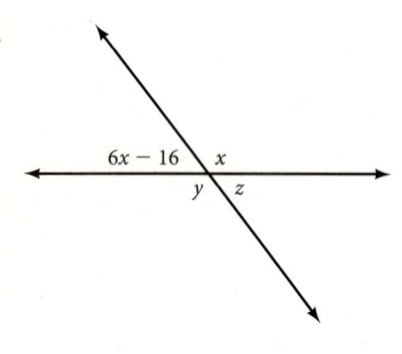

46. $2x - 3$, x, y, z

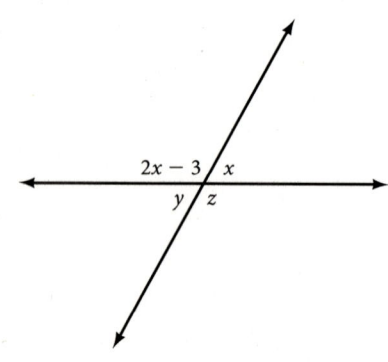

47. $2y - 6$, $2x$, y, z

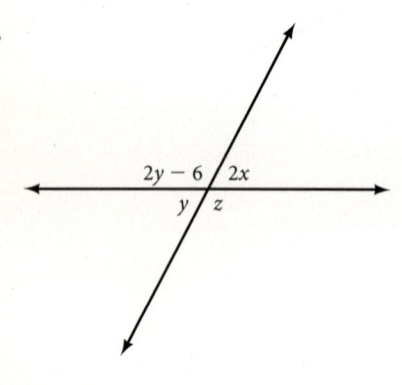

48. $5z + 15$, x, y, $6z - 1$

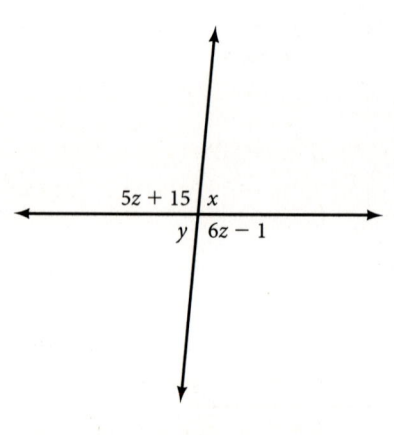

49. $7x - 12$, x, y, z

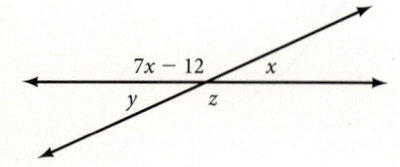

50. $x + 16$, x, y, z

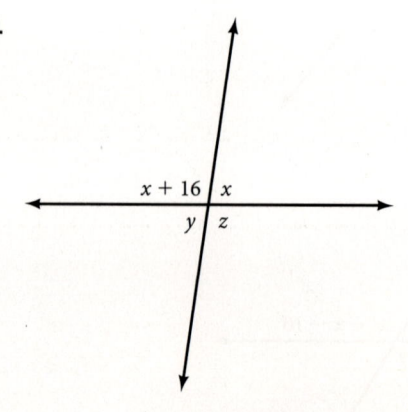

5.2 Parallel Lines

Objectives
- **A** Classify the angles formed from parallel lines.
- **B** Classify angles formed by parallel lines cut by a transversal.

A Parallel Lines

A plane is a two-dimensional surface that extends indefinitely in all directions. Two lines in the same plane that never intersect are called parallel lines.

> **Definition**
> Two or more lines that lie in the same plane are **parallel lines** if they have no points in common.

That is, parallel lines never intersect no matter how far they are extended in the plane. The following lines ℓ_1 and ℓ_2 are parallel and denoted by $\ell_1 \parallel \ell_2$, which is read "ℓ one is parallel to ℓ two." The symbol $>$ can also be used to show two or more lines are parallel, see Figure 1.

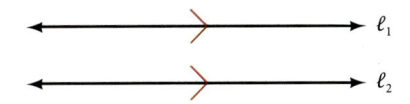

FIGURE 1 Parallel lines ℓ_1 and ℓ_2

Lines in the same plane that intersect at exactly one point are called intersecting lines. That is, if two different lines in the same plane are not parallel, then they are intersecting lines.

When two or more parallel lines are intersected by a third line, the third line is called a transversal. In this section we will solve problems about the angles formed when two parallel lines are cut by a transversal.

> **Definition**
> A **transversal** is a line that intersects two or more lines in the same plane at different points.

In Figure 2, line t intersects both lines ℓ_1 and ℓ_2. In a case like this, t is called a *transversal*. We have labeled all the angles formed by these intersections with the numbers 1 through 8.

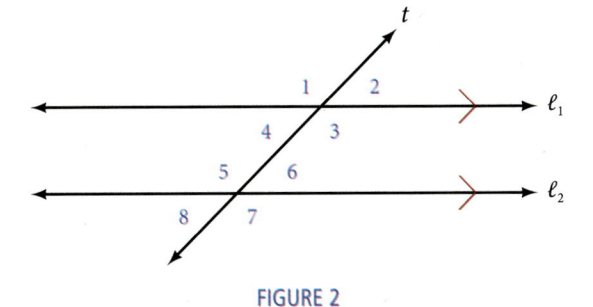

FIGURE 2

Chapter 5 Geometry

B Classifying the Angles Formed

Once two parallel lines are cut by a transversal, certain angle relationships are formed between the interior and exterior angles. The angles that lie between ℓ_1 and ℓ_2, are called *interior angles* and the angles that lie outside ℓ_1 and ℓ_2 are called *exterior angles*. See Figure 3.

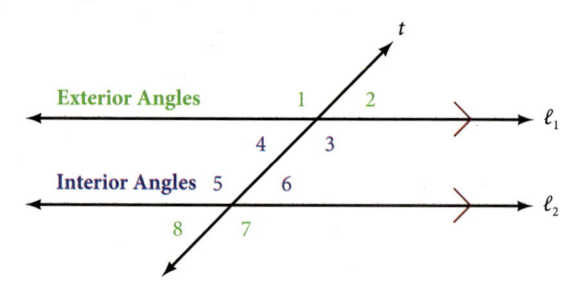

FIGURE 3

These angles are further classified in pairs:

    Alternate interior angles: ∠4 and ∠6
                                         ∠3 and ∠5

    Alternate exterior angles: ∠1 and ∠7
                                         ∠2 and ∠8

    Corresponding angles: ∠1 and ∠5
                                  ∠2 and ∠6
                                  ∠3 and ∠7
                                  ∠4 and ∠8

If two parallel lines are cut by a transversal (as shown in Figure 3) all the angle pairs noted above are congruent.

    Consecutive Interior angles: ∠3 and ∠6
                                             ∠4 and ∠5

These angles are supplementary.

EXAMPLE 1   Name two pairs of alternate exterior angles in Figure 4.

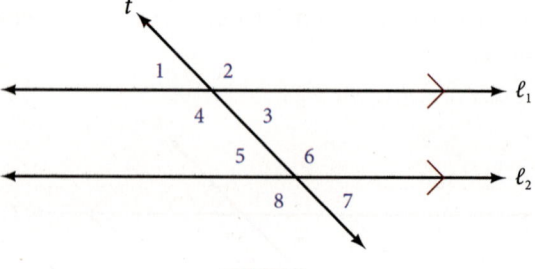

FIGURE 4

SOLUTION   Notice that four exterior angles exist, ∠1, ∠2, ∠7, and ∠8. Also notice that ∠1 and ∠7 are located on opposite sides of transversal *t*. Because these angles alternate from side to side of the transversal, we refer to ∠1 and ∠7 as alternate exterior angles. ∠2 and ∠8 have the same relationship. Therefore, two pairs of

    Alternate Exterior Angles

        ∠1 and ∠7
        ∠2 and ∠8

5.2 Parallel Lines

EXAMPLE 2 Name two pairs of alternate interior angles in Figure 5.

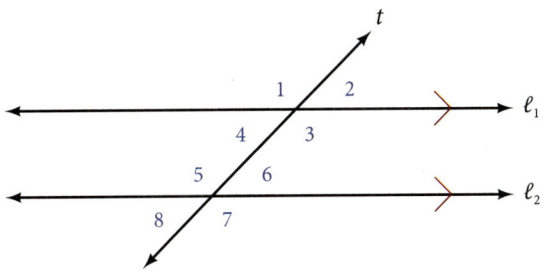

FIGURE 5

SOLUTION The four interior angles in the figure are ∠3, ∠4, ∠5, and ∠6. Notice that ∠4 and ∠6 are on opposite (or alternating) sides of transversal t. Therefore, ∠4 and ∠6 are alternate interior angles. ∠3 and ∠5 have the same relationship. Therefore, two pairs of

Alternate Interior Angles

∠4 and ∠6
∠3 and ∠5

Hint: Another method of identifying alternate interior angles is to look for a "Z" pattern in the figure. Following the green line below, follow across ℓ_1, down transversal t, and across ℓ_2. ∠4 and ∠6 are the included angles as are ∠3 and ∠5.

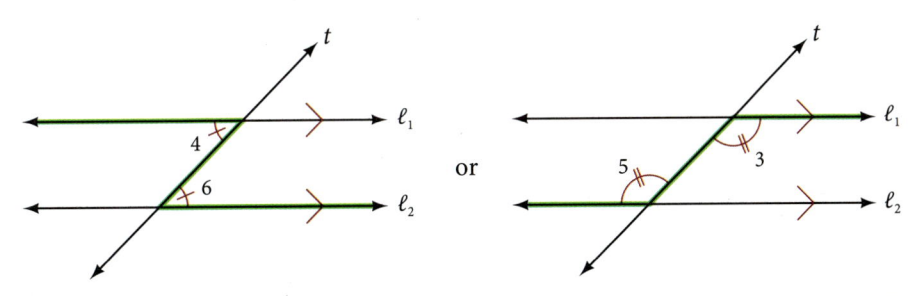

FIGURE 6 Alternate interior angles as identified using the "Z" pattern

EXAMPLE 3 Name two pairs of consecutive interior angles in Figure 7.

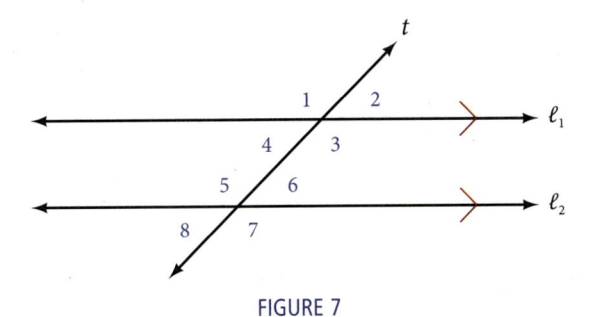

FIGURE 7

SOLUTION If two parallel lines are cut by a transversal, consecutive interior angles are supplementary. Consecutive interior angles are interior angles located on the same side of the transversal. In Figure 7, ∠4 and ∠5 are a pair of consecutive interior angles. ∠3 and ∠6 are also consecutive interior angles.

Consecutive Interior Angles

∠4 and ∠5
∠3 and ∠6

Chapter 5 Geometry

EXAMPLE 4   Name two pairs of corresponding angles in Figure 8.

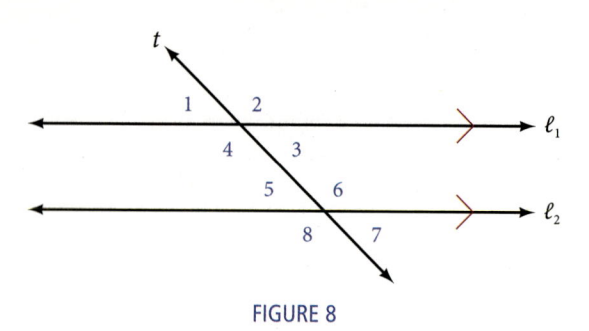

FIGURE 8

SOLUTION   Corresponding angles are angles that correspond to or "match" each other. Notice that ∠1 corresponds to ∠5 as if these angles were pieces in a puzzle with the same measure. You could "cut" ∠1 and "paste" it over ∠5. Two pairs of corresponding angles are ∠1 and ∠5, as well as ∠4 and ∠8. Can you find the other two pairs of corresponding angles? Here are all pairs of corresponding angles:

Corresponding Angles

∠1 and ∠5
∠2 and ∠6
∠3 and ∠7
∠4 and ∠8

In summary, if two parallel lines are cut by a transversal, we have the following theorems:

Theorems for Parallel Lines
For lines ℓ_1 and ℓ_2 cut by transversal t, and the eight angles formed, we have the following theorems:

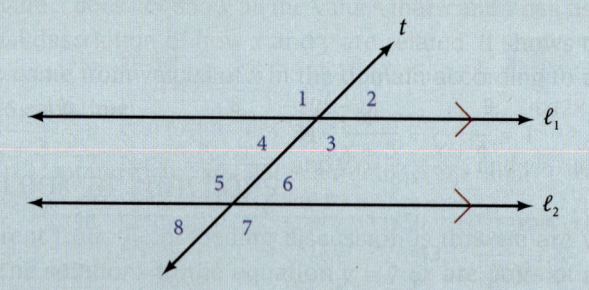

Corresponding Angle Theorem
$\ell_1 \parallel \ell_2$, ⇔ Corresponding angles are congruent

∠1 ≅ ∠5
∠4 ≅ ∠8
∠2 ≅ ∠6
∠3 ≅ ∠7

Alternate Interior Angle Theorem
$\ell_1 \parallel \ell_2$, ⇔ Alternate interior angles are congruent

∠4 ≅ ∠6
∠5 ≅ ∠3

5.2 Parallel Lines

Alternate Exterior Angle Theorem
$\ell_1 \parallel \ell_2, \Leftrightarrow$ Alternate exterior angles are equal

$$\angle 1 \cong \angle 7$$
$$\angle 2 \cong \angle 8$$

Consecutive Interior Angle Theorem
$\ell_1 \parallel \ell_2, \Leftrightarrow$ Consecutive interior angles are supplementary

$$m\angle 4 + m\angle 5 = 180°$$
$$m\angle 3 + m\angle 6 = 180°$$

EXAMPLE 5 If $\ell_1 \parallel \ell_2$, and $m\angle 1 = 155°$, find $m\angle 6$.

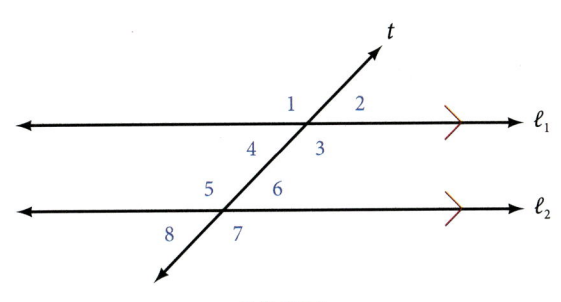

FIGURE 9

SOLUTION If two parallel lines are cut by a transversal, corresponding angles are congruent and $\angle 1 \cong \angle 5$. If the angles are congruent, their measures are equal. Therefore, $m\angle 5 = 155°$. Knowing that $\angle 5$ and $\angle 6$ are supplementary, $m\angle 5 + m\angle 6 = 180°$ and through substitution:

$$155° + m\angle 6 = 180°$$
$$m\angle 6 = 25°$$

EXAMPLE 6 If $\ell_1 \parallel \ell_2$, and $m\angle 2 = 53°$, find $m\angle 8$.

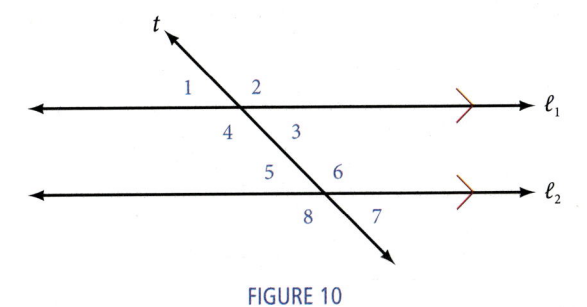

FIGURE 10

SOLUTION If two parallel lines are cut by a transversal, alternate exterior angles are congruent. In this example, $\angle 2$ and $\angle 8$ are a pair of alternate exterior angles. Therefore $m\angle 2 = m\angle 8$. Therefore $m\angle 8 = 53°$

PRACTICE PROBLEMS

5. Use the information in Example 5 to find $m\angle 7$.

6. Use the information in Example 6 to find $m\angle 5$.

Answer
5. 155°
6. 127°

7. Find x for Example 7 if:
   ∠6 = 3x + 47
   ∠8 = 7x − 1

Chapter 5 Geometry

EXAMPLE 7 If $\ell_1 \parallel \ell_2$, find x.

∠2 = 2x + 10
∠7 = 5x + 30

FIGURE 11

SOLUTION   There are multiple ways to solve this problem. Notice that ∠2 corresponds to ∠6, but is also an alternate exterior angle to ∠8. Let's solve this problem using alternate exterior angles.

Because two parallel lines are cut by a transversal, $2x + 10 = m\angle 8$. Since they are congruent, the measures are equal and $m\angle 8 = m\angle 2$.

We also know that ∠8 and ∠7 are supplementary angles, so $m\angle 8 + m\angle 7 = 180°$.

Using substitution:

$$(2x + 10) + (5x + 30) = 180$$
$$7x + 40 = 180$$
$$7x = 140$$
$$x = 20$$

EXAMPLE 8   In the figure below, ℓ_1 is parallel to ℓ_2 and t is a transversal. Find the measure of angle 2 if $m\angle ACD = 40°$.

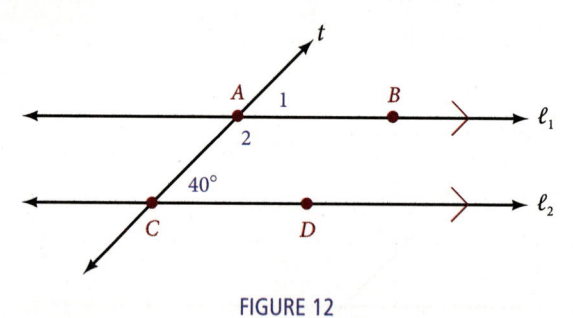

8. Find angle 2 if m∠ACD in Figure 12 is 52°.

FIGURE 12

SOLUTION   We know that ∠1 and ∠ACD are corresponding angles. Since the two lines are parallel, these angles must be equal. Therfore:

$$m\angle 1 = 40°$$

On the other hand, angles 1 and 2 are supplementary angles, giving us

$$m\angle 1 + m\angle 2 = 180°$$

Answer
7. x = 12
8. 128°

By substituting 40° for ∠1, we have

$$40° + m\angle 2 = 180°$$

$$m\angle 2 = 140° \quad \text{Subtract 40° from each side}$$

Notice that ∠2 and ∠ACD are supplementary because their sum is 180°.

In general, when two lines cut by a transversal are parallel, interior angles on the same side of the transversal are supplementary, as are exterior angles on the same side of the transversal. That is,

$$\ell_1 \text{ and } \ell_2 \text{ parallel} \Leftrightarrow \begin{cases} \angle 1 + \angle 7 = 180° \\ \angle 3 + \angle 5 = 180° \\ \angle 8 + \angle 2 = 180° \\ \angle 4 + \angle 6 = 180° \end{cases}$$

Transitive Property
The transitive property says that if $m\angle A \cong m\angle B$ and $m\angle B \cong m\angle C$ then $m\angle A \cong m\angle C$.

EXAMPLE 9 Given the diagram in Figure 13 with $m\angle 3 = m\angle 14$, show that lines t_1 and t_2 are parallel.

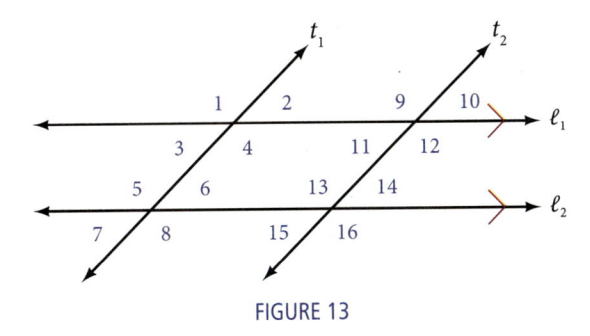

FIGURE 13

SOLUTION We reason like this: Since ℓ_1 and ℓ_2 are parallel, the measures of angles 3 and 7 are equal because they are corresponding angles for those two parallel lines. The problem states that angles 3 and 14 are equal. Through the transitive property, $\angle 7 \cong \angle 14$. If these alternate exterior angles are congruent, $t_1 \parallel t_2$. Here is a summary:

$$m\angle 3 = m\angle 7 \quad \text{Corresponding angles}$$
$$m\angle 3 = m\angle 14 \quad \text{Given in the problem}$$

Therefore, $m\angle 7 = m\angle 14$ Transitive Property

Perpendicular Lines

You are probably already familiar with perpendicular lines to some extent. Here is the formal definition:

Definition
Two lines are **perpendicular** if and only if a right angle is formed at their intersection.

---

9. Show that t_1 and t_2 are parallel if $m\angle 8 = m\angle 9$.

Answers
9. See Solutions to Selected Practice Problems

Chapter 5 Geometry

EXAMPLE 10 In Figure 14, $\ell_1 \parallel \ell_2$ and $\angle ADC$ is a right angle. Prove that ℓ_2 is perpendicular to t.

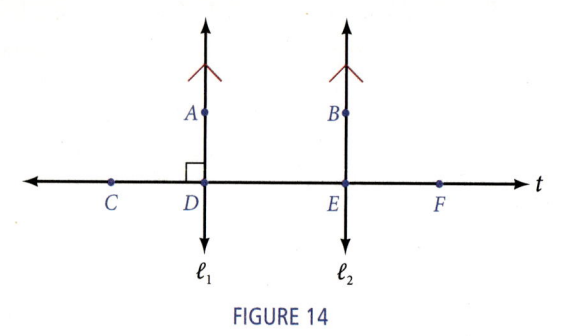

FIGURE 14

SOLUTION Since ℓ_1 and ℓ_2 are parallel, $\angle ADC \cong \angle BEC$ because they are corresponding angles. Further, since $\angle ADC = 90°$, then $\angle BEC = 90°$. Therefore ℓ_2 is perpendicular to t because a right angle is formed at the intersection.

> **Theorem**
> Two lines perpendicular to the same line are parallel to each other.

Getting Ready for Class

After reading through the preceding section, respond in your own words and in complete sentences.

1. With respect to Figure 2, show that $\angle 2$ and $\angle 5$ are supplementary angles.
2. With respect to Figure 2, what can be said about the sum of $m\angle 2 + m\angle 5$? Prove it.
3. Write the relationship between alternate interior angles if ℓ_1 and ℓ_2 are parallel.
4. Name the pairs of vertical angles in Figure 2.

Problem Set 5.2

For each problem, solve for the angle using the given information and the figure below.

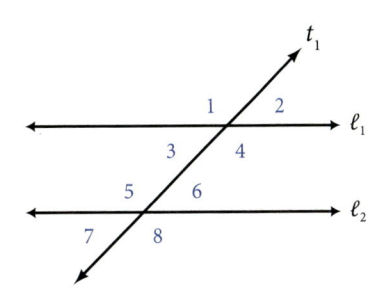

FIGURE 15

1. $m\angle 6 = 60°$
   $m\angle 1 =$

2. $m\angle 3 = 45°$
   $m\angle 6 =$

3. $m\angle 2 = 65°$
   $m\angle 7 =$

4. $m\angle 8 = 125°$
   $m\angle 3 =$

5. $m\angle 1 = 102°$
   $m\angle 7 =$

6. $m\angle 4 = 3(m\angle 2)$
   $m\angle 4 =$

7. Angle 1 is twice the size of angle 2. Find angle 2.

8. Angle 3 is $\frac{2}{3}$ of angle 5. Find the two angles.

9. Angle 6 is $\frac{1}{4}$ of angle 1. Find the two angles.

10. Angle 8 is 40° more than $\angle 6$. Find the two angles.

Using the figure below and the information, determine if ℓ_1 and ℓ_2 are parallel, not parallel, or not enough information.

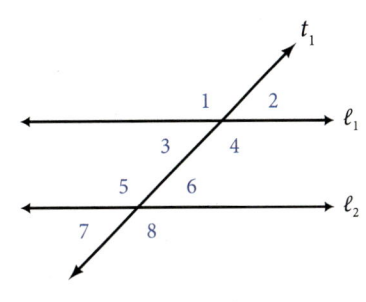

FIGURE 16

11. $m\angle 4 + m\angle 6 = 180°$

12. $m\angle 2 = m\angle 7$

13. $m\angle 5 = m\angle 8$

14. $m\angle 2 + m\angle 6 = 100°$

15. $m\angle 4 = m\angle 7$; $m\angle 4 = 60°$

16. $m\angle 1 + m\angle 2 \neq m\angle 4 + m\angle 6$

Chapter 5 Geometry

Use the properties of parallel lines to solve for x in each of the following.

17. $6x + 12$; $4x - 2$

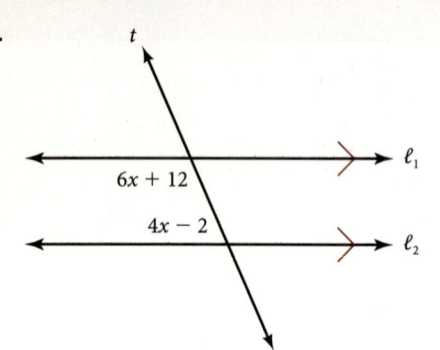

18. $3x - 2$; $8x + 6$

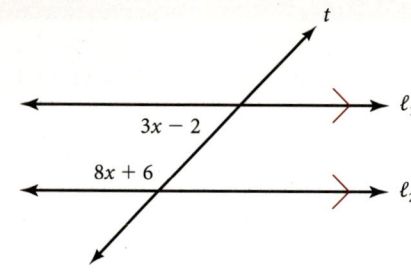

19. $12x - 17$; $9x - 13$

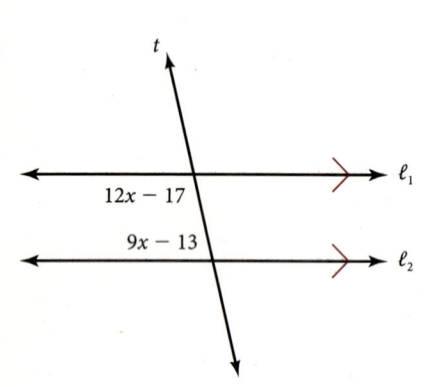

20. $4x + 18$; $7x + 8$

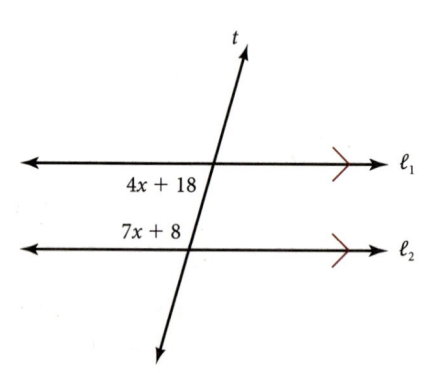

21. $11x - 2$; $8x + 13$

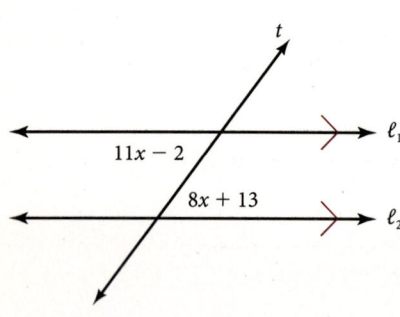

22. $10x + 4$; $12x - 18$

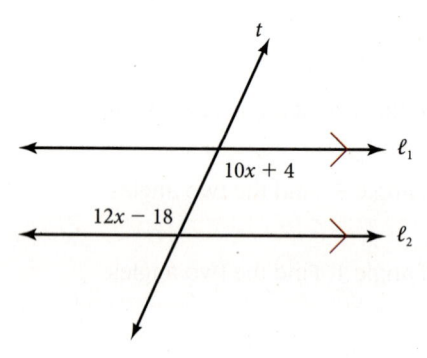

23. $4x - 14$; $3x + 27$

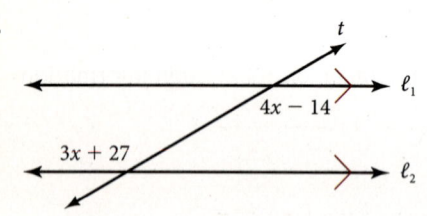

24. $x + 19$; $3x - 29$

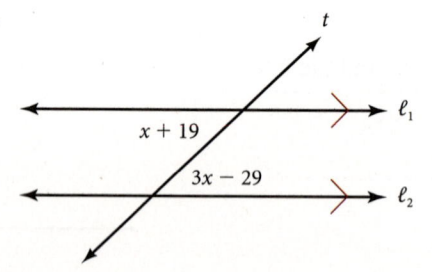

25. x^2 ; $8x + 33$

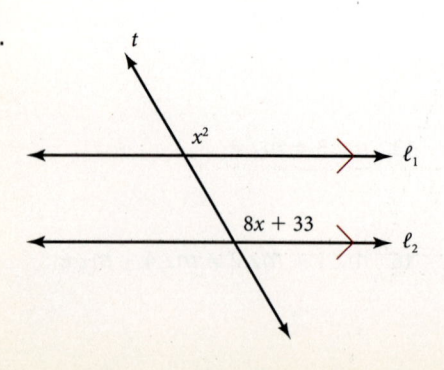

26. $x^2 - x$; $5x$

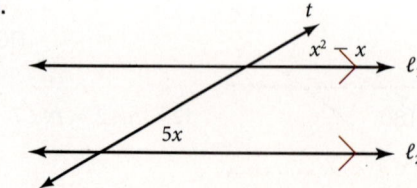

27.

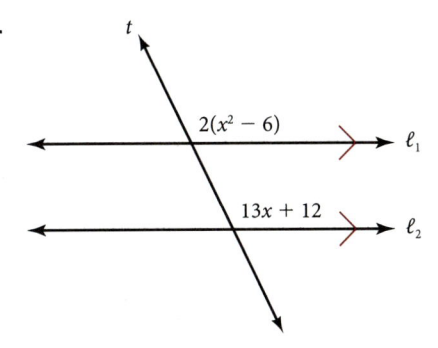

$2(x^2 - 6)$ — ℓ_1
$13x + 12$ — ℓ_2

28.

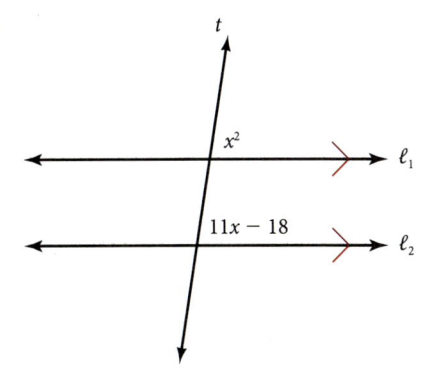

x^2 — ℓ_1
$11x - 18$ — ℓ_2

29.

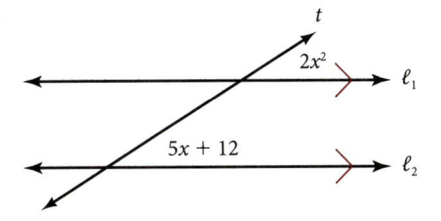

$2x^2$ — ℓ_1
$5x + 12$ — ℓ_2

30.

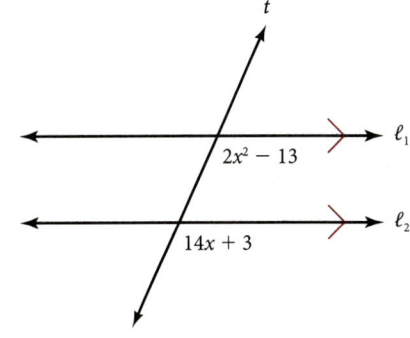

$2x^2 - 13$ — ℓ_1
$14x + 3$ — ℓ_2

31.

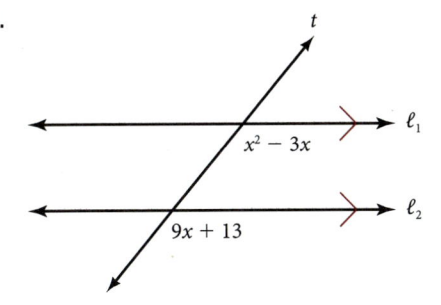

$x^2 - 3x$ — ℓ_1
$9x + 13$ — ℓ_2

32.

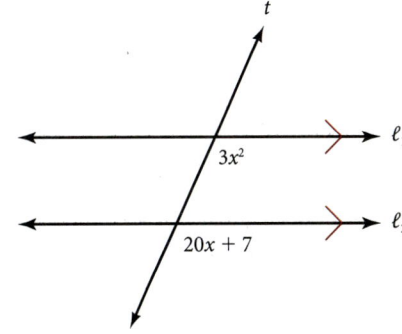

$3x^2$ — ℓ_1
$20x + 7$ — ℓ_2

33.

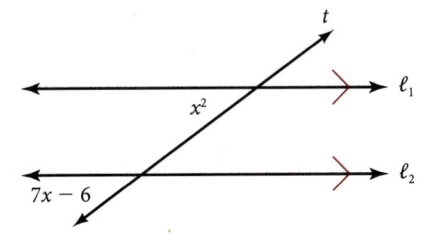

x^2 — ℓ_1
$7x - 6$ — ℓ_2

34.

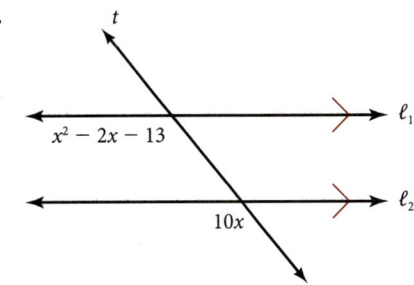

$x^2 - 2x - 13$ — ℓ_1
$10x$ — ℓ_2

35.

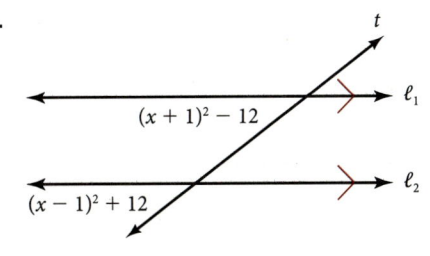

$(x + 1)^2 - 12$ — ℓ_1
$(x - 1)^2 + 12$ — ℓ_2

36.

$3x^2$ — ℓ_1
$15x + 18$ — ℓ_2

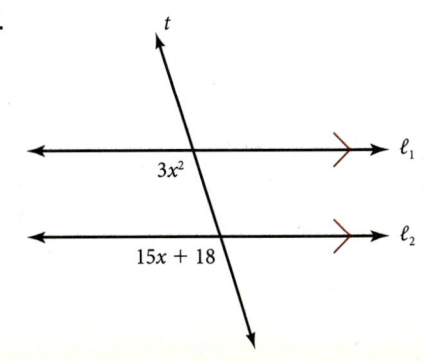

37.

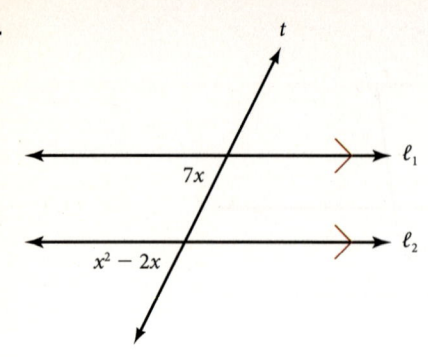

$7x$

$x^2 - 2x$

38.

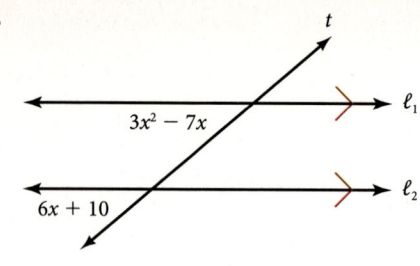

$3x^2 - 7x$

$6x + 10$

39.

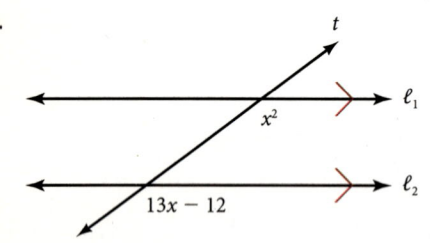

x^2

$13x - 12$

40.

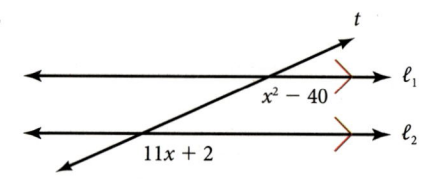

$x^2 - 40$

$11x + 2$

41.

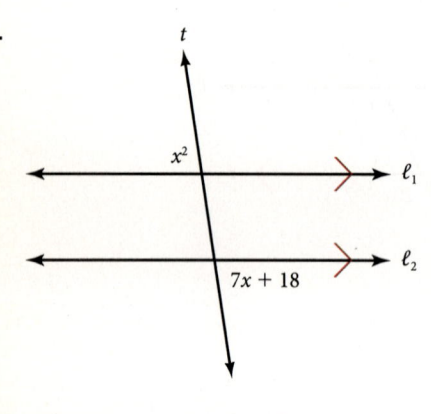

x^2

$7x + 18$

42.

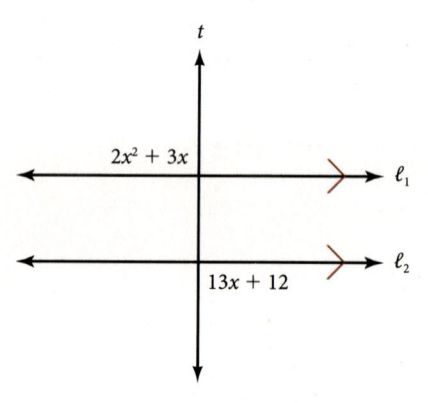

$2x^2 + 3x$

$13x + 12$

43.

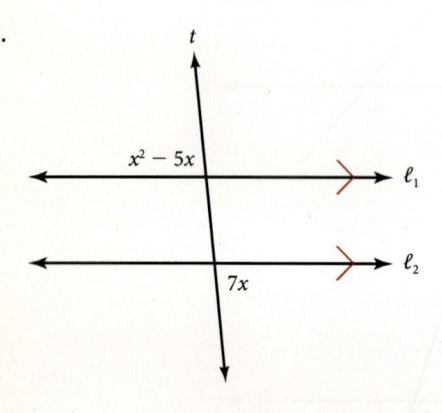

$x^2 - 5x$

$7x$

44.

$5x^2$

$20x + 25$

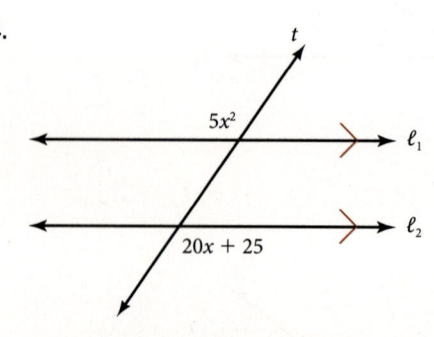

Classification of Triangles

5.3

Objectives
A Classification of triangles by sides or angles.
B Using the sum of the angles in a triangle to solve for unknown angles.

Triangles

A triangle is a three-sided polygon. Every triangle has three sides and three angles. We denote the angles or vertices with uppercase letters and the lengths of the sides with lowercase letters, as shown in Figure 1. It is standard practice in mathematics to label the sides and angles so that a is opposite A, b is opposite B, and c is opposite C.

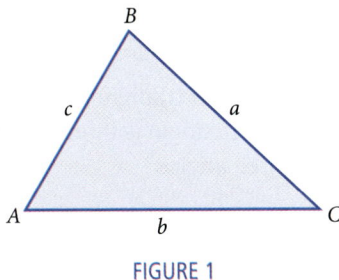

FIGURE 1

A Classification of Triangles

There are different types of triangles that are named according to the relative lengths of their sides or angles (Figure 2). Therefore we classify triangles by sides or angles. In an *equilateral triangle*, all three sides are of equal length measure. An *isosceles triangle* has two congruent sides, and the base angles, angles opposite the congrunet sides, are also congruent. If all the sides and angles have different measure, the triangle is called *scalene*. In an *acute triangle*, all three angles are acute (or measure less than 90°). An *obtuse triangle* has exactly one obtuse angle (or measure greater than 90°), and a *right triangle* has one right angle.

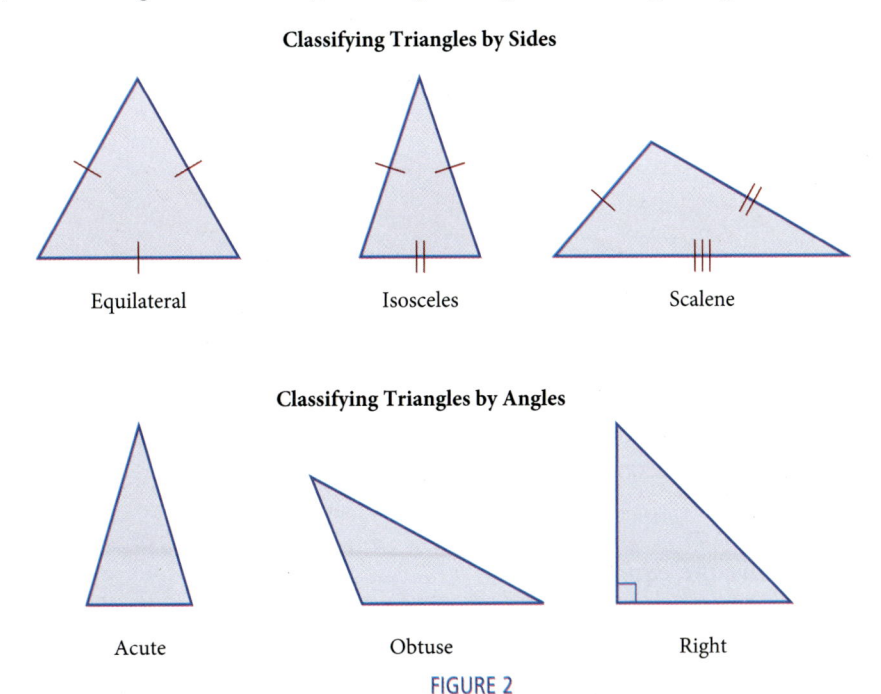

Classifying Triangles by Sides

Equilateral   Isosceles   Scalene

Classifying Triangles by Angles

Acute   Obtuse   Right

FIGURE 2

A *right triangle* is a triangle in which one of the angles is a right angle. In every right triangle, the longest side is called the *hypotenuse*, and it is always opposite

5.3 Classification of Triangles

Chapter 5 Geometry

the right angle. The other two sides are called the *legs* of the right triangle. Since the sum of the angles in any triangle is 180°, the other two angles in a right triangle must be complementary, acute angles.

B The Sum of the Angles in a Triangle

In any triangle, the sum of the interior angles is 180°. For the triangle shown in Figure 1, the relationship is written

$$A + B + C = 180°$$

PRACTICE PROBLEMS

1. The angles in a triangle are such that one angle is three times the smallest angle, whereas the largest angle is five times the smallest angle. Find the measure of all three angles.

EXAMPLE 1   The angles in a triangle are such that one angle is twice the smallest angle, whereas the third angle is three times as large as the smallest angle. Find the measure of all three angles.

SOLUTION

Step 1: **Read and list.**
  Known items: The sum of all three angles is 180°; one angle is twice the smallest angle; the largest angle is three times the smallest angle.

  Unknown items: The measure of each angle

Step 2: **Assign a variable, and translate information.**
  Let x be the smallest angle, then $2x$ will be the measure of another angle and $3x$ will be the measure of the largest angle.

Step 3: **Reread, and write an equation.**
  When working with geometric objects, drawing a generic diagram sometimes will help us visualize what it is that we are asked to find. In Figure 3, we draw a triangle with angles A, B, and C.

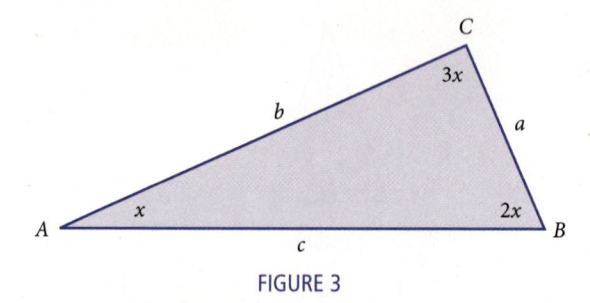

FIGURE 3

We can let the value of $A = x$, the value of $B = 2x$, and the value of $C = 3x$. We know that the sum of angles A, B, and C will be 180°, so our equation becomes

$$x + 2x + 3x = 180°$$

Step 4: **Solve the equation.**

$$x + 2x + 3x = 180°$$
$$6x = 180°$$
$$x = 30°$$

Step 5: **Write the answer.**
  The smallest angle A measures 30°.
  Angle B measures $2x$, or $2(30°) = 60°$.
  Angle C measures $3x$, or $3(30°) = 90°$.

Answer
1. 20°, 60°, and 100°

Step 6: Reread, and check.

The angles must add to 180°:

$$A + B + C = 180°$$
$$30° + 60° + 90° \stackrel{?}{=} 180°$$
$$180° = 180° \quad \text{Our answers check}$$

EXAMPLE 2 Solve for x using the properties of isosceles triangles.

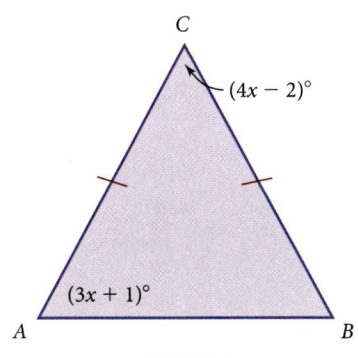

FIGURE 4

SOLUTION Since $\overline{AC} = \overline{CB}$,

$m\angle CAB = m\angle CBA$, which are both $(3x + 1)°$.

The sum of the interior angles is 180°, so we have

$$(3x + 1) + (3x + 1) + (4x - 2) = 180$$
$$10x + 2 - 2 = 180$$
$$10x = 180$$
$$x = 18$$

2. Solve for x using the properties of isosceles triangles.

The Exterior Angle of a Triangle

In any triangle, the angle formed by extending one side of a triangle an its adjacent angle is called an exterior angle, shown in Figure 4. The exterior angle is equal to the sum of the two remote interior angles, the two angles not adjacent to the exterior angle.

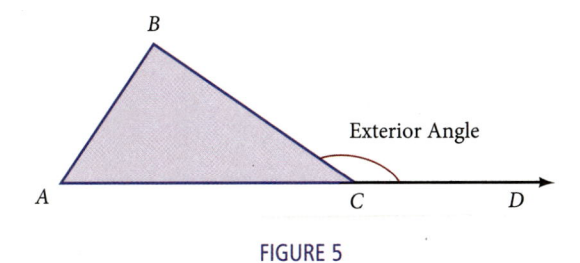

FIGURE 5

$$m\angle BCD = m\angle A + m\angle B$$

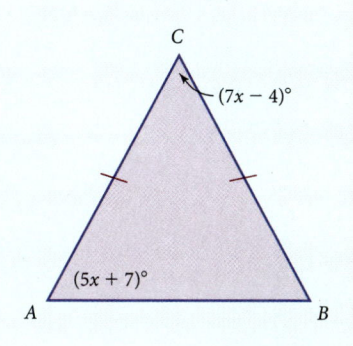

Answer
2. $x = 10$

3. Find the measure of the two remote angles in the figure.

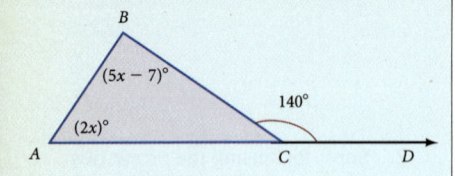

EXAMPLE 3 Find the measure of the two remote angles in the figure.

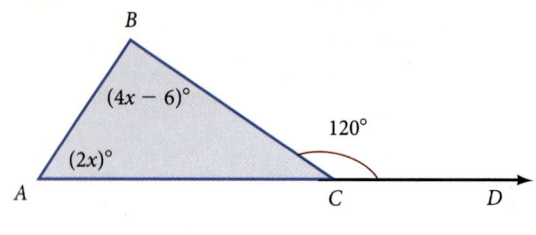

FIGURE 6

SOLUTION   Start by writing the sum of the remote angles is 120°.

$2x + 4x - 6 = 120°$
$6x - 6 = 120°$ Simplify
$6x = 126°$ Add 6
$x = 21°$ Divide by 6

$2(21) = 42°$

$4(21) - 6 = 78°$

The two remote angles are 42° and 78°.

Answer
3. 21°

Problem Set 5.3

Classify each triangle by their angles.

1.

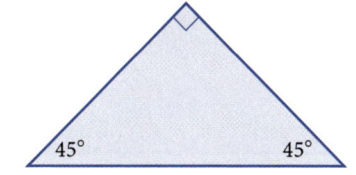

2.

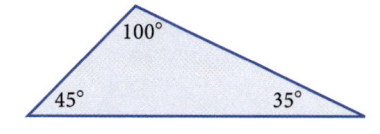

3.

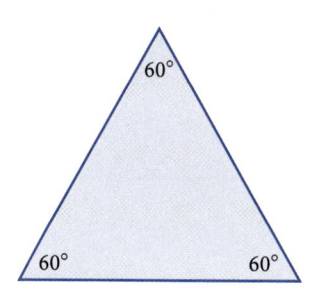

Classify each triangle by their sides.

4.

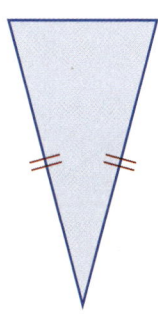

5.

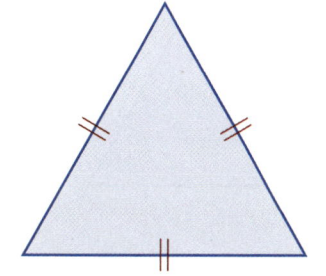

6.

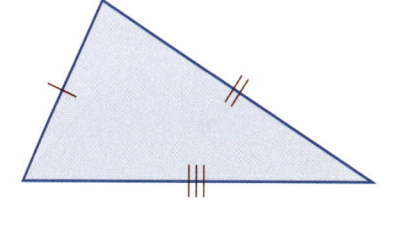

A Triangle Problems [Example 1]

7. The smallest angle in a triangle is $\frac{1}{5}$ as large as the largest angle. The third angle is twice the smallest angle. Find the three angles.

8. One angle in a triangle is half the largest angle but three times the smallest. Find all three angles.

9. A right triangle has one 37° angle. Find the other two angles.

10. In a right triangle, one of the acute angles is twice as large as the other acute angle. Find the measure of the two acute angles.

11. One angle of a triangle measures 20° more than the smallest, while a third angle is twice the smallest. Find the measure of each angle.

12. One angle of a triangle measures 50° more than the smallest, while a third angle is three times the smallest. Find the measure of each angle.

332 Chapter 5 Geometry

For problems 13–18 the expressions give the measure of each interior angle. Solve for x.

13. Triangle ABC: angle $B = x^2 - 4$, angle $A = x^2 - 4$, angle $C = x^2 - 4$.

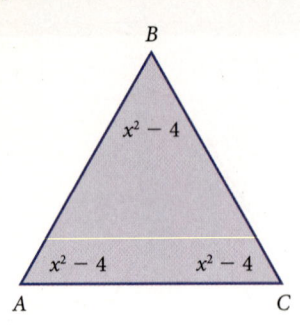

14. Triangle ABC: angle $B = x^2$, angle $A = 7x + 2$, angle $C = 7x + 2$.

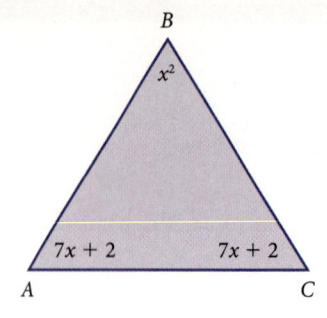

15. Triangle ABC: angle $B = x^2 - 11$, angle $A = 6(x + 1)$, angle $C = 5(x + 1)$.

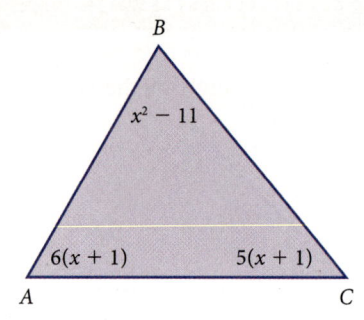

16. Triangle ABC: angle $B = 2x^2$, angle $A = x^2 + 4x$, angle $C = x^2 + 12$.

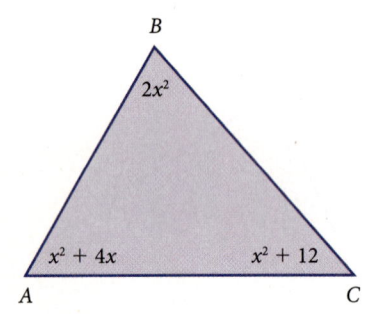

17. Triangle ABC: angle $B = 6x^2$, angle $A = 7x^2$, angle $C = 7x^2$.

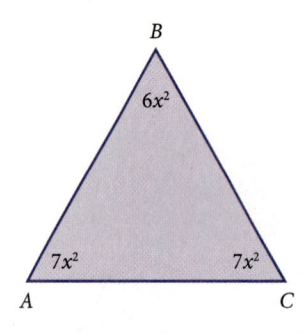

18. Triangle ABC: angle $B = x^2$, angle $A = 3x^2$, angle $C = x^2$.

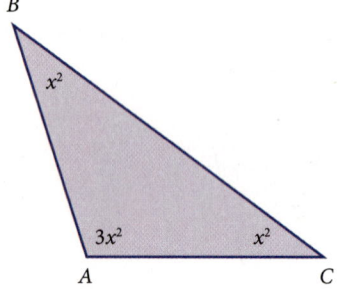

Solve for x.

19. Triangle with exterior angle at $B = 93°$, angle $D = 6x - 7$, angle $C = 2x + 4$.

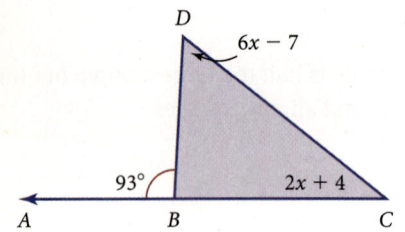

20. Triangle with exterior angle at $B = 72°$, angle $D = 4x + 3$, angle $C = 5x - 12$.

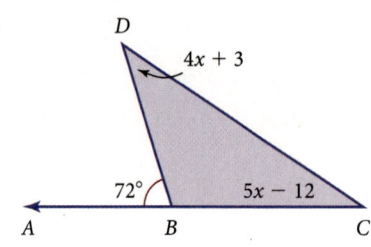

21. Triangle with exterior angle at $B = 90°$, angle $D = 4x - 3$, angle $C = 3x + 9$.

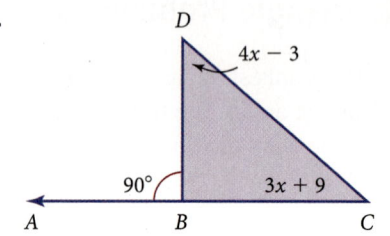

22. Triangle with exterior angle at $B = 136°$, angle $D = 5x + 5$, angle $C = 3x + 11$.

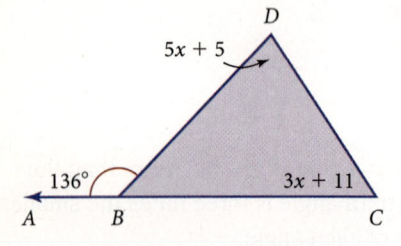

23. Triangle with exterior angle at $B = 128°$, angle $D = 3x + 5$, angle $C = 6x - 12$.

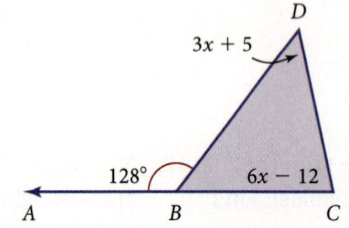

24. Triangle with exterior angle at $B = 140°$, angle $D = 7x + 6$, angle $C = 5x + 2$.

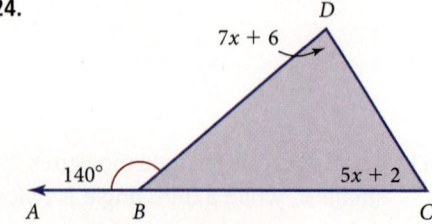

5.3 Problem Set 333

25. Triangle with angle x at B, angle x^2 at A, exterior angle $110°$ at C.

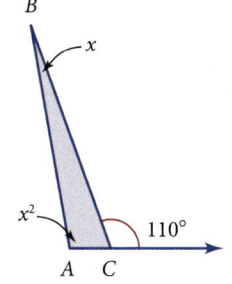

26. Triangle with angle x^2 at B, angle x^2 at A, exterior angle $98°$ at C.

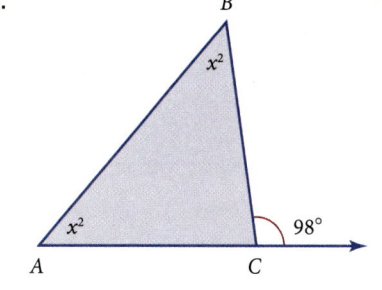

27. Triangle with angle x^2 at B, angle $2x^2$ at A, exterior angle $84°$ at C.

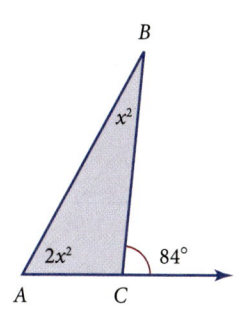

28. Triangle with angle $2x^2 + 9x$ at B, angle x^2 at A, exterior angle $120°$ at C.

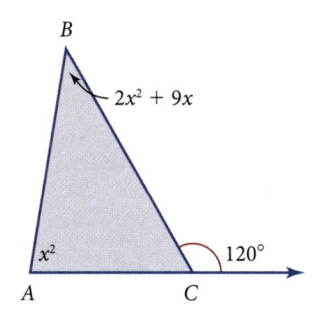

29. Triangle with angle x at B, angle x^2 at A, exterior angle $72°$ at C.

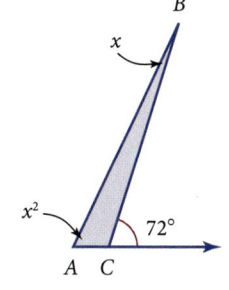

30. Triangle with angle $3x$ at B, angle x^2 at A, exterior angle $154°$ at C.

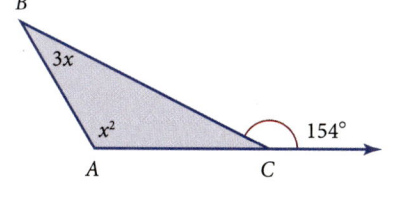

Use the exterior angle theorem to solve for x.

31. Triangle with angle $x - 4$ at B, angle $3x - 7$ at A, exterior angle $3x + 21$ at C.

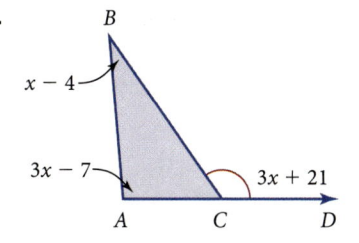

32. Triangle with angle $\frac{3}{2}x$ at B, angle $3x$ at A, exterior angle $5x - 7$ at C.

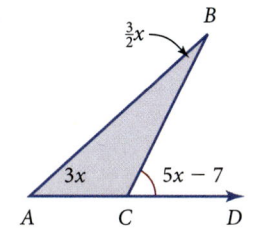

33. Triangle with angle $\frac{5}{4}x$ at B, angle $\frac{3}{2}x$ at A, exterior angle $3x - 11$ at C.

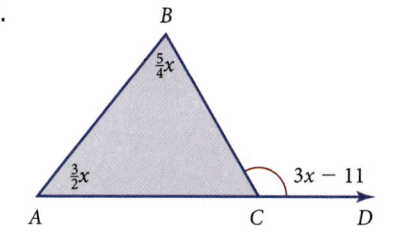

34. Triangle with angle $\frac{4}{3}x$ at B, angle $\frac{5}{2}x$ at A, exterior angle $4x - 4$ at C.

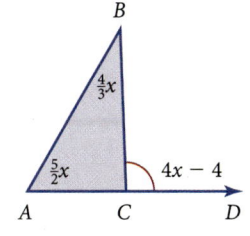

35. Triangle with angle $\frac{7}{5}x$ at B, angle $\frac{5}{3}x$ at A, exterior angle $3x + 3$ at C.

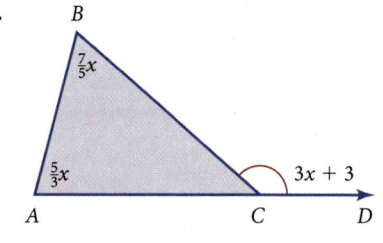

36. Triangle with angle $4x$ at B, angle $\frac{7}{2}x$ at A, exterior angle $8x - 6$ at C.

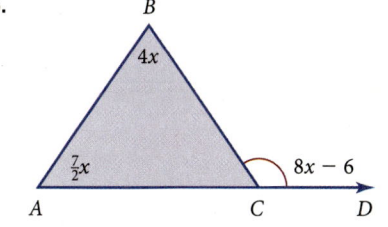

Chapter 5 Geometry

Use the properties of isosceles triangles to solve for *x*.

37. Triangle with C at top (7x + 5), A bottom-left (4x + 5), B bottom-right. Tick marks on AC and BC.

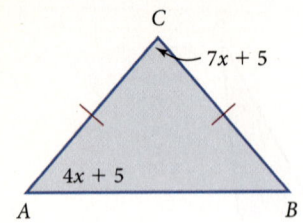

38. Triangle with C at top (9x − 4), A bottom-left (3x + 17), B bottom-right. Tick marks on AC and BC.

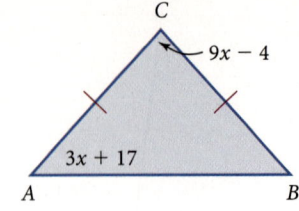

39. Triangle with C at top (4x), A bottom-left (2x − 10), B bottom-right. Tick marks on AC and BC.

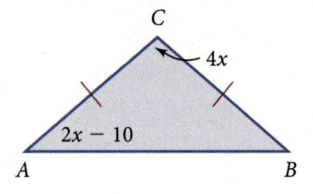

40. Triangle with C at top (6x − 10), A bottom-left, B bottom-right (4x + 4). Tick marks on AC and BC.

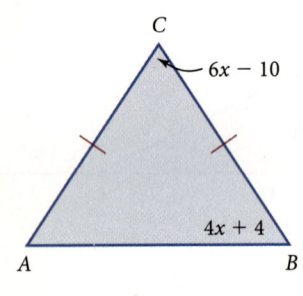

41. Triangle with C at top (x − 8), A bottom-left, B bottom-right (2x − 6). Tick marks on AC and BC.

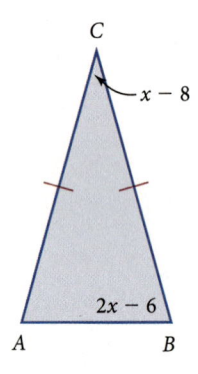

42. Triangle with C at top (9x + 13), A bottom-left (5x − 2), B bottom-right. Tick marks on AC and BC.

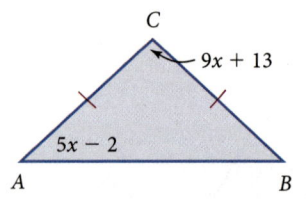

43. Triangle with B at top, AB = 4x + 3, BC = 7x − 9. Tick marks at A and C.

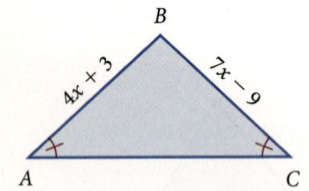

44. Triangle with B at top, AB = 6x − 8, BC = 3x + 4. Tick marks at A and C.

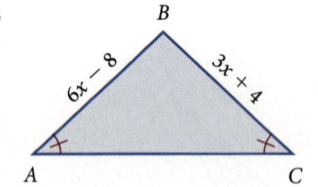

45. Triangle with B at top, AB = x − 18, BC = x/3 + 6. Tick marks at A and C.

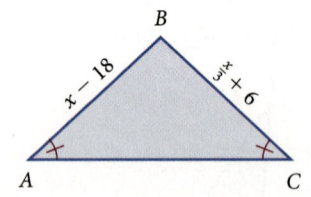

46. Triangle with B at top, AB = 3x − 14, BC = 2x + 1. Tick marks at A and C.

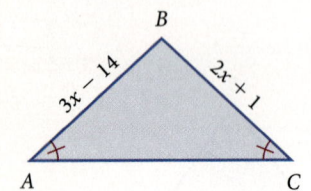

47. Triangle with B at top, AB = 2x + 14, BC = 5x + 2. Tick marks at A and C.

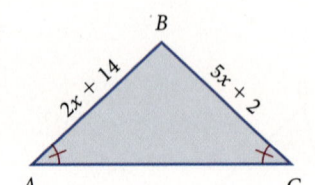

48. Triangle with B at top, AB = 4x + 3, BC = 9x − 7. Tick marks at A and C.

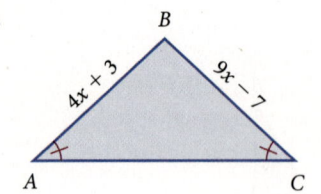

Find the value of x and the measure of angle A, B, and C.

49. Triangle ABC with angle A = 4x, angle B = 6x, angle C = 8x.

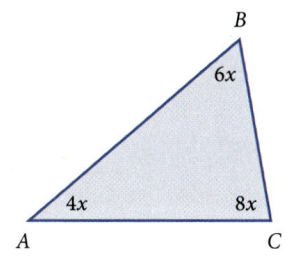

50. Triangle ABC with angle A = 2x + 4, angle B = 3x + 8, angle C = 4x + 6.

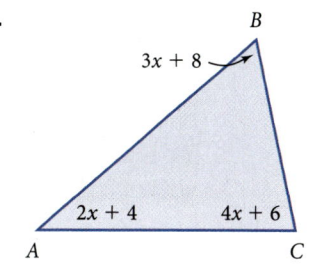

51. Triangle ABC with angle A = 7x + 5, angle B = 11x + 10, angle C = 14x + 5.

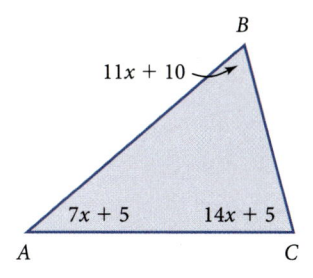

52. Triangle ABC with angle A = 12x + 11, angle B = 15x − 9, angle C = 19x − 6.

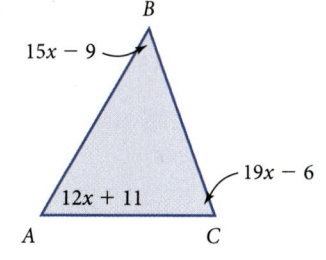

53. Triangle ABC with angle A = 2x + 6, angle B = 6x − 10, angle C = 4x + 4.

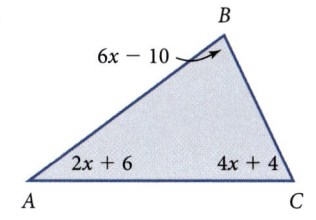

54. Triangle ABC with angle A = 2x, angle B = 5x, angle C = 3x.

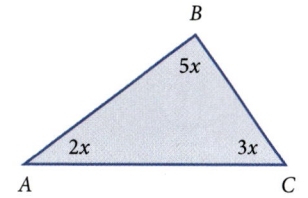

55. Triangle ABC with angle A = x, angle B = 11x, angle C = 3x.

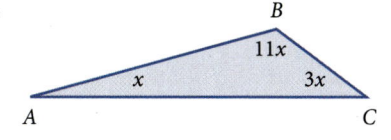

56. Triangle ABC with angle A = 4x, angle B = 9x, angle C = 7x.

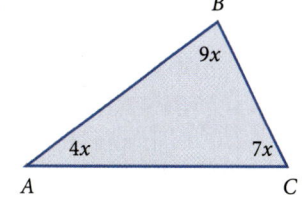

Find the value of x and the measure of ∠A and ∠B.

57. Triangle with ∠B = 3x + 11, ∠A = 4x + 7, exterior angle at C = 102°

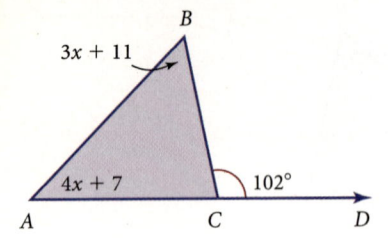

58. Triangle with ∠B = 4x + 11, ∠A = 8x + 5, exterior angle at C = 124°

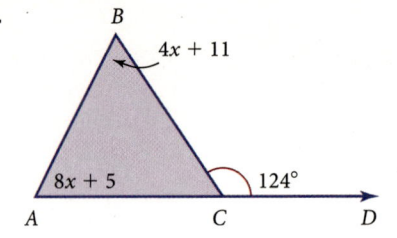

59. Triangle with ∠B = 9x + 7, ∠A = 11x + 8, exterior angle at C = 115°

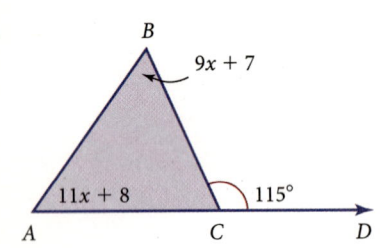

60. Triangle with ∠B = 6x − 3, ∠A = 5x + 5, exterior angle at C = 90°

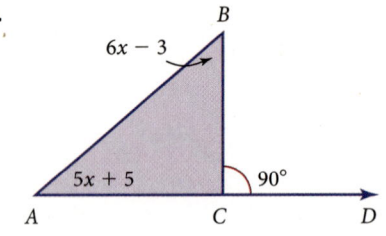

61. Triangle with ∠B = 7x + 11, ∠A = 3x + 6, exterior angle at C = 77°

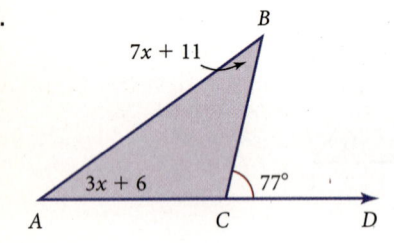

62. Triangle with ∠B = 7x − 8, ∠A = 4x + 6, exterior angle at C = 130°

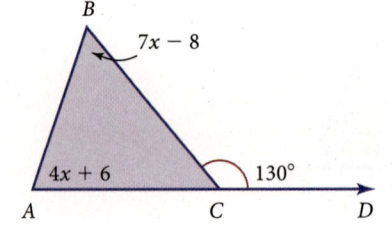

63. Triangle with ∠B = 3(5x + 3), ∠A = 17x − 5, exterior angle at C = 164°

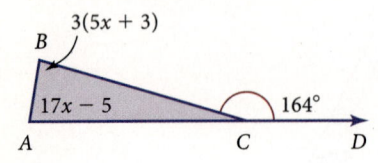

64. Triangle with ∠B = 3x + 1, ∠A = 9x − 2, exterior angle at C = 59°

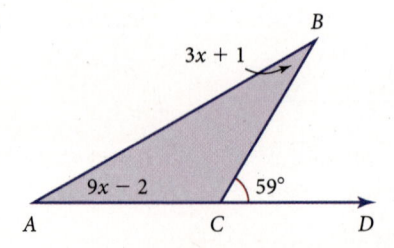

Congruent Triangles

5.4

Two triangles that are the same size and shape are congruent. If the three sides and angles of one triangle are congruent to the same six parts of another triangle, those triangles are congruent.

Objectives
A Corresponding parts of triangles.
B Reflexive, Symetric, and Transitive Properties in Triangles.
C Using tests for congruent triangles to solve triangles.

Definition
Corresponding parts of **congruent triangles** are congruent.

A Corresponding Parts

Given: $\triangle ABC \cong \triangle DEF$

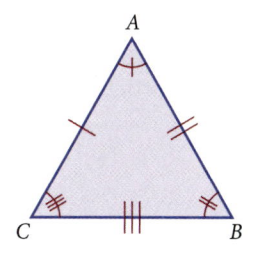

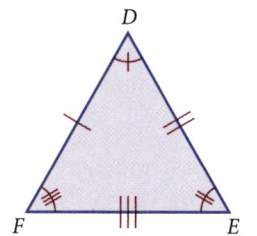

FIGURE 1

$\angle A \cong \angle D$ $\qquad$ $\overline{AB} \cong \overline{DE}$
$\angle B \cong \angle E$ $\qquad$ $\overline{BC} \cong \overline{EF}$
$\angle C \cong \angle F$ $\qquad$ $\overline{AC} \cong \overline{DF}$ $\qquad$ Parts match

EXAMPLE 1 Given: $\triangle ABC \cong \triangle DEF$

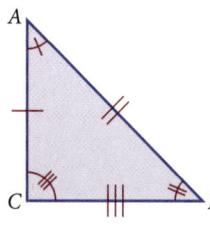

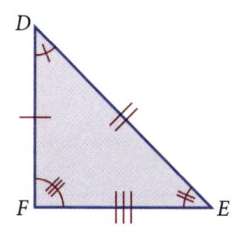

FIGURE 2

Given the following angles and sides, name the corresponding part

a. $\angle A$ $\qquad$ **b.** $\angle C$ $\qquad$ **c.** $\overline{BC}$ $\qquad$ **d.** $\overline{DF}$

SOLUTION Since its given that $\triangle ABC$ is congruent to $\triangle DEF$, we know that the corresponding (or matching) parts of these triangles are congruent.

Therefore,

a. $\angle A \cong \angle D$ $\qquad$ **b.** $\angle C \cong \angle F$ $\qquad$ **c.** $\overline{BC} \cong \overline{EF}$ $\qquad$ **d.** $\overline{DF} \cong \overline{AC}$

PRACTICE PROBLEMS

1. Refer to Figure 2 and name the corresponding part.
   a. $\angle B$
   b. $\overline{DE}$

Answer
1. **a.** $\angle E$ **b.** $\overline{AB}$

2. Refer to Figure 3 and complete the congruence statement.

   a. △DBC ≅ _____
   b. △BAD ≅ _____

Chapter 5 Geometry

EXAMPLE 2 Complete the congruence statement.

a. △ABD ≅ _____
b. △BCD ≅ _____

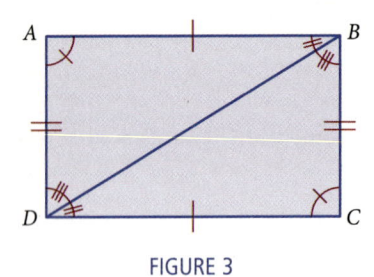

FIGURE 3

SOLUTION By examining the figure, we can match corresponding angles and sides by the marks shown on the triangles. When completing a congruence statement it is important that the letters of the vertices are listed in the correct order.

The correct answer for **a** and **b** are:

△ABD ≅ △CDB
△BCD ≅ △DAB

B Reflexive, Symmetric, and Transitive Properties in Triangles

Triangles can be said to have reflexive, symmetric, and transitive properties. These properties are shown in Figure 4.

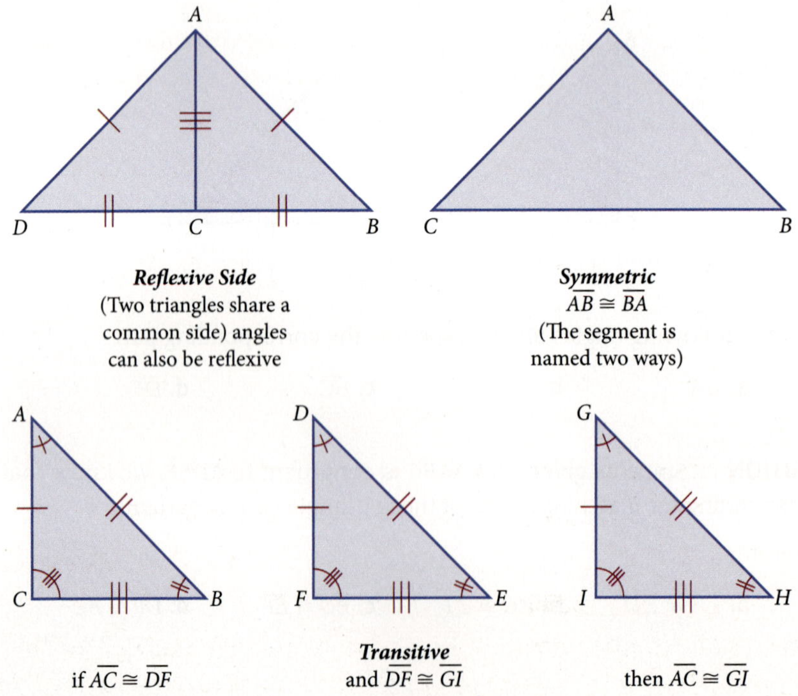

Reflexive Side
(Two triangles share a common side) angles can also be reflexive

Symmetric
$\overline{AB} \cong \overline{BA}$
(The segment is named two ways)

if $\overline{AC} \cong \overline{DF}$

Transitive
and $\overline{DF} \cong \overline{GI}$

then $\overline{AC} \cong \overline{GI}$

FIGURE 4

Answer
2. a. △BDA   b. △DCB

5.4 Congruent Triangles

EXAMPLE 3 Refer to Figure 5 and complete the congruence statement and name the property.

a. $\overline{BF} \cong$ _____
b. If $\overline{AF} \cong \overline{CF}$, and $\overline{CF} \cong \overline{CE}$, then $\overline{AF} \cong$ _____

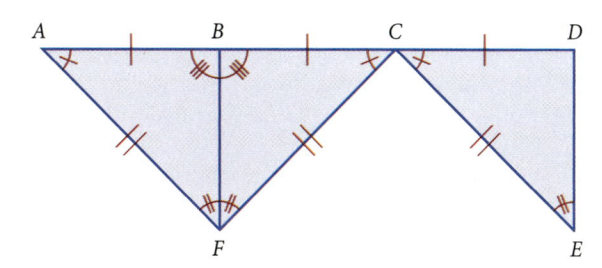

FIGURE 5

SOLUTION $\triangle ABF$ and $\triangle CBF$ share a commons side, $\overline{BF}$. Therefore, the solution to **a** is

$\overline{BF} \cong \overline{BF}$ because of the reflexive property

Because congruence of triangles is transitive, so are their corresponding parts. Therefore, if $\overline{AF} \cong \overline{CF}$, and $\overline{CF} \cong \overline{CE}$, then $\overline{AF} \cong \overline{CE}$ because of the transitive property.

C Tests for Congruent Triangles

If two triangles are congruent, it is not necessary to show all three sides and all three angles are congruent to the parts of the second triangle. The following four postulates are shortcuts to test triangle congruence. Before we go on we should understand the meaning of an included side and an included angle. An included side is located between two adjacent angles. An included angle is the angle between (or formed) by two adjacent sides. For example:

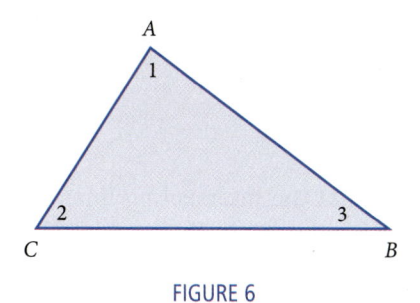

FIGURE 6

$\overline{AB}$ is the included side of $\angle 1$ and $\angle 3$

$\angle 2$ is the included angle of the sides $\overline{AC}$ and $\overline{BC}$

> **Postulate (ASA)**
> If two angles and the included side in one triangle are congruent to two angles and the included side in another triangle, then the triangles must be congruent.

3. Refer to Figure 5 and complete the congruence statement.
   a. $\angle ABF \cong$ _____
   b. If $\overline{AB} \cong \overline{BC}$ and $\overline{BC} \cong \overline{CD}$ then $\overline{AB} \cong$ _____.

Answer
3. a. $\angle CBF$ b. $\overline{CD}$

Chapter 5 Geometry

> **Postulate (AAS)**
> If two angles and the non-included side in one triangle are congruent to the corresponding two angles and non-included side in another triangle, then the two triangles are congruent.

> **Postulate (SAS)**
> If two sides and the included angle in one triangle are congruent to two sides and the included angle in another triangle, then the triangles must be congruent.

> **Postulate (SSS)**
> If three sides of one triangle are congruent to the three sides in another triangle, then the triangles must be congruent.

EXAMPLE 4 Determine which theorem can be used to prove the triangles are congruent.

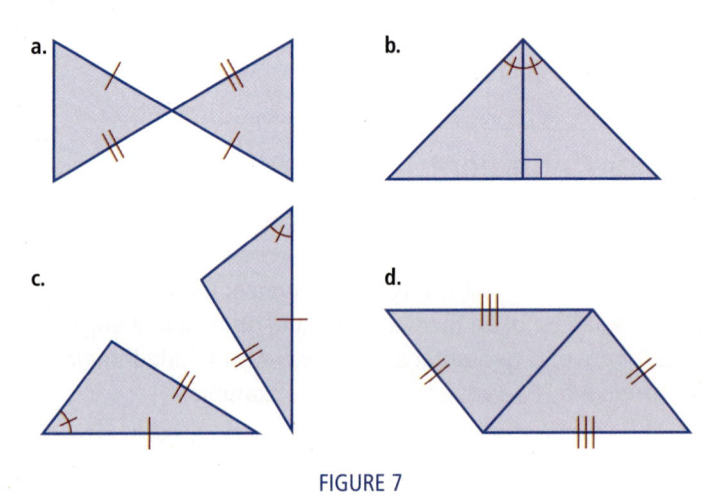

a.

b.

c.

d.

FIGURE 7

SOLUTION

a. In Example **4a** notice that the intersecting lines form vertical angles. All vertical angles are congruent. Since their angles are included between two pairs of congruent sides, the triangles are congruent through **SAS**.

b. In Example **4b** we have perpendicular lines which form two right angles. We also have a reflexive side included between two congruent angles. These triangles are congruent through **ASA**.

c. Notice the marked angle in either triangle is not the included angle. Therefore, the answer is not SAS. Furthermore, there is no such postulate as SSA. For these reasons, 4c is **no solution**.

d. Because of the reflexive side, the triangles shown by **4d** are congruent through **SSS**.

5.4 Congruent Triangles

EXAMPLE 5 Determine if the triangles are congruent, then find x and y.

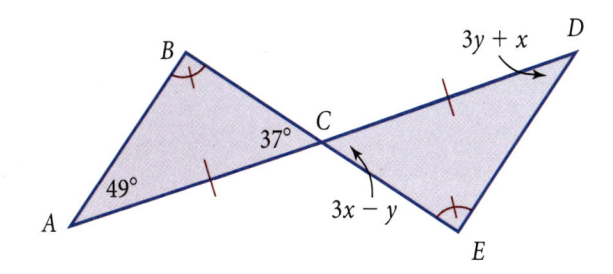

SOLUTION The intersecting lines form vertical angles which are congruent by definition. Since the angle is between a congruent angle and a congruent side, the triangles are congruent through **AAS**.

To find x and y we can write

    1. $3x - y = 37$ Vertical angles congruent
    2. $3y + x = 49$ Corresponding parts congruent

Solve equation 1 for y.

    3. $y = 3x - 37$

Plug equation 3 into equation 2 and solve for x.

$$3(3x - 37) + x = 49$$
$$x = 16$$

Plug $x = 16$ into equation 3.

$$y = 3(16) - 37$$
$$y = 11$$

So, $x = 16$ and $y = 11$.

Left Blank Intentionally

Problem Set 5.4

Given the following figures, name the corresponding parts.

1.

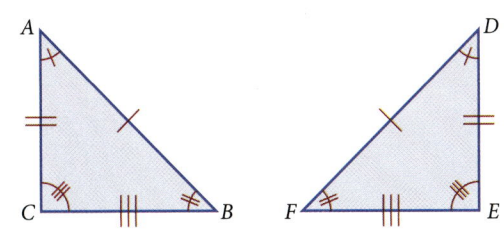

a. ∠B ≅ ____   b. ∠E ≅ ____
c. $\overline{BC}$ ≅ ____   d. $\overline{DE}$ ≅ ____

2.

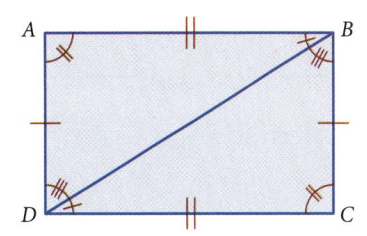

a. ∠A ≅ ____   b. ∠DBC ≅ ____
c. $\overline{CD}$ ≅ ____   d. $\overline{BD}$ ≅ ____

3.

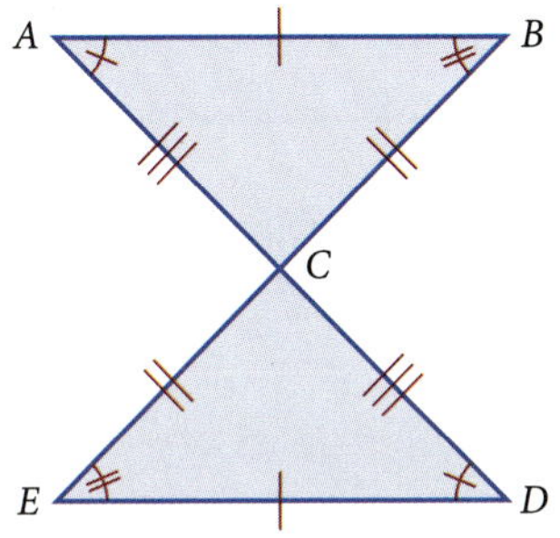

a. ∠B ≅ ____   b. ∠E ≅ ____
c. $\overline{DE}$ ≅ ____   d. $\overline{AC}$ ≅ ____

4.

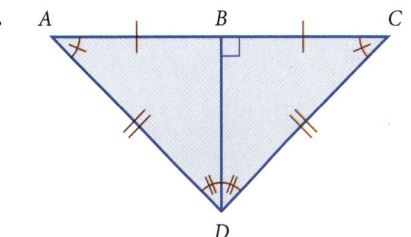

a. ∠ADB ≅ ____   b. ∠CBD ≅ ____
c. $\overline{BD}$ ≅ ____   d. $\overline{AB}$ ≅ ____

Complete the congruence statement using the figures.

5.

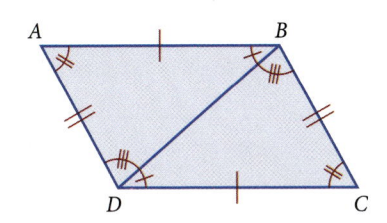

△ABD ≅ ____

6.

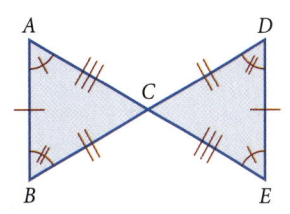

△ABC ≅ ____

7.

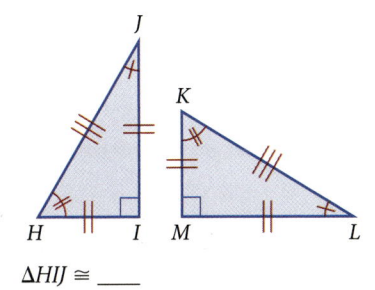

△HIJ ≅ ____

8.

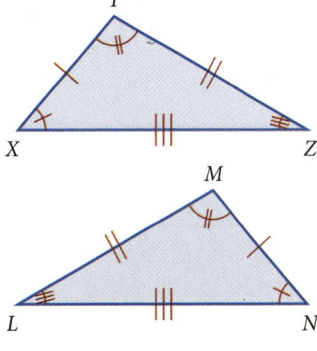

△XYZ ≅ ____

Complete the congruence statement and name the property.

9.

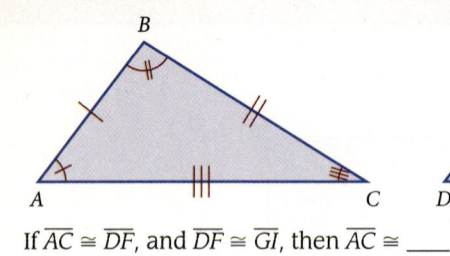

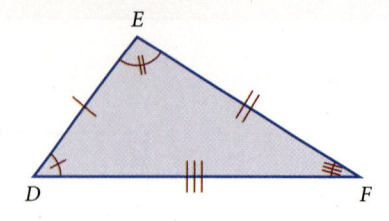

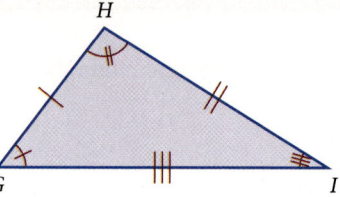

If $\overline{AC} \cong \overline{DF}$, and $\overline{DF} \cong \overline{GI}$, then $\overline{AC} \cong$ ___

10.

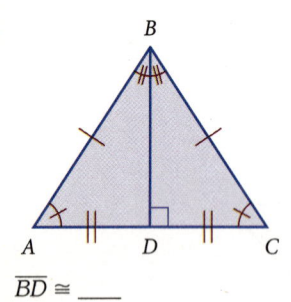

$\overline{BD} \cong$ ___

11.

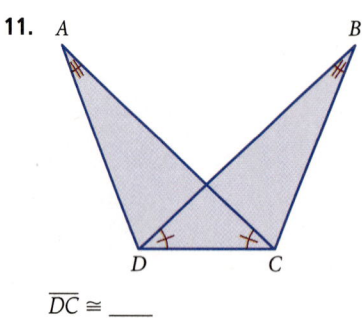

$\overline{DC} \cong$ ___

12.

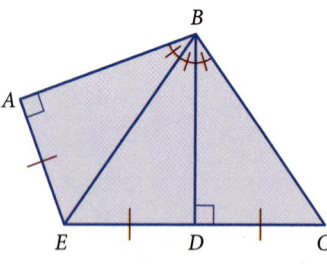

If $\overline{AE} \cong \overline{ED}$, and $\overline{ED} \cong \overline{DC}$, then $\overline{AE} \cong$ ___

△ABC is congruent to △DEF. Find x and $\overline{AC}$ and $\overline{DF}$.

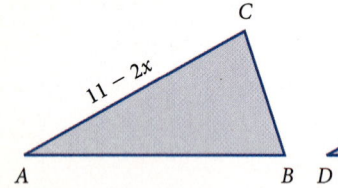

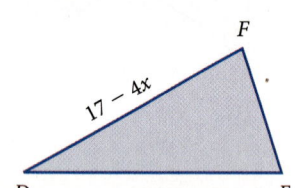

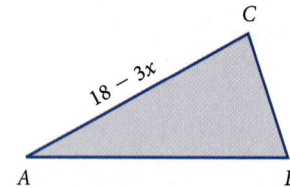

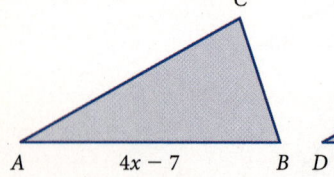

13. $6x + 4$; $3x + 10$

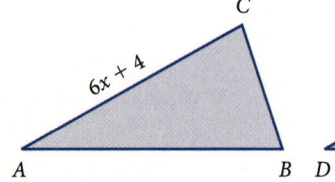

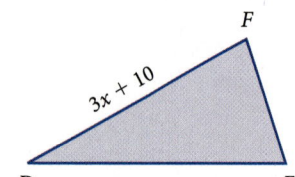

14. $11x - 9$; $6x + 6$

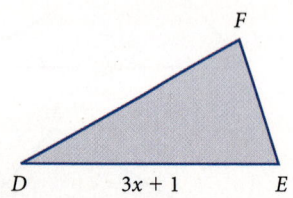

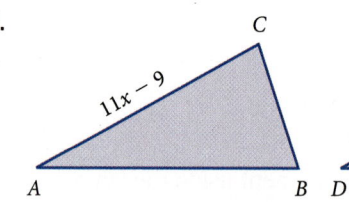

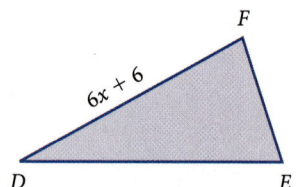

15. $11 - 2x$; $17 - 4x$

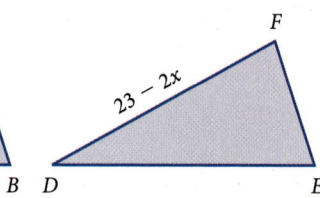

16. $18 - 3x$; $23 - 2x$

17. $4x - 7$; $3x + 1$

18. $4x + 11$; $11x - 10$

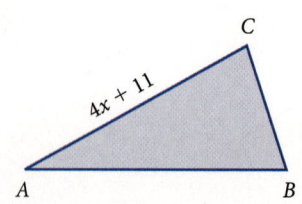

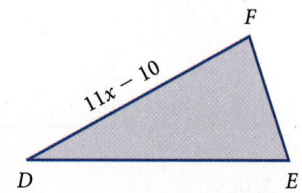

5.4 Problem Set

Determine which postulate can be used to prove the triangles are congruent.

19.

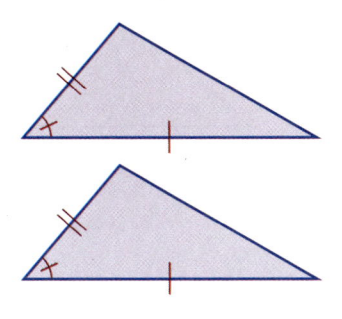

20.

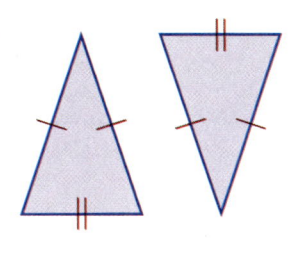

21.

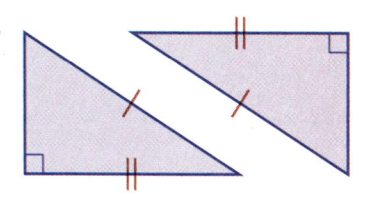

22.

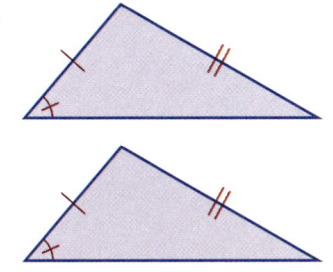

23.

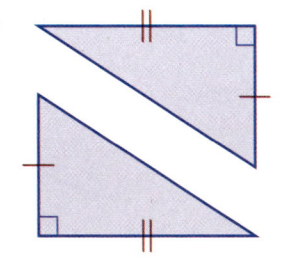

24.

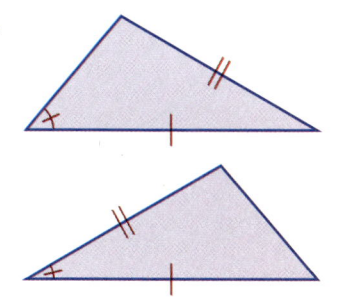

For each pair of triangles, give the reason for congruence and find x.

25. $7x + 3$ ... $11x - 9$

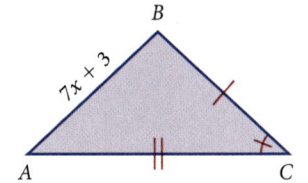

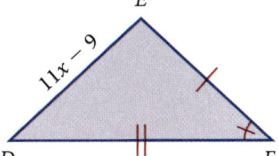

26. $8x + 5$ ... $11x - 4$

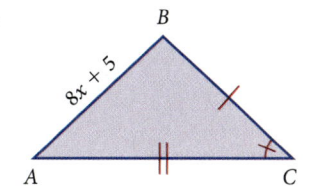

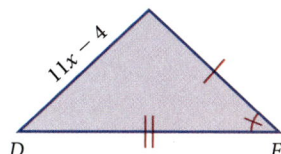

27. $4x - 9$ ... $2x + 1$

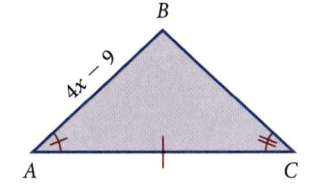

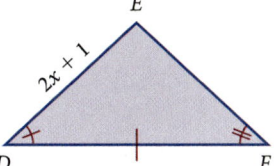

28. $3x + 11$ ... $4x - 4$

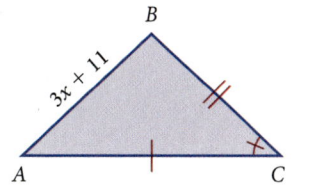

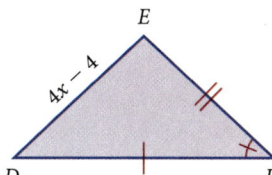

29. $4x + 48$ ... $9x - 37$

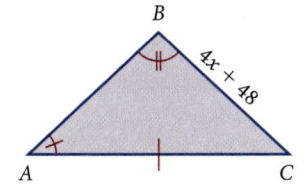

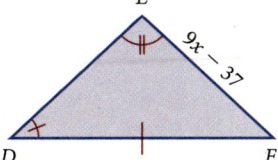

30. $6x - 10$ ... $3x + 17$

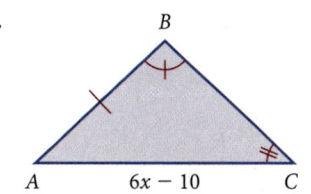

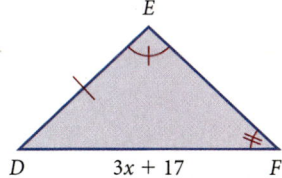

346 Chapter 5 Geometry

31.

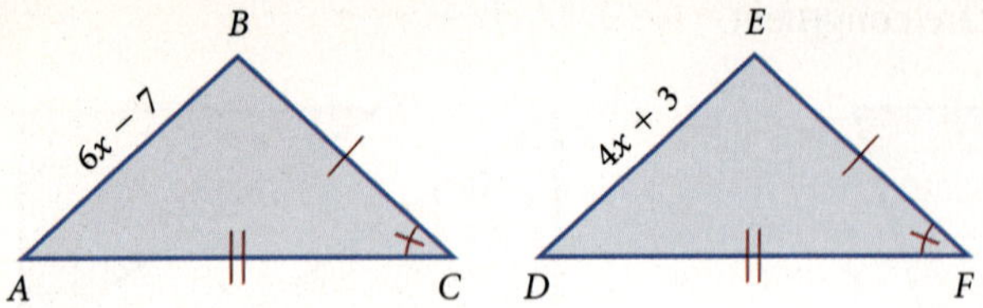

32.

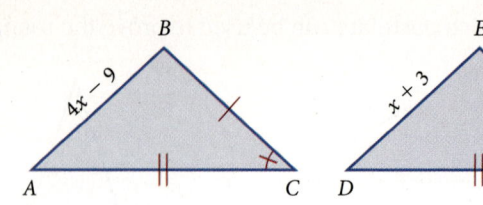

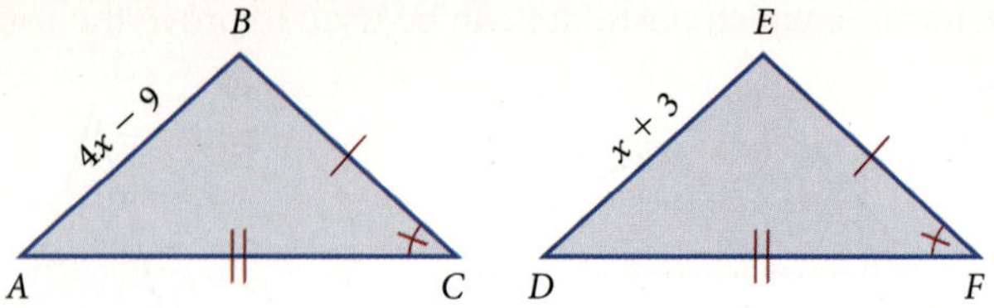

For each pair of triangles, give the reason for congruence and find x and y.

33.

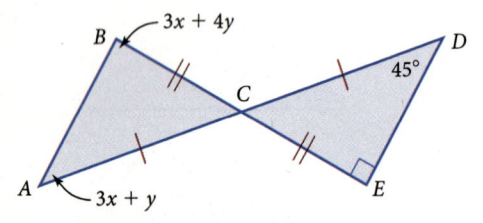

34.

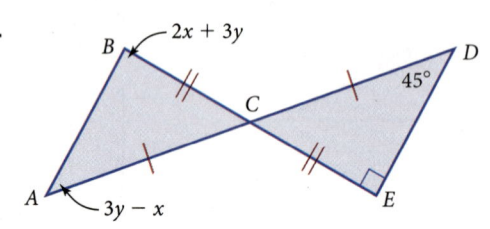

35.

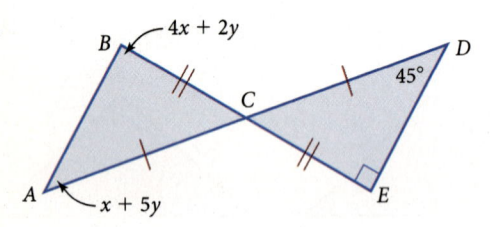

36.

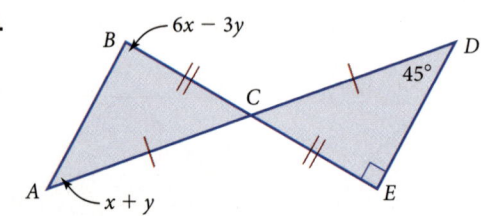

37.

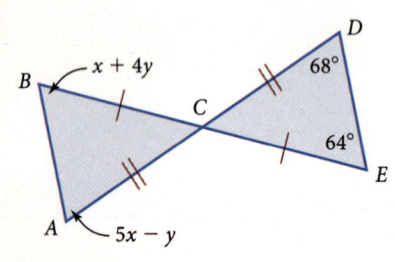

38.

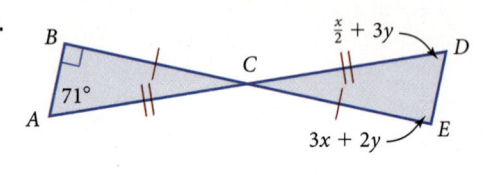

39.

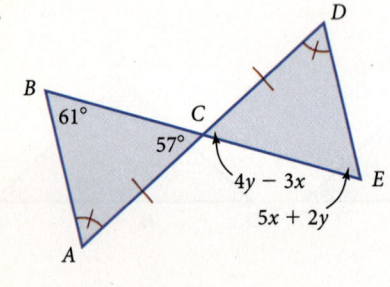

40.

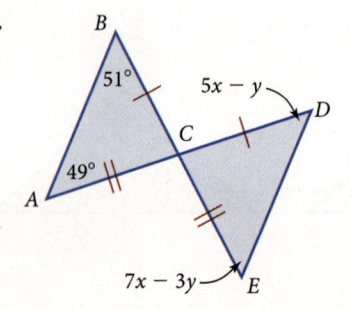

Parallelograms 5.5

Objectives
A Define polygons.
B Find interior angles of a polygon.
C Find missing sides and angles of parallelograms.

A Polygons

A polygon is a closed-sided figure with segments that intersect at the endpoints. Polygons can be regular or not regular. A regular polygon is a polygon whose sides and interior angles are congruent.

Examples of polygons:

Pentagon (not regular)   Pentagon

Quadrilateral (not regular)   Quadrilateral

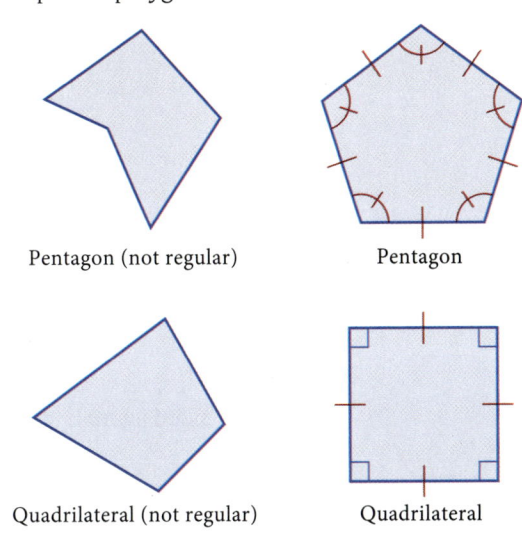

FIGURE 1

B Interior Angles of a Polygon

We know the sum of the interior angles of a triangle is 180°. We can use this fact to find the sum of the interior angles of other polygons as well. For example, the quadrilaterals here are drawn using triangles, with the triangles having the same vertices as the quadrilaterals.

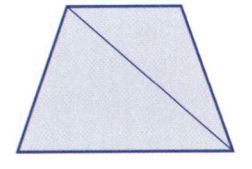

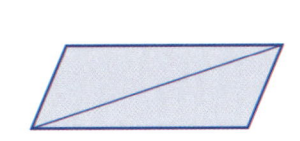

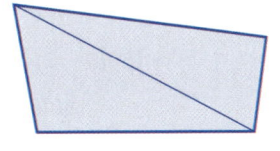

FIGURE 2

From here we can see the sum of the angles in any quadrilateral is 360°. In fact we can use this same method of drawing triangles in any polygon to find the measure of the interior angles

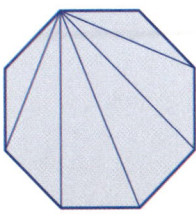

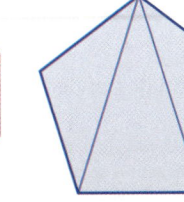

$180(6 \triangle\text{'s}) = 1080°$ $180(3 \triangle\text{'s}) = 540°$

FIGURE 3

5.5 Parallelograms

Chapter 5 Geometry

We can see a pattern develop with the number of sides of a polygon and the sum of the interior angles. That is, the sum of the interior angles of a polygon is given by the formula

$$180(n - 2)$$

given in degrees where n is the number of sides of the polygon.

Polygon	Number of Sides	Number of $\triangle$'s	Sum of Interior $\angle$'s
Triangle	3	1	1(180) = 180°
Quadrilateral	4	2	2(180) = 360°
Pentagon	5	3	3(180) = 540°
Hexagon	6	4	4(180) = 720°
Heptagon	7	5	5(180) = 900°
Octagon	8	6	6(180) = 1080°
Nonagon	9	7	7(180) = 1260°
Decagon	10	8	8(180) = 1440°

PRACTICE PROBLEMS

1. Find the sum of the interior angles for the figures given below.
    a. A seven sided figure
    b. A nine sided figure
    c. A twelve sided figure

EXAMPLE 1 Find the sum of the interior angles for the figures given below and list the polygon's name based on the number of sides.

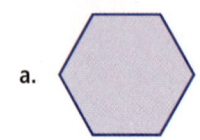

a. [hexagon figure] b. An eight sided figure c. A ten sided figure

SOLUTION We begin by counting the number of sides for each figure, and then use our formula, $180(n - 2)$ for the sum of the interior angles.

a. Since we count six sides on this figure, we have a hexagon. We have:

$$180(6 - 2)$$
$$= 180(4)$$
$$= 720°$$

The sum of the interior angles of a hexagon is 720°.

b. Since we have an eight-sided figure, which is an octagon, we have:

$$180(8 - 2)$$
$$= 180(6)$$
$$= 1,080°$$

The sum of the interior angles of the octagon is 1,080°.

c. Since we have a ten-sided figure, which is a decagon, we have:

$$180(10 - 2)$$
$$= 180(8)$$
$$= 1,440°$$

The sum of the interior angles of decagon is 1,440°.

Answer
4. a. 900° b. 1260° c. 1800°

5.5 Parallelograms

EXAMPLE 2 Find the measure of the missing angle in each figure.

a. [quadrilateral with angles 75°, 100°, 115°, x]

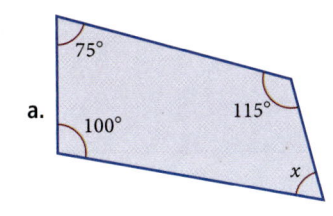

b. [pentagon with angles 90°, 95°, 120°, 115°, x]

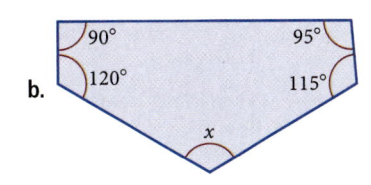

SOLUTION By first knowing the sum of the interior angles for the figure, and subtracting the sum of the given angles, we will have the measure of the missing angle.

a. The figure here is a quadrilateral. The sum of the interior angles of a quadrilateral is 360°.
The sum of given angles is 75 + 100 + 115 = 290. Subtracting 290 from 360 we have

$$360 - 290 = 70$$

The missing angle has a measure of 70°.

b. The figure here is a pentagon, or five-sided polygon. The sum of the interior angles of a pentagon is given here:

$$180(5 - 2)$$
$$= 180(3)$$
$$= 540°$$

The sum of given angles is 95 + 120 + 115 + 90 = 420. Subtracting 420 from 540 we have

$$540 - 420 = 120$$

The missing angle has a measure of 120°.

C Parallelograms

Parallelograms are special quadrilaterals with both pairs of opposite sides parallel. Note the symbols on the segments below to designate the parallel sides. The symbol for parallelograms is ▱. Thus, the figure below is noted: ▱ABCD. Notice the vertices are labeled in order.

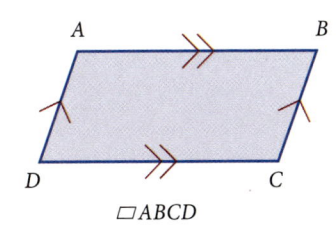

[Parallelogram ABCD with A top-left, B top-right, C bottom-right, D bottom-left]

▱ABCD

FIGURE 4

Parallelograms also have the following characteristics:

1. Opposite sides are congruent.
2. Opposite angles are congruent.
3. The diagonals bisect each other.

---

2. Find the measure of the missing angle in each figure.

a. [figure with angles 75°, 113°, 140°, 46°, x]

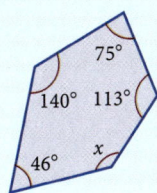

b. [figure with angles 103°, 114°, 56°, x]

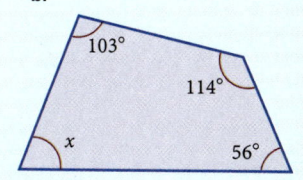

Answer
5. a. 166° b. 87°

Chapter 5 Geometry

EXAMPLE 3   Given: ▱ABCD

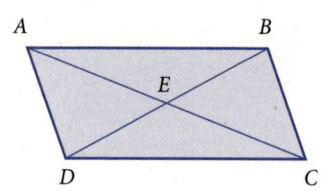

a. $\overline{AD} \cong$ ____
b. $\overline{DE} \cong$ ____
c. $\angle ABC \cong \angle$ ____
d. $2(AE) =$ ____

SOLUTION
a. $\overline{AD} \cong \overline{BC}$
b. $\overline{DE} \cong \overline{BE}$
c. $\angle ABC \cong \angle ADC$
d. $2(AE) = AC$

Also note that properties from the section on parallel lines can be used to find angle measures and congruence. For example: Given ▱ABCD

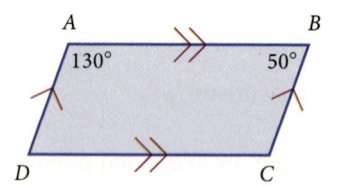

Segments $\overline{AD}$ and $\overline{BC}$ (two parallel lines) are being cut by transversal $\overline{AB}$. We have consecutive interior angles which we know are supplementary.

Problem Set 5.5

Find the sum of the interior angles for each of the following figures.

1.

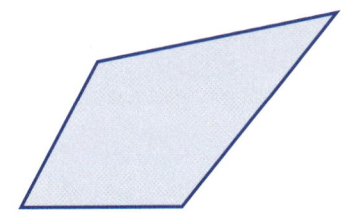

2.

3.

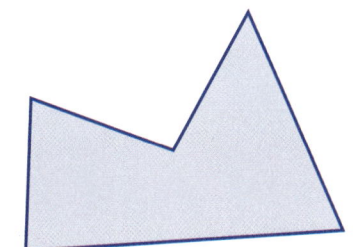

4.

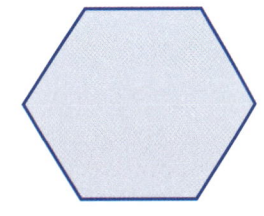

5.

6.

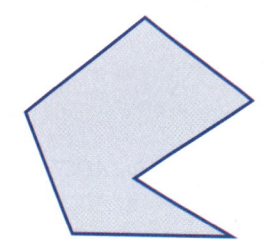

Figure 5 shows ▱ABCD. Use it for problems 7–18.

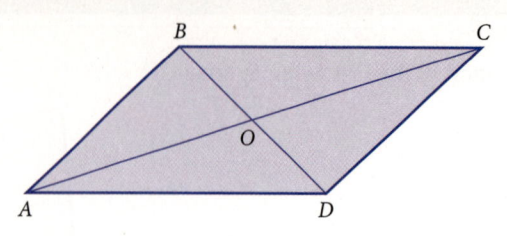

FIGURE 5

7. $AB = 13$
    $CD = $ ___

8. $m\angle BOA = 36°$
    $m\angle AOD = $ ___

9. $AC = 27$
    $OC = $ ___

10. $m\angle ABC = 112°$
    $m\angle ADC = $ ___

11. $BO = 63$
    $DO = $ ___

12. $m\angle BAD = 15°$
    $m\angle ABC = $ ___

13. $AO = 14$
    $m\angle ABC = 100°$
    $BD = 7$
    $OD = $ ___
    $m\angle BCD = $ ___
    $AC = $ ___

14. $m\angle CBO = 32°$
    $m\angle BCD = 90°$
    $CD = 6$
    $m\angle $ ___ $= 32°$
    $m\angle ABO = $ ___
    $AB = $ ___

15. $m\angle AOD = 145°$
    $BD = 17$
    $m\angle ABO = 82°$
    $m\angle $ ___ $= 145°$
    $m\angle ODC = $ ___
    $OD = $ ___

16. $AB = 32$
    $m\angle OAD = 32°$
    $AC = 16$
    $m\angle $ ___ $= 32°$
    $AO = $ ___
    ___ $= 32$

17. $m\angle ADC = 112°$
    $m\angle ABO = 90°$
    $AO = 9$
    ___ $= 18$
    $m\angle ODA = $ ___
    $m\angle $ ___ $= 112°$

18. $BD = 12$
    $AO = 7$
    $m\angle ABO = 90°$
    $BO = $ ___
    $AC = $ ___
    $m\angle ODC = $ ___

5.5 Problem Set

Use Figure 6 to name the component of the parallelogram that is congruent to the given part.

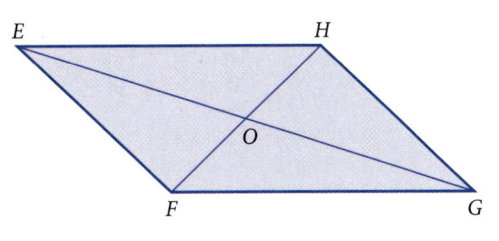

FIGURE 6

19. $EH \cong$ _____

20. $\angle HOG \cong \angle$ _____

21. $\angle EFO \cong \angle$ _____

22. $HO \cong$ _____

23. $\angle HGF \cong \angle$ _____

24. $\angle HEO \cong \angle$ _____

25. $EF \cong$ _____

26. $\angle OEF \cong \angle$ _____

Use the given information to find the congruent part of ▱ $WXYZ$ with diagonals intersecting at O. There will be more than one answer.

27. $WX \cong XY$
$m\angle WXY = 90°$
$ZO \cong$ _____

28. $m\angle WZY = 90°$
$\angle WXY \cong$ _____

29. $m\angle WZO \cong m\angle OZY = 45°$
$WX \cong$ _____

30. $WY \cong XZ$
$\angle XYZ \cong$ _____

31. $\angle WOZ \cong \angle WOX$
$WX \cong$ _____

32. $\angle OXY \cong \angle XYO$
$XO \cong$ _____

For each pair of triangles, give the reason for congruence and find x and y.

33.

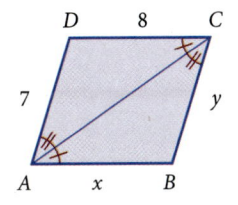

34.

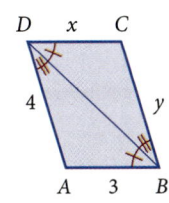

35.

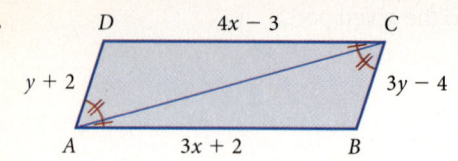

D $4x - 3$ C
$y + 2$ $3y - 4$
A $3x + 2$ B

36.

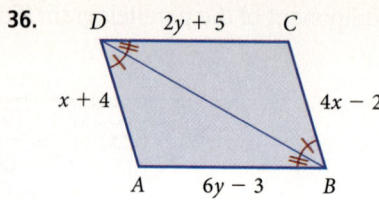

D $2y + 5$ C
$x + 4$ $4x - 2$
A $6y - 3$ B

37.

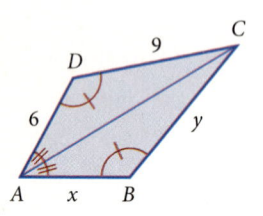

D, 9, C
6, y
A, x, B

38.

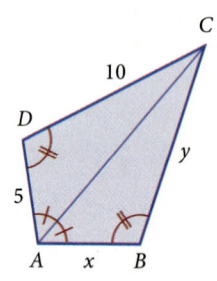

C
10
D
5 y
A x B

39.

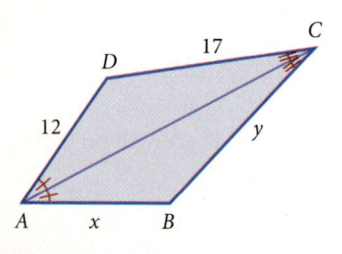

D 17 C
12 y
A x B

40.

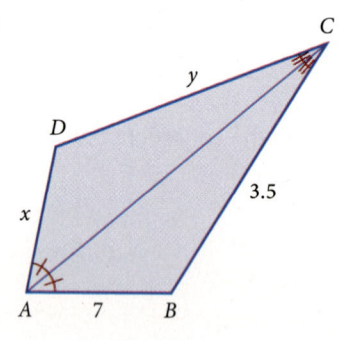

C
y
D
x 3.5
A 7 B

41.

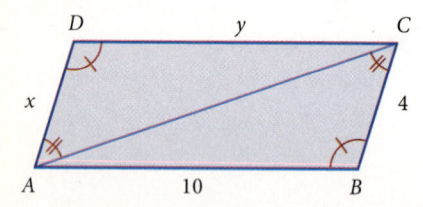

D y C
x 4
A 10 B

42.

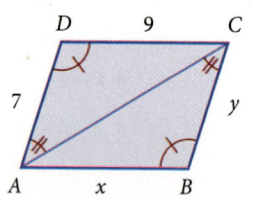

D 9 C
7 y
A x B

43.

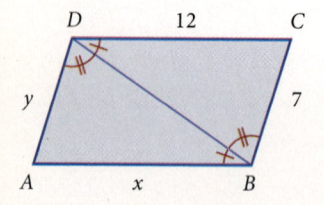

D 12 C
y 7
A x B

44.

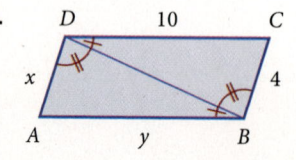

D 10 C
x 4
A y B

Similar Figures and Proportions

5.6

Objectives

A Name the terms in a proportion.
B Use the fundamental property of proportions to solve a proportion.
C Solve for missing parts of similar triangles.
D Solve for similar parts of other similar figures.
E Draw similar figures.
F Solve application problems involving similar figures.

This 8-foot-high bronze sculpture "Cellarman" in Napa, California, is an exact replica of the smaller, 12-inch sculpture. Both pieces are the product of artist Tim Lloyd of Arroyo Grande, California.

In mathematics, when two or more objects have the same shape, but are different sizes, we say they are similar. If two figures are similar, then their corresponding sides are proportional.

Definition

A statement that two ratios are equal is called a **proportion**. If $\frac{a}{b}$ and $\frac{c}{d}$ are two equal ratios, then the statement

$$\frac{a}{b} = \frac{c}{d}$$

is called a proportion.

A Terms of a Proportion

Each of the four numbers in a proportion is called a *term* of the proportion. We number the terms of a proportion as follows:

First term ⟶ $\frac{a}{b} = \frac{c}{d}$ ⟵ Third term
Second term   ⟵ Fourth term

The first and fourth terms of a proportion are called the *extremes,* and the second and third terms of a proportion are called the *means.*

Means ⟶ $\frac{a}{b} = \frac{c}{d}$ ⟵ Extremes

EXAMPLE 1 In the proportion $\frac{3}{4} = \frac{6}{8}$, name the four terms, the means, and the extremes.

SOLUTION The terms are numbered as follows:

First term = 3   Third term = 6
Second term = 4   Fourth term = 8

The means are 4 and 6; the extremes are 3 and 8.

1. In the proportion $\frac{2}{3} = \frac{6}{9}$, name the four terms, the means, and the extremes.

Answer
1. See Solutions to Selected Practice Problems.

5.6 Similar Figures and Proportions

Chapter 5 Geometry

The final thing we need to know about proportions is expressed in the following property.

B The Fundamental Property of Proportions

Fundamental Property of Proportions
In any proportion, the product of the extremes is equal to the product of the means. This property is also referred to as the means/extremes property, and in symbols, it looks like this:

If $\dfrac{a}{b} = \dfrac{c}{d}$ then $ad = bc$

EXAMPLE 2 Verify the fundamental property of proportions for the following proportions.

a. $\dfrac{3}{4} = \dfrac{6}{8}$ b. $\dfrac{17}{34} = \dfrac{1}{2}$

SOLUTION We verify the fundamental property by finding the product of the means and the product of the extremes in each case.

Proportion	Product of the Means	Product of the Extremes
a. $\dfrac{3}{4} = \dfrac{6}{8}$	$4 \cdot 6 = 24$	$3 \cdot 8 = 24$
b. $\dfrac{17}{34} = \dfrac{1}{2}$	$34 \cdot 1 = 34$	$17 \cdot 2 = 34$

For each proportion the product of the means is equal to the product of the extremes.

We can use the fundamental property of proportions to solve an equation that has the form of a proportion.

EXAMPLE 3 Solve for x.

$$\dfrac{2}{3} = \dfrac{4}{x}$$

SOLUTION Applying the fundamental property of proportions, we have

If $\dfrac{2}{3} = \dfrac{4}{x}$

then $2 \cdot x = 3 \cdot 4$ The product of the extremes equals the product of the means

$2x = 12$ Multiply

The result is an equation. We know from previous work that we can divide both sides of an equation by the same nonzero number without changing the solution to the equation. In this case we divide both sides by 2 to solve for x:

$2x = 12$

$\dfrac{2x}{2} = \dfrac{12}{2}$ Divide both sides by 2

$x = 6$ Simplify each side

---

2. Verify the fundamental property of proportions for the following proportions.

   a. $\dfrac{5}{6} = \dfrac{15}{18}$

   b. $\dfrac{13}{39} = \dfrac{1}{3}$

   c. $\dfrac{\frac{2}{3}}{\frac{5}{3}} = \dfrac{2}{5}$

   d. $\dfrac{0.12}{0.18} = \dfrac{2}{3}$

3. Find the missing term:

   a. $\dfrac{3}{4} = \dfrac{9}{x}$

   b. $\dfrac{5}{8} = \dfrac{3}{x}$

Note In some of these problems you will be able to see what the solution is just by looking the problem over. In those cases it is still best to show all the work involved in solving the proportion. It is good practice for the more difficult problems.

Answer
2. See Solutions to Selected Practice Problems.

5.6 Similar Figures and Proportions

The solution is 6. We can check our work by using the fundamental property of proportions:

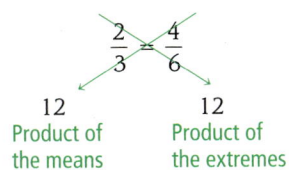

$$\frac{2}{3} \times \frac{4}{6}$$

12 Product of the means   12 Product of the extremes

Because the product of the means and the product of the extremes are equal, our work is correct.

EXAMPLE 4 Solve for y: $\frac{5}{y} = \frac{10}{13}$

SOLUTION We apply the fundamental property and solve as we did in Example 3:

If $\quad \frac{5}{y} = \frac{10}{13}$

then $\quad 5 \cdot 13 = y \cdot 10 \quad$ The product of the extremes equals the product of the means

$\quad\quad\quad 65 = 10y \quad$ Multiply $5 \cdot 13$

$\quad\quad\quad \frac{65}{10} = \frac{10y}{10} \quad$ Divide both sides by 10

$\quad\quad\quad 6.5 = y \quad 65 \div 10 = 6.5$

The solution is 6.5. We could check our result by substituting 6.5 for y in the original proportion and then finding the product of the means and the product of the extremes.

C Similar Triangles

Two triangles that have the same shape are similar when their corresponding sides are proportional, or have the same ratio. The triangles below are similar.

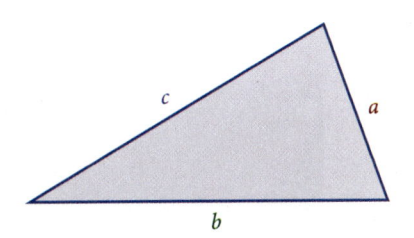

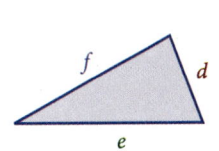

Corresponding Sides	Ratio
side a corresponds with side d	$\frac{a}{d}$
side b corresponds with side e	$\frac{b}{e}$
side c corresponds with side f	$\frac{c}{f}$

Because their corresponding sides are proportional, we write

$$\frac{a}{d} = \frac{b}{e} = \frac{c}{f}$$

4. Solve for y: $\frac{2}{y} = \frac{8}{19}$

Answer
3. a. 12 b. 4.8
4. 4.75

5. The two triangles below are similar. Find the missing side, x.

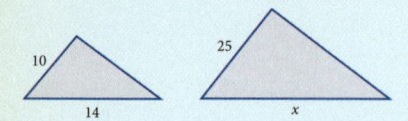

Chapter 5 Geometry

EXAMPLE 5 The two triangles below are similar. Find side x.

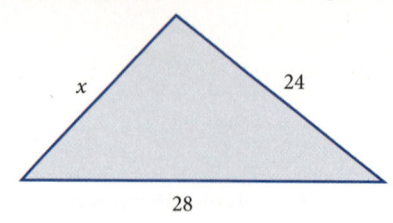

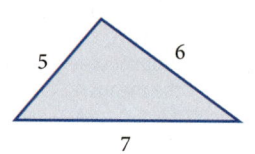

SOLUTION To find the length x, we set up a proportion of equal ratios. The ratio of x to 5 is equal to the ratio of 24 to 6 and to the ratio of 28 to 7. Algebraically we have

$$\frac{x}{5} = \frac{24}{6} \quad \text{and} \quad \frac{x}{5} = \frac{28}{7}$$

We can solve either proportion to get our answer. The first gives us

$$\frac{x}{5} = 4 \qquad \frac{24}{6} = 4$$
$x = 4 \cdot 5$ Multiply both sides by 5
$x = 20$ Simplify

B Other Similar Figures

When one shape or figure is either a reduced or enlarged copy of the same shape or figure, we consider them similar. For example, video viewed over the Internet was once confined to a small "postage stamp" size. Now it is common to see larger video over the Internet. Although the width and height have increased, the shape of the video has not changed.

EXAMPLE 6 The width and height of the two video clips are proportional. Find the height, h, in pixels of the larger video window.

6. Find the height, h, in pixels of a video clip proportional to those in Example 6 with a width of 360 pixels.

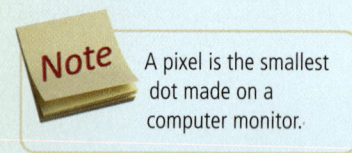

Note A pixel is the smallest dot made on a computer monitor.

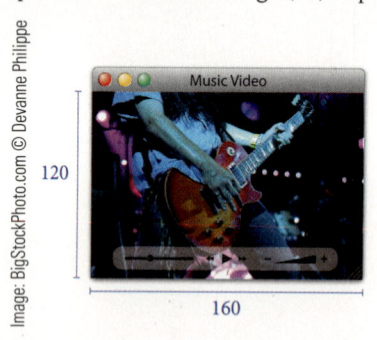

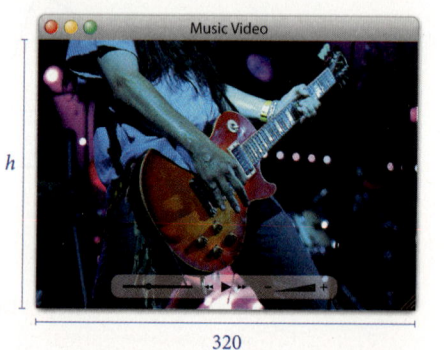

SOLUTION We write our proportion as the ratio of the height of the new video to the height of the old video is equal to the ratio of the width of the new video to the width of the old video:

$$\frac{h}{120} = \frac{320}{160}$$
$$\frac{h}{120} = 2$$
$$h = 2 \cdot 120$$
$$h = 240$$

The height of the larger video is 240 pixels.

Answers
5. 35
6. 270

C Drawing Similar Figures

EXAMPLE 7 Draw a triangle similar to triangle *ABC*, if *AC* is proportional to *DF*. Make *E* the third vertex of the new triangle.

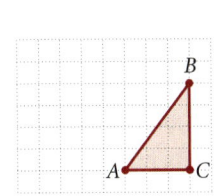

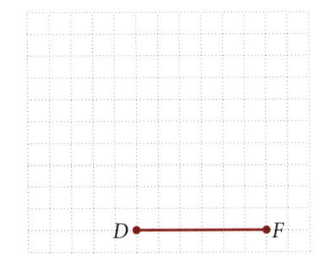

SOLUTION We see that *AC* is 3 units in length and *BC* has a length of 4 units. Since *AC* is proportional to *DF*, which has a length of 6 units, we set up a proportion to find the length *EF*.

$$\frac{EF}{BC} = \frac{DF}{AC}$$

$$\frac{EF}{4} = \frac{6}{3}$$

$$\frac{EF}{4} = 2$$

$$EF = 8$$

Now we can draw *EF* with a length of 8 units, then complete the triangle by drawing line *DE*.

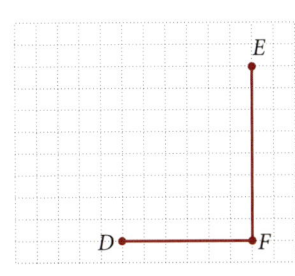

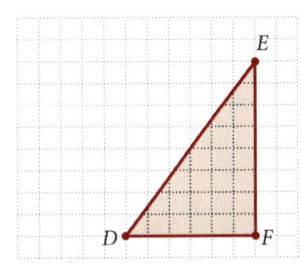

We have drawn triangle *DEF* similar to triangle *ABC*.

D Applications

EXAMPLE 8 A building casts a shadow of 105 feet while a 21-foot flagpole casts a shadow that is 15 feet. Find the height of the building.

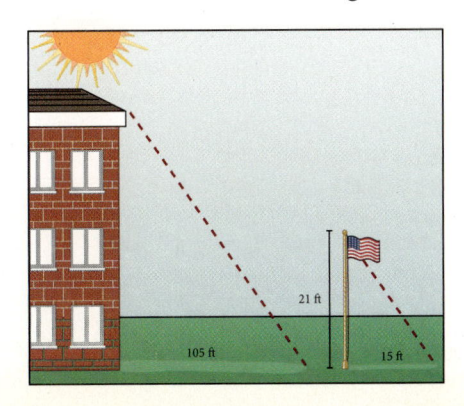

---

7. Draw a triangle similar to triangle *ABC*, if *AC* is proportional to *GI*.

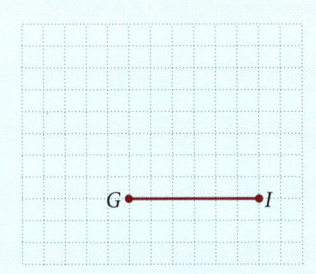

8. A building casts a shadow of 42 feet, while an 18-foot flagpole casts a shadow that is 12 feet. Find the height of the building.

Answer
7. See Solutions to Selected Practice Problems.

Chapter 5 Geometry

SOLUTION The figure shows both the building and the flagpole, along with their respective shadows. From the figure it is apparent that we have two similar triangles. Letting x = the height of the building, we have

$$\frac{x}{21} = \frac{105}{15}$$

$15x = 2205$ Cross Multiply

$x = 147$ Divide both sides by 15

The height of the building is 147 feet.

THE VIOLIN FAMILY The instruments in the violin family include the bass, cello, viola, and violin. These instruments can be considered similar figures because the entire length of each instrument is proportional to its body length.

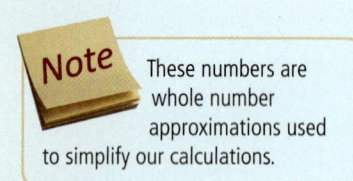

> **Note** These numbers are whole number approximations used to simplify our calculations.

9. Find the body length of an instrument proportional to the violin family that has a total length of 32 inches.

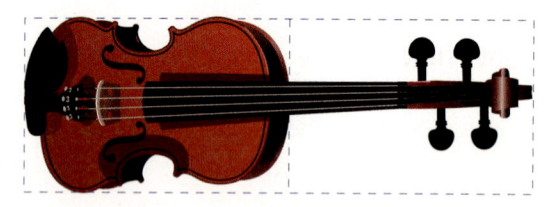

EXAMPLE 9 The entire length of a violin is 24 inches, while the body length is 15 inches. Find the body length of a cello if the entire length is 48 inches.

SOLUTION Let b equal the body length of the cello, and set up the proportion.

$$\frac{b}{15} = \frac{48}{24}$$

$$\frac{b}{15} = 2$$

$b = 2 \cdot 15$

$b = 30$

The body length of a cello is 30 inches.

Getting Ready for Class

After reading through the preceding section, respond in your own words and in complete sentences.

1. What are similar figures?
2. How do we know if corresponding sides of two triangles are proportional?
3. When labeling a triangle *ABC*, how do we label the sides?
4. How are proportions used when working with similar figures?

Answers
8. 63 ft
9. 20 in.

Problem Set 5.6

B Find the missing term in each of the following proportions. Write your answers as fractions in lowest terms.
[Examples 1–3]

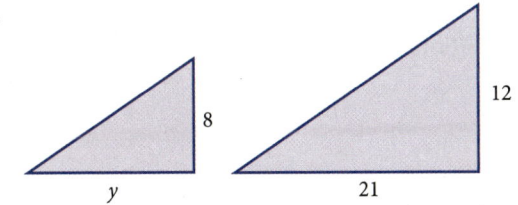

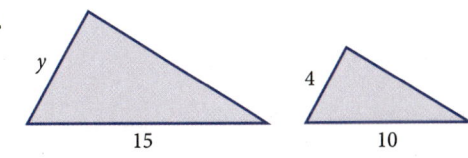

1. $\dfrac{2}{5} = \dfrac{4}{x}$
2. $\dfrac{3}{8} = \dfrac{9}{x}$
3. $\dfrac{1}{y} = \dfrac{5}{12}$
4. $\dfrac{2}{y} = \dfrac{6}{10}$
5. $\dfrac{x}{4} = \dfrac{3}{8}$
6. $\dfrac{x}{5} = \dfrac{7}{10}$

7. $\dfrac{5}{9} = \dfrac{x}{2}$
8. $\dfrac{3}{7} = \dfrac{x}{3}$
9. $\dfrac{3}{7} = \dfrac{3}{x}$
10. $\dfrac{2}{9} = \dfrac{2}{x}$
11. $\dfrac{x}{2} = 7$
12. $\dfrac{x}{3} = 10$

13. $\dfrac{\frac{1}{2}}{y} = \dfrac{\frac{1}{3}}{12}$
14. $\dfrac{\frac{2}{3}}{y} = \dfrac{\frac{1}{3}}{5}$
15. $\dfrac{n}{12} = \dfrac{\frac{1}{4}}{\frac{1}{2}}$
16. $\dfrac{n}{10} = \dfrac{\frac{3}{5}}{\frac{3}{8}}$
17. $\dfrac{10}{20} = \dfrac{20}{n}$
18. $\dfrac{8}{4} = \dfrac{4}{n}$

C In problems 19–22, for each pair of similar triangles, set up a proportion in order to find the unknown. [Example 5]

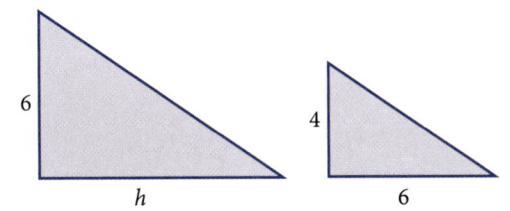

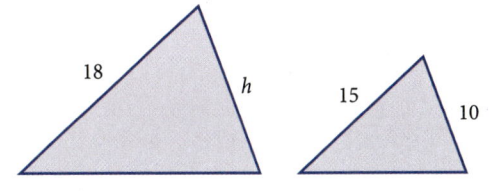

19. Triangles with sides 6, h and 4, 6.

20. Triangles with sides 18, h and 15, 10.

21. Triangles with sides y, 8 and 21, 12.

22. Triangles with sides y, 15 and 4, 10.

362 Chapter 5 Geometry

D In problems 23–28, for each pair of similar figures, set up a proportion in order to find the unknown. [Example 6]

23. 16, 12, x, 9

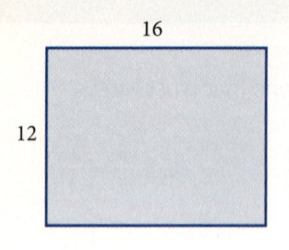

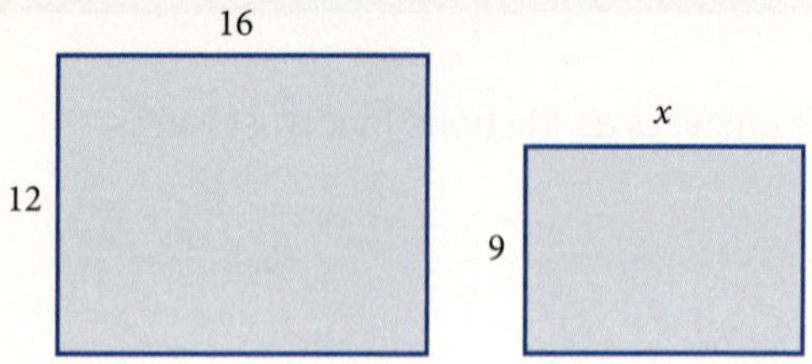

24. x, 40, 9, 24

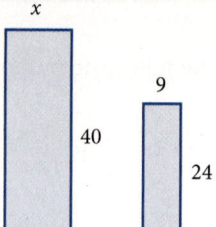

25. 5, a, 3, 15

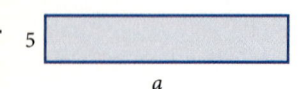

26. 48, 54, a, 36

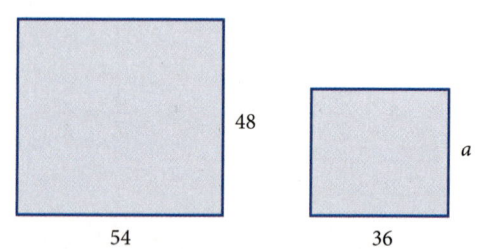

27. 50, 40, 40, y

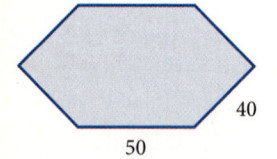

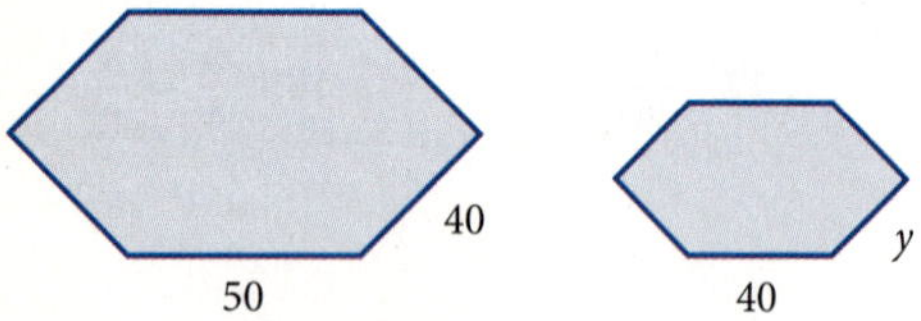

28. 42, 30, 28, y

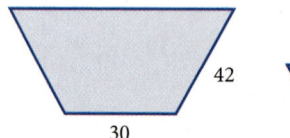

E For each problem, draw a figure on the grid on the right that is similar to the given figure. [Example 7]

29. AC is proportional to DF.

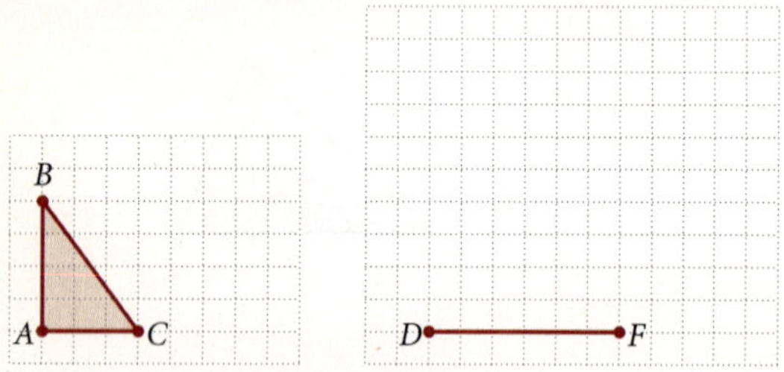

30. AC is proportional to DF.

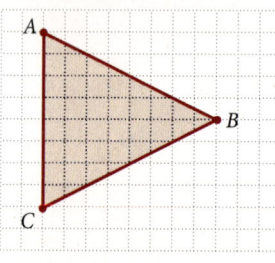

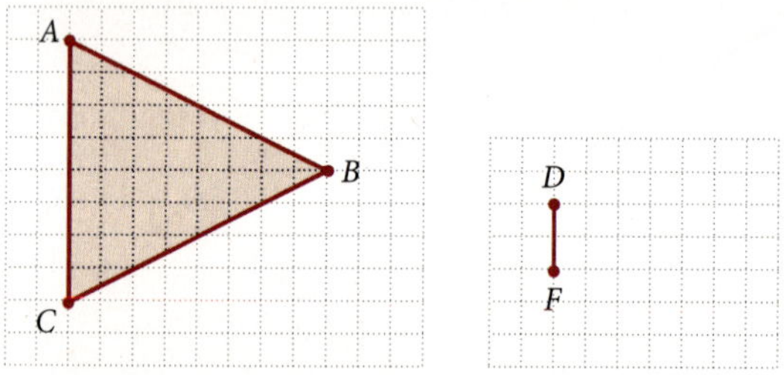

31. DC is proportional to HG.

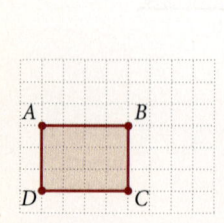

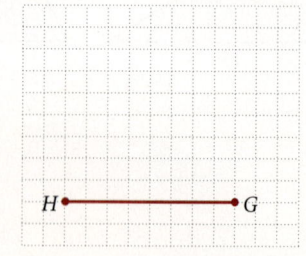

32. AB is proportional to FG.

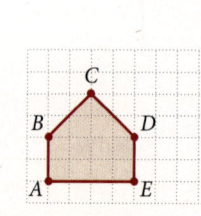

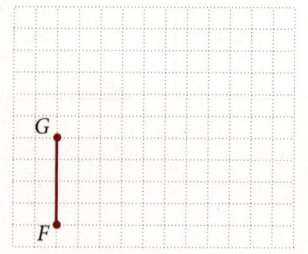

Triangle ABC is similar to △DEF. Find the value of x.

33.

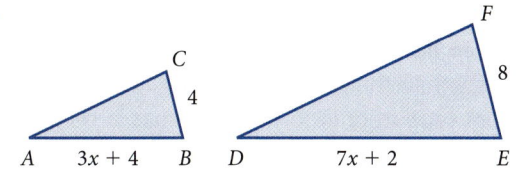

C, 4, A, 3x + 4, B
F, 8, D, 7x + 2, E

34.

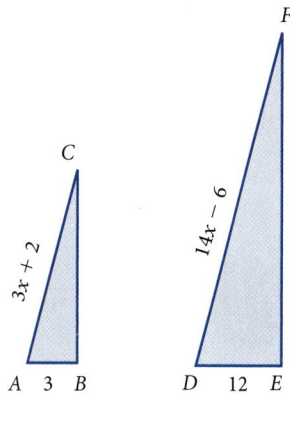

C, 3x + 2, A, 3, B
F, 14x − 6, D, 12, E

35.

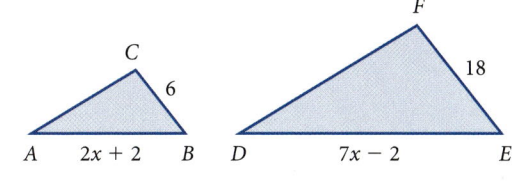

C, 6, A, 2x + 2, B
F, 18, D, 7x − 2, E

36.

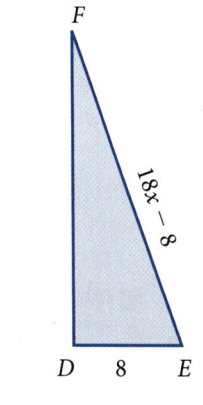

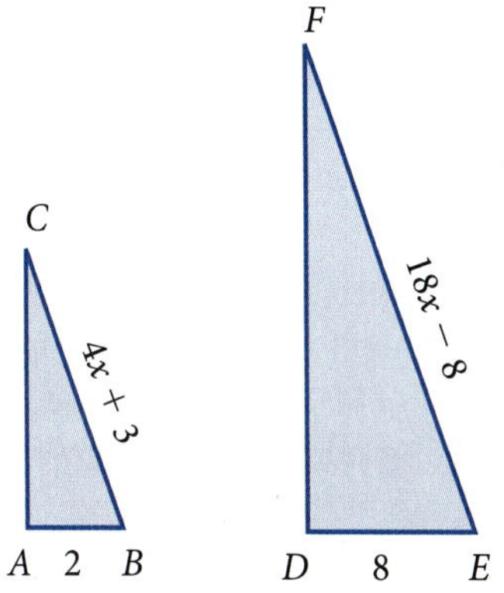

C, 4x + 3, A, 2, B
F, 18x − 8, D, 8, E

37.

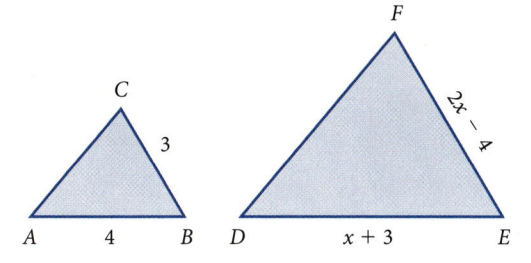

C, 3, A, 4, B
F, 2x − 4, D, x + 3, E

38.

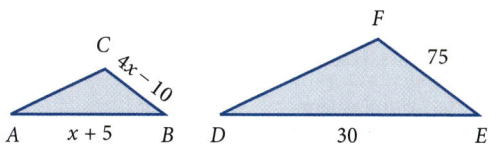

C, 4x − 10, A, x + 5, B
F, 75, D, 30, E

39.

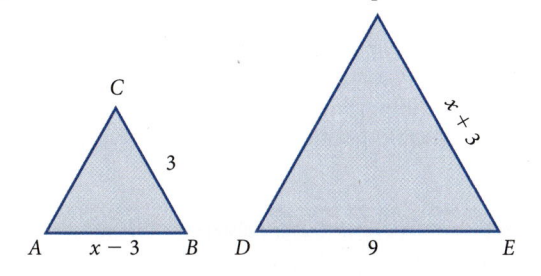

C, 3, A, x − 3, B
F, x + 3, D, 9, E

40.

C, 4, A, x − 5, B
F, x + 5, D, 6, E

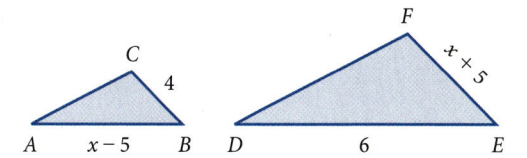

F Applying the Concepts [Examples 8–9]

41. Length of a Bass The entire length of a violin is 24 inches, while its body length is 15 inches. The bass is an instrument proportional to the violin. If the total length of a bass is 72 inches, find its body length.

42. Length of an Instrument The entire length of a violin is 24 inches, while the body length is 15 inches. Another instrument proportional to the violin has a body length of 25 inches. What is the total length of this instrument?

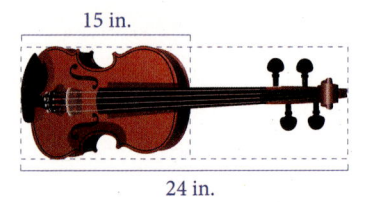

43. Video Resolution A new graphics card can increase the resolution of a computer's monitor. Suppose a monitor has a horizontal resolution of 800 pixels and a vertical resolution of 600 pixels. By adding a new graphics card, the resolutions remain in the same proportions, but the horizontal resolution increases to 1,280 pixels. What is the new vertical resolution?

44. Screen Resolution The display of a 20" computer monitor is proportional to that of a 23" monitor. A 20" monitor has a horizontal resolution of 1,680 pixels and a vertical resolution of 1,050 pixels. If a 23" monitor has a horizontal resolution of 1,920 pixels, what is its vertical resolution?

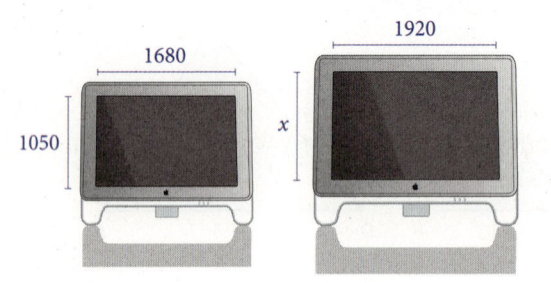

45. Eiffel Tower At the Paris Las Vegas Hotel is a replica of the Eiffel Tower in France. The heights of the tower in Las Vegas and the tower in France are 460 feet and 1,063 feet respectively. The base of the Eiffel Tower in France is 410 feet wide. What is the width of the base of the tower in Las Vegas? Round to the nearest foot.

46. Pyramids The Luxor Hotel in Las Vegas is almost an exact model of the pyramid of Khafre, the second largest Egyptian pyramid. The heights of the Luxor hotel and the pyramid of Khafre are 350 feet and 470 feet respectively. If the base of the pyramid in Khafre was 705 feet wide, what is the width of the base of the Luxor Hotel?

5.7 The Pythagorean Theorem

Objectives
- **A** Using the Pythagorean Theorem to solve right triangles.
- **B** Solving 30°–60°–90° Triangles.
- **C** Solving 45°–45°–90° Triangles.

A Pythagorean Theorem

In any right triangle, the square of the length of the longest side (called the hypotenuse) is equal to the sum of the squares of the lengths of the other two sides (called legs).

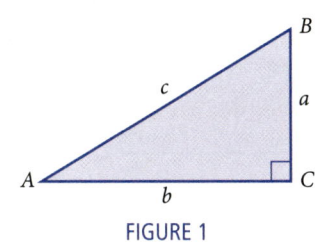

If $C = 90°$,
then $c^2 = a^2 + b^2$

FIGURE 1

Next we will prove the Pythagorean Theorem. Part of the proof involves finding the area of a triangle. In any triangle, the area is given by the formula

$$\text{Area} = \frac{1}{2}(\text{base})(\text{height})$$

For the right triangle shown in Figure 1, the base is b, and the height is a. Therefore the area is $A = \frac{1}{2}ab$.

There are many ways to prove the Pythagorean Theorem. The method that we are offering here is based on the diagram shown in Figure 2 and the formula for the area of a triangle.

Figure 2 is constructed by taking the right triangle in the lower right corner and repeating it three times so that the final diagram is a square in which each side has length $a + b$.

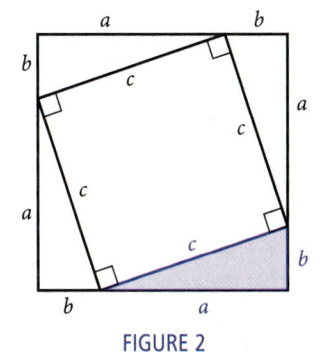

FIGURE 2

To derive the relationship between a, b, and c, we simply notice that the area of the large square is equal to the sum of the areas of the four triangles and the inner square. In symbols we have

Area of large square		Area of four triangles		Area of inner square
$(a + b)^2$	$=$	$4\left(\dfrac{1}{2}ab\right)$	$+$	c^2

We expand the left side using the formula from algebra for the square of a binomial. We simplify the right side by multiplying 4 with $\frac{1}{2}$.

$$a^2 + 2ab + b^2 = 2ab + c^2$$

Chapter 5 Geometry

Adding $-2ab$ to each side, we have the relationship we are after:

$$a^2 + b^2 = c^2$$

PRACTICE PROBLEMS

1. Solve for x:

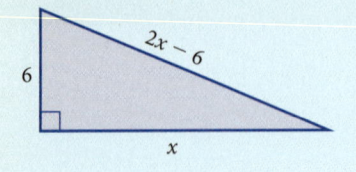

   (Triangle with legs 6 and x, hypotenuse $2x - 6$)

EXAMPLE 1 Solve for x in the right triangle in Figure 3.

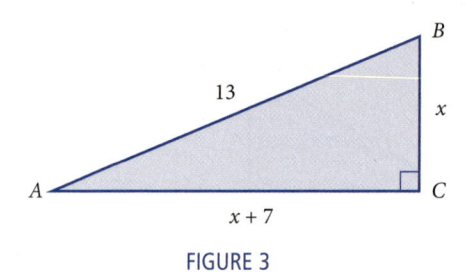

FIGURE 3 (Right triangle with sides 13, x, and $x + 7$)

SOLUTION Applying the Pythagorean Theorem gives us a quadratic equation to solve.

$$(x + 7)^2 + x^2 = 13^2$$
$$x^2 + 14x + 49 + x^2 = 169 \quad \text{Expand } (x + 7)^2 \text{ and } 13^2$$
$$2x^2 + 14x + 49 = 169 \quad \text{Combine similar terms}$$
$$2x^2 + 14x - 120 = 0 \quad \text{Add } -169 \text{ to both sides}$$
$$x^2 + 7x - 60 = 0 \quad \text{Divide both sides by 2}$$
$$(x - 5)(x + 12) = 0 \quad \text{Factor the left side}$$
$$x - 5 = 0 \quad \text{or} \quad x + 12 = 0 \quad \text{Set each factor to 0}$$
$$x = 5 \quad \text{or} \quad x = -12$$

Our only solution is $x = 5$. We cannot use $x = -12$ since x is the length of a side of triangle ABC and therefore cannot be negative.

NOTE The lengths of the sides of the triangle in Example 1 are 5, 12, and 13. Whenever the three sides in a right triangle are natural numbers, those three numbers are called a *Pythagorean triple*.

EXAMPLE 2 The vertical rise of the Forest Double chair lift (Figure 4) is 1,170 feet and the length of the chair lift as 5,750 feet. To the nearest foot, find the horizontal distance covered by a person riding this lift.

FIGURE 4

2. Another chair lift has a vertical rise of 960 feet and the length of the chair is 4,520 feet. To the nearest foot, find the horizontal distance covered by a person riding the lift.

Answer
1. 8

SOLUTION Figure 5 is a model of the Forest Double chair lift. A rider gets on the lift at point A and exits at point B. The length of the lift is AB.

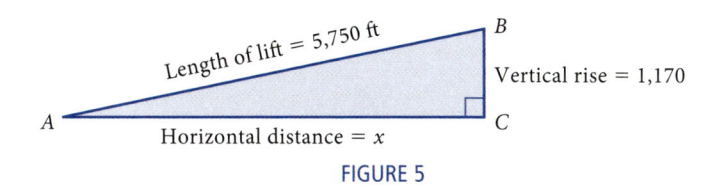

FIGURE 5

To find the horizontal distance covered by a person riding the chair lift we use the Pythagorean Theorem:

$5{,}750^2 = x^2 + 1{,}170^2$ Pythagorean Theorem

$33{,}062{,}500 = x^2 + 1{,}368{,}900$ Simplify squares

$x^2 = 33{,}062{,}500 - 1{,}368{,}900$ Solve for x^2

$x^2 = 31{,}693{,}600$ Simplify the right side

$x = \sqrt{31{,}693{,}600}$

$x = 5{,}630$ ft To the nearest foot

A rider getting on the lift at point A and riding to point B will cover a horizontal distance of approximately 5,630 feet.

Before leaving the Pythagorean Theorem we should mention something about Pythagoras and his followers, the Pythagoreans. They established themselves as a secret society around the year 540 B.C. The Pythagoreans kept no written record of their work; everything was handed down by spoken word. Their influence was not only in mathematics, but also in religion, science, medicine, and music. Among other things, they discovered the correlation between musical notes and the reciprocals of counting numbers, $\frac{1}{2}, \frac{1}{3}, \frac{1}{4}$, and so on. In their daily lives they followed strict dietary and moral rules to achieve a higher rank in future lives. The British philosopher Bertrand Russell has referred to Pythagoras as "intellectually one of the most important men that ever lived."

B The 30°–60°–90° Triangle

In any right triangle in which the two acute angles are 30° and 60°, the longest side (the hypotenuse) is always twice the shortest side (the side opposite the 30° angle), and the side of medium length (the side opposite the 60° angle) is always $\sqrt{3}$ times the shortest side (Figure 6).

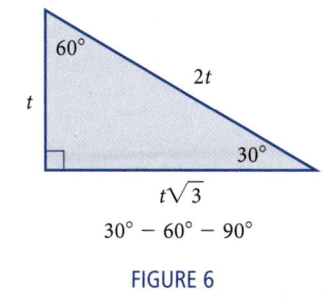

30° – 60° – 90°

FIGURE 6

Answer
2. 4,417 feet

NOTE The shortest side t is opposite the smallest angle 30°. The longest side $2t$ is opposite the largest angle 90°. To verify the relationship between the sides in this triangle, we draw an equilateral triangle (one in which all three sides are equal) and label half the base with t (Figure 7).

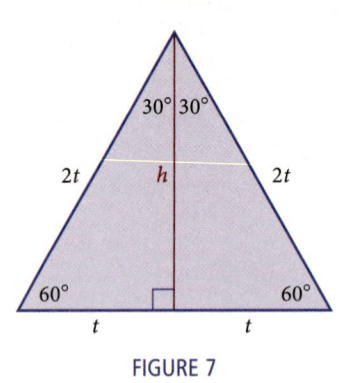

FIGURE 7

The altitude h bisects the base. We have two 30°–60°–90° triangles. The longest side in each is $2t$. We find that h is $t\sqrt{3}$ by applying the Pythagorean Theorem.

$$t^2 + h^2 = (2t)^2$$
$$h = \sqrt{4t^2 - t^2}$$
$$= \sqrt{3t^2}$$
$$= t\sqrt{3}$$

EXAMPLE 3 If the shortest side of a 30°–60°–90° triangle is 5, find the other two sides.

SOLUTION The longest side is 10 (twice the shortest side), and the side opposite the 60° angle is $5\sqrt{3}$ (Figure 8).

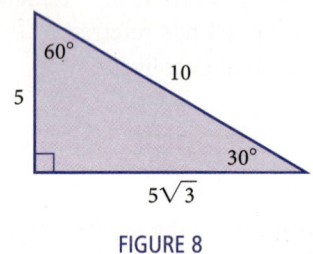

FIGURE 8

3. If the hypotenuse of a 30°–60°–90° triangle is 12, find the other 2 sides.

Algebra Review: Rationalizing the Denominator

Radical expressions that are in simplified form are generally easier to work with. Recall that a radical expression is in simplified form if it has three special characteristics.

> **Definition**
> A radical expression is in **simplified form** if
> 1. There are no perfect squares that are factors of the quantity under the square root sign, no perfect cubes that are factors of the quantity under the cube root sign, and so on. We want as little as possible under the radical sign.
> 2. There are no fractions under the radical sign.
> 3. There are no radicals in the denominator.

Answer
3. $6, 6\sqrt{3}$

5.7 The Pythagorean Theorem

A radical expression that has these three characteristics is said to be in simplified form. Remember that simplified form is not always the least complicated expression. In many cases, the simplified expression looks more complicated than the original expression. The important thing about simplified form for radicals is that simplified expressions are easier to work with.

The tools we will use to put radical expressions into simplified form are the properties of radicals that we covered previously. We list the properties again for clarity.

> **Note** Simplified form for radicals is the form that we work toward when simplifying radicals. The properties of radicals are the tools we use to get us to simplified form.

Properties of Radicals

If a and b represent any two nonnegative real numbers, then it is always true that

1. $\sqrt{a}\,\sqrt{b} = \sqrt{a \cdot b}$
2. $\dfrac{\sqrt{a}}{\sqrt{b}} = \sqrt{\dfrac{a}{b}}$ $b \neq 0$
3. $\sqrt{a}\,\sqrt{a} = (\sqrt{a})^2 = a$ This property comes directly from the definition of radicals

The following examples review how we put a radical expression into simplified form using the three properties of radicals. Although the properties are stated for square roots only, they hold for all roots. [Property 3 written for cube roots would be $\sqrt[3]{a}\,\sqrt[3]{a}\,\sqrt[3]{a} = (\sqrt[3]{a})^3 = a$.]

EXAMPLE 4   Put $\sqrt{\dfrac{1}{2}}$ into simplified form.

SOLUTION   The expression $\sqrt{\dfrac{1}{2}}$ is not in simplified form because there is a fraction under the radical sign. We can change this by applying Property 2 for radicals:

$$\sqrt{\dfrac{1}{2}} = \dfrac{\sqrt{1}}{\sqrt{2}} \qquad \text{Property 2 for radicals}$$

$$= \dfrac{1}{\sqrt{2}} \qquad \sqrt{1} = 1$$

The expression $\dfrac{1}{\sqrt{2}}$ is not in simplified form because there is a radical sign in the denominator. If we multiply the numerator and denominator of $\dfrac{1}{\sqrt{2}}$ by $\sqrt{2}$, the denominator becomes $\sqrt{2} \cdot \sqrt{2} = 2$:

$$\dfrac{1}{\sqrt{2}} = \dfrac{1}{\sqrt{2}} \cdot \dfrac{\sqrt{2}}{\sqrt{2}} \qquad \text{Multiply numerator and denominator by } \sqrt{2}$$

$$= \dfrac{\sqrt{2}}{2} \qquad \begin{array}{l} 1 \cdot \sqrt{2} = \sqrt{2} \\ \sqrt{2} \cdot \sqrt{2} = \sqrt{4} = 2 \end{array}$$

If we check the expression $\dfrac{\sqrt{2}}{2}$ against our definition of simplified form for radicals, we find that all three rules hold. There are no perfect squares that are factors of 2. There are no fractions under the radical sign. No radicals appear in the denominator. The expression $\dfrac{\sqrt{2}}{2}$, therefore, must be in simplified form.

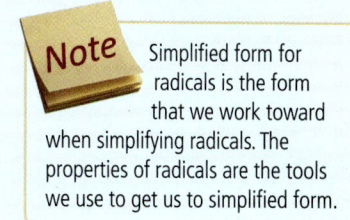

4. Put $\sqrt{\dfrac{1}{3}}$ in simplified form.

Answer
4. $\dfrac{\sqrt{3}}{3}$

Chapter 5 Geometry

5. Write $\sqrt{\dfrac{3}{5}}$ in simplified form.

EXAMPLE 5 Write $\sqrt{\dfrac{2}{3}}$ in simplified form.

SOLUTION We proceed as we did in Example 4:

$$\sqrt{\dfrac{2}{3}} = \dfrac{\sqrt{2}}{\sqrt{3}} \quad \text{Use Property 2 to separate radicals}$$

$$= \dfrac{\sqrt{2}}{\sqrt{3}} \cdot \dfrac{\sqrt{3}}{\sqrt{3}} \quad \text{Multiply by } \dfrac{\sqrt{3}}{\sqrt{3}} \text{ to remove the radical from the denominator}$$

$$= \dfrac{\sqrt{6}}{3} \quad \begin{array}{l}\sqrt{2} \cdot \sqrt{3} = \sqrt{6} \\ \sqrt{3} \cdot \sqrt{3} = \sqrt{9} = 3\end{array}$$

6. Another ladder is leaning against a wall. The top of the ladder is 7 feet above the ground, and the bottom of the ladder makes an angle of 60° with the ground. How long is the ladder and how far from the wall is the bottom of the ladder?

EXAMPLE 6 A ladder is leaning against a wall. The top of the ladder is 4 feet above the ground and the bottom of the ladder makes an angle of 60° with the ground (Figure 9). How long is the ladder, and how far from the wall is the bottom of the ladder?

SOLUTION The triangle formed by the ladder, the wall, and the ground is a 30°–60°–90° triangle. If we let x represent the distance from the bottom of the ladder to the wall, then the length of the ladder can be represented by $2x$. The distance from the top of the ladder to the ground is $x\sqrt{3}$, since it is opposite the 60° angle (Figure 10). It is also given as 4 feet. Therefore,

$$x\sqrt{3} = 4$$

$$x = \dfrac{4}{\sqrt{3}}$$

$$= \dfrac{4\sqrt{3}}{3} \quad \text{Rationalize the denominator by multiplying the numerator and denominator by } \sqrt{3}.$$

FIGURE 9

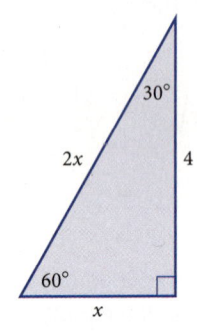

FIGURE 10

The distance from the bottom of the ladder to the wall, x, is $\dfrac{4\sqrt{3}}{3}$ feet, so the length of the ladder, $2x$, must be $\dfrac{8\sqrt{3}}{3}$ feet. Note that these lengths are given in exact values. If we want a decimal approximation for them, we can replace $\sqrt{3}$ with 1.732 to obtain

$$\dfrac{4\sqrt{3}}{3} \approx \dfrac{4(1.732)}{3} = 2.309 \text{ ft}$$

$$\dfrac{8\sqrt{3}}{3} \approx \dfrac{8(1.732)}{3} = 4.619 \text{ ft}$$

Answers

5. $\dfrac{\sqrt{15}}{5}$

6. 4.04 ft., 8.08 ft.

CALCULATOR NOTE On a scientific calculator, this last calculation could be done as follows:

$$8 \; \boxed{\times} \; 3 \; \boxed{\sqrt{}} \; \boxed{\div} \; 3 \; \boxed{=}$$

On a graphing calculator, the calculation is done like this:

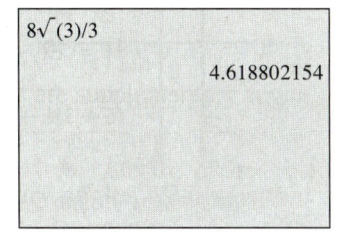

$$8 \; \boxed{\times} \; \boxed{\sqrt{}} \; \boxed{(} \; 3 \; \boxed{)} \; \boxed{\div} \; 3 \; \boxed{\text{ENTER}}$$

```
8√(3)/3
              4.618802154
```

FIGURE 11

Some graphing calculators use parentheses with certain functions, such as the square root function. For example, the TI-83/84 will automatically insert a left parenthesis, so TI-83/84 users should skip this key. Other models do not require them. For the sake of clarity, we will often include parentheses throughout this book. You may be able to omit one or both parentheses with your model.

C The 45°–45°–90° Triangle

If the two acute angles in a right triangle are both 45°, then the two shorter sides (the legs) are equal and the longest side (the hypotenuse) is $\sqrt{2}$ times as long as the shorter sides. That is, if the shorter sides are of length t, then the longest side has length $t\sqrt{2}$ (Figure 12).

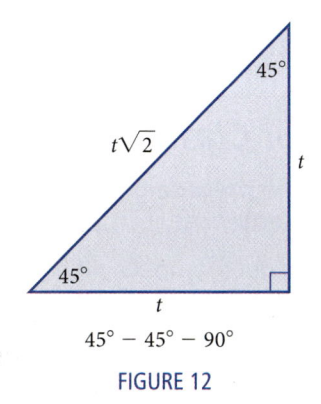

45° − 45° − 90°

FIGURE 12

To verify this relationship, we simply note that if the two acute angles are equal, then the sides opposite them are also equal. We apply the Pythagorean Theorem to find the length of the hypotenuse.

$$\begin{aligned} \text{hypotenuse} &= \sqrt{t^2 + t^2} \\ &= \sqrt{2t^2} \\ &= t\sqrt{2} \end{aligned}$$

7. If the rope in Example 7 was 12 feet long, find the length of the tent pole.

Chapter 5 Geometry

EXAMPLE 7 A 10-foot rope connects the top of a tent pole to the ground. If the rope makes an angle of 45° with the ground, find the length of the tent pole (Figure 13).

SOLUTION Assuming that the tent pole forms an angle of 90° with the ground, the triangle formed by the rope, tent pole, and the ground is a 45° – 45° – 90° triangle (Figure 14).

If we let x represent the length of the tent pole, then the length of the rope, in terms of x, is $x\sqrt{2}$. It is also given as 10 feet. Therefore,

$$x\sqrt{2} = 10$$

$$x = \frac{10}{\sqrt{2}} = 5\sqrt{2}$$

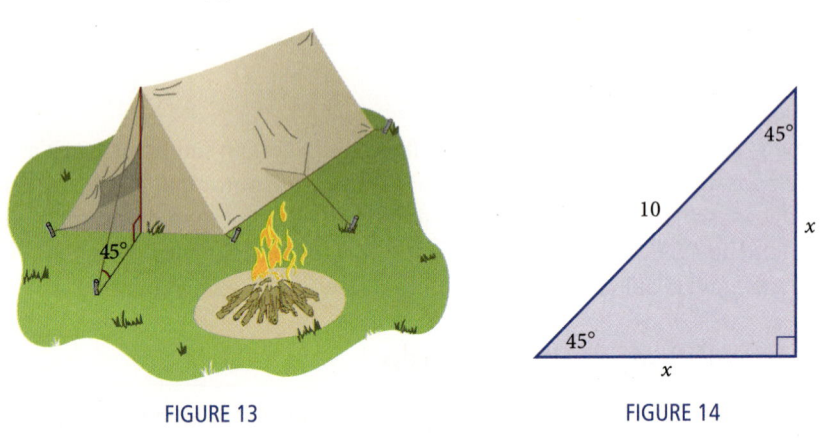

FIGURE 13

FIGURE 14

The length of the tent pole is $5\sqrt{2}$ feet. Again, $5\sqrt{2}$ is the exact value of the length of the tent pole. To find a decimal approximation, we replace $\sqrt{2}$ with 1.414 to obtain

$$5\sqrt{2} \approx 5(1.414) = 7.07 \text{ ft}$$

Getting Ready for Class

After reading through the preceding section, respond in your own words and in complete sentences.

1. Describe how you would put $\sqrt{\frac{1}{2}}$ in simplified form.
2. What is simplified form for an expression that contains a square root?
3. What does it mean to rationalize the denominator in an expression?
4. Why is it important to recognize 30° – 60° – 90° and 45° – 45° – 90° triangles?

Answer
7. $6\sqrt{2} \approx 8.49$ ft.

Problem Set 5.7

Problems 1 through 6 refer to right triangle ABC with $C = 90°$.

1. If $a = 4$ and $b = 3$, find c.
2. If $a = 6$ and $b = 8$, find c.
3. If $a = 8$ and $c = 17$, find b.
4. If $a = 2$ and $c = 6$, find b.
5. If $b = 12$ and $c = 13$, find a.
6. If $b = 10$ and $c = 26$, find a.

Find the value of x in the following special triangles.

7. 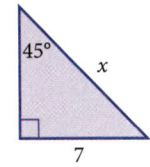
8. 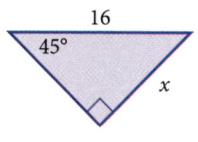
9. 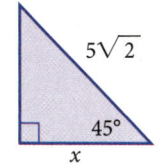
10. 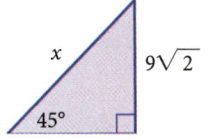
11. 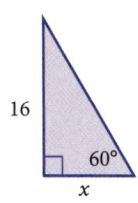
12. 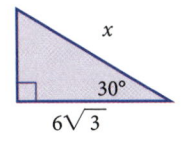
13. 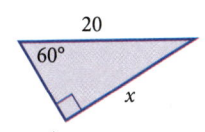
14.

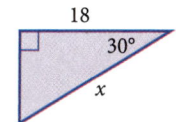

Solve for x in each of the following right triangles.

15. 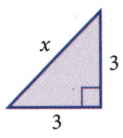
16. 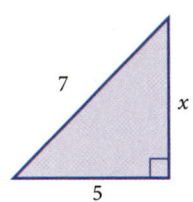
17. 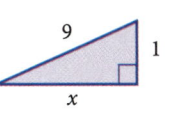
18. 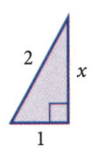
19. 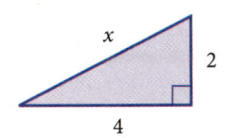
20. 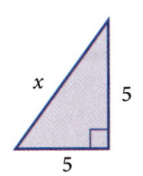
21. 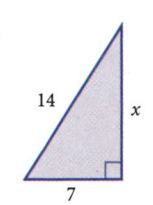
22.

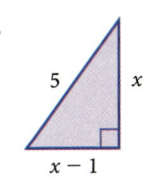

Chapter 5 Geometry

In right △ABC, find x, $\overline{AB}$, $\overline{BC}$, and $\overline{AC}$.

23. Triangle with BC = x, CA = x + 1, BA = x + 2

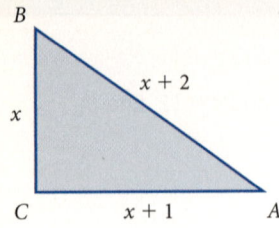

24. Triangle with BC = x + 2, CA = 4x + 4, BA = 5x

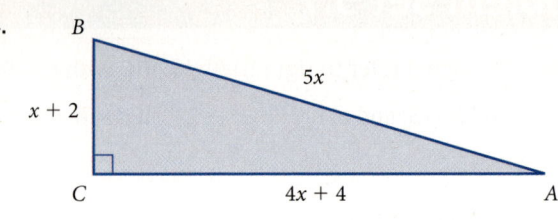

25. Triangle with BC = 2x, CA = 3x − 1, BA = 4x − 2

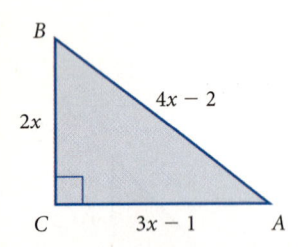

26. Triangle with BC = 2x + 2, CA = 6x, BA = 7x − 2

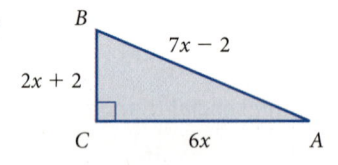

27. Triangle with BC = x − 5, CA = 2x, BA = 2x + 1

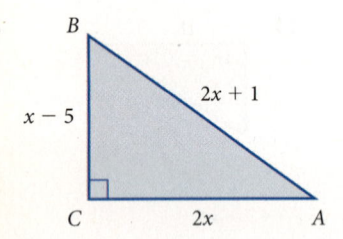

28. Triangle with BC = x, CA = x + 2, BA = x + 4

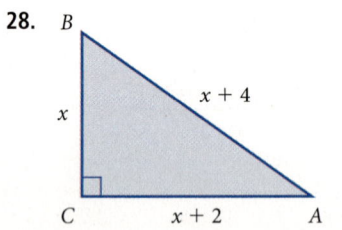

29. Triangle with BC = x, CA = 4x + 4, BA = 5x − 4

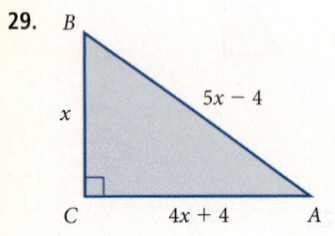

30. Triangle with BC = $\frac{x}{2}$, CA = 4x + 8, BA = 4x + 9

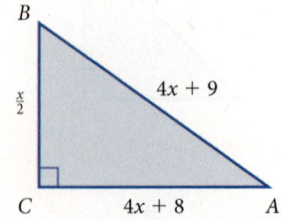

Find the remaining sides of a 30° – 60° – 90° triangle if

31. the longest side is 8.

32. the longest side is 5.

33. the side opposite 60° is 6.

34. the side opposite 60° is 4.

35. Escalator An escalator in a department store is to carry people a vertical distance of 20 feet between floors. How long is the escalator if it makes an angle of 30° with the ground?

36. Escalator What is the length of the escalator in Problem 27 if it makes an angle of 60° with the ground?

37. Tent Design A two-person tent is to be made so that the height at the center is 4 feet. If the sides of the tent are to meet the ground at an angle of 60°, and the tent is to be 6 feet in length, how many square feet of material will be needed to make the tent? (Figure 15; assume that the tent has a floor and is closed at both ends, and give your answer to the nearest tenth of a square foot.)

38. Tent Design If the height at the center of the tent in Problem 29 is to be 3 feet, how many square feet of material will be needed to make the tent?

FIGURE 15

Find the remaining sides of a 45° – 45° – 90° triangle if

39. the shorter sides are each $\frac{4}{5}$.

40. the shorter sides are each $\frac{1}{2}$.

41. the longest side is $8\sqrt{2}$.

42. the longest side is $5\sqrt{2}$.

43. the longest side is 4.

44. the longest side is 12.

45. Distance a Bullet Travels A bullet is fired into the air at an angle of 45°. How far does it travel before it is 1,000 feet above the ground? (Assume that the bullet travels in a straight line, neglect the forces of gravity, and give your answer to the nearest foot.)

46. Time a Bullet Travels If the bullet in Problem 37 is traveling at 2,828 feet per second, how long does it take for the bullet to reach a height of 1,000 feet?

Geometry: Characteristics of a Cube The object shown in Figure 16 is a cube (all edges are equal in length).

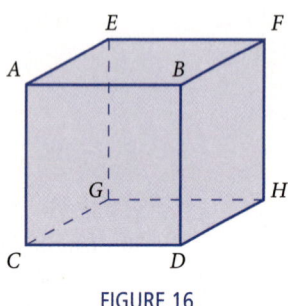

FIGURE 16

47. If the length of each edge of the cube shown in Figure 16 is 1 inch, find
  a. the length of diagonal *CH*.
  b. the length of diagonal *CF*.

48. If the length of each edge of the cube shown in Figure 16 is 5 centimeters, find
  a. the length of diagonal *GD*.
  b. the length of diagonal *GB*.

49. If the length of each edge of the cube shown in Figure 16 is unknown, we can represent it with the variable *x*. Then we can write formulas for the lengths of any of the diagonals. Finish each of the following statements:
  a. If the length of each edge of a cube is *x*, then the length of the diagonal of any face of the cube will be _____.
  b. If the length of each edge of a cube is *x*, then the length of any diagonal that passes through the center of the cube will be _____.

50. What is the measure of ∠*GDH* in Figure 16?

Right Triangle Trigonometry

5.8

Objectives

A Finding values for the six trigonometric functions.

B Using cofunctions and the Cofunction Theorem.

C Using special triangles to find exact trigonometric values.

D Find exact values.

The word *trigonometry* is derived from two Greek words: *tri'gonon*, which translates as *triangle*, and *met'ron*, which means *measure*. Trigonometry, then, is triangle measure. In this section, we will give a second definition for the trigonometric functions that is, in fact, based on "triangle measure." We will define the trigonometric functions as the ratios of sides in right triangles.

A Using The Six Trigonometric Functions

Definition
If triangle ABC is a right triangle with $C = 90°$ (Figure 1), then the six trigonometric functions for A are defined as follows:

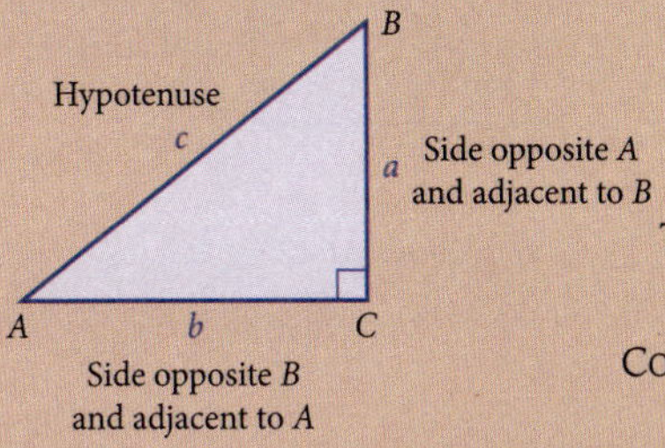

$$\text{Sine } A \ (\sin A) = \frac{\text{side opposite } A}{\text{hypotenuse}} = \frac{a}{c}$$

$$\text{Cosine } A \ (\cos A) = \frac{\text{side adjacent } A}{\text{hypotenuse}} = \frac{b}{c}$$

$$\text{Tangent } A \ (\tan A) = \frac{\text{side opposite } A}{\text{side adjacent } A} = \frac{a}{b}$$

$$\text{Cotangent } A \ (\cot A) = \frac{\text{side adjacent } A}{\text{side opposite } A} = \frac{b}{a}$$

$$\text{Secant } A \ (\sec A) = \frac{\text{hypotenuse}}{\text{side adjacent } A} = \frac{c}{b}$$

$$\text{Cosecant } A \ (\csc A) = \frac{\text{hypotenuse}}{\text{side opposite } A} = \frac{c}{a}$$

FIGURE 1

EXAMPLE 1
Triangle ABC is a right triangle with $C = 90°$. If $a = 6$ and $c = 10$, find the six trigonometric functions of A.

SOLUTION We begin by making a diagram of ABC, and then use the given information and the Pythagorean Theorem to solve for b (Figure 2).

$$b = \sqrt{c^2 - a^2}$$
$$= \sqrt{100 - 36}$$
$$= \sqrt{64}$$
$$= 8$$

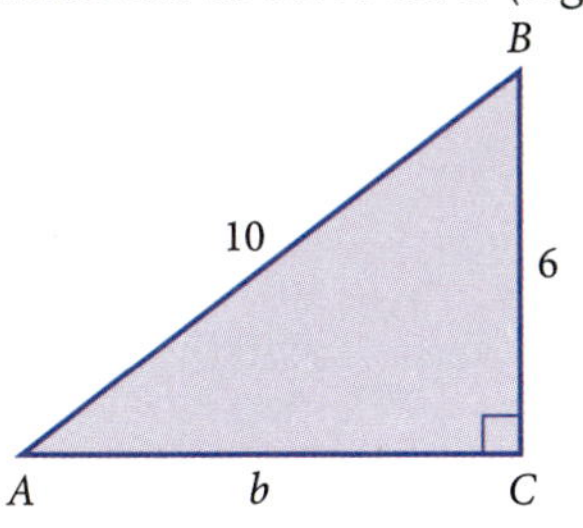

FIGURE 2

Now we write the six trigonometric functions of A using $a = 6$, $b = 8$, and $c = 10$.

$$\sin A = \frac{a}{c} = \frac{6}{10} = \frac{3}{5} \qquad \csc A = \frac{c}{a} = \frac{5}{3}$$

$$\cos A = \frac{b}{c} = \frac{8}{10} = \frac{4}{5} \qquad \sec A = \frac{c}{b} = \frac{5}{4}$$

$$\tan A = \frac{a}{b} = \frac{6}{8} = \frac{3}{4} \qquad \cot A = \frac{b}{a} = \frac{4}{3}$$

Here is another definition that we will need before we can take the definition of trigonometric functions any further.

PRACTICE PROBLEMS

1. Find the six trigonometric functions of B in Figure 2.

Answers
1. $\sin B = \frac{4}{5}$ $\csc B = \frac{5}{4}$
   $\cos B = \frac{3}{5}$ $\sec B = \frac{5}{3}$
   $\tan B = \frac{4}{3}$ $\cot B = \frac{3}{4}$

Chapter 5 Geometry

B Cofunctions and the Cofunction Theorem

Definition
Sine and cosine are cofunctions, as are tangent and cotangent, and secant and cosecant. We say sine is the cofunction of cosine, and cosine is the cofunction of sine.

Now let's see what happens when we apply the definition of trigonometric functions to B in right triangle ABC.

$$\sin B = \frac{\text{side opposite } B}{\text{hypotenuse}} = \frac{b}{c} = \cos A$$

$$\cos B = \frac{\text{side adjacent } B}{\text{hypotenuse}} = \frac{a}{c} = \sin A$$

$$\tan B = \frac{\text{side opposite } B}{\text{side adjacent } B} = \frac{b}{a} = \cot A$$

$$\cot B = \frac{\text{side adjacent } B}{\text{side opposite } B} = \frac{a}{b} = \tan A$$

$$\sec B = \frac{\text{hypotenuse}}{\text{side adjacent } B} = \frac{c}{a} = \csc A$$

$$\csc B = \frac{\text{hypotenuse}}{\text{side opposite } B} = \frac{c}{b} = \sec A$$

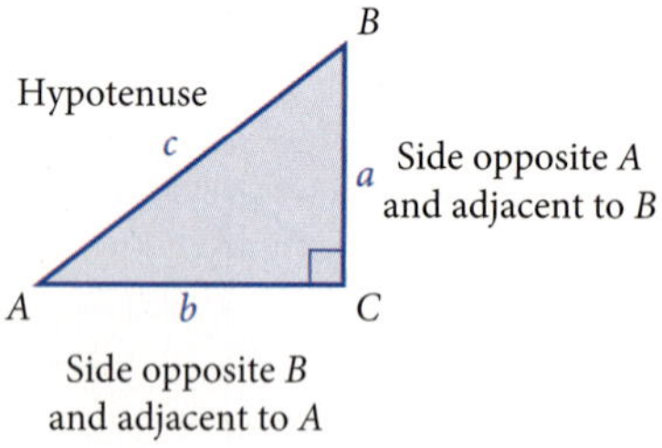

FIGURE 3

As you can see in Figure 3, every trigonometric function of A is equal to the cofunction of B. That is, $\sin A = \cos B$, $\sec A = \csc B$, and $\tan A = \cot B$, to name a few. Since A and B are the acute angles in a right triangle, they are always complementary angles; that is, their sum is always 90°. What we actually have here is another property of trigonometric functions: The sine of an angle is the cosine of its complement, the secant of an angle is the cosecant of its complement, and the tangent of an angle is the cotangent of its complement. Or, in symbols,

if $A + B = 90°$, then $\begin{cases} \sin A = \cos B \\ \sec A = \csc B \\ \tan A = \cot B \end{cases}$

and so on.
We generalize this discussion with the following theorem.

Cofunction Theorem
A trigonometric function of an angle is always equal to the cofunction of the complement of the angle.

To clarify this further, if two angles are complementary, such as 40° and 50°, then a trigonometric function of one is equal to the cofunction of the other. That is, $\sin 40° = \cos 50°$, $\sec 40° = \csc 50°$, and $\tan 40° = \cot 50°$.

EXAMPLE 2 Fill in the blanks so that each expression becomes a true statement.
a. $\sin \_\_\_ = \cos 30°$ **b.** $\tan y = \cot \_\_\_$ **c.** $\sec 75° = \csc \_\_\_$

Note The prefix *co-* in *co*sine, *co*secant, and *co*tangent is a reference to the complement. Around the year 1463, Regiomontanus used the term *sinus rectus complementi*, presumably referring to the cosine as the sine of the complementary angle. In 1620, Edmund Gunter shortened this to *co.sinus*, which was further abbreviated as *cosinus* by John Newton in 1658.

2. **a.** csc ___ = sec 70°
   b. sin x = cos ___
   c. cot 32° = tan ___

SOLUTION Using the theorem on cofunctions of complementary angles, we fill in the blanks as follows:

a. sin $\underline{60°}$ = cos 30°   since sine and cosine are cofunctions and 60° + 30° = 90°

b. tan y = cot $\underline{(90° - y)}$ since tangent and cotangent are cofunctions and $y + (90° - y) = 90°$

c. sec 75° = csc $\underline{15°}$ since secant and cosecant are cofunctions and 75° + 15° = 90°

C Special Triangles and their Trigonometric Values

For our next application, we need to recall the two special triangles we introduced previously. They are the 30° – 60° – 90° triangle and the 45° – 45° – 90° triangle (Figure 4).

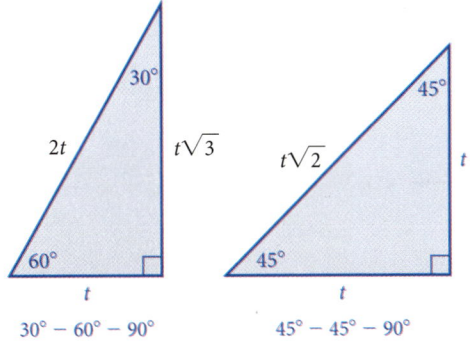

FIGURE 4

We can use these two special triangles to find the trigonometric functions of 30°, 45°, and 60°. For example,

$$\sin 60° = \frac{\text{side opposite } 60°}{\text{hypotenuse}} = \frac{t\sqrt{3}}{2t} = \frac{\sqrt{3}}{2}$$

$$\tan 30° = \frac{\text{side opposite } 30°}{\text{side adjacent } 30°} = \frac{t}{t\sqrt{3}} = \frac{\sqrt{3}}{3}$$

Dividing out the common factor t

We could go on in this manner—using the definition or trigonometric functions and the two special triangles—to find the six trigonometric ratios for 30°, 45°, and 60°. Instead, let us vary things a little and use the information just obtained for sin 60° and tan 30° and the theorem on cofunctions of complementary angles to find cos 30° and cot 60°.

$$\cos 30° = \sin 60° = \frac{\sqrt{3}}{2}$$

$$\cot 60° = \tan 30° = \frac{\sqrt{3}}{3}$$

Cofunction Theorem

To vary things even more, we can use some reciprocal identities to find csc 60° and cot 30°.

$$\csc 60° = \frac{1}{\sin 60°} = \frac{2}{\sqrt{3}}$$

$$= \frac{2\sqrt{3}}{3}$$ If the denominator is rationalized

$$\cot 30° = \frac{1}{\tan 30°} = \sqrt{3}$$

Answers
2. a. 20° **b.** 90° − x **c.** 58°

Chapter 5 Geometry

The idea behind using the different methods listed here is not to make things confusing. We have a number of tools at hand, and it does not hurt to show the different ways they can be used.

D Finding Exact Values

If we were to continue finding the sine, cosine, and tangent for these special angles, we would obtain the results summarized in Table 1 and illustrated in Figure 5.

Table 1 is called a table of exact values to distinguish it from a table of approximate values. Later in this chapter we will work with a calculator to obtain tables of approximate values.

FIGURE 5

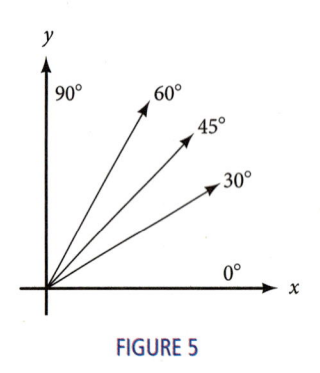

TABLE 1 EXACT VALUES

θ	$\sin \theta$	$\cos \theta$	$\tan \theta$
30°	$\dfrac{1}{2}$	$\dfrac{\sqrt{3}}{2}$	$\dfrac{1}{\sqrt{3}} = \dfrac{\sqrt{3}}{3}$
45°	$\dfrac{1}{\sqrt{2}} = \dfrac{\sqrt{2}}{2}$	$\dfrac{1}{\sqrt{2}} = \dfrac{\sqrt{2}}{2}$	1
60°	$\dfrac{\sqrt{3}}{2}$	$\dfrac{1}{2}$	$\sqrt{3}$

EXAMPLE 3 Use the exact values from Table 1 to show that the following are true.
a. $\cos^2 30° + \sin^2 30° = 1$
b. $\cos^2 45° + \sin^2 45° = 1$

SOLUTION

a. $\cos^2 30° + \sin^2 30° = \left(\dfrac{\sqrt{3}}{2}\right)^2 + \left(\dfrac{1}{2}\right)^2 = \dfrac{3}{4} + \dfrac{1}{4} = 1$

b. $\cos^2 45° + \sin^2 45° = \left(\dfrac{1}{\sqrt{2}}\right)^2 + \left(\dfrac{1}{\sqrt{2}}\right)^2 = \dfrac{1}{2} + \dfrac{1}{2} = 1$

EXAMPLE 4 Let $x = 30°$ and $y = 45°$ in each of the expressions that follow, and then simplify each expression as much as possible.
a. $2 \sin x$ **b.** $\sin 2x$ **c.** $4 \sin (3y - 90°)$

SOLUTION

a. $2 \sin x = 2 \sin 30° = 2\left(\dfrac{1}{2}\right) = 1$

b. $\sin 2x = \sin 2(30°) = \sin 60° = \dfrac{\sqrt{3}}{2}$

c. $4 \sin (3y - 90°) = 4 \sin [3(45°) - 90°] = 4 \sin 45° = 4\left(0\right) = \dfrac{\sqrt{2}}{2}$

---

3. Use the exact values from Table 1 to show that the following are true.
   a. $\dfrac{2 \tan 45°}{1 + \tan^2 45°} = 1$
   b. $\cos^2 30° - \sin^2 30° = \dfrac{1}{2}$

4. Let $x = 60°$ and $y = 45°$ and simplify each expression.
   a. $2 \sin^2 x$
   b. $\cos \dfrac{x}{2}$
   c. $6 \tan (2y - x)$

Answers
3. See Solutions to Selected Practice Problems
4. a. $\dfrac{3}{2}$ b. $\dfrac{\sqrt{3}}{2}$ c. $2\sqrt{3}$

Problem Set 5.8

Problems 1 through 6 refer to right triangle ABC with C = 90°. In each case, use the given information to find the six trigonometric functions of A. Do not rationalize your answers.

1. $b = 3, c = 5$

2. $b = 5, c = 13$

3. $a = 2, b = 1$

4. $a = 3, b = 2$

5. $a = 2, b = \sqrt{5}$

6. $a = 3, b = \sqrt{7}$

In each right triangle below, find sin A, cos A, tan A, and sin B, cos B, tan B. Do not rationalize your answers.

7.
B, 6, 5, A, C

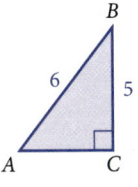

8.
B, 4, A, 3, C

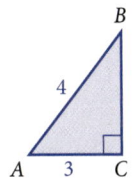

9.
C, 1, 1, A, B

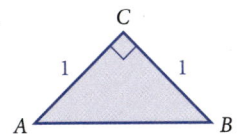

10.
C, 1, 2, A, B

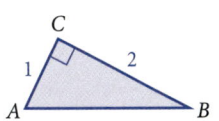

11.
A, C, 10, 6, B

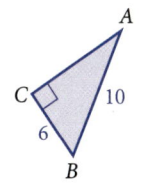

12.
A, $\sqrt{5}$, C, 1, B

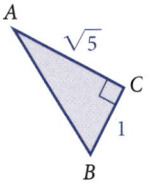

13.
B, 2x, C, x, A

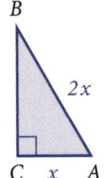

14.
B, x, C, x, A

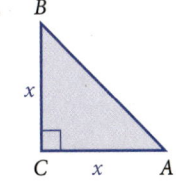

Chapter 5 Geometry

Find x to the nearest tenth.

15. Right triangle with legs x (vertical) and 12 (horizontal); angle 40° between hypotenuse and base of length 12.

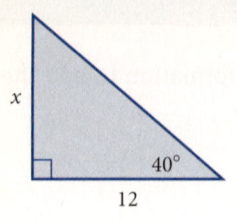

16. Right triangle with top angle 32°, hypotenuse 8, and horizontal leg x at bottom.

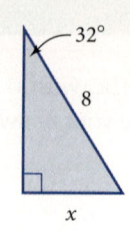

17. Right triangle with top angle 28°, hypotenuse 15, and vertical leg x on the left.

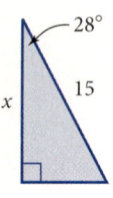

18. Right triangle with top angle 45°, hypotenuse x, and horizontal leg 12.

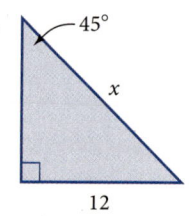

19. Right triangle with left side 13, top side x, and bottom-right angle 47°.

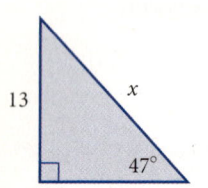

20. Right triangle with vertical side 27 and bottom-right side x.

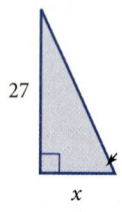

21. Right triangle with left side x, top side 21, and bottom-right angle 31°.

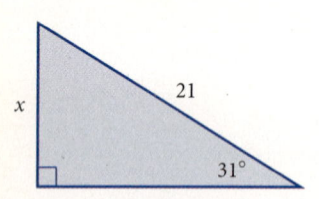

22. Right triangle with side 28 and angle 85°, bottom side x.

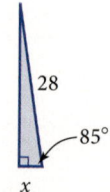

In each diagram below, angle A is at the origin. In each case, find the coordinates of point B and then find sin A, cos A, and tan A.

23. Coordinate grid with point B above point C on the x-axis; A at origin.

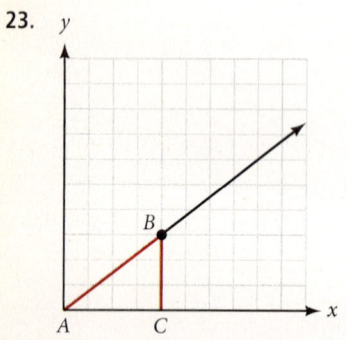

24. Coordinate grid with point B above point C on the x-axis; A at origin.

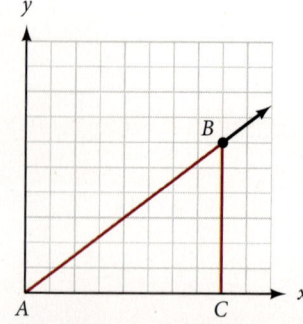

Use the Cofunction Theorem to fill in the blanks so that each expression becomes a true statement.

25. $\tan 8° = \cot$ _____

26. $\cot 12° = \tan$ _____

27. $\sin x = \cos$ _____

28. $\sin y = \cos$ _____

29. $\tan (90° - x) = \cot$ _____

30. $\tan (90° - y) = \cot$ _____

Complete the following tables using exact values. Do not rationalize any denominators.

31.

x	sin x	csc x
30°	$\frac{1}{2}$	
45°	$\frac{1}{\sqrt{2}}$	
60°	$\frac{\sqrt{3}}{2}$	

32.

x	cos x	sec x
30°	$\frac{\sqrt{3}}{2}$	
45°	$\frac{1}{\sqrt{2}}$	
60°	$\frac{1}{2}$	

Simplify each expression by first substituting values from the table of exact values and then simplifying the resulting expression.

33. $4 \sin 30°$

34. $5 \sin^2 30°$

35. $(\sin 60° + \cos 60°)^2$

36. $\sin^2 60° + \cos^2 60°$

37. $\sin^2 45° - 2 \sin 45° \cos 45° + \cos^2 45°$

38. $(\sin 45° - \cos 45°)^2$

39. $(\tan 45° + \tan 60°)^2$

40. $\tan^2 45° + \tan^2 60°$

For each expression that follows, replace x with 30°, y with 45°, and z with 60°, and then simplify as much as possible.

41. $2 \sin x$

42. $4 \cos y$

43. $-3 \sin 2x$

44. $3 \sin 2x$

45. $2 \cos (3x - 45°)$

46. $2 \sin (90° - z)$

Find exact values for each of the following:

47. sec 30°

48. csc 30°

49. cot 45°

50. cot 30°

51. sec 45°

52. csc 45°

Problems 53 through 56 refer to right triangle ABC with $C = 90°$. In each case, use a calculator to find sin A, cos A, sin B, and cos B. Round your answers to the nearest hundredth.

53. $a = 3.42, c = 5.70$

54. $b = 8.88, c = 9.62$

55. $a = 19.44, b = 5.67$

56. $a = 11.28, b = 8.46$

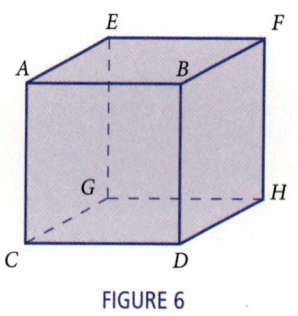

FIGURE 6

57. Suppose each edge of the cube shown in Figure 6 is 5 inches long. Find the sine and cosine of the angle formed by diagonals CF and CH.

58. Suppose each edge of the cube shown in Figure 6 is 3 inches long. Find the sine and cosine of the angle formed by diagonals DE and DG.

59. Suppose each edge of the cube shown in Figure 6 is x inches long. Find the sine and cosine of the angle formed by diagonals CF and CH.

60. Suppose each edge of the cube shown in Figure 6 is x inches long. Find the sine and cosine of the angle formed by diagonals DE and DG.

Solving Right Triangles 5.9

Objectives
A Introduce significant digits.
B Use trigonometric functions to solve right triangles.
C Solve applications using trigonometric functions.

The first Ferris wheel was designed and built by American engineer George W. G. Ferris in 1893. The diameter of this wheel was 250 feet. It had 36 cars, each of which held 40 passengers. The top of the wheel was 264 feet above the ground. It took 20 minutes to complete one revolution. As you will see trigonometric functions can be used to model the motion of a rider on a Ferris wheel. The model can be used to give information about the position of the rider at any time during a ride. For instance, in Example 5 in this section, we will use the definition for the trigonometric functions to find the height a rider is above the ground at certain positions on a Ferris wheel.

In this section, we will use our definition for trigonometric functions of an acute angle, along with our calculators, to find the missing parts to some right triangles.

A Significant Digits

> **Definition**
> The number of **significant digits** (or figures) in a number is found by counting all the digits from left to right beginning with the first nonzero digit on the left.

Before we begin, however, we need to talk about significant digits. According to this definition,

> 0.042 has two significant digits
> 0.005 has one significant digit
> 20.5 has three significant digits
> 6.000 has four significant digits
> 9,200 has four significant digits
> 700 has three significant digits

NOTE In actual practice it is not always possible to tell how many significant digits an integer like 700 has. For instance, if exactly 700 people signed up to take history at your school, then 700 has three significant digits. On the other hand, if 700 is the result of a calculation in which the answer, 700, has been rounded to the nearest ten, then it has two significant digits. There are ways to write integers like 700 so that the number of significant digits can be determined exactly. One way is with scientific notation. However, to simplify things, we will assume that integers have the greatest possible number of significant digits. In the case of 700, that number is three.

Sometimes, with angles, the nearest degree is not accurate enough. Some professions, such as Civil Engineering and Land Surveying, use a more precise system of measureing angles that further divides a degree into minutes, denoted by the symbol '. There are 60 minutes in a degree. For example, 36°21' is 36 degrees and 21 minutes.

The relationship between the accuracy of the sides of a triangle and the accuracy of the angles in the same triangle is shown in Table 1.

Chapter 5 Geometry

TABLE 1

Accuracy of sides	Accuracy of Angles
Two significant digits	Nearest degree
Three significant digits	Nearest 10 minutes or $\frac{1}{6}$ of a degree
Four significant digits	Nearest minute or $\frac{1}{60}$ of a degree

B Solving Right Triangles

We solve a right triangle by using the information given about it to find all of the missing sides and angles. In all of the examples and in the Problem Set that follows, we will assume C is the right angle in all of our right triangles, unless otherwise noted.

Unless stated otherwise, we round our answers so that the number of significant digits in our answers matches the number of significant digits in the least significant number given in the original problem. Also, we round our answers only but not any of the numbers in the intermediate steps. Finally, we are showing the values of the trigonometric functions to four significant digits simply to avoid cluttering the page with long decimal numbers. This does not mean that you should stop halfway through a problem and round the values of trigonometric functions to four significant digits before continuing.

EXAMPLE 1 In right triangle ABC, $A = 40°$ and $c = 12$ centimeters. Find a, b, and B.

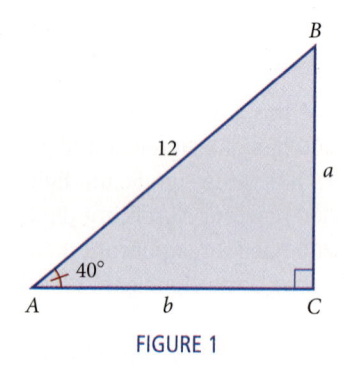

FIGURE 1

SOLUTION We begin by making a diagram of the situation (Figure 1). The diagram is very important because it lets us visualize the relationship between the given information and the information we are asked to find.

To find B, we use the fact that the sum of the two acute angles in any right triangle is 90°.

$$B = 90° - A$$
$$= 90° - 40°$$
$$B = 50°$$

To find a, we can use the formula for sin A.

$$\sin A = \frac{a}{c}$$

PRACTICE PROBLEMS

1. In right triangle ABC, $A = 35°$ and $a = 4$ inches. Find b, c, and B

5.9 Solving Right Triangles

Multiplying both sides of this formula by c and then substituting in our given values of A and c we have

$a = c \sin A$
$ = 12 \sin 40°$
$ = 12(0.6428)$ sin 40° = 0.6428
$a = 7.7$ cm   Answer rounded to two significant digits

There is more than one way to find b.

Using $\cos A = \dfrac{b}{c}$, we have   Using the Pythagorean Theorem, we have

$b = c \cos A$ $c^2 = a^2 + b^2$
$ = 12 \cos 40°$ $b = \sqrt{c^2 - a^2}$
$ = 12(0.7660)$ $ = \sqrt{12^2 - (7.7)^2}$
$b = 9.2$ cm $ = \sqrt{144 - 59.29}$
$ = \sqrt{84.71}$
$ b = 9.2$

In Example 2, we are given two sides and asked to find the remaining parts of a right triangle.

EXAMPLE 2 In right triangle ABC, $a = 2.73$ and $b = 3.41$. Find the remaining side and angles.

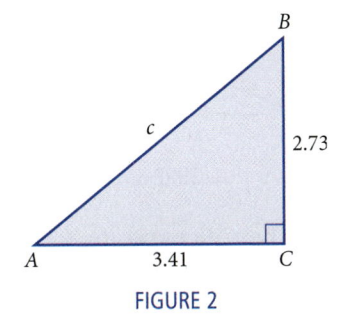

FIGURE 2

SOLUTION Figure 2 is a diagram of the triangle. We can find A by using the formula for $\tan A$.

$\tan A = \dfrac{a}{b}$
$ = \dfrac{2.73}{3.41}$
$\tan A = 0.8006$

Now, to find A, we use a calculator.

$A = \tan^{-1}(0.8006) = 38.7°$

Next we find B.

$B = 90.0° - A$
$ = 90.0° - 38.7°$
$B = 51.3°$

2. If $a = 2.32$ and $c = 4.10$, find the remaining side and angles

Answer
1. $b = 5.7$ in., $c = 7.0$ in., $B = 55°$

Chapter 5 Geometry

Notice we are rounding each angle to the nearest tenth of a degree since the sides we were originally given have three significant digits.

We can find c using the Pythagorean Theorem or one of our trigonometric functions. Let's start with a trigonometric function.

If $\quad \sin A = \dfrac{a}{c}$

then $\quad c = \dfrac{a}{\sin A}$ Multiply each side by c, then divide each side by $\sin A$

$\quad\quad = \dfrac{2.73}{\sin 38.7°}$

$\quad\quad = \dfrac{2.73}{0.6252}$

$\quad\quad = 4.37$ To three significant digits

Using the Pythagorean Theorem, we obtain the same result.

If $\quad c^2 = a^2 + b^2$

then $\quad c = \sqrt{a^2 + b^2}$

$\quad\quad = \sqrt{(2.73)^2 + (3.41)^2}$

$\quad\quad = \sqrt{19.081}$

$\quad\quad = 4.37$

EXAMPLE 3 The circle in Figure 3 has its center at C and a radius of 18 inches. If triangle ADC is a right triangle and A is 35°, find x, the distance from A to B.

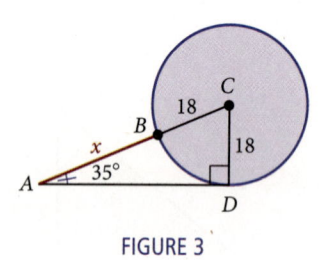

FIGURE 3

SOLUTION In triangle ADC, the side opposite A is 18 and the hypotenuse is $x + 18$. We can use $\sin A$ to write an equation that will allow us to solve for x.

$\sin 35° = \dfrac{18}{x + 18}$

$(x + 18) \sin 35° = 18$ Multiply each side by $x + 18$

$x + 18 = \dfrac{18}{\sin 35°}$ Divide each side by $\sin 35°$

$x = \dfrac{18}{\sin 35°} - 18$ Subtract 18 from each side

$\quad = \dfrac{18}{0.5736} - 18$

$\quad = 13$ inches To two significant digits

3. Find x if the circle in Example 3 has a radius of 6 feet and A is 40°.

Answers
2. $A = 34°, B = 56°, b = 3.4$
3. 3.3 feet

5.9 Solving Right Triangles

EXAMPLE 4 In Figure 4, the distance from A to D is 32 feet. Use the information in Figure 4 to solve for x, the distance between D and C.

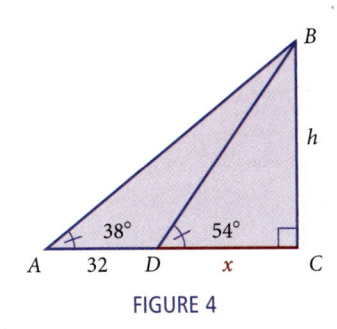

FIGURE 4

SOLUTION To find x we write two equations, each of which contains the variables x and h. Then we solve each equation for h and set the two expressions for h equal to each other.s

Two equations involving both x and h
$$\begin{cases} \tan 54° = \dfrac{h}{x} \Rightarrow h = x \tan 54° \\ \tan 38° = \dfrac{h}{x+32} \Rightarrow h = (x+32) \tan 38° \end{cases}$$ Solve each equation for h

Setting the two expressions for h equal to each other gives us an equation that involves only x.

Since $h = h$

we have $x \tan 54° = (x + 32) \tan 38°$

$x \tan 54° = x \tan 38° + 32 \tan 38°$ Distributive property

$x \tan 54° - x \tan 38° = 32 \tan 38°$ Subtract $x \tan 38°$ from each side

$x(\tan 54° - \tan 38°) = 32 \tan 38°$ Factor x from each term on the left side

$x = \dfrac{32 \tan 38°}{\tan 54° - \tan 38°}$ Divide each side by the coefficient of x

$= \dfrac{32(0.7813)}{1.3764 - 0.7813}$

$= 42$ ft   To two significant digits

EXAMPLE 5 In the introduction to this section, we gave some of the facts associated with the first Ferris wheel. Figure 5 is a simplified model of that Ferris wheel. If θ is the central angle formed as a rider moves from position P_0 to position P_1, find the rider's height above the ground h when θ is 45.0°.

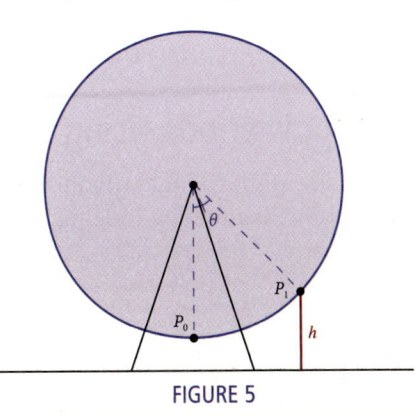

FIGURE 5

4. Find x in Example 4 if the distance from A to D is 26 feet and the given angles are the same.

5. Using the information from Example 5, find the rider's height above the ground when θ is 72.0°.

Answer
4. 34 feet

SOLUTION We know from the introduction to this section that the diameter of the first Ferris wheel was 250 feet, which means the radius was 125 feet. Since the top of the wheel was 264 feet above the ground, the distance from the ground to the bottom of the wheel was 14 feet (the distance to the top minus the diameter of the wheel). To form a right triangle, we draw a horizontal line from P_1 to the vertical line connecting the center of the wheel with P_0. This information is shown in Figure 6.

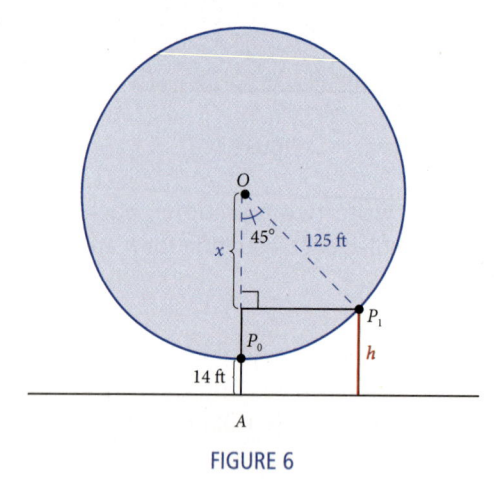

FIGURE 6

The key to solving this problem is recognizing that x is the difference between OA (the distance from the center of the wheel to the ground) and h. Since OA is 139 feet (the radius of the wheel plus the distance between the bottom of the wheel and the ground: $125 + 14 = 139$), we have

$$x = 139 - h$$

We use a cosine ratio to write an equation that contains h.

$$\cos 45.0° = \frac{x}{125}$$

$$= \frac{139 - h}{125}$$

Solving for h we have

$$125 \cos 45.0° = 139 - h$$

$$h = 139 - 125 \cos 45.0°$$

$$= 139 - 125(0.7071)$$

$$= 139 - 88.4$$

$$= 50.6 \text{ ft}$$

If $\theta = 45.0°$, a rider at position P_1 is $\frac{1}{8}$ of the way around the wheel. At that point, the rider is approximately 50.6 feet above the ground.

C Applications Using Trigonometric Functions

We can use right triangle trigonometry to solve a variety of problems. We will see how this is done by looking at a number of applications of right triangle trigonometry.

Answer
5. 100 feet

5.9 Solving Right Triangles

EXAMPLE 6 The two equal sides of an isosceles triangle are each 24 centimeters. If each of the two equal angles measures 52°, find the length of the base and the altitude.

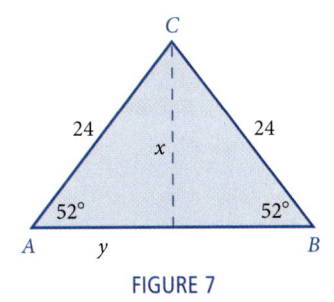

FIGURE 7

SOLUTION An isosceles triangle is any triangle with two equal sides. The angles opposite the two equal sides are called the base angles, and they are always equal. Figure 7 is a picture of our isosceles triangle.

We have labeled the altitude x. We can solve for x using a sine ratio.

If $\quad \sin 52° = \dfrac{x}{24}$

then $\quad x = 24 \sin 52°$

$\quad\quad\quad = 24(0.7880)$

$\quad\quad\quad = 19$ cm   Rounded to 2 significant digits

We have labeled half the base with y. To solve for y, we can use a cosine ratio.

If $\quad \cos 52° = \dfrac{y}{24}$

then $\quad y = 24 \cos 52°$

$\quad\quad\quad = 24(0.6157)$

$\quad\quad\quad = 14.8$ cm

The base is $2y = 2(14.8) = 30$ cm to the nearest centimeter.

6. Find the base and altitude of the triangle in Example 6, if the equal sides are 14 inches and the equal angles are 16°.

Answer
6. altitude = 3.9 inches, base = 27 inches

Left Blank Intentionally

Problem Set 5.9

Problems 1 through 14 refer to right triangle ABC with C = 90°. Begin each problem by drawing a picture of the triangle with both the given and asked for information labeled appropriately. Also, write your answers for angles in decimal degrees.

1. If $A = 42°$ and $c = 15$ ft, find a.

2. If $A = 42°$ and $c = 89$ cm, find b.

3. If $A = 34°$ and $a = 22$ m, find c.

4. If $A = 34°$ and $b = 55$ m, find c.

5. If $B = 24.5°$ and $c = 2.34$ ft, find a.

6. If $B = 16.9°$ and $c = 7.55$ cm, find b.

7. If $B = 55.33°$ and $b = 12.34$ yd, find a.

8. If $B = 77.66°$ and $a = 43.21$ inches, find b.

9. If $a = 16$ cm and $b = 26$ cm, find A.

10. If $a = 42.3$ inches and $b = 32.4$ inches, find B.

11. If $b = 6.7$ m and $c = 7.7$ m, find A.

12. If $b = 9.8$ mm and $c = 12$ mm, find B.

13. If $c = 45.54$ ft and $a = 23.32$ ft, find B.

14. If $c = 5.678$ ft and $a = 4.567$ ft, find A.

Problems 15 through 32 refer to right triangle ABC with C = 90°. In each case, solve for all the missing parts using the given information. (In Problems 27 through 32, write your angles in decimal degrees.)

15. $A = 25°$, $c = 24$ m

16. $A = 41°$, $c = 36$ m

17. $A = 32.6°$, $a = 43.4$ inches

18. $A = 48.3°$, $a = 3.48$ inches

19. $A = 10° 42'$, $b = 5.932$ cm

20. $A = 66° 54'$, $b = 28.28$ cm

21. $B = 76°$, $c = 5.8$ ft

22. $B = 21°$, $c = 4.2$ ft

23. $B = 26° 30'$, $b = 324$ mm

24. $B = 53° 30'$, $b = 725$ mm

25. $B = 23.45°, a = 5.432$ mi

26. $B = 44.44°, a = 5.555$ mi

27. $a = 37$ ft, $b = 87$ ft

28. $a = 91$ ft, $b = 85$ ft

29. $a = 2.75$ cm, $c = 4.05$ cm

30. $a = 62.3$ cm, $c = 73.6$ cm

31. $b = 12.21$ inches, $c = 25.52$ inches

32. $b = 377.3$ inches, $c = 588.5$ inches

In Problems 33 and 34, use the information given in the diagram to find A to the nearest degree.

33.

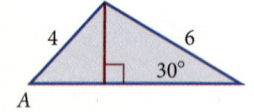

34.

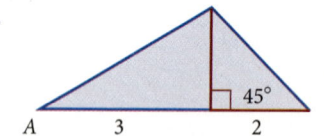

The circle in Figure 18 has a radius of r and center at C. The distance from A to B is x. For Problems 35 through 38, redraw Figure 18, label it as indicated in each problem, and then solve the problem.

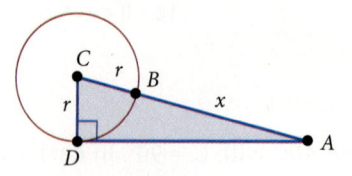

FIGURE 18

35. If $A = 31°$ and $r = 12$, find x.

36. If $C = 26°$ and $r = 20$, find x.

37. If $A = 45°$ and $x = 15$, find r.

38. If $C = 65°$ and $x = 22$, find r.

Figure 19 shows two right triangles drawn at 90° to each other. For Problem 39 through 42, redraw Figure 19, label it as the problem indicates, and then solve the problem.

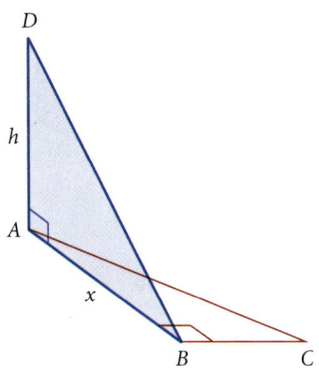

FIGURE 19

39. If $\angle ABD = 27°$, $C = 62°$, and $BC = 42$, find x and then find h.

40. If $\angle ABD = 53°$, $C = 48°$, and $BC = 42$, find x and then find h.

41. If $AC = 32$, $h = 19$, and $C = 41°$, find $\angle ABD$.

42. If $AC = 19$, $h = 32$, and $C = 49°$, find $\angle ABD$.

In Figure 20, the distance from A to D is y, the distance from D to C is x, and the distance from C to B is h. Use Figure 20 to solve Problems 43 through 48.

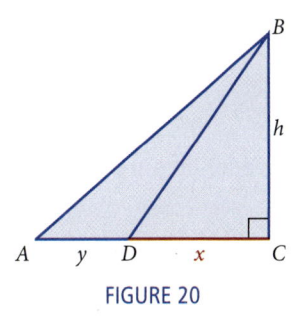

FIGURE 20

43. If $A = 41°$, $\angle BDC = 58°$, $AB = 18$, and $DB = 14$, find x, then y.

44. If $A = 32°$, $\angle BDC = 48°$, $AB = 17$, and $DB = 12$, find x, then y.

45. If $A = 41°$, $\angle BDC = 58°$, and $AB = 28$, find h, then x.

46. If $A = 32°$, $\angle BDC = 48°$, and $AB = 56$, find h, then x.

47. If $A = 43°$, $\angle BDC = 57°$, and $y = 10$, find x.

48. If $A = 32°$, $\angle BDC = 41°$, and $y = 14$, find x.

FIGURE 21

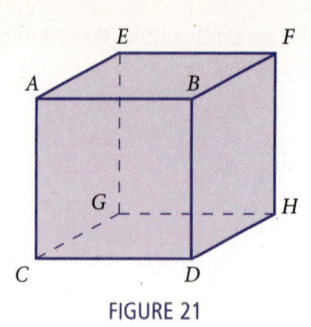

49. Suppose each edge of the cube shown in Figure 21 is 5 inches long. Find the measure of the angle formed by diagonals CF and CH. Round your answer to the nearest tenth of a degree.

50. Suppose each edge of the cube shown in Figure 21 is 3 inches long. Find the measure of the angle formed by diagonals DE and DG. Round your answer to the nearest tenth of a degree.

51. Suppose each edge of the cube shown is x inches long. Find the measure of the angle formed by diagonals CF and CH in Figure 21. Round your answer to the nearest tenth of a degree.

52. Suppose each edge of the cube shown is y inches long. Find the measure of the angle formed by diagonals DE and DG in Figure 21. Round your answer to the nearest tenth of a degree.

Repeat Example 5 from this section for the following values of θ.

53. $\theta = 30°$ **54.** $\theta = 60°$ **55.** $\theta = 120°$ **56.** $\theta = 135°$

C Applying the Concepts

Solve each of the following problems. In each case, be sure to make a diagram of the situation with all the given information labeled.

57. Geometry The two equal sides of an isosceles triangle are each 42 centimeters. If the base measures 30 centimeters, find the height and the measure of the two equal angles.

58. Geometry An equilateral triangle (one with all sides the same length) has an altitude of 4.3 inches. Find the length of the sides.

59. Geometry The height of a right circular cone is 25.3 centimeters. If the diameter of the base is 10.4 centimeters, what angle does the side of the cone make with the base (Figure 22)?

60. Geometry The diagonal of a rectangle is 348 millimeters, while the longer side is 278 millimeters. Find the shorter side of the rectangle and the angles the diagonal makes with the sides.

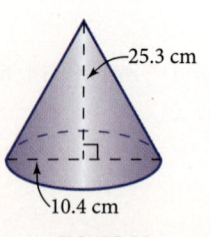

FIGURE 22

61. Length of an Escalator How long should an escalator be if it is to make an angle of 33° with the floor and carry people a vertical distance of 21 feet between floors?

62. Height of a Hill A road up a hill makes an angle of 5.1° with the horizontal. If the road from the bottom of the hill to the top of the hill is 2.5 miles long, how high is the hill?

63. Length of a Rope A 72.5-foot rope from the top of a circus tent pole is anchored to the ground 43.2 feet from the bottom of the pole. What angle does the rope make with the pole? (Assume the pole is perpendicular to the ground.)

64. Angle of a Ladder A ladder is leaning against the top of a 7.0-foot wall. If the bottom of the ladder is 4.5 feet from the wall, what is the angle between the ladder and the wall?

Left Blank Intentionally

Area and Perimeter

5.10

Objectives
A Find the perimeter of a polygon.
B Find the circumference of a circle.
C Find the area of a polygon.
D Find the area of a circle.

A Perimeter

We begin this section by reviewing the definition of a polygon, and the definition of perimeter.

> **Definition**
> A **polygon** is a closed geometric figure, with at least three sides, in which each side is a straight line segment.

The most common polygons are squares, rectangles, and triangles.

> **Definition**
> The **perimeter** of any polygon is the sum of the lengths of the sides, and it is denoted with the letter P.

To find the perimeter of a polygon we add all the lengths of the sides together.
  Here are the most common polygons, along with the formula for the perimeter of each.

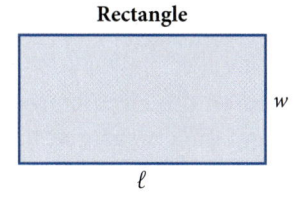

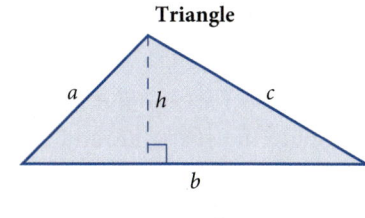

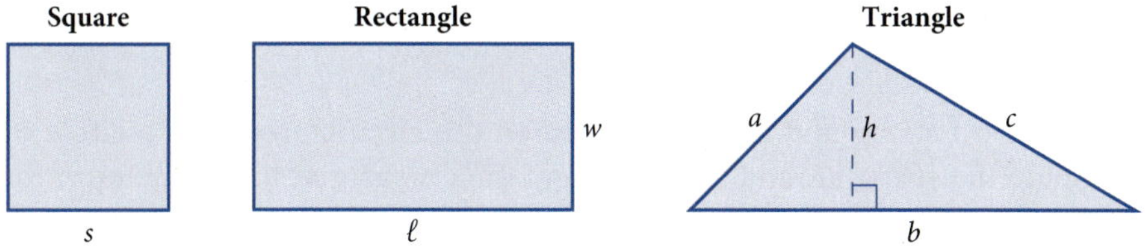

Square
s
$P = 4s$

Rectangle
ℓ, w
$P = 2\ell + 2w$

Triangle
a, h, b, c
$P = a + b + c$

We can justify our formulas as follows. If each side of a square is s units long, then the perimeter is found by adding all four sides together:

$$\text{Perimeter} = P = s + s + s + s = 4s$$

Likewise, if a rectangle has a length of l and a width of w, then to find the perimeter we add all four sides together:

$$\text{Perimeter} = P = \ell + \ell + w + w = 2\ell + 2w$$

EXAMPLE 1
Find the perimeter of the given rectangle.

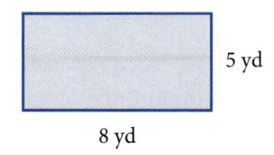

5 yd
8 yd

SOLUTION The given rectangle has a width of 5 yards and a length of 8 yards. We can use the formula for $P = 2\ell + 2w$ to find the perimeter.

We have: $P = 2(8) + 2(5)$
$= 16 + 10$
$= 26$ yards

PRACTICE PROBLEMS
1. Find the perimeter of the rectangle in Example 1 if the length and width are each increased by 1 yard.

Answer
1. 30 yards

5.10 Area and Perimeter

B Circumference

The *circumference* of a circle is the distance around the outside, just as the perimeter of a polygon is the distance around the outside. The circumference of a circle can be found by measuring its radius or diameter and then using the appropriate formula. The *radius* of a circle is the distance from the center of the circle to the circle itself. The radius is denoted by the letter r. The *diameter* of a circle is the distance from one side to the other, through the center. The diameter is denoted by the letter d. In the figure below we can see that the diameter is twice the radius, or

$$d = 2r$$

The relationship between the circumference and the diameter or radius is not as obvious. As a matter of fact, it takes some fairly complicated mathematics to show just what the relationship between the circumference and the diameter is.

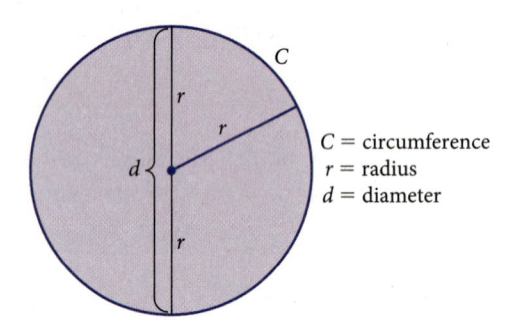

C = circumference
r = radius
d = diameter

If you took a string and actually measured the circumference of a circle by wrapping the string around the circle and then measured the diameter of the same circle, you would find that the ratio of the circumference to the diameter, $\frac{C}{d}$, would be approximately equal to 3.14. The actual ratio of C to d in any circle is an irrational number. It can't be written in decimal form. We use the symbol π (Greek *pi*) to represent this ratio. In symbols, the relationship between the circumference and the diameter in any circle is

$$\frac{C}{d} = \pi$$

Knowing what we do about the relationship between division and multiplication, we can rewrite this formula as

$$C = \pi d$$

This is the formula for the circumference of a circle. When we do the actual calculations, we will use the approximation 3.14 for π.

Because $d = 2r$, the same formula written in terms of the radius is

$$C = 2\pi r$$

EXAMPLE 2 Find the circumference of each coin.

a. 1 Euro coin (Round to the nearest whole number.)

Diameter = 23.25 millimeters

2. Find the circumference for each coin in Example 2 from the dimensions given below. Round answers to the nearest hundredth.

   a. Diameter = 0.92 inches
   b. Radius = 13.20 millimeters

b. Susan B. Anthony dollar (Round to the nearest hundredth.)

Radius = 0.52 inch

SOLUTION Applying our formulas for circumference we have:

a. $C = \pi d \approx (3.14)(23.25) \approx 73$ mm

b. $C = 2\pi r \approx 2(3.14)(0.52) \approx 3.27$ in.

Circles and the Earth

There are many circles modeled on the surface of the earth. The most familiar are the latitude and longitude lines. Of these circles, the ones with the largest circumference are called *great circles*. All of the longitude lines are great circles. Of the lattitude lines, only the equator is a great circle.

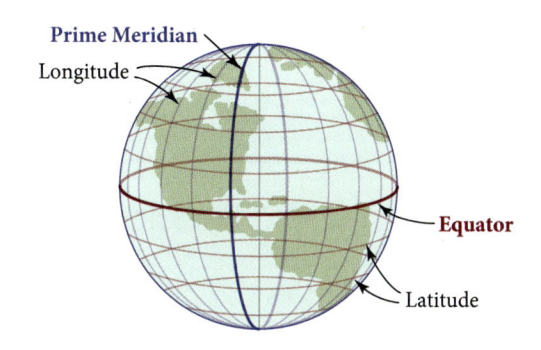

EXAMPLE 3 If the circumference of the earth is approximately 24,900 miles at the equator, what is the diameter of the earth to the nearest 10 miles?

SOLUTION We substitute 24,900 for C in the formula $C = \pi d$, and then we solve for d.

$24,900 = \pi d$

$24,900 \approx 3.14d$ Substitute 3.14 for π

$d \approx \dfrac{24,900}{3.14}$ Divide each side by 3.14

$d \approx 7,930$ miles

3. The circumference of the moon at the equator is 6,790 miles. To the nearest 10 miles, find the diameter of the moon

Answers
2. **a.** 2.89 in.   **b.** 82.90 mm
3. Approximately 2,160 miles

C Area of a Polygon

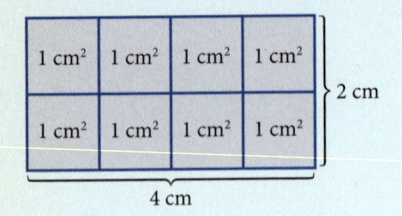

Recall that the area of a flat object is a measure of the amount of surface the object has. The area of the rectangle in the margin is 8 square centimeters, because it takes 8 square centimeters to cover it.

As we have noted previously, the area of this rectangle can also be found by multiplying the length and the width.

$$\begin{aligned} \text{Area} &= (\text{length}) \cdot (\text{width}) \\ &= (4 \text{ centimeters}) \cdot (2 \text{ centimeters}) \\ &= (4 \cdot 2) \cdot (\text{centimeters} \cdot \text{centimeters}) \\ &= 8 \text{ square centimeters} \end{aligned}$$

From this example, and others, we conclude that the area of any rectangle is the product of the length and width.

Here are the most common geometric figures along with the formula for the area of each one. The only formulas that are new to us are the ones that accompany the parallelogram and the circle.

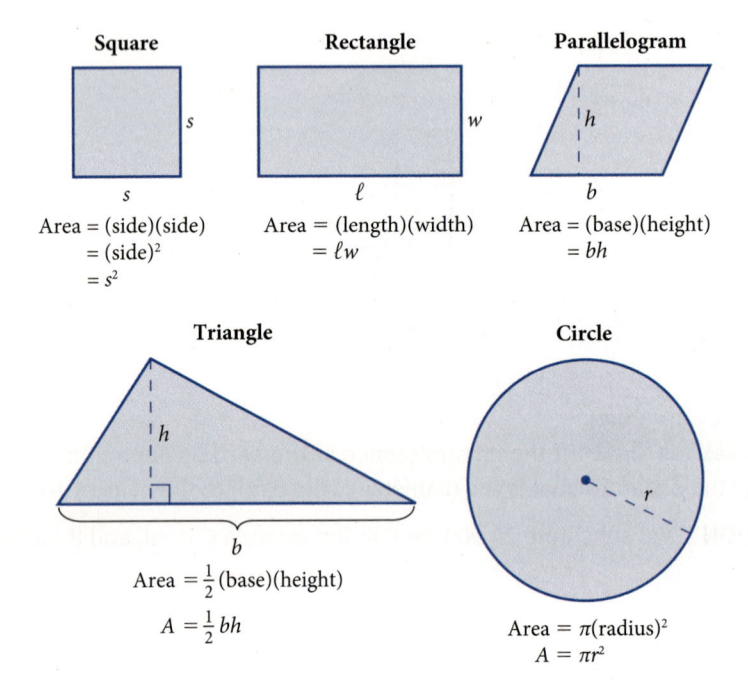

Square
Area = (side)(side)
    = (side)^2
    = s^2

Rectangle
Area = (length)(width)
    = ℓw

Parallelogram
Area = (base)(height)
    = bh

Triangle
Area = $\frac{1}{2}$(base)(height)
$A = \frac{1}{2}bh$

Circle
Area = π(radius)^2
$A = \pi r^2$

EXAMPLE 4 The parallelogram below has a base of 5 centimeters and a height of 2 centimeters. Find the area.

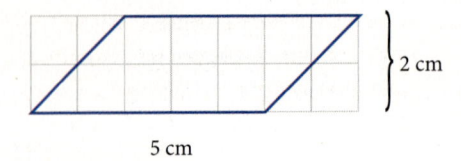

SOLUTION If we apply our formula we have

$$\begin{aligned} \text{Area} &= (\text{base})(\text{height}) \\ A &= bh \\ &= 5 \cdot 2 \\ &= 10 \text{ cm}^2 \end{aligned}$$

4. Find the area.

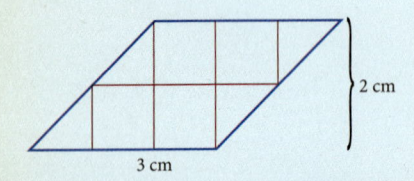

5.10 Area and Perimeter   403

Or, we could simply count the number of square centimeters it takes to cover the object. There are 8 complete squares and 4 half-squares, giving a total of 10 squares for an area of 10 square centimeters. Counting the squares in this manner helps us see why the formula for the area of a parallelogram is the product of the base and the height.

To justify our formula in general, we simply rearrange the parts to form a rectangle.

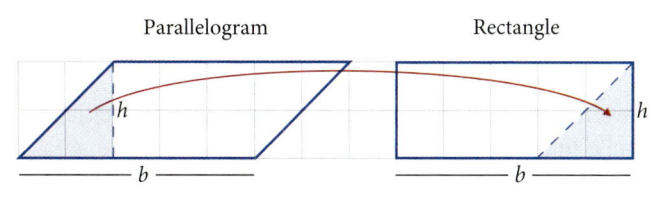

Parallelogram   Rectangle

Move triangle to right side

EXAMPLE 5 The triangle below has a base of 6 centimeters and a height of 3 centimeters. Find the area.

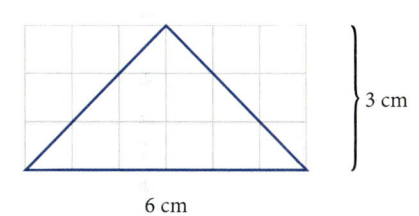

SOLUTION If we apply our formula we have

$$\text{Area} = \frac{1}{2}(\text{base})(\text{height})$$

$$A = \frac{1}{2}bh$$

$$= \frac{1}{2} \cdot 6 \cdot 3$$

$$= 9 \text{ cm}^2$$

As was the case in Example 4, we can also count the number of square centimeters it takes to cover the triangle. There are 6 complete squares and 6 half-squares, giving a total of 9 squares for an area of 9 square centimeters.

D Area of a Circle

EXAMPLE 6 A circle has a radius of 14.5 millimeters. Find the area of the circle to the nearest whole number.

SOLUTION Using our formula for the area of a circle, and using 3.14 for π, we have

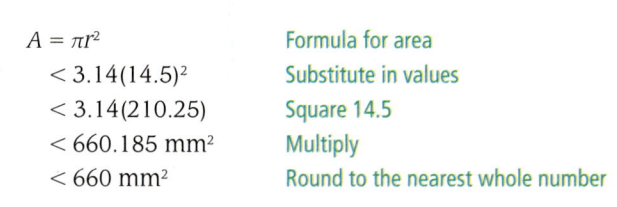

$A = \pi r^2$ Formula for area
$ < 3.14(14.5)^2$ Substitute in values
$ < 3.14(210.25)$ Square 14.5
$ < 660.185 \text{ mm}^2$ Multiply
$ < 660 \text{ mm}^2$ Round to the nearest whole number

5. Find the area.

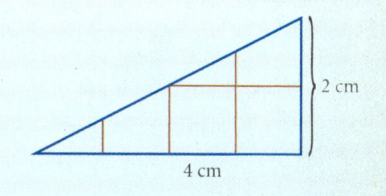

6. A circle has a radius of 5 feet. Find the area.

Answers
4. 6 cm²
5. 4 cm²
6. 78.5 ft²

7. Find the area of the shaded portion of this figure.

10 cm

Chapter 5 Geometry

EXAMPLE 7 Find the area of the shaded portion of this figure.

12 ft

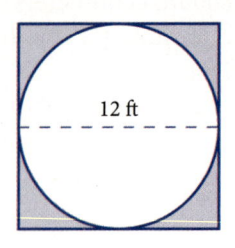

SOLUTION We have a circle inscribed in a square. We notice the diameter of the circle is the same length as one side of the square. To find the area of the shaded region, we subtract the area of the circle from the area of the square as follows:

$$A = 12^2 - 6^2\pi$$
$$= 144 - 36\pi$$
$$< 30.96 \text{ ft}^2$$

Getting Ready for Class

After reading through the preceding section, respond in your own words and in complete sentences.

1. What is the perimeter of a polygon?
2. How are perimeter and circumference related?
3. What does π represent?
4. How do you find the area of a circle?

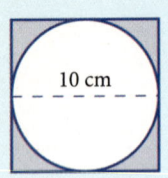

Answers
7. 21.5 cm² to the nearest tenth

Problem Set 5.10

Use the fomulas for area and perimeter of a rectangle to complete the table.

	Length	Width	Area	Perimeter
1.	5 mm		40 mm²	
2.		7 cm		18 cm
3.	6 in.			24 in.
4.	14 cm	3.5 cm		

For problems 5–8, find the area of $\square MNQT$, and state if it is a rectangle and/or square.

5. $\overline{TM} \perp \overline{MN}$, $QT = 9$ cm, $QN = 6$ cm

6. $\overline{TM} \perp \overline{MN}$, $\overline{TM} \cong \overline{MN}$, $QN = 12$ in.

7. $\overline{MQ} \cong \overline{NT}$, $\overline{MQ} \perp \overline{NT}$, and $QT = 4.5$ cm

8. $\overline{MQ} \cong \overline{NT}$, $\overline{MQ} \perp \overline{NT}$, the perimeter is 52 mm

For problems 9–12 refer to rectangle RECT, not drawn to scale.

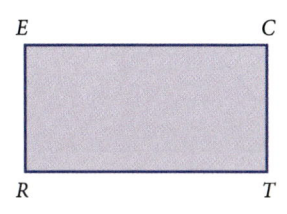

9. If $RE = (2x - 8)$ mm and $RT = (2x + 4)$ mm, and the area of RECT is 64 mm², find RE and RT.

10. If $RT = (4x + 5)$ cm and $RE = (6x - 7)$ cm, and the perimeter of RECT is $(22x - 10)$ cm, find the area, and the perimeter.

11. If $CT = (2x + 3)$ cm and $CE = x$ cm, and RECT has an area of $(7x + 16)$ cm². Find x, the area, and the perimeter.

12. If $CT = x$ in. and $CE = (3x - 5)$ in., and RECT has an area of $(8x - 4)$ in², Find x, the area, and the perimeter of RECT.

For problems 13–16, use the formulas for the area and perimeter of a square and refer to square SQAR, not drawn to scale.

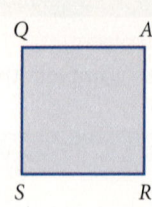

13. If $QA = (2x + 6)$ cm and the perimeter of SQAR is $(11x + 3)$ cm. Find x, the area, and the perimeter of SQAR.

14. If $SR = (3x - 5)$ mm and the perimeter of SQAR is $(9x + 4)$ mm. Find x, the area, and the perimeter of SQAR.

15. If $AR = (x + 5)$ cm and the area of SQAR is $(14x + 21)$ cm². Find x, the area, and the perimeter of SQAR.

16. If $QA = (x - 4)$ in. and the area of SQAR is $(2x + 7)$ in². Find x, the area, and the perimeter of SQAR.

Find the perimeter of each figure.

17.

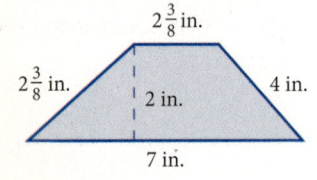

18.

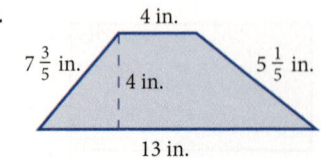

19.

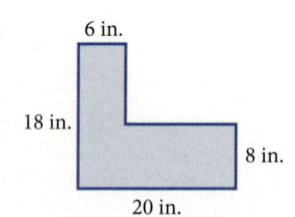

20.

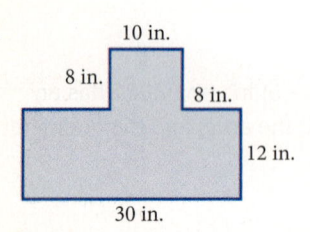

21.

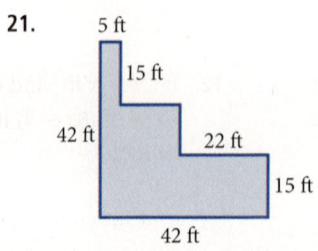

22.

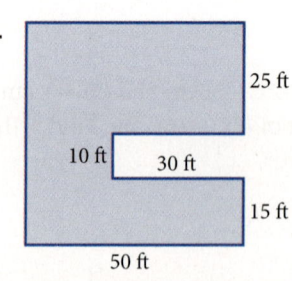

Find the perimeter and area of each figure. Use 3.14 for π.

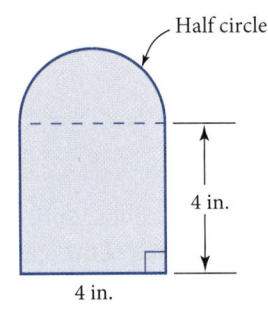

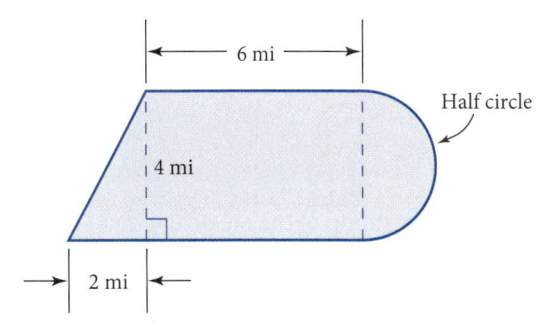

23. Half circle, 4 in., 4 in.

24. 6 mi, Half circle, 4 mi, 2 mi

Find the circumference and area of each circle. Use 3.14 for π.

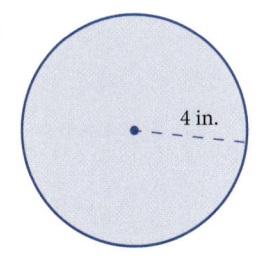

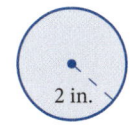

25. 4 in.

26. 2 in.

Applying the Concepts

27. **Art at the Getty.** The painting below has been displayed at the Getty Museum. Find the perimeter.

   $33\frac{3}{4}$ in.
   $55\frac{1}{2}$ in.

28. **Art at the Getty.** The painting below has been displayed at the Getty Museum. Find the perimeter.

   $18\frac{7}{8}$ in.
   $24\frac{7}{8}$ in.

29. **Circumference** A dinner plate has a radius of 6 inches. Find the circumference.

30. **Circumference** A salad plate has a radius of 3 inches. Find the circumference.

31. **Circumference** The radius of the earth is approximately 3,900 miles. Find the circumference of the earth at the equator. (The equator is a circle around the earth that divides the earth into two equal halves.)

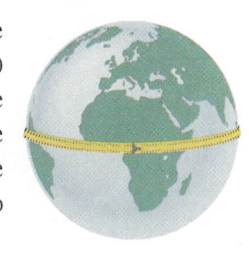

32. **Circumference** The radius of the moon is approximately 1,100 miles. Find the circumference of the moon around its equator.

33. Area A swimming pool is 20 feet wide and 40 feet long. If it is surrounded by square tiles, each of which is 1 foot by 1 foot, how many tiles are there surrounding the pool?

34. Area A garden is rectangular with a width of 8 feet and a length of 12 feet. If it is surrounded by a walkway 2 feet wide, how many square feet of area does the walkway cover?

35. Area of a Stamp A stamp of the Mexican artist Frida Kahlo was issued in 2001. The image area of the stamp has a width of 0.84 inches and a length of 1.41 inches. Find the area of the image. Round to the nearest hundredth.

36. Area of a Stamp A stamp of the Italian scientist Enrico Fermi was issued in 2001. The image area of the stamp has a width of 21.4 millimeters and a length of 35.8 millimeters. Find the area of the image. Round to the nearest whole number.

37. Area of a Euro The 10-euro banknote shown below has a width of 67 millimeters and a length of 127 millimeters. Find the area.

38. Area of a Dollar The $10 bill shown below has a width of 2.56 inches and a length of 6.14 inches. Find the area. Round to the nearest hundredth.

39. Comparing Areas The side of a square is 5 feet long. If all four sides are increased by 2 feet, by how much is the area increased?

40. Comparing Areas The length of a side in a square is 20 inches. If all four sides are decreased by 4 inches, by how much is the area decreased?

41. The circle here is said to be *inscribed* in the square. If the area of the circle is 64π square centimeters, find the length of one of the diagonals of the square (the distance from *A* to *D*). Round to the nearest tenth.

42. The painting below is from the Getty Museum. Find the diameter, circumference, and area of the circle that is inside of the square.

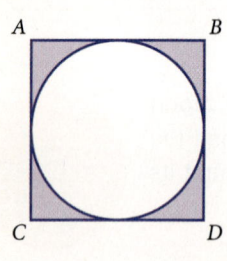

$17\frac{7}{8}$ in.

Chapter 5 Summary

■ Complementary and Supplementary Angles [5.1]

Two angles that add to 90° are called *complementary angles*. Two angles that add to 180° are called *supplementary angles*.

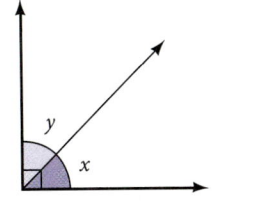

Complementary angles: $x + y = 90°$ 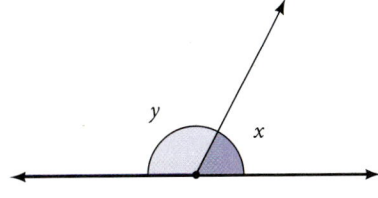   Supplementary angles: $x + y = 180°$

■ Vertical and Adjacent Angles [5.1]

Two angles are *vertical angles* if and only if they are non-adjacent and formed by intersecting lines.
   Two angles are *adjacent angles* if and only if they share a common ray and a common vertex.

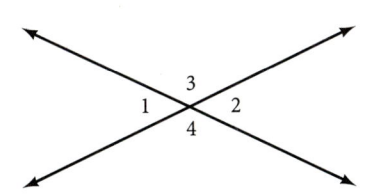

$\angle 1$ and $\angle 2$ are verical angles.
$\angle 1$ and $\angle 3$ are adjacent angles.

■ Parallel and Perpendicular Lines [5.2]

Two or more lines that lie in the same plane are *parallel lines* if they have no points in common.
   Two lines are *perpendicular* if and only if a right angle is formed at their intersection.

■ Theorem for Parallel lines [5.2]

If $\ell_1 \parallel \ell_2$, then,

    Corresponding angles are congruent
    Alternate interior angles are congruent
    Alternate exterior angles are congruent
    Consecutive interior angles are supplementary

Triangles [5.3]

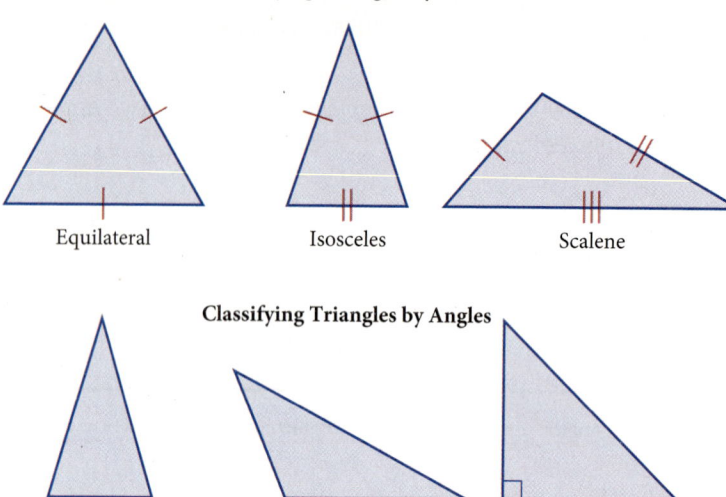

Classifying Triangles by Sides

Equilateral   Isosceles   Scalene

Classifying Triangles by Angles

Acute   Obtuse   Right

For all triangles, the sum of the interior angles is 180°.

Congruent Triangles [5.4]

Two triangles are *congruent* if all of their corresponding parts are the same size.

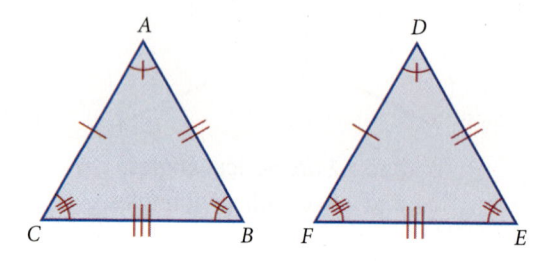

Test for Congruent Triangles [5.4]

Postulate (ASA)
If two angles and the included side in one triangle are congruent to two angles and the included side in another triangle, then the triangles must be congruent.

Postulate (AAS)
If two angles and the non-included side in one triangle are congruent to the corresponding two angles and non-included side in another triangle, then the two triangles are congruent.

Postulate (SAS)
If two angles and the included angle in one triangle are congruent to two sides and the included angle in another triangle, then the triangles must be congruent.

Postulate (SSS)
If three sides of one triangle are congruent to the three sides in another triangle, then the triangles must be congruent.

Polygons [5.5]

The sum of the interior angles of a polygon is given by the formula
$$180(n - 2)$$

Parallelograms [5.5]

Parallelograms are quadrilaterals with the following characteristics:
1. Both pairs of opposite sides are parallel.
2. Opposite sides are congruent.
3. Opposite angles are congruent.
4. The diagonals bisect each other.

Similar Figures [5.6]

Two triangles that have the same shape are similar when their corresponding sides are proportional, or have the same ratio. The triangles below are similar.

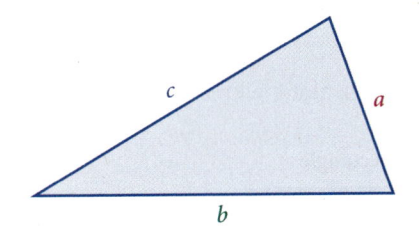

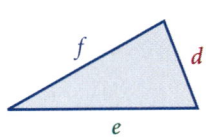

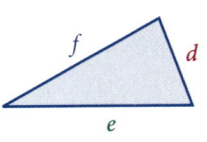

Find x.

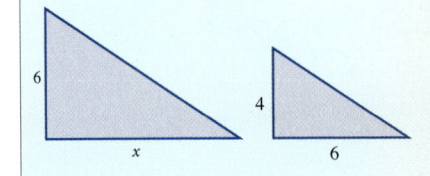

$$\frac{4}{6} = \frac{6}{x}$$
$$36 = 4x$$
$$9 = x$$

Corresponding Sides   **Ratio**

side a corresponds with side d $\frac{a}{d}$

side b corresponds with side e $\frac{b}{e}$

side c corresponds with side f $\frac{c}{f}$

Because their corresponding sides are proportional, we write
$$\frac{a}{d} = \frac{b}{e} = \frac{c}{f}$$

Pythagorean Theorem [5.7]

In any right triangle, the length of the longest side (the hypotenuse) is equal to the square root of the sum of the squares of the two shorter sides.

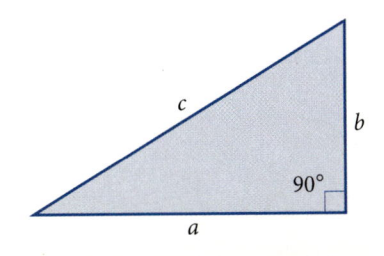

$$c = \sqrt{a^2 + b^2}$$

Trigonometric Functions [5.8]

Definition
If triangle ABC is a right triangle with C = 90° (Figure 1), then the six trigonometric functions for A are defined as follows:

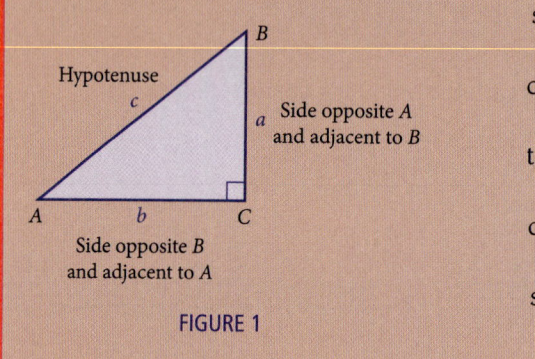

$$\sin A = \frac{\text{side opposite } A}{\text{hypotenuse}} = \frac{a}{c}$$

$$\cos A = \frac{\text{side adjacent } A}{\text{hypotenuse}} = \frac{b}{c}$$

$$\tan A = \frac{\text{side opposite } A}{\text{side adjacent } A} = \frac{a}{b}$$

$$\cot A = \frac{\text{side adjacent } A}{\text{side opposite } A} = \frac{b}{a}$$

$$\sec A = \frac{\text{hypotenuse}}{\text{side adjacent } A} = \frac{c}{b}$$

$$\csc A = \frac{\text{hypotenuse}}{\text{side opposite } A} = \frac{c}{a}$$

FIGURE 1

Area and Perimeter [5.10]

Below are two common geometric figures, along with the formulas for their areas.

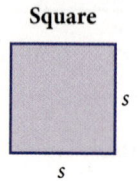

Square

Area = (side)(side)
= (side)^2
= s^2

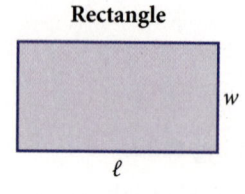

Rectangle

Area = (length)(width)
= ℓw

Perimeter = 2(length) + 2(width)
Perimeter = 4(side) = $2\ell + 2w$
Circumference = 2π(radius) = $4s = 2\pi r$

Circle

Area = π(radius)^2
= πr^2

Chapter 5 Review

B is the midpoint of $\overline{AC}$. Find x. [5.1]

1. B is the midpoint of $\overline{AC}$. Find x.

   A —— $x+10$ —— B —— $-3(x+2)$ —— C

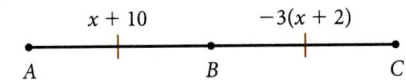

2. B is the midpoint of $\overline{AC}$. Find x.

   A —— $4x$ —— B —— $6x-10$ —— C

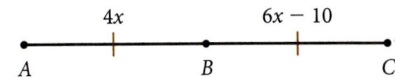

Solve for x, and find the two angles. [5.1]

3. Find x and the measure of angles 1 and 2 given the following:

   $\angle 1 = 11x + 13$
   $\angle 2 = 14x - 5$

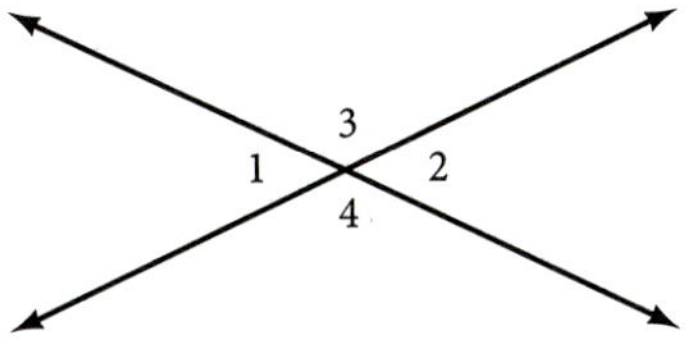

4. $\angle ABC$ is a right angle.
   Find x and the measure of $\angle ABD$ and $\angle DBC$.

   $8x + 2$
   $3x + 11$

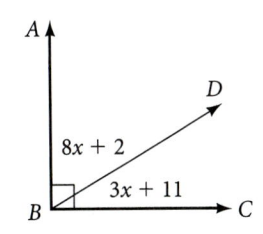

[5.2]

5. $4x + 4$, $6x - 12$, lines ℓ_1 and ℓ_2 with transversal t.

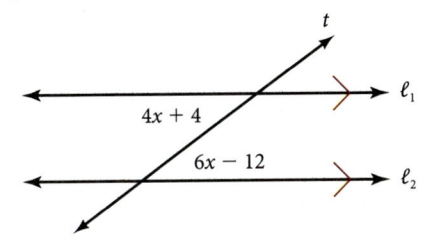

6. $2x^2$, $16x + 18$, lines ℓ_1 and ℓ_2 with transversal t.

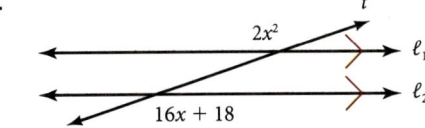

Solve for x. [5.3]

7. Triangle with $x^2 + 3$ at B, $7x + 3$ at A, and 84° at C.

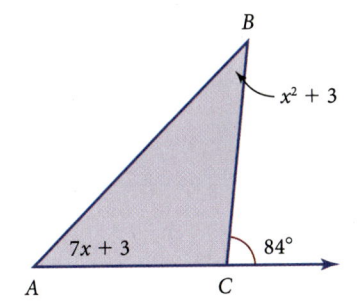

8. Triangle with 98° at A, $6x + 3$ at D, $5x - 4$ at C.

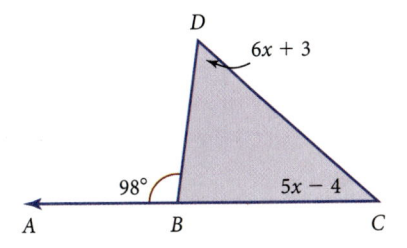

9. Triangle ABC with $5x - 6$ at C, $3x + 5$ at A, sides CA and CB marked equal.

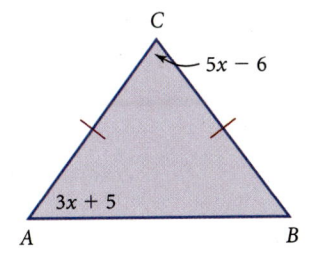

10. Triangle with $4x - 11$ at A side, $2x + 5$ at C side, sides AB and CB marked equal.

Chapter 5 Review   413

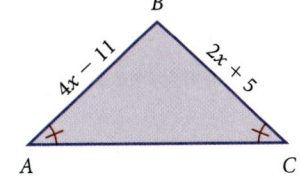

Chapter 5 Geometry

Give the reason for congruence and find x and y. [5.4]

11. Triangle ABC with AB = 3x + 10, and triangle DEF with DE = 5x − 6.

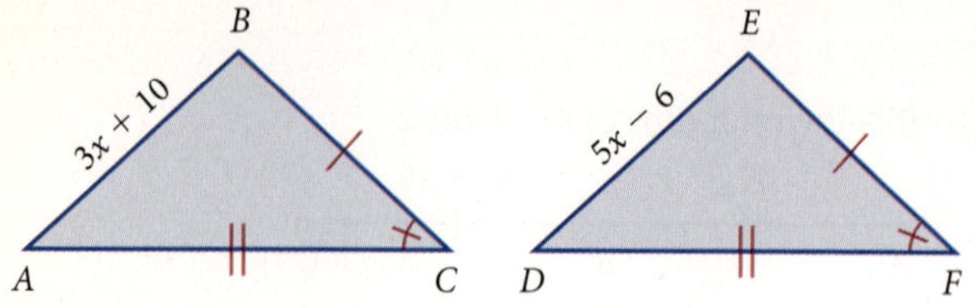

12. Quadrilateral with DC = 12, DA = 10, AB = x, CB = y.

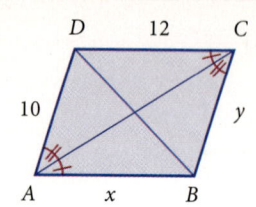

Triangle ABC is similar to △DEC. Solve for x. [5.6]

13. Small triangle with CB = 4, AB = x + 3. Large triangle with CE = 12, DE = 4x + 3.

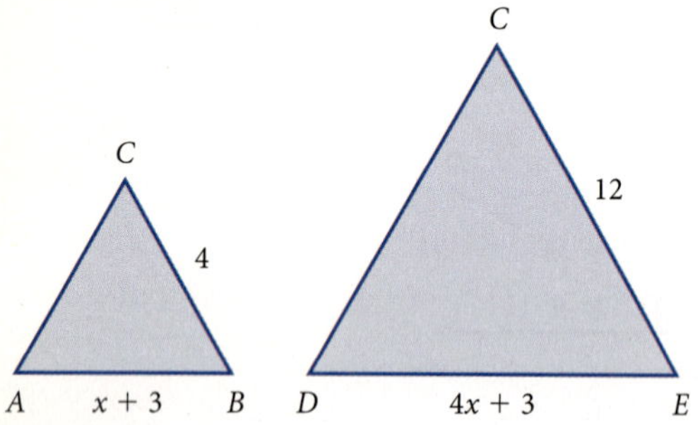

Find the length of the hypotenuse in each right triangle. Round your answers to the nearest tenth. [5.7]

14. Right triangle with legs 5 ft and 2 ft.

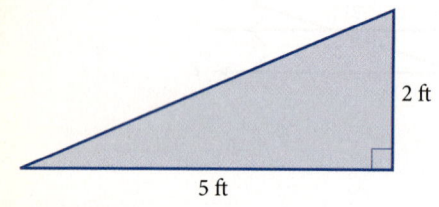

15. Right triangle with legs 5 ft and 4 ft.

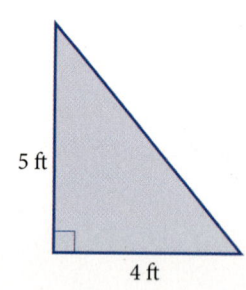

Solve for x. Round to the nearest tenth. [5.8]

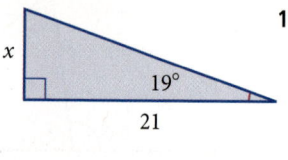

16. Right triangle with hypotenuse 13, angle 56°, opposite side x.

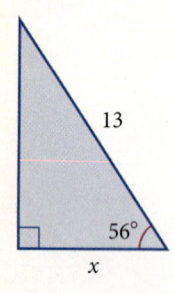

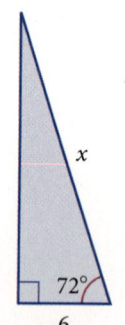

17. Right triangle with side x, hypotenuse 21, angle 19°.

18. Right triangle with side x, base 6, angle 72°.

19. Right triangle with side 4, hypotenuse x, angle 28°.

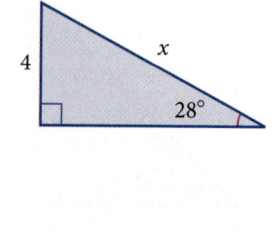

Find x and the area if the perimeter is 46. [5.10]

20. Rectangle with width 2x and length 3x + 3.

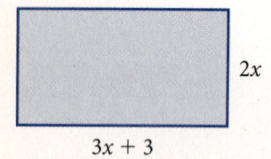

Find the perimeter and area of the shaded portion. [5.10]

21. Square with inscribed circle of radius 3 in.

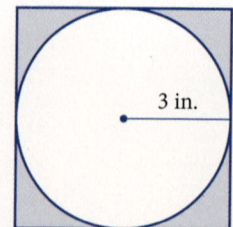

Chapter 5 Cumulative Review

Simplify each of the following.

1. $-(-3)^3$
2. $-|-7|$
3. $4^3 - 6^2$
4. $2(9-4)^2$
5. $42 \div 7 \cdot 3$
6. $64 \div 16 \cdot 2$
7. $42 - 24 \div 6 \cdot 3$
8. $57 - 36 \div (4+2)$

Find the value of each expression when x is 6.

9. $x^2 - 3x + 9$
10. $x^2 - 16$

Reduce the following fractions to lowest terms.

11. $\dfrac{216}{312}$
12. $\dfrac{252}{396}$

Subtract.

13. $\dfrac{3}{4} - \dfrac{7}{12}$
14. $\dfrac{7}{36} + \dfrac{17}{24}$

15. Add $\dfrac{1}{2}$ to the product of -8 and $\dfrac{7}{16}$.

16. Subtract $-\dfrac{3}{4}$ from the product of -4 and $\dfrac{3}{16}$.

Simplify each of the following expressions.

17. $16\left(\dfrac{5}{8}x - \dfrac{3}{4}y\right)$
18. $6\left(\dfrac{1}{3}x + \dfrac{1}{2}y\right)$

Solve the following equations.

19. $\dfrac{3}{4}x - 8 = 4$
20. $\dfrac{3}{2}(4x - 6) - 7 = 2$
21. $-6 + 2(3x + 4) = 7$
22. $5x - (x + 5) = 19$
23. $|3x + 2| - 6 = 2$
24. $|2x - 4| = 3x + 6$

Solve the following expressions for y and simplify.

25. $2x + 3y = 18$
26. $y - 4 = \dfrac{3}{4}(8x - 4)$

If $f(x) = x^2 - 6x$, $g(x) = 2x + 4$ and $h(x) = x - 2$, find the following.

27. $g(-3)$
28. $f(4) + h(9)$
29. $(f \circ g)(-3)$
30. $(f + g)(x)$

31. Specify the domain and range for the relation $\{(-2, 3), (6, 1), (-2, 7)\}$. Is the relation also a function?

Find x and the two angles.

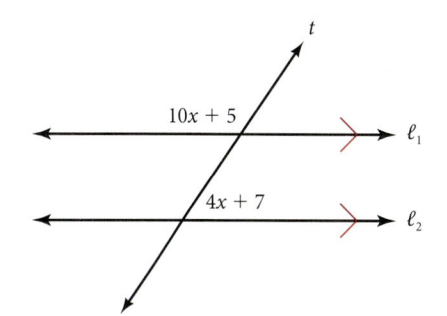

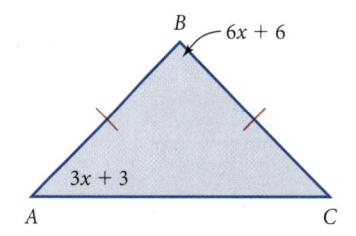

32. [figure: two parallel lines ℓ_1 and ℓ_2 cut by transversal t, with angles $10x + 5$ and $4x + 7$]

33. [figure: triangle ABC with sides labeled $6x + 6$ and $3x + 3$, with tick marks showing congruent sides]

34. Triangle ABC and △DEF are similar. Find x.

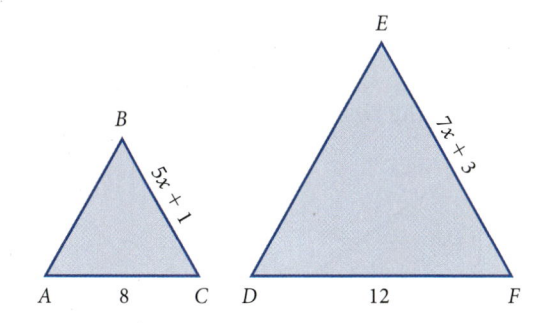

[figure: triangle ABC with AC = 8, side $5x + 1$; triangle DEF with DF = 12, side $7x + 3$]

35. State why the triangles are congruent, find x, and the angles.

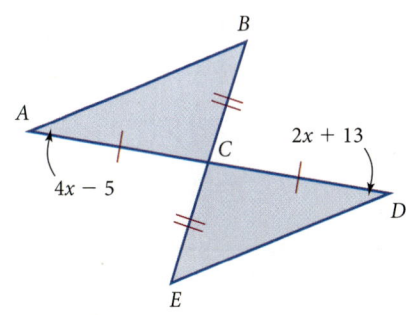

[figure: two triangles sharing vertex C; sides $4x - 5$ and $2x + 13$ with tick marks]

Chapter 5   Test

1. B is the midpoint of $\overline{AC}$. Find x. [5.1]

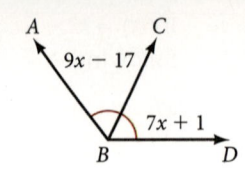

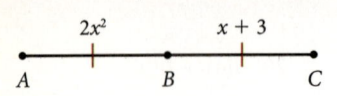

   A —— $2x^2$ —— B —— $x+3$ —— C

2. Angle at B with rays to A, C (labeled $9x-17$), and D (labeled $7x+1$).

3. C above point B on line AD; $\angle A = 2x-10$, $\angle CBD = 9x+3$.

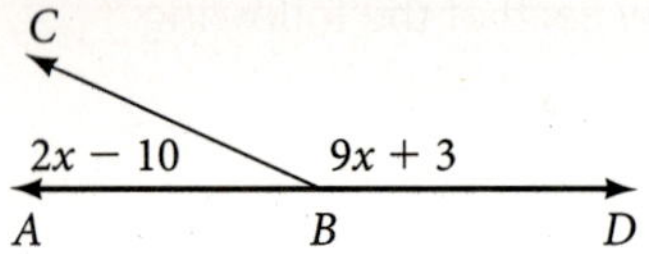

[5.2]

4. Two intersecting lines with angles $8x+6$ and $2x^2-4$.

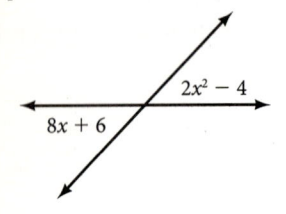

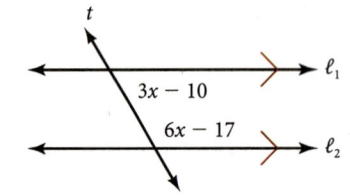

5. Parallel lines $\ell_1 \parallel \ell_2$ cut by transversal t; angles $3x-10$ and $6x-17$.

6. Parallel lines $\ell_1 \parallel \ell_2$ cut by transversal t; angles $6x^2$ and $9x+6$.

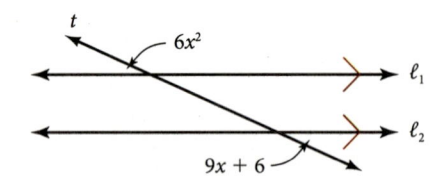

[5.3]

7. Triangle ABC with angle at $A = x^2+2$, exterior angle at $C = 34°$, and angle at $B = 4x$.

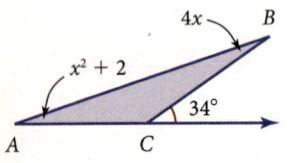

8. Triangle with angles $5x$, $3x-2$, and exterior angle $7x+16$ at C on line ACD.

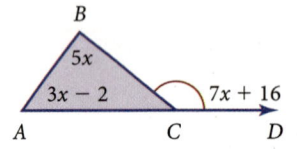

Give the reason for congruence and x and y. [5.4]

9. Triangles sharing side; $\angle DCE$ (or labeled region) $= 101°$, $\angle E = 24°$, $CB = 4y-3x$, $AC = 9y+5x$.

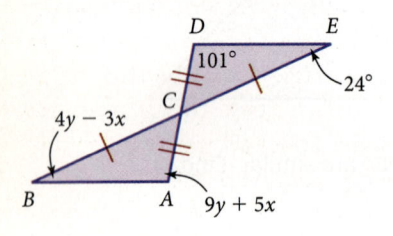

10. Quadrilateral $ABCD$ with diagonal; $DC = 3y+2$, $CB = 8x-4$, $AB = 5y-6$, $AD = 3x+6$.

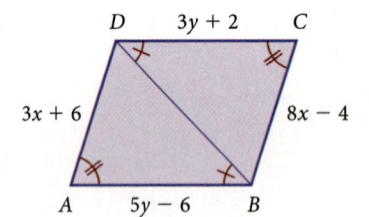

Triangle ABC is similar to $\triangle DEC$. Find x. [5.6]

11. Small triangle ABC with $BC = 10$, $AC = 3x-7$; large triangle DEC with $CE = 15$, $DE = 2x+2$.

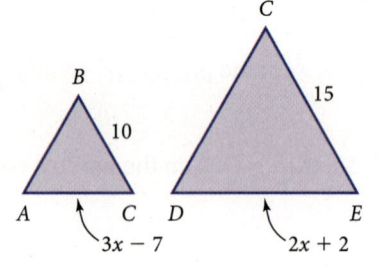

Find x. Round to the nearest tenth if necessary. [5.7]

12. Right triangle with legs $3x-3$ and $2x+2$, hypotenuse $4x-2$.

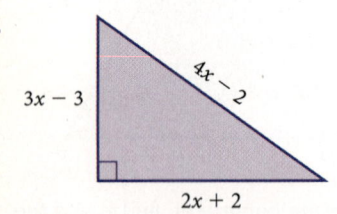

13. Right triangle with leg x opposite $29°$, hypotenuse 36.

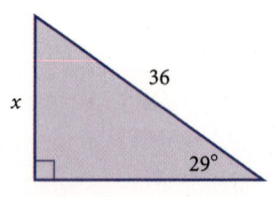

14. Find the area and the perimeter. Round to the nearest hundredth (use 3.14 for π). [5.10]

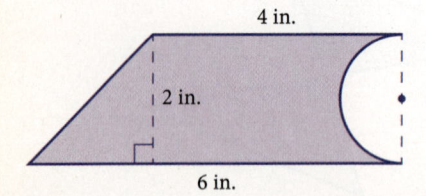

    Figure with top 4 in., height 2 in., bottom 6 in.

15. Find x, the perimeter, and the area if the perimeter is $(10x+4)$ in. (use 3.14 for π). [5.10]

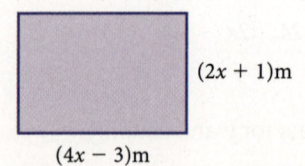

    Rectangle with sides $(2x+1)$ m and $(4x-3)$ m.

416   Chapter 5 Geometry

Chapter 5 Projects

GEOMETRY

GROUP PROJECT

Constructing the Spiral of Roots

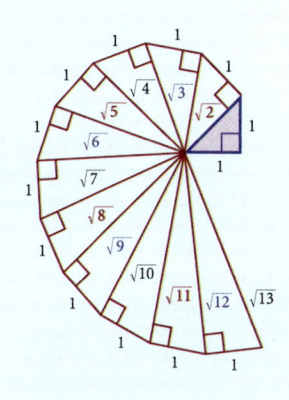

Number of People 3

Time Needed 20 minutes

Equipment Two sheets of graph paper (4 or 5 squares per inch) and pencils.

Background The spiral of roots gives us a way to visualize the positive square roots of the counting numbers, and in so doing, we see many line segments whose lengths are irrational numbers.

Procedure You are to construct a spiral of roots from a line segment 1 inch long. The graph paper you have contains either 4 or 5 squares per inch, allowing you to accurately draw 1-inch line segments. Because the lines on the graph paper are perpendicular to one another, if you are careful, you can use the graph paper to connect one line segment to another so that they form a right angle.

1. Fold one of the pieces of graph paper so it can be used as a ruler.

2. Use the folded paper to draw a line segment 1-inch long, just to the right of the middle of the unfolded paper. On the end of this segment, attach another segment of 1-inch length at a right angle to the first one. Connect the end points of the segments to form a right triangle. Label each side of this triangle. When you are finished, your work should resemble Figure 1.

3. On the end of the hypotenuse of the triangle, attach a 1-inch line segment so that the two segments form a right angle. (Use the folded paper to do this.) Draw the hypotenuse of this triangle. Label all the sides of this second triangle. Your work should resemble Figure 2.

4. Continue to draw a new right triangle by attaching 1-inch line segments at right angles to the previous hypotenuse. Label all the sides of each triangle.

5. Stop when you have drawn a hypotenuse $\sqrt{8}$ inches long.

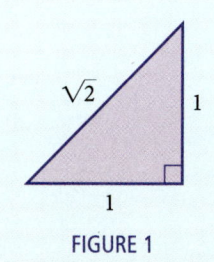

FIGURE 1

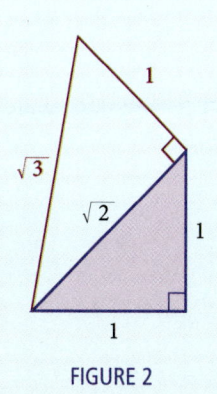

FIGURE 2

RESEARCH PROJECT

Maria Gaetana Agnesi (1718–1799)

January 9, 1999, was the 200th anniversary of the death of Maria Agnesi, the author of *Instituzioni analitiche ad uso della gioventu italiana* (1748), a calculus textbook considered to be the best book of its time and the first surviving mathematical work written by a woman. Maria Agnesi is also famous for a curve that is named for her. The curve is called the Witch of Agnesi in English, but its actual translation is the Locus of Agnesi. The foundation of the curve is shown in the figure. Research the Witch of Agnesi and then explain, in essay form, how the diagram in the figure is used to produce the Witch of Agnesi. Include a rough sketch of the curve, starting with a circle of diameter a as shown in the figure. Then, for comparison, sketch the curve again, starting with a circle of diameter $2a$.

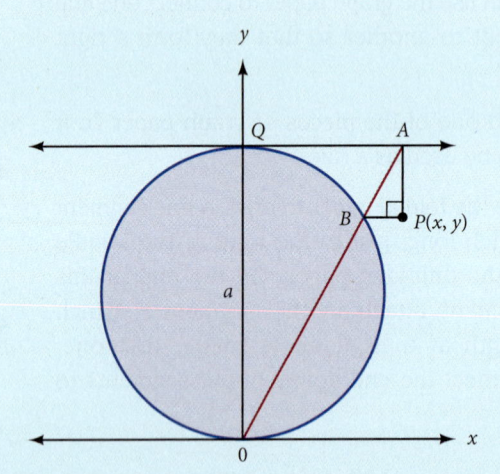

Answers to Odd-Numbered Problems

Chapter 1

Pretest for Chapter 1
1. 103   2. 35   3. 12   4. 0   5. 3   6. 0   7. $\frac{26}{21}$ 8. $-3x$ 9. $-14x - 35$ 10. $5y - 3$ 11. $12x - 11$
12. $16a - 22$ 13. -2 14. $\frac{3}{2}$ 15. -1 16. $-\frac{2}{5}$ 17. $\frac{3}{2}$ 18. $\frac{1}{5}$

Getting Ready for Chapter 1
1. 3.8   2. 0.5   3. 1.8   4. 3   5. 4   6. 44   7. 10   8. 21   9. 210   10. 231   11. 630   12. 13,500   13. 3.6
14. 0.35   15. 215.6   16. 0.2052   17. 785   18. 87   19. 320   20. 3   21. -4 22. -9 23. -12 24. $-\frac{7}{8}$
25. $-\frac{7}{8}$ 26. $-\frac{3}{8}$ 27. x 28. x 29. $7y - 3$ 30. $-7x + 14$

Problem Set 1.1
1. $4, -4, \frac{1}{4}$ 3. $-\frac{1}{2}, \frac{1}{2}, -2$ 5. $5, -5, \frac{1}{5}$ 7. $\frac{3}{8}, -\frac{3}{8}, \frac{8}{3}$ 9. $-\frac{1}{6}, \frac{1}{6}, -6$ 11. $3, -3, \frac{1}{3}$ 13. $-1, 1$ 15. 0   17. 2
19. $\frac{3}{4}$ 21. π 23. -4 25. -2 27. $-\frac{3}{4}$ 29. $\frac{21}{40}$ 31. 2   33. $\frac{8}{27}$ 35. $\frac{1}{10,000}$ 37. $\frac{72}{385}$ 39. 1   41. $6 + x$
43. $a + 8$ 45. $15y$ 47. x 49. a 51. x 53. $3x + 18$ 55. $12x + 8$ 57. $15a + 10b$ 59. $\frac{4}{3}x + 2$ 61. $2 + y$
63. $40t + 8$ 65. $9x + 3y - 6z$ 67. $3x + 7y$ 69. $6x + 7y$ 71. $3x + 1$ 73. $2x - 1$ 75. $x + 2$ 77. $a - 3$ 79. $x + 24$
81. $3x - 2y$ 83. $3x + 8y$ 85. $8x + 5y$ 87. $15x + 10$ 89. $8y + 32$ 91. $15t + 9$ 93. $28x + 11$ 95. $\frac{7}{15}$ 97. $\frac{29}{35}$ 99. $\frac{35}{144}$
101. $\frac{949}{1,260}$ 103. $\frac{47}{105}$ 105. 15   107. $14a + 7$ 109. $6y + 6$ 111. $12x + 2$ 113. $8y + 11$ 115. $24a + 15$
117. $11x + 20$ 119. Commutative property of addition   121. Commutative property of multiplication   123. Additive inverse
125. Commutative property of addition   127. Associative and commutative properties of multiplication
129. Commutative and associative properties of addition   131. Distributive property   133. $7(x + 2) = 7x + 14; 7x + 7(2) = 7x + 14$
135. $x(y + 4) = xy + 4x; xy + 4x$ 137. $12 + \frac{1}{4}(12) = 15$

Problem Set 1.2
1. 4   3. -4 5. -10 7. -4 9. $\frac{19}{12}$ 11. $-\frac{32}{105}$ 13. -8 15. -12 17. $-7x$ 19. 13   21. -14 23. $6a$
25. -15 27. 15   29. -24 31. $-10x$ 33. x 35. y 37. $-8x + 6$ 39. $-3a + 4$ 41. -14 43. 18   45. 16
47. -19 49. 50   51. 20   53. -2 55. 1   57. -30 59. 18   61. 277   63. -73 65. 39   67. 11   69. -5
71. 11   73. 7   75. -44 77. 2   79. $\frac{4}{3}$ 81. 0   83. Undefined   85. $-\frac{2}{3}$ 87. $x - 5$ 89. $x - 7$ 91. $-6x + 9y$
93. $3x - 18$ 95. $3a + 6$ 97. $3x + 60$ 99. $11y$ 101. $14x + 12$ 103. $7m - 15$ 105. $-2x + 9$ 107. $7y + 10$
109. $-20x + 5$ 111. $-11x + 10$ 113. -2 115. Undefined   117. 0   119. $-\frac{2}{3}$ 121. 32   123. 64   125. $-\frac{1}{18}$
127. $\frac{5}{3}$ 129. -1 131. 12   133.

a	b	Sum $a + b$	Difference $a - b$	Product ab	Quotient $\frac{a}{b}$
3	12	15	-9	36	$\frac{1}{4}$
-3	12	9	-15	-36	$-\frac{1}{4}$
3	-12	-9	15	-36	$-\frac{1}{4}$
-3	-12	-15	9	36	$\frac{1}{4}$

135.

x	$3(5x - 2)$	$15x - 6$	$15x - 2$
-2	-36	-36	-32
-1	-21	-21	-17
0	-6	-6	-2
1	9	9	13
2	24	24	28

137. a. 1 b. $\frac{3}{2}$ c. -1 d. 135   139. 1.5283   141. -0.0794
143. -0.0714 145. 3.4   147. 1.6   149. 1,200   151. 190   153. Santa Fe 10:10 P.M., Detroit 3:30 A.M.

Problem Set 1.3

1. 4.25×10^5 3. 6.78×10^6 5. 1.1×10^4 7. 8.9×10^7 9. 38,400 11. 57,100,000 13. 3,300 15. 89,130,000
17. 3.5×10^{-4} 19. 7×10^{-4} 21. 6.035×10^{-2} 23. 1.276×10^{-1} 25. 0.00083 27. 0.0625
29. 0.3125 31. 0.005 33. 4.0×10^4; 51,180,000 35. 1.2×10^6; 87,870,000 37.
39. 1.024×10^3 41. 1.074×10^9 43. $4x$ 45. $-3x$ 47. $9x$ 49. $-6a$
51. 6×10^{10} 53. 1.5×10^{11} 55. 6.84×10^{-2} 57. 1.62×10^{10} 59. 3.78×10^{-1}
61. 2×10^3 63. 4×10^{-9} 65. 5×10^{14} 67. 6×10^1 69. 2×10^{-8} 71. 4×10^6
73. 4×10^{-4} 75. 2.1×10^{-2} 77. 3×10^{-5} 79. 6.72×10^{-1} 81. 7.81×10^{-3}
83. 1.37×10^{-2} 85. 2,500,000 stones or 2.5×10^6 87. 7 89. -14 91. -14 93. 21

Jupiter's Moon		Period (Seconds)
Io	153,000	1.53×10^5
Europa	307,000	3.07×10^5
Ganymede	618,000	6.18×10^5
Callisto	1,440,000	1.44×10^6

Problem Set 1.4

1. 8 3. 5 5. 2 7. -7 9. $-\frac{9}{2}$ 11. -4 13. $-\frac{4}{3}$ 15. -4 17. -2 19. $\frac{3}{4}$ 21. 12 23. -10 25. 7
27. -7 29. 3 31. $\frac{4}{5}$ 33. 1 35. 4 37. 17 39. 2 41. 6 43. $-\frac{4}{3}$ 45. $-\frac{3}{2}$ 47. $\frac{5}{3}$ 49. $\frac{3}{2}$ 51. 1
53. No solution 55. All real numbers are solutions. 57. No solution 59. $-\frac{2}{3}$ 61. $-\frac{7}{640}$ 63. 2 65. 20 67. 0.7
69. 2,400 71. 24 73. 6 75. 3 77. 8.5 79. 2 81. 5 83. 18 85. 36 87. 4 89. 30 91. 6,000
93. 5,000 95. $\frac{8}{15}$ 97. $-\frac{5}{11}$ 99. a. $\frac{5}{8}$ b. 0 c. $10x - 10$ d. 0 e. $8x - 40$ f. 5 101. a. $\$6.60 = \$0.40n + \$1.80$ b. 12 miles
103. a. $3,937,000 = 1125A$ b. 3,500 square miles 105. 50 miles per hour 107. Commutative 109. Associative
111. Commutative and associative 113. Multiplicative identity 115. Commutative 117. Additive identity 119. 0.5
121. 62.5 123. 0 125. 1.25 127. 13 129. a. 3 b. 5 c. 9 131. 2 133. $-\frac{3}{2}$ 135. $-\frac{21}{2}$ 137. No solution

Problem Set 1.5

1. 10 feet by 20 feet 3. 7 feet 5. 5 inches 7. 4 meters 9. $92.00 11. $200.00 13. $3,260.66 per month
15. $99.6 million 17. $6,000 at 8%, $3,000 at 9% 19. $5,000 at 12%, $10,000 at 10% 21. $4,000 at 8%, $2,000 at 9%
23. 16 adults, 22 children 25. $54 27.

t	0	$\frac{1}{4}$	1	$\frac{7}{4}$	2
h	0	7	16	7	0

29. **HORSE RACING**

Year	Bets (millions of dollars)
1985	20.2
1990	34.4
1995	44.8
2000	65.4
2005	104

31. **HOT COFFEE SALES**

Year	Sales (billions of dollars)
2005	7
2006	7.5
2007	8
2008	8.6
2009	9.2

33. **DISTANCE**

Speed (mph)	Distance (miles)
20	10
30	15
40	20
50	25
60	30
70	35

35.

Time (hours)	Distance Upstream (miles)	Distance Downstream (miles)
1	6	14
2	12	28
3	18	42
4	24	56
5	30	70
6	36	84

37.

Age (years)	Maximum Heart Rate (beats per minute)
18	202
19	201
20	200
21	199
22	198
23	197

39.

w	l	A
2	22	44
4	20	80
6	18	108
8	16	128
10	14	140
12	12	144

Chapter 1 Review

1. $-2, \frac{1}{2}$ 2. $\frac{2}{5}, -\frac{5}{2}$ 3. 3 4. -5 5. 4 6. 6 7. a 8. c 9. a 10. b, d 11. a, c 12. f 13. e
14. g 15. $-6x + 10$ 16. $-6x + 21$ 17. $-x + 3$ 18. $-15x + 3$ 19. $2y$ 20. $3x$ 21. $8x + 5$ 22. $y + 2$
23. $-2y + 9$ 24. $-18x - 14$ 25. $5a - 22$ 26. 2 27. -2 28. 1 29. 17 30. 3 31. -6 32. 2
33. 66 34. $-\frac{5}{6}$ 35. $-\frac{5}{6}$ 36. 1 37. $\frac{27}{64}$ 38. 2 39. -42 40. 30 41. $21x$ 42. $-6x$ 43. $-\frac{5}{6}$
44. -36 45. $\frac{1}{10}$ 46. $-\frac{2}{7}$ 47. 17 48. 4 49. 13 50. 7 51. 9 52. 32 53. 30 54. 43 55. -13
56. -17 57. 0 58. -1 59. -36 60. -34 61. 16 62. -24 63. 2 64. 39 65. 0 66. 7.38×10^7
67. 2.935×10^{-3} 68. 4,400,000,000 69. 0.000016 70. 10 71. 2 72. 2 73. -3 74. -3 75. 0 76. $\frac{2}{3}$
77. $\frac{10}{13}$ 78. $\frac{5}{2}$ 79. 1 80. $-\frac{5}{11}$ 81. $-\frac{2}{9}$ 82. 4 feet by 12 feet 83. 3 meters, 4 meters, 5 meters

Answers to Odd-Numbered Problems   A-3

Chapter 1 Test

1. a. 3   **b.** -2 **2.** Commutative property of addition   **3.** Multiplicative identity property   **4.** Associative property of multiplication   **5.** Associative property of addition   **6.** 2   **7.** 0   **8.** $\frac{59}{72}$ **9.** $-4x$ **10.** $-5x - 8$ **11.** $4y - 10$ **12.** $3x - 17$ **13.** $11a - 10$ **14.** $-\frac{7}{3}$ **15.** $-\frac{5}{2}$ **16.** 57   **17.** 10   **18.** 16   **19.** 0   **20.** 3.85×10^{-5} **21.** 6,750,000   **22.** 1.701×10^3 **23.** 5.0×10^2 **24.** 4.0×10^{-4} **25.** 2.56×10^{-3} **26.** 12   **27.** $-\frac{4}{3}$ **28.** $-\frac{7}{4}$ **29.** 2   **30.** 6 cm by 24 cm   **31.** 6 m   **32.** 12 music videos

Chapter 2

Pretest for Chapter 2

1. $\frac{125}{8}$ **2.** $\frac{1}{81}$ **3.** a^4 **4.** x^2 **5.** $128a^{11}b^7$ **6.** $\frac{2}{25x^2y^6}$ **7.** $\frac{6}{5}x^3 - \frac{6}{5}x^2 - \frac{8}{5}x - \frac{6}{5}$ **8.** $-14x - 58$ **9.** $-5x^2 - 31x + 28$ **10.** $6x^3 + 14x^2 - 27x + 10$ **11.** $9a^8 - 42a^4 + 49$ **12.** $4x^2 - 9$ **13.** $3x^3 - 17x^2 - 28x$ **14.** $14x^2 - \frac{1}{14}$ **15.** $(x-1)(x-5)$ **16.** $3(5x^2 - 4)(x^2 + 3)$ **17.** $(3x + 2y)(3x - 2y)(9x^2 + 4y^2)$ **18.** $(6x - y)(a + 3b^2)$ **19.** $\left(y + \frac{1}{3}\right)\left(y^2 - \frac{1}{3}y + \frac{1}{9}\right)$ **20.** $3x^2y^4(x - 3y)(x + 8y)$ **21.** $(a - 6 - b)(a + 6 - b)$ **22.** $(2 - x)(2 + x)(4 + x^2)$ **23.** $-10, -\frac{1}{2}$ **24.** $0, 3$ **25.** $-6, 3$ **26.** $-5, -3, 3$ **27.** $-3, 5$ **28.** $-5, 4$

Getting Ready for Chapter 2

1. 3   **2.** 2   **3.** -16 **4.** -7 **5.** -32 **6.** -6 **7.** $\frac{1}{2}$ **8.** $\frac{1}{2}$ **9.** 81   **10.** 8   **11.** 1,000   **12.** 10,000   **13.** -8 **14.** -27 **15.** $\frac{1}{25}$ **16.** $\frac{1}{9}$ **17.** $-\frac{1}{8}$ **18.** $-\frac{1}{125}$ **19.** $\frac{9}{4}$ **20.** $\frac{16}{9}$ **21.** 4,520   **22.** 39,100   **23.** 3.76   **24.** 2.74

Problem Set 2.1

1. 16   **3.** -16 **5.** -0.027 **7.** 32   **9.** $\frac{1}{8}$ **11.** $-\frac{25}{36}$ **13.** x^9 **15.** 64   **17.** $-\frac{8}{27}x^6$ **19.** $-6a^6$ **21.** a^{14} **23.** 4^8 **25.** $9x^8$ **27.** $8b^7$ **29.** $-\frac{1}{16}a^{12}$ **31.** $\frac{1}{9}$ **33.** $-\frac{1}{32}$ **35.** $\frac{16}{9}$ **37.** 17   **39.** x^3 **41.** $\frac{a^6}{b^{15}}$ **43.** $\frac{8}{125y^{18}}$ **45.** $\frac{1}{5}$ **47.** $\frac{24a^{12}c^6}{b^3}$ **49.** $\frac{8x^{22}}{81y^{23}}$ **51.** $\frac{1}{x^{10}}$ **53.** a^{10} **55.** $\frac{1}{t^6}$ **57.** x^{12} **59.** x^{18} **61.** $\frac{1}{x^{22}}$ **63.** $\frac{a^3b^7}{4}$ **65.** $\frac{y^{38}}{x^{16}}$ **67.** $\frac{16y^{16}}{x^8}$ **69.** x^4y^6 **71.** x^2y **73.** $3ab^2$ **75.** $2a$ **77.** $4xy^4$ **79. a.** 32   **b.** 64   **c.** 32   **d.** 64   **81. a.** 8   **b.** 8   **c.** $\frac{1}{16}$ **d.** $\frac{1}{16}$ **83. a.** $\frac{1}{7}$ **b.** $\frac{1}{11}$ **c.** $\frac{1}{2x}$ **d.** $\frac{1}{8x^2}$ **85.** 2^n **87.** r^3 **89.** $80x^3y^6$ **91.** $72x^3y^3$ **93.** $15xy$ **95.** $12x^7y^6$ **97.** $54a^6b^2c^4$ **99.** $2x^3$ **101.** $4x$ **103.** $-4xy^4$ **105.** $-2x^2y^2$ **107.** 275,625 square feet   **109.** $5x$ **111.** $8x^2$ **113.** $2x^3$ **115.** -5 **117.** $-2x + 3$ **119.** $9x + 6$ **121.** 1,200   **123.** $\frac{1}{x^3}$ **125.** y^3 **127.** x^5

Problem Set 2.2

1. Trinomial, 2, 5   **3.** Binomial, 1, 3   **5.** Trinomial, 2, 8   **7.** Polynomial, 3, 4   **9.** Monomial, 0, $-\frac{3}{4}$ **11.** Trinomial, 3, 6   **13.** $7x + 1$ **15.** $2x^2 + 7x - 15$ **17.** $12a^2 - 7ab - 10b^2$ **19.** $x^2 - 13x + 3$ **21.** $\frac{1}{4}x^2 - \frac{7}{12}x - \frac{1}{4}$ **23.** $-y^3 - y^2 - 4y + 7$ **25.** $2x^3 + x^2 - 3x - 17$ **27.** $\frac{1}{14}x^2 + \frac{1}{7}xy + \frac{5}{7}y^2$ **29.** $-3a^3 + 6a^2b - 5ab^2$ **31.** $-3x$ **33.** $3x^2 - 12xy$ **35.** $17x^5 - 12$ **37.** $14a^2 - 2ab + 8b^2$ **39.** $2 - x$ **41.** $10x - 5$ **43.** $9x - 35$ **45.** $9y - 4x$ **47.** $9a + 2$ **49.** -2 **51. a.** 208   **b.** 103   **53. a.** 51   **b.** -15 **55. a.** 110   **b.** -120 **57. a.** $-5,000$ **b.** 3,000   **59.** 1st year $51,568x + 67,073y$; 2nd year $51,568(2x) + 67,073y$; total $154,709x + 134,146y$ **61.** $2x^2 + 7x - 15$ **63.** $6x^3 - 11x^2y + 11xy^2 - 12y^3$ **65.** $-12x^4$ **67.** $20x^5$ **69.** a^6 **71.** 650   **73.** $x - 5$ **75.** $x - 7$ **77.** $5x - 1$ **79.** $x - 1$ **81.** $3x + 8y$

Problem Set 2.3

1. $12x^3 - 10x^2 + 8x$ **3.** $-3a^5 + 18a^4 - 21a^2$ **5.** $2a^5b - 2a^3b^2 + 2a^2b^4$ **7.** $x^2 - 2x - 15$ **9.** $6x^4 - 19x^2 + 15$ **11.** $x^3 + 9x^2 + 23x + 15$ **13.** $a^3 - b^3$ **15.** $8x^3 + y^3$ **17.** $2a^3 - a^2b - ab^2 - 3b^3$ **19.** $x^2 + x - 6$ **21.** $6a^2 + 13a + 6$ **23.** $20 - 2t - 6t^2$ **25.** $x^6 - 2x^3 - 15$ **27.** $y^2 - 2y - 15$ **29.** $12 + a - a^2$ **31.** $6x^2 + 11x + 4$ **33.** $y^2 - 10y + 24$ **35.** $20x^2 - 9xy - 18y^2$ **37.** $18t^2 - \frac{2}{9}$ **39.** $25x^2 + 20xy + 4y^2$ **41.** $25 - 30t^3 + 9t^6$ **43.** $4a^2 - 9b^2$ **45.** $9r^4 - 49s^2$ **47.** $y^2 + 3y + \frac{9}{4}$ **49.** $a^2 - a + \frac{1}{4}$ **51.** $x^2 + \frac{1}{2}x + \frac{1}{16}$ **53.** $t^2 + \frac{2}{3}t + \frac{1}{9}$ **55.** $\frac{1}{9}x^2 - \frac{4}{25}$ **57.** $x^3 - 6x^2 + 12x - 8$ **59.** $x^3 - \frac{3}{2}x^2 + \frac{3}{4}x - \frac{1}{8}$ **61.** $3x^3 - 18x^2 + 33x - 18$ **63.** $a^2b^2 + b^2 + 8a^2 + 8$ **65.** $3x^2 + 12x + 14$ **67.** $24x$ **69.** $x^2 + 4x - 5$ **71.** $4a^2 - 30a + 56$ **73.** $32a^2 + 20a - 18$ **75.** $2^4 - 3^4 = -65$; $(2-3)^4 = 1$; $(2^2 + 3^2)(2+3)(2-3) = -65$

77. $x(6{,}200 - 25x) = 6{,}200x - 25x^2$; $18,375   **79.** $A = 100 + 400r + 600r^2 + 400r^3 + 100r^4$ **81.** $8a^2$ **83.** -48 **85.** $2a^3b$
87. $-3b^2$ **89.** $-y^4$ **91.** $(x+y)^2 + (x+y) - 20 = x^2 + 2xy + y^2 + x + y - 20$ **93.** $x^{2n} - 5x^n + 6$ **95.** $10^{2n} + 13x^n - 3$
97. $x^{2n} + 10x^n + 25$ **99.** $x^{3n} + 1$

Problem Set 2.4

1. $5x^2(2x - 3)$ **3.** $9y^3(y^3 + 2)$ **5.** $3ab(3a - 2b)$ **7.** $7xy^2(3y^2 + x)$ **9.** $3xy^2(4x - 3y)$
11. $5xy(4y - 1)$ **13.** $2x^3y(9xy^4 - 7)$ **15.** $3(a^2 - 7a + 10)$ **17.** $4x(x^2 - 4x - 5)$
19. $10x^2y^2(x^2 + 2xy - 3y^2)$ **21.** $xy(-x + y - xy)$ **23.** $2xy^2z(2x^2 - 4xz + 3z^2)$
25. $5abc(4abc - 6b + 5ac)$ **27.** $(a - 2b)(5x - 3y)$ **29.** $3(x+y)^2(x^2 - 2y^2)$
31. $(x + 5)(2x^2 + 7x + 6)$ **33.** $(x + 1)(3y + 2a)$ **35.** $(xy + 1)(x + 3)$
37. $(x - 2)(3y^2 + 4)$ **39.** $(x - a)(x - b)$ **41.** $(b + 5)(a - 1)$ **43.** $(b^2 + 1)(a^4 - 5)$
45. $(x + 3)(x^2 - 4)$ **47.** $(x + 2)(x^2 - 25)$ **49.** $(2x + 3)(x^2 - 4)$ **51.** $(x + 3)(4x^2 - 9)$
53. 6 **55.** $P(1 + r) + P(1 + r)r = (1 + r)(P + Pr) = (1 + r)P(1 + r) = P(1 + r)^2$
57. $3x^2(x^2 - 3xy - 6y^2)$ **59.** $(x - 3)(2x^2 - 4x - 3)$ **61.** $3x^2 + 5x - 2$
63. $3x^2 - 5x + 2$ **65.** $x^2 + 5x + 6$ **67.** $6y^2 + y - 35$ **69.** $20 - 19a + 3a^2$
71. (graph to the right)

Two Numbers a and b	Their Product ab	Their Sum $a + b$
1, −24	−24	−23
−1, 24	−24	23
2, −12	−24	−10
−2, 12	−24	10
3, −8	−24	−5
−3, 8	−24	5
4, −6	−24	−2
−4, 6	−24	2

Problem Set 2.5

1. $(x - 3)^2$ **3.** $(a - 6)^2$ **5.** $(y - 9)^2$ **7.** $(y - 7)^2$ **9.** $(5 - t)^2$ **11.** $\left(\frac{1}{3}x + 3\right)^2$ **13.** $(2y^2 - 3)^2$ **15.** $(4a + 5b)^2$
17. $\left(\frac{1}{5} + \frac{1}{4}t^2\right)^2$ **19.** $\left(y + \frac{3}{2}\right)^2$ **21.** $\left(a - \frac{1}{2}\right)^2$ **23.** $\left(x - \frac{1}{4}\right)^2$ **25.** $\left(t + \frac{1}{3}\right)^2$ **27.** $4(2x - 3)^2$ **29.** $3a(5a + 1)^2$
31. $(x + 2 + 3)^2 = (x + 5)^2$ **33.** $(x + 3)(x - 3)$ **35.** $(7x + 8y)(7x - 8y)$ **37.** $\left(2a + \frac{1}{2}\right)\left(2a - \frac{1}{2}\right)$ **39.** $\left(x + \frac{3}{5}\right)\left(x - \frac{3}{5}\right)$
41. $(3x + 4y)(3x - 4y)$ **43.** $10(5 + t)(5 - t)$ **45.** $(x^2 + 9)(x + 3)(x - 3)$ **47.** $(3x^3 + 1)(3x^3 - 1)$ **49.** $(4a^2 + 9)(2a + 3)(2a - 3)$
51. $\left(\frac{1}{9} + \frac{y^2}{4}\right)\left(\frac{1}{3} + \frac{y}{2}\right)\left(\frac{1}{3} - \frac{y}{2}\right)$ **53.** $(x - y)(x + y)(x^2 + xy + y^2)(x^2 - xy + y^2)$ **55.** $2a(a - 2)(a + 2)(a^2 + 2a + 4)(a^2 - 2a + 4)$
57. $(x + 1)(x - 5)$ **59.** $y(y + 8)$ **61.** $(x - 5 + y)(x - 5 - y)$ **63.** $(a + 4 + b)(a + 4 - b)$ **65.** $(x + y + a)(x + y - a)$
67. $(x + 3)(x + 2)(x - 2)$ **69.** $(x + 2)(x + 5)(x - 5)$ **71.** $(2x + 3)(x + 2)(x - 2)$ **73.** $(x + 3)(2x + 3)(2x - 3)$ **75.** $(2x - 15)(2x + 5)$
77. $(a - 3 - 4b)(a - 3 + 4b)$ **79.** $(a - 3 - 4b)(a - 3 + 4b)$ **81.** $(x + 4)(x - 3)^2$ **83.** $(x - y)(x^2 + xy + y^2)$ **85.** $(a + 2)(a^2 - 2a + 4)$
87. $(3 + x)(9 - 3x + x^2)$ **89.** $(y - 1)(y^2 + y + 1)$ **91.** $10(r - 5)(r^2 + 5r + 25)$ **93.** $(4 + 3a)(16 - 12a + 9a^2)$
95. $(2x - 3y)(4x^2 + 6xy + 9y^2)$ **97.** $\left(t + \frac{1}{3}\right)\left(t^2 - \frac{1}{3}t + \frac{1}{9}\right)$ **99.** $\left(3x - \frac{1}{3}\right)\left(9x^2 + x + \frac{1}{9}\right)$ **101.** $(4a + 5b)(16a^2 - 20ab + 25b^2)$
103. 30 and −30   **105.** $y(y^2 + 25)$ **107.** $2ab^3(b^2 + 4b + 1)$ **109.** $(2x - 3)(2x + a)$ **111.** $(x + 2)(x - 2)$ **113.** $(x - 3)^2$
115. $(3a - 4)(2a - 1)$ **117.** $(x + 2)(x^2 - 2x + 4)$ **119.** $(a - b + 3)(a + b - 3)$ **121.** $(x - y - 8)(x + y + 2)$ **123.** $k = 144$
125. $k = \pm 126$

Problem Set 2.6

1. $(x + 3)(x + 4)$ **3.** $(x + 1)(x + 2)$ **5.** $(a + 3)(a + 7)$ **7.** $(x - 2)(x - 5)$ **9.** $(y - 3)(y - 7)$ **11.** $(x - 4)(x + 3)$ **13.** $(y + 4)(y - 3)$
15. $(x + 7)(x - 2)$ **17.** $(r - 9)(r + 1)$ **19.** $(x - 6)(x + 5)$ **21.** $(a + 7)(a + 8)$ **23.** $(y + 6)(y - 7)$ **25.** $(x + 6)(x + 7)$
27. $2(x + 1)(x + 2)$ **29.** $3(a + 4)(a - 5)$ **31.** $100(x - 2)(x - 3)$ **33.** $100(p - 5)(p - 8)$ **35.** $x^2(x + 3)(x - 4)$ **37.** $2r(r + 5)(r - 3)$
39. $2y^2(y + 1)(y - 4)$ **41.** $x^3(x + 2)(x + 2)$ **43.** $3y^2(y + 1)(y - 5)$ **45.** $4x^2(x - 4)(x - 9)$ **47.** $(2x - 5)(x + 3)$ **49.** $(2x - 3)(x - 5)$
51. $(2x - 5)(x - 3)$ **53.** Prime   **55.** $(2 + 3a)(1 + 2a)$ **57.** $15(4y + 3)(y - 1)$ **59.** $x^2(3x - 2)(2x + 1)$ **61.** $10r(2r - 3)^2$
63. $(4x + y)(x - 3y)$ **65.** $(2x - 3a)(5x + 6a)$ **67.** $(3a + 4b)(6a - 7b)$ **69.** $200(1 + 2t)(3 - 2t)$ **71.** $y^2(3y - 2)(3y + 5)$
73. $2a^2(3 + 2a)(4 - 3a)$ **75.** $4x^2 - 9$ **77.** $4x^2 - 12x + 9$ **79.** $8x^3 - 27$ **81.** $\frac{5}{8}$ **83.** x^3 **85.** $4x^2$ **87.** $\frac{1}{2}$ **89.** x^2 **91.** $3x$
93. $2y$ **95.** $(2x^3 + 5y^2)(4x^3 + 3y^2)$ **97.** $(3x - 5)(x + 100)$ **99.** $\left(\frac{1}{4}x + 1\right)\left(\frac{1}{2}x + 2\right)$ **101.** $(2x + 0.5)(x + 0.5)$

Problem Set 2.7

1. $(x + 9)(x - 9)$ 3. $(x - 3)(x + 5)$ 5. $(x + 2)(x + 3)^2$ 7. $(x^2 + 2)(y^2 + 1)$ 9. $2ab(a^2 + 3a + 1)$ 11. Prime
13. $3(2a + 5)(2a - 5)$ 15. $(3x - 2y)^2$ 17. $(5 - t)^2$ 19. $4x(x^2 + 4y^2)$ 21. $2y(y + 5)^2$ 23. $a^4(a + 2b)(a^2 - 2ab + 4b^2)$
25. $(t + 3 + x)(t + 3 - x)$ 27. $(x + 5)(x + 3)(x - 3)$ 29. $5(a + b)^2$ 31. Prime   33. $3(x + 2y)(x + 3y)$ 35. $\left(3a + \frac{1}{3}\right)^2$
37. $(x - 3)(x - 7)^2$ 39. $(x + 8)(x - 8)$ 41. $(2 - 5x)(4 + 3x)$ 43. $a^5(7a + 3)(7a - 3)$ 45. $\left(r + \frac{1}{5}\right)\left(r - \frac{1}{5}\right)$ 47. Prime
49. $100(x - 3)(x + 2)$ 51. $a(5a + 3)(5a + 1)$ 53. $(3x^2 + 1)(x^2 - 5)$ 55. $3a^2b(2a - 1)(4a^2 + 2a + 1)$ 57. $(4 - r)(16 + 4r + r^2)$
59. $5x^2(2x + 3)(2x - 3)$ 61. $100(2t + 3)(2t - 3)$ 63. $2x^3(4x - 5)(2x - 3)$ 65. $(y + 1)(y - 1)(y^2 - y + 1)(y^2 + y + 1)$
67. $2(5 + a)(5 - a)$ 69. $3x^2y^2(2x + 3y)^2$ 71. $(x - 2 + y)(x - 2 - y)$ 73. $\left(a - \frac{2}{3}b\right)^2$ 75. $\left(x - \frac{2}{5}y\right)^2$ 77. $\left(a - \frac{5}{6}b\right)^2$
79. $\left(x - \frac{4}{5}y\right)^2$ 81. $(2x - 3)(x - 5)(x + 2)$ 83. $(x - 4)^3(x - 3)$ 85. $2(y - 3)(y^2 + 3y + 9)$ 87. $2(a - 4b)(a^2 + 4ab + 16b^2)$
89. $2(x + 6y)(x^2 - 6xy + 36y^2)$ 91. 60 geese; 48 ducks   93. 150 oranges; 144 apples   95. $2x^2 + 2x + 1$ 97. $t^2 - 4t + 3$
99. $(x - 6)(x + 4)$ 101. $x(2x + 1)(x - 3)$ 103. $(x + 2)(x - 3)(x + 3)$ 105. 6   107. $-\frac{1}{2}$

Problem Set 2.8

1. $6, -1$ 3. $-1, 4$ 5. $-2, -4$ 7. $-5, 3$ 9. $0, 2, 3$ 11. $\frac{1}{3}, -4$ 13. $\frac{2}{3}, \frac{3}{2}$ 15. $5, -5$ 17. $0, -3, 7$ 19. $-4, \frac{5}{2}$
21. $0, \frac{4}{3}$ 23. $-\frac{1}{5}, \frac{1}{3}$ 25. $0, -\frac{4}{3}, \frac{4}{3}$ 27. $-10, 0$ 29. $-5, 1$ 31. $1, 2$ 33. $-2, 3$ 35. $-2, \frac{1}{4}$ 37. $-3, -2, 2$
39. $-2, -5, 5$ 41. $-\frac{3}{2}, -2, 2$ 43. $-3, -\frac{3}{2}, \frac{3}{2}$ 45. $-2, \frac{5}{3}$ 47. $-1, 9$ 49. $0, -3$ 51. $-4, -2$ 53. $-4, 2$ 55. $-\frac{3}{2}$
57. $-2, 8$ 59. -3 61. $-7, 1$ 63. $0, 5$ 65. $-1, 8$ 67. a. $\frac{25}{9}$ b. $-\frac{5}{3}, \frac{5}{3}$ c. $-3, 3$ d. $\frac{5}{3}$ 69. 3 hours   71. $-5, -4$ or $4, 5$
73. 24 feet   75. $6, 8, 10$ 77. 2 feet, 8 feet   79. 18 inches, 4 inches   81. 0 and 2 seconds   83. 1 and 2 seconds
85. 0 and $\frac{3}{2}$ seconds   87. 2 and 3 seconds

Chapter 2 Review

1. x^{10} 2. $25x^6$ 3. $-32x^{18}y^8$ 4. $\frac{1}{8}$ 5. $\frac{9}{4}$ 6. $\frac{1}{2}$ 7. $\frac{1}{a^9}$ 8. $\frac{x^{12}}{4}$ 9. x^2 10. $2x^2 - 5x + 7$ 11. $2x^3 - 2x^2 - 2x - 4$
12. $x^2 - 2x - 3$ 13. $30x + 12$ 14. 15   15. $12x^3 - 6x^2 + 3x$ 16. $2a^4b^3 + 4a^3b^4 + 2a^2b^5$ 17. $18 - 9y + y^2$ 18. $6x^4 + 5x^2 - 4$
19. $2t^3 - 4t^2 - 6t$ 20. $x^3 + 27$ 21. $8x^3 - 27$ 22. $a^4 - 4a^2 + 4$ 23. $9x^2 + 30x + 25$ 24. $16x^2 - 24xy + 9y^2$ 25. $x^2 - \frac{1}{9}$
26. $4a^2 - b^2$ 27. $x^3 - 3x^2 + 3x - 1$ 28. $x^{2m} - 4$ 29. $3xy(2x^3 - 3y^3 + 6x^2y^2)$ 30. $4(x + y)^2(x^2 - 2y^2)$ 31. $(4x^2 + 5)(2 - y)$
32. $(1 - y)(1 + y)(x^3 + 8b^2)$ 33. $(x - 2)(x - 3)$ 34. $2x(x + 5)(x - 3)$ 35. $(5a - 4b)(4a - 5b)$ 36. $x^2(3x + 2)(2x - 5)$
37. $3y(4x + 5)(2x - 3)$ 38. $(x^2 + 4)(x + 2)(x - 2)$ 39. $3(a^2 + 3)^2$ 40. $(a - 2)(a^2 + 2a + 4)$ 41. $5x(x + 3y)^2$
42. $3ab(a - 3b)(a + 3b)$ 43. $(x - 5 + y)(x - 5 - y)$ 44. $(6 - 5a)(6 + 5a)$ 45. $(x + 3)(x - 3)(x + 4)$ 46. $-3, -2$ 47. $-\frac{1}{2}, \frac{4}{5}$
48. $-\frac{5}{3}, \frac{5}{3}$ 49. $0, -2$ 50. $-3, 6$ 51. $-4, 3, -3$ 52. $-10, -8$ or $8, 10$ 53. $-5, -4$ or $4, 5$

Chapter 2 Cumulative Review

1. -125 2. $-\frac{1}{125}$ 3. 60   4. 48   5. 48   6. 121   7. -1 8. 2   9. 11.49   10. -11.51 11. 22   12. 16
13. $6x - 1$ 14. $20x - 9y$ 15. $-2x - 4$ 16. $5a - 2$ 17. $8x^3 - 7x^2 - 14$ 18. $\frac{1}{3}x^2 - \frac{17}{6}x$ 19. $81x^8y^4$ 20. $-15a^5b^5$
21. $-2xy^4$ 22. $\frac{81a^2}{b^4}$ 23. $15x^2 - 4x - 4$ 24. $15x^3 + 10x^2 + 35x$ 25. $x^2 - \frac{4}{3}x + \frac{4}{9}$ 26. $t^3 + 2t^2 + \frac{4}{3}t + \frac{8}{27}$ 27. 2.8×10^{10}
28. 4×10^2 29. $(x - a)(x - y)$ 30. $(3a - 7)(2a + 5)$ 31. $(x - 2)(x + 8)$ 32. $\left(\frac{1}{2} + t\right)\left(\frac{1}{4} - \frac{1}{2}t + t^2\right)$ 33. 2   34. $-1, 5$
35. -36 36. $-\frac{3}{4}$ 37. $0, -\frac{6}{5}, \frac{6}{5}$ 38. -4 39. 11   40. -21 41. -13 42. 8   43. 25   44. 25   45. 1.765×10^8
46. 9.61×10^7 47. 5×10^{-3} 48. 1×10^{-2}

Chapter 2 Test

1. $\frac{16}{9}$ 2. $\frac{1}{32}$ 3. x^8 4. a^2 5. $32x^{12}y^{11}$ 6. $\frac{a^{14}}{4b^{18}}$ 7. 149   8. -151 9. $\frac{3}{4}x^3 - \frac{5}{4}x^2 - 2x - 1$ 10. $4x + 75$
11. $6y^2 + y - 35$ 12. $2x^3 + 3x^2 - 26x + 15$ 13. $64 - 48t^3 + 9t^6$ 14. $1 - 36y^2$ 15. $4x^3 - 2x^2 - 30x$ 16. $10t^4 - \frac{1}{10}$
17. $(x+4)(x-3)$ 18. $2(3x^2 - 1)(2x^2 + 5)$ 19. $(2a - 3y)(2a + 3y)(4a^2 + 9y^2)$ 20. $(7a - b^2)(x^2 - 2y)$ 21. $\left(t + \frac{1}{2}\right)\left(t^2 - \frac{1}{2}t + \frac{1}{4}\right)$
22. $4a^3b(a - 8b)(a + 2b)$ 23. $(x - 5 - b)(x - 5 + b)$ 24. $(3 - x)(3 + x)(9 + x^2)$ 25. $-\frac{1}{3}, 2$ 26. $0, 5$ 27. $-5, 2$
28. $-4, -2, 4$ 29. $2, 3$ 30. $2, 4$ 31. 9 32. 5 33. 6 34. $25x - .2x^2$ 35. $P(x) = -.2x^2 + 23x - 100$ 36. $500
37. $300   38. $200

Chapter 3

Pretest for Chapter 3

1. $x + y$ 2. $\frac{2x-3}{x-1}$ 3. $3(a+2)$ 4. $4(a-2)$ 5. $\frac{x+1}{2}$ 6. $\frac{27}{50}$ 7. $\frac{17}{32}$ 8. $\frac{1}{a+5}$ 9. $\frac{2(4x+1)}{(x+1)(2x-1)}$ 10. $3x^2y\sqrt{3xy}$
11. $4y^2\sqrt[3]{2x^2y}$ 12. $\frac{\sqrt{15}}{5}$ 13. $\frac{2ab^3\sqrt{14bc}}{7c}$ 14. $x = -\frac{10}{19}$ 15. no solution   16. $y = -\frac{12}{67}$ 17. $x = 3, 7$ 18. $\frac{9}{2}, -\frac{3}{2}$
19. $\frac{-1 \pm 2\sqrt{3}}{4}$ 20. $-4 \pm \sqrt{7}$

Getting Ready for Chapter 3

1. $\frac{3}{4}$ 2. $\frac{2}{5}$ 3. 40 4. 40 5. 785 6. 518.1 7. $x - 5$ 8. $b - a$ 9. $8x^2 - 16x$ 10. -2 11. 0 12. Undefined
13. $-4xy^4$ 14. x^5 15. y^6 16. $243r^{40}s^{100}$ 17. 9 18. 4 19. $\frac{25}{4}$ 20. $(x+a)(x-a)$ 21. $(x-3)(x-3)$ 22. $x^2(x-y)$
23. $y(x-1)$ 24. $2(y-1)(y+1)$ 25. $(2y-3)(4y^2 + 6y + 9)$

Problem Set 3.1

1. a. $\frac{1}{5}$ b. $\frac{1}{x-3}; x \neq \pm 3$ c. $\frac{1}{x+3}; x \neq \pm 3$ d. $\frac{x^2 + 3x + 9}{x+3}; x \neq \pm 3$ 3. $\frac{x-4}{6}$ 5. $\frac{4x-3}{x(x+y)}$ 7. $(a^2 + 9)(a+3)$ 9. $\frac{a-6}{a+6}$ 11. $\frac{2y+3}{y+1}$
13. $\frac{a^2 - ab + b^2}{a-b}$ 15. $\frac{x+3}{3}$ 17. $(x^2 + 4)(x+2)$ 19. $\frac{x-5}{x+6}$ 21. $\frac{2x-3}{x-4}$ 23. $\frac{2(x-1)}{x}$ 25. $\frac{2x+3y}{2x+y}$ 27. $\frac{x+3}{y-4}$ 29. $\frac{x+b}{x-2b}$
31. $x + 2$ 33. $\frac{2x^2 - 5}{3x-2}$ 35. $\frac{x^2 + 2x + 4}{x+2}$ 37. $4 + t$ 39. $\frac{4x^2 + 6x + 9}{2x+3}$ 41. -1 43. $-(y+6)$ 45. $-\frac{3a+1}{3a-1}$ 47. 3
49. $x + a$ 51. $2; 2$ 53. Undefined; 4 55. $1; 1$ 57. $3; 3$
59. The graph of $y = x + 2$ includes the point $(2, 4)$ while the other graph does not.

61.
Weeks	Weight (pounds)
x	W
0	200
1	194
4	184
12	173
24	168

63. 6.8 feet per second

65. $3,768$ inches per minute; $2,826$ inches per minute   67. $3x^2 - 7x + 4$ 69. 9 71. $8x^2$ 73. $2x^3$ 75. $-2x^2y^2$ 77. 185.12
79. $4x^3 - 8x^2$ 81. $4x^3 - 6x - 20$ 83. $-3x + 9$ 85. $(x+a)(x-a)$ 87. $(x+y)(x-7y)$
89. a. 2 b. -4 c. Undefined   d. 2 e. 1 f. -6

Problem Set 3.2

1. $\frac{1}{6}$ 3. $\frac{9}{4}$ 5. $\frac{1}{2}$ 7. $\frac{8}{3}$ 9. 2 11. $\frac{5y^5}{3x^2}$ 13. $\frac{5x^3}{8y^2}$ 15. $\frac{15y}{x^2}$ 17. $\frac{b}{a}$ 19. $\frac{2y^5}{z^3}$ 21. $\frac{x+3}{x+2}$ 23. $y+1$
25. $\frac{3(x+4)}{x-2}$ 27. 1 29. $\frac{(a-2)(a+2)}{a-5}$ 31. $\frac{9t^2-6t+4}{4t^2-2t+1}$ 33. $\frac{x+3}{x+4}$ 35. $\frac{5a-b}{9a^2+15ab+25b^2}$ 37. 2 39. $\frac{x(x-1)}{x^2+1}$ 41. $\frac{(a+4b)(a-3b)}{(a-4b)(a+5b)}$
43. $\frac{2y-1}{2y-3}$ 45. $\frac{(y-2)(y+1)}{(y+2)(y-1)}$ 47. $\frac{x-1}{x+1}$ 49. $\frac{x-2}{x+3}$ 51. a. $\frac{5}{21}$ b. $\frac{5x+3}{25x^2+15x+9}$ c. $\frac{5x-3}{25x^2+15x+9}$ d. $\frac{5x+3}{5x-3}$
53. a. $\frac{(x+2)^2}{(x-1)^2}$ b. $(x-3)^2$ 55. $\frac{(x+2)(x-4)}{4(x+9)}$ 57. $(x+3)(x+4)$ 59. $3x$ 61. $2(x+5)$ 63. $x-2$ 65. $-(y-4)$
67. $(a-5)(a+1)$ 69. $\frac{(x-4)^2}{x-3}$ 71. $-\frac{x-1}{x+3}$ 73. $(y-2)(x-7)$ 75. $\frac{(x+3)^2}{x-4}$ 77. $\frac{(2x-3)(x-1)}{x-2}$
79.

Number of Copies	Price per Copy ($)
1	20.33
10	9.33
20	6.40
50	4.00
100	3.05

81. $305.00 83. x^2-6x+9 85. $10x^5+8x^3-6x^2$ 87. $12a^2+11a-5$
89. $12xy-6x+28y-14$ 91. $9-6t^2+t^4$ 93. $3x^3+18x^2+33x+18$ 95. $\frac{2}{3}$
97. $\frac{47}{105}$ 99. $x-7$ 101. $(x+1)(x-1)$ 103. $2(x+5)$ 105. $(a-b)(a^2+ab+b^2)$
107. $\frac{x^4-x^2y^2+y^4}{x^2+y^2}$ 109. $\frac{(a+5)(a-1)}{3a^2-2a+1}$ 111. $\frac{a(c-1)}{a-b}$

Problem Set 3.3

1. $\frac{5}{4}$ 3. $\frac{1}{3}$ 5. $\frac{41}{24}$ 7. $\frac{19}{144}$ 9. $\frac{31}{24}$ 11. 1 13. -1 15. $\frac{1}{x+y}$ 17. 1 19. $\frac{a^2+2a-3}{a^3}$ 21. 1
23. a. $\frac{1}{16}$ b. $\frac{9}{4}$ c. $\frac{13}{24}$ d. $\frac{5x+15}{(x-3)^2}$ e. $\frac{x+3}{5}$ f. $\frac{x-2}{x-3}$ 25. $\frac{4-3t}{2t^2}$ 27. $\frac{1}{2}$ 29. $\frac{x+3}{2(x+1)}$ 31. $\frac{a-b}{a^2+ab+b^2}$ 33. $\frac{2y-3}{4y^2+6y+9}$
35. $\frac{2(2x-3)}{(x-3)(x-2)}$ 37. $\frac{1}{2t-7}$ 39. $\frac{4}{(a-3)(a+1)}$ 41. $\frac{-4x^2}{(2x+1)(2x-1)(4x^2+2x+1)}$ 43. $\frac{2}{(2x+3)(4x+3)}$ 45. $\frac{a}{(a+4)(a+5)}$ 47. $\frac{x+1}{(x-2)(x+3)}$
49. $\frac{x-1}{(x+1)(x+2)}$ 51. $\frac{1}{(x+2)(x+1)}$ 53. $\frac{1}{(x+2)(x+3)}$ 55. $\frac{4x+5}{2x+1}$ 57. $\frac{22-5t}{4-t}$ 59. $\frac{2x^2+3x-4}{2x+3}$ 61. $\frac{2x-3}{2x}$ 63. $\frac{1}{2}$
65. $-\frac{(2x-11)}{4(x-4)}$ 67. $\frac{3}{x+4}$ 69. $\frac{(2x+1)(x+5)}{(x-2)(x+1)(x+3)}$ 71. $\frac{(x+2)(x-1)}{(3x-2)(3x+2)(x-2)}$ 73. $\frac{3(x-2)(2x+1)}{(2x+1)(x-1)(x-3)}$ 75. $\frac{2x}{(x-9)(x-7)}$
77. $x+\frac{4}{x}=\frac{x^2+4}{x}$ 79. $\frac{51}{10}=5.1$ 81. a. $T=120$ months b. The two objects will never meet. 83. 5.4×10^4 85. 3.4×10^{-4}
87. 6,440 89. 0.00644 91. 1.2×10^4 93. $\frac{6}{5}$ 95. $x+2$ 97. $3-x$ 99. $(x+2)(x-2)$ 101. $\frac{x-1}{x+3}$ 103. $\frac{a^2+ab+a-v}{u+v}$
105. $\frac{6x+5}{(4x-1)(3x-4)}$ 107. $\frac{y(y^2+1)}{(y+1)^2(y-1)^2}$

Problem Set 3.4

1. $-\frac{35}{3}$ 3. $-\frac{18}{5}$ 5. $\frac{36}{11}$ 7. 2 9. 5 11. 2 13. $-3, 4$ 15. $1, -\frac{4}{3}$ 17. $-\frac{9}{5}, 5$ 19. $\frac{9}{2}$ 21. $\emptyset$ 23. $3, -\frac{4}{3}$ 25. $-\frac{4}{3}$
27. $\emptyset$ 29. a. $\frac{1}{3}$ b. 3 c. 9 d. 4 e. $\frac{1}{3}, 3$ 31. a. $\frac{6}{(x-4)(x+3)}$ b. $\frac{x-3}{x-4}$ c. possible $-2, 5$; only 5 checks
33. Possible solution -1, does not check; $\emptyset$ 35. 5 37. $-\frac{1}{2}, \frac{5}{3}$ 39. $\frac{2}{3}$ 41. 18 43. Possible solution 4, does not check; $\emptyset$
45. Possible solutions 3 and -4; only -4 checks; 47. -6 49. -5 51. $\frac{53}{17}$ 53. Possible solutions 1 and 2; only 2 checks;
55. Possible solution 3, does not check; $\emptyset$ 57. $\frac{22}{3}$ 59. 2 61. 1, 5
63. $x=\frac{ab}{a-b}$ 65. $R=\frac{R_1R_2}{R_1+R_2}$ 67. $y=\frac{x-3}{x-1}$ 69. $y=\frac{1-x}{3x-2}$
73. (Table to the right) 75. 5 77. $-6, -5$ or $5, 6$ 79. $3, 4, 5$ 81. 2,358
83. 12.3 85. 3 87. $9, -1$ 89. 60 91. $-\frac{2}{3}, -\frac{5}{2}, \frac{5}{2}$ 93. $-2, 2, 3$
95. $-\frac{a}{5}$ 97. $v=\frac{16t^2+s}{t}$ 99. $f=\frac{pg}{g-p}$

Time t(sec)	Speed of Kayak Relative to the Water v(m/sec)	Current of the River c(m/sec)
240	4	1
300	4	2
514	4	3
338	3	1
540	3	2
N/A	3	3

Problem Set 3.5

1. $2\sqrt{2}$ 3. $7\sqrt{2}$ 5. $12\sqrt{2}$ 7. $4\sqrt{5}$ 9. $4\sqrt{3}$ 11. $15\sqrt{3}$ 13. $3\sqrt[3]{2}$ 15. $4\sqrt[3]{2}$ 17. $6\sqrt[3]{2}$ 19. $2\sqrt[5]{2}$ 21. $3x\sqrt{2x}$
23. $5x^2\sqrt{2x}$ 25. $2a\sqrt{5a}$ 27. $4y^3\sqrt{3}$ 29. $2y\sqrt[4]{2y^3}$ 31. $2xy^2\sqrt[3]{5xy}$ 33. $4abc^2\sqrt[3]{3b}$ 35. $2bc\sqrt[3]{6a^2c}$ 37. $2xy^2\sqrt[5]{2x^3y^2}$
39. $3xy^2z\sqrt[5]{x^2}$ 41. $2\sqrt{3}$ 43. $\sqrt{-20}$, which is not a real number 45. $\frac{\sqrt{11}}{2}$ 47. $\frac{2\sqrt{3}}{3}$ 49. $\frac{5\sqrt{6}}{6}$ 51. $\frac{\sqrt{2}}{2}$ 53. $\frac{\sqrt{5}}{5}$
55. $2\sqrt[3]{4}$ 57. $\frac{2\sqrt[3]{3}}{3}$ 59. $\frac{\sqrt[3]{24x^2}}{2x}$ 61. $\frac{\sqrt[3]{8y^3}}{y}$ 63. $\frac{\sqrt[3]{36xy^2}}{3y}$ 65. $\frac{\sqrt[3]{6xy^2}}{3y}$ 67. $\frac{\sqrt[3]{2x}}{2x}$ 69. $\frac{3x\sqrt{15xy}}{5y}$ 71. $\frac{5xy\sqrt{6xz}}{2z}$
73. $\frac{2ab\sqrt[3]{6ac^2}}{3c}$ 75. $\frac{2xy^2\sqrt[3]{3z^2}}{3z}$ 77. $\sqrt{x}$ 79. $\sqrt[6]{xy}$ 81. $\sqrt[12]{a}$ 83. $x\sqrt[9]{6x}$ 85. $ab^2c\sqrt[6]{c}$ 87. $abc^2\sqrt[15]{3a^2b}$ 89. $2b^2\sqrt[9]{a}$
91. $5|x|$ 93. $3|xy|\sqrt{3x}$ 95. $|x-5|$ 97. $|2x+3|$ 99. $2|a(a+2)|$
101. $2|x|\sqrt{x-2}$ 103. $\sqrt{9+16}=\sqrt{25}=5$; $\sqrt{9}+\sqrt{16}=3+4=7$ 105. $5\sqrt{13}$ feet
107. a. 13 feet b. $2\sqrt{14}\approx 7.5$ feet 109. $7x$ 111. $27xy^2$ 113. $\frac{5}{6}x$ 115. $3\sqrt{2}$
117. $5y\sqrt{3xy}$ 119. $2a\sqrt[3]{3ab^2}$ 121. $\frac{y^3}{x^2}$ 123. 1 125. $\frac{4x^2-6x+9}{9x^2-3x+1}$ 127. $12\sqrt[3]{5}$
129. $6\sqrt[3]{49}$ 131. $\frac{\sqrt[10]{a^7}}{a}$ 133. $\frac{\sqrt[20]{a^9}}{a}$ 135. (Graph to the right)
137. About $\frac{3}{4}$ of a unit 139. $x=0$

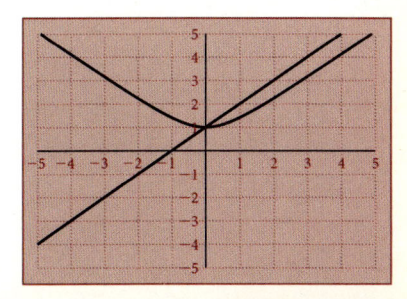

Problem Set 3.6

1. a. $\frac{-2 \pm \sqrt{10}}{3}$ b. $\frac{2 \pm \sqrt{10}}{3}$ c. $\frac{-2 \pm \sqrt{10}}{2}$ d. $\frac{-3 \pm \sqrt{15}}{2}$ 3. a. $\frac{-1 \pm \sqrt{17}}{4}$ b. $\frac{4 \pm \sqrt{10}}{3}$ c. $\frac{5 \pm \sqrt{17}}{4}$ d. $\frac{1 \pm \sqrt{19}}{3}$ e. $\frac{-1 \pm \sqrt{10}}{3}$ 5. 1, 2
7. $\frac{3 \pm \sqrt{22}}{2}$ 9. 0, 5 11. $0, -\frac{4}{5}$ 13. $\frac{3 \pm \sqrt{5}}{4}$ 15. $-3 \pm \sqrt{17}$ 17. $\frac{-1 \pm \sqrt{7}}{2}$ 19. 1 21. $-1, \frac{3}{4}$ 23. $4 \pm \sqrt{2}$ 25. $\frac{1}{2}, 1$
27. $-\frac{1}{2}, 3$ 29. $-2, 1$ 31. $1 \pm \sqrt{2}$ 33. $\frac{-3 \pm \sqrt{5}}{2}$ 35. $3, -5$ 37. $0, \frac{-1 - \sqrt{7}}{2}$ 39. $0, 1 \pm \sqrt{3}$
41. $-1, 0, \frac{1}{3}$ 43. a and b 45. a. $\frac{5}{3}, 0$ b. $\frac{5}{3}, 0$ 47. No, $2 \pm \sqrt{11}$ 49. Yes 51. a. $-1, 3$ b. $1 \pm \sqrt{15}$ c. ± 2 d. $2 \pm 2\sqrt{2}$
53. a. $-2, \frac{5}{3}$ b. $\pm \frac{3\sqrt{6}}{2}$ c. $\emptyset$ d. ± 1 55. 2 seconds 57. 20 or 60 items 59. 0.49 centimeter (8.86 cm is impossible)
61. a. $\ell + w = 10, \ell w = 15$ b. 8.16 yards, 1.84 yards
c. Two answers are possible because either dimension (long or short) may be considered the length.
63. $4y + 1 + \frac{2}{2y - 7}$ 65. $x^2 + 7x + 12$ 67. 169 69. 0 71. ± 12 73. $x^2 - x - 6$ 75. $x^3 - 4x^2 - 3x + 18$ 77. $-2\sqrt{3}, \sqrt{3}$
79. $\frac{-1 \pm \sqrt{3}}{\sqrt{2}} = \frac{-\sqrt{2} \pm \sqrt{6}}{2}$ 81. $\frac{2\sqrt{2} \pm \sqrt{5}}{\sqrt{3}} = \frac{2\sqrt{6} \pm \sqrt{15}}{3}$

Chapter 3 Review

1. $\frac{25x^2}{7y^3}$ 2. $\frac{a(a-b)}{4}$ 3. $\frac{x-5}{x+5}$ 4. $\frac{a+1}{a-1}$ 5. $\frac{9}{5}$ 6. $\frac{3x}{4y^2}$ 7. $\frac{x-1}{x^2+1}$ 8. 1 9. $\frac{x+2}{x-2}$ 10. $(2x-3)(x+3)$ 11. $\frac{31}{30}$
12. -1 13. $\frac{x^2+x+1}{x^3}$ 14. $\frac{1}{(y+4)(y+3)}$ 15. $\frac{x-1}{2(x+1)(x+2)}$ 16. $\frac{15x-2}{5x-2}$ 17. 6 18. 1 19. -6
20. Possible solution -3, which does not check; $\emptyset$ 21. $\frac{22}{3}$ 22. Possible solutions 4 and -5; only 4 checks 23. $2\sqrt{3}$ 24. $5\sqrt{2}$
25. $2\sqrt[3]{2}$ 26. $3x\sqrt{2}$ 27. $4ab^2c\sqrt{5a}$ 28. $2abc\sqrt[3]{2bc^2}$ 29. $\frac{3\sqrt{2}}{2}$ 30. $3\sqrt[3]{4}$ 31. $\frac{4x\sqrt{21xy}}{7y}$ 32. $\frac{2y\sqrt[4]{45x^2z^2}}{3z}$ 33. $-5, 2$
34. $0, \frac{9}{4}$ 35. $\frac{2 \pm \sqrt{23}}{2}$ 36. $1 \pm \sqrt{2}$ 37. $-\frac{4}{3}, \frac{1}{2}$ 38. 3 39. $5 \pm \sqrt{7}$ 40. $0, \frac{5 \pm \sqrt{21}}{2}$ 41. $-1, \frac{3}{5}$ 42. $-2 \pm 2\sqrt{3}$
43. $-\frac{1}{3}, 1$ 44. $\frac{3 \pm \sqrt{29}}{2}$

Chapter 3 Cumulative Review

1. $-\frac{27}{8}$ 2. $\frac{36}{25}$ 3. 47 4. -55 5. 16 6. 25 7. $\frac{3}{5}$ 8. 2 9. $\frac{x^3}{y^7}$ 10. $\frac{a^8}{b^2}$ 11. $2\sqrt[3]{4}$ 12. $5x\sqrt{2x}$ 13. $\frac{9}{20}$
14. $\frac{9}{4}$ 15. -2 16. $5a - 7b > 5a + 7b$ 17. 159 18. -161 19. 90 20. 80 21. $0, \frac{5 \pm \sqrt{37}}{6}$ 22. $12x$
23. $\frac{1}{(y+3)(y+2)}$ 24. $-\frac{1}{x+y}$ 25. $12t^4 - \frac{1}{12}$ 26. $x + 2$ 27. $2x^{2n} - 3x^{3n}$ 28. $a^3 - a^2 + 2a - 4 + \frac{7}{a+2}$ 29. x^3 30. $\frac{x^3}{y^7}$
31. $x + 3$ 32. $x - 6y$ 33. $(5a - 2b)(5a + 2b)(25a^2 + 4b^2)$ 34. $(x + 7)(x - 10)$ 35. $(x + 5 - y)(x + 5 + y)$
36. $\left(y + \frac{2}{3}x\right)\left(y^2 - \frac{2}{3}xy + \frac{4}{9}x^2\right)$ 37. -20 38. 3 39. 1 40. $\frac{5}{6}$ 41. 5 42. $\frac{3}{2}, 4$ 43. $3\sqrt[4]{2}$ 44. $\frac{\sqrt[3]{2x^2y}}{x}$ 45. 27,300,000
46. 19,700,000

Chapter 3 Test

1. $x + y$ 2. $\frac{x-1}{x+1}$ 3. 5.9 feet/second 4. $2(a + 4)$ 5. $4(a + 3)$ 6. $x + 3$ 7. $\frac{38}{105}$ 8. $\frac{7}{8}$ 9. $\frac{1}{a-3}$ 10. $\frac{3(x-1)}{x(x-3)}$
11. $\frac{x}{(x+4)(x+5)}$ 12. $\frac{x+4}{(x+1)(x+2)}$ 13. $-\frac{3}{5}$ 14. $\emptyset$ 15. $\frac{3}{13}$ 16. $-2, 3$ 17. $5xy^2\sqrt{5xy}$ 18. $2x^2y^2\sqrt{5xy^2}$ 19. $\frac{\sqrt{6}}{3}$
20. $\frac{2a^2b\sqrt{15bc}}{5c}$ 21. $-\frac{9}{2}, \frac{1}{2}$ 22. $3 \pm \sqrt{2}$ 23. $5 \pm \sqrt{46}$ 24. $1 \pm \sqrt{10}$ 25. $\frac{3 \pm \sqrt{17}}{4}$ 26. $-1 \pm \sqrt{7}$
27. $\frac{1}{2}$ second, $\frac{3}{2}$ seconds 28. 15 or 100 cups

Chapter 4

Pretest for Chapter 4

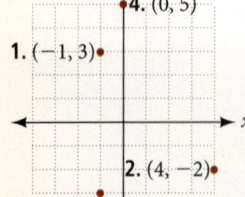

5. yes 6. no 7. no 8. yes 9. -2 10. 0 11. 17 12. 18 13. 12 14. -9
15. 1 16. 19

1. $(-1, 3)$
2. $(4, -2)$
3. $(-1, -3)$
4. $(0, 5)$

Answers to Odd-Numbered Problems   A-9

Getting Ready for Chapter 4

1. 324   2. 0.0005   3. 3   4. −9   5. 3   6. 0   7. 113   8. 7   9. 3   10. 4   11. $-\frac{3}{50}$ 12. $-\frac{3}{5}$ 13. $y = \frac{2}{3}x - \frac{5}{3}$
14. $y = -\frac{1}{2}x + \frac{7}{2}$ 15. (5, 0), (4, 3)   16. (0, 0), (2, 0)   17.

t	0	1	2	3	4
h	0	48	64	48	0

18.

t	4	6	8	10
s	15	10	7.5	6

19. 135   20. $\frac{1}{2}$

Problem Set 4.1

1. *(graph with points: a. (1, 2), b. (−1, −2), c. (5, 0), d. (0, 2), e. (−5, −5), f. ($\frac{1}{2}$, 2))*

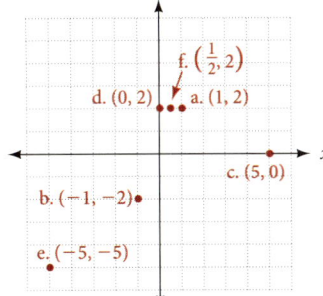

3. A. (4, 1)   B. (−4, 3)   C. (−2, −5)   D. (2, −2)   E. (0, 5)   F. (−4, 0)   G. (1, 0)

5. *(graph: $2x - 3y = 6$, through (3, 0) and (0, −2))*

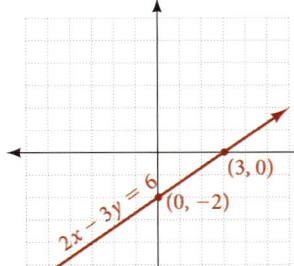

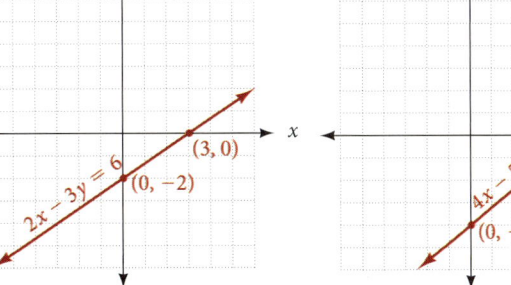

7. *(graph: $4x - 5y = 20$, through (5, 0) and (0, −4))*

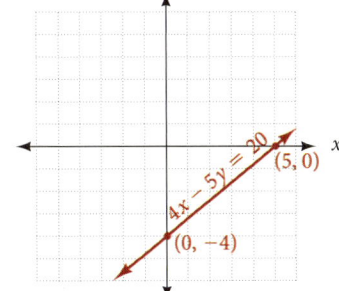

9. *(graph: $y = 2x + 3$, through $(-\frac{3}{2}, 0)$ and (0, 3))*

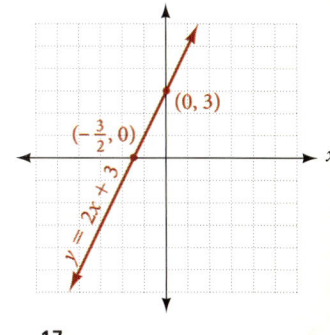

11. *(graph: $-3x + 2y = 12$, through (−4, 0) and (0, 6))*

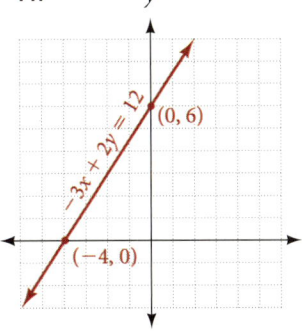

13. *(graph: $6x - 5y - 20 = 0$, through $(\frac{10}{3}, 0)$ and (0, −4))*

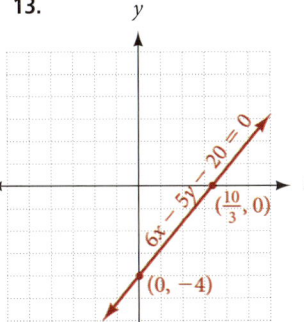

15. *(graph: $y = 3x - 5$, through $(\frac{5}{3}, 0)$ and (0, −5))*

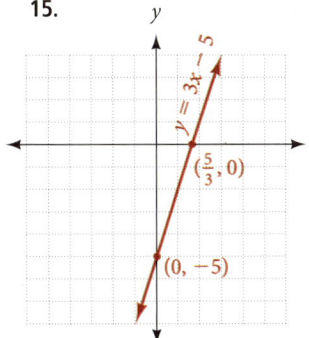

17. *(graph: $\frac{x}{2} + \frac{y}{3} = 1$, through (2, 0) and (0, 3))*

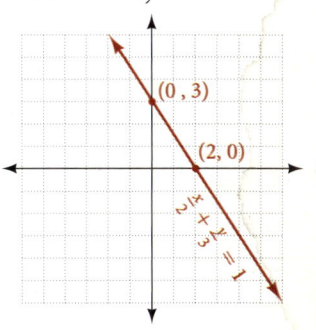

19. b.

21. *(graph: $y = \frac{1}{3}x$, through (−3, −1), (0, 0), (3, 1))*

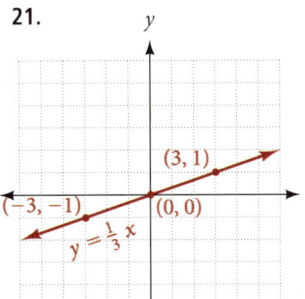

23. *(graph: $-2x + y = -3$, through $(\frac{3}{2}, 0)$ and (0, −3))*

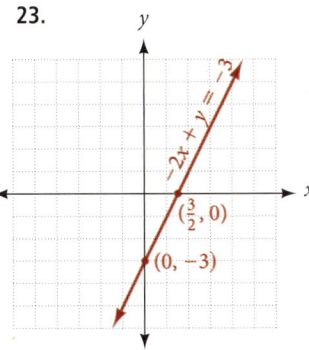

25. *(graph: $y = -\frac{2}{3}x + 1$, through (−3, 3), (0, 1), (3, −1))*

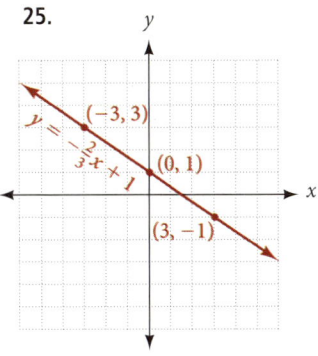

27. *(graph: $\frac{x}{3} + \frac{y}{4} = 1$, through (0, 4) and (3, 0))*

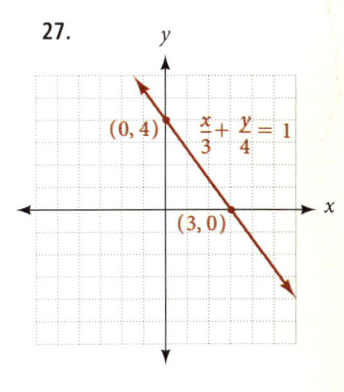

29. b.

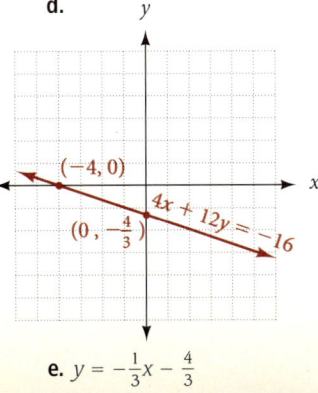

31. a. −7   b. −4   c. $-\frac{4}{3}$
   d. *(graph: $4x + 12y = -16$, through (−4, 0) and $(0, -\frac{4}{3})$)*
   e. $y = -\frac{1}{3}x - \frac{4}{3}$

33. a. *(graph: $y = 2x$, through (−2, −4) and (1, 2))*

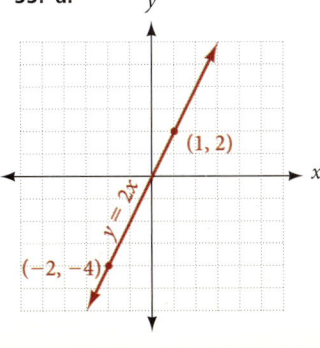

b. *(graph: $x = -3$)*

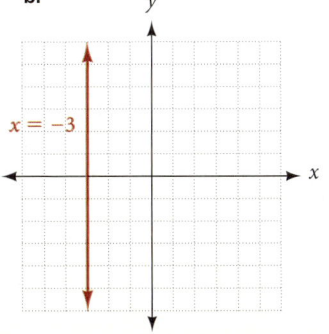

c. *(graph: $y = 2$)*

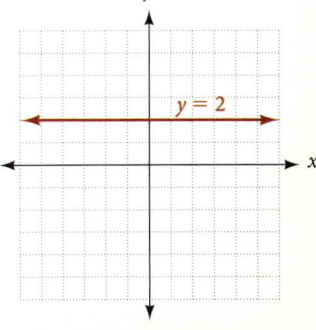

35. a.

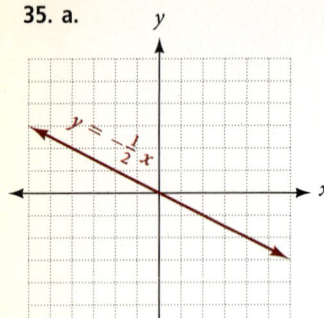

b. $x = 4$

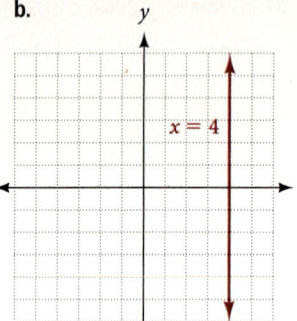

c. $y = -3$

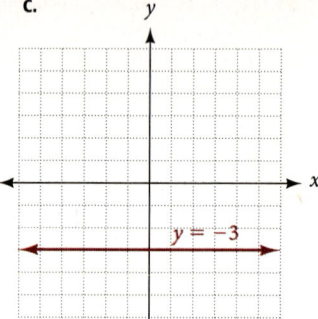

37. $0.02x + 0.03y = 0.06$

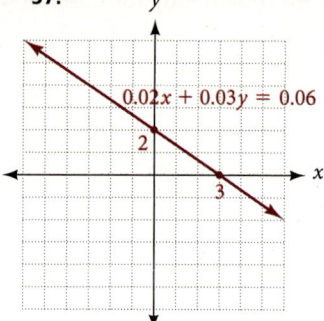

39. $y = 3x$; points $(-2, -6)$, $(-1, -3)$, $(0, 0)$, $(1, 3)$, $(2, 6)$

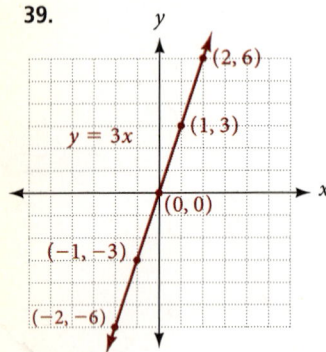

41. a. (5, 40), (10, 80), (20, 160)
b. $320
c. 30 hours
d. No, if she works 35 hours she should be paid $280

43. Projected Non-Camera Phone Sales — 2006: 300, 2007: 250, 2008: 175, 2009: 150, 2010: 125 (Phones in millions)

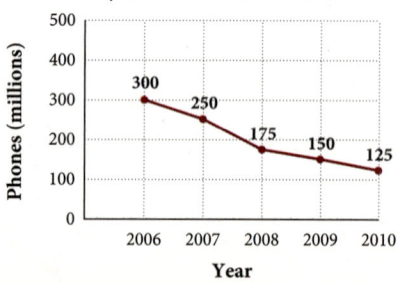

45. (1985, 20.2), (1990, 34.4), (1995, 44.8), (2000, 65.4), (2004, 99.4)

47.

x (year)	1970	1975	1980	1985	1990	1995	2000	2005
	21.5	21.75	22.75	23.75	24.25	24.5	24.75	25

49. $A = (1, 2)$, $B = (6, 7)$ **51.** $A = (3, 3)$, $B = (3, 6)$, $C = (8, 6)$ **53. a.** Yes **b.** No **c.** Yes   **55.** 2   **57.** -2 **59.** 1,200
61. 2   **63.** $\frac{8}{15}$ **65.** $-\frac{6}{100}$ **67.** 1   **69.** $-\frac{4}{3}$ **71.** Undefined   **73. a.** $\frac{3}{2}$ **b.** $-\frac{3}{2}$
75. x-intercept = $\frac{c}{a}$, y-intercept = $\frac{b}{a}$,   **77.** x-intercept = a, y-intercept = b,

Problem Set 4.2

1. Domain = {1, 2, 4}; Range = {3, 5, 1}; a function   **3.** Domain = {−1, 1, 2}; Range = {3, −5}; a function
5. Domain = {7, 3}; Range = {−1, 4}; not a function   **7.** Domain = {a, b, c, d}; Range = {3, 4, 5}; a function
9. Domain = {a}; Range = {1, 2, 3, 4}; not a function   **11.** Yes   **13.** No   **15.** No   **17.** Yes   **19.** Yes
21. Domain = {x | −5 ≤ x ≤ 5}; Range = {y | 0 ≤ y ≤ 5}   **23.** Domain = {x | −5 ≤ x ≤ 3}; Range = {y | y = 4}
25. Domain = All real numbers; Range = {y | y ≥ −1}; a function $y = x^2 - 1$
27. Domain = All real numbers; Range = {y | y ≥ 4}; a function $y = x^2 + 4$
29. Domain = {x | x ≥ −1}; Range = All real numbers; not a function $x = y^2 - 1$
31. Domain = {x | x ≥ 4}; Range = All real numbers; not a function $x = y^2 + 4$

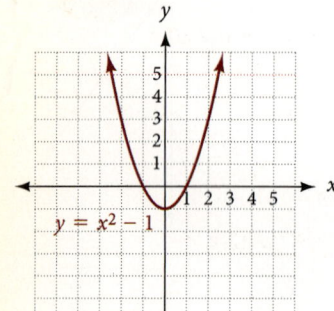

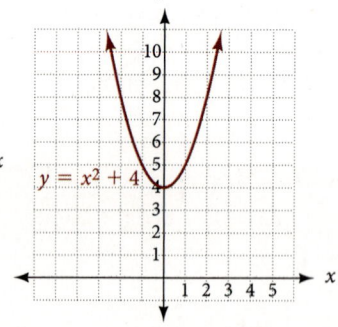

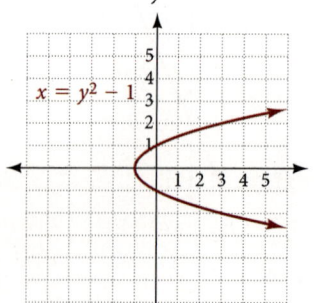

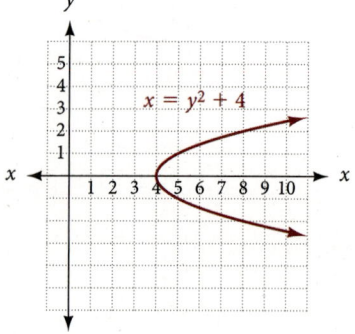

33. Domain = All real numbers; Range = $\{y \mid y \geq 0\}$; a function

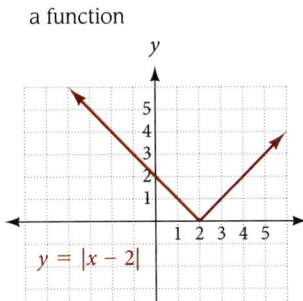

$y = |x - 2|$

35. Domain = All real numbers; Range = $\{y \mid y \geq -2\}$; a function

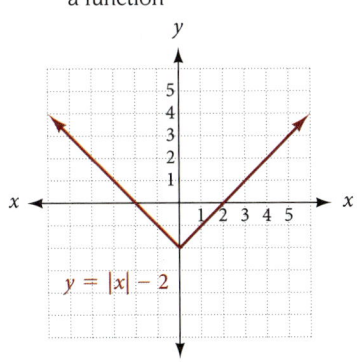

$y = |x| - 2$

37. Domain = {2004–2010}; Range = {680, 730, 800, 900, 920, 990, 1030};

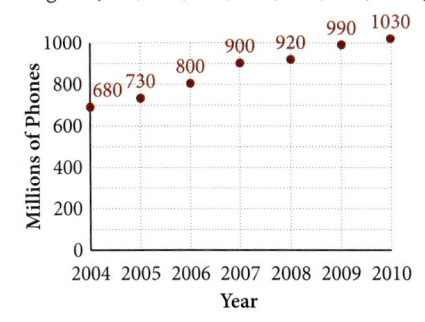

39. a. $y = 8.5x$ for $10 \leq x \leq 40$ **b.** (table below) **c.**

Hours Worked	Function Rule	Gross Pay ($)
x	$y = 8.5x$	y
10	$y = 8.5(10) = 85$	85
20	$y = 8.5(20) = 170$	170
30	$y = 8.5(30) = 255$	255
40	$y = 8.5(40) = 340$	340

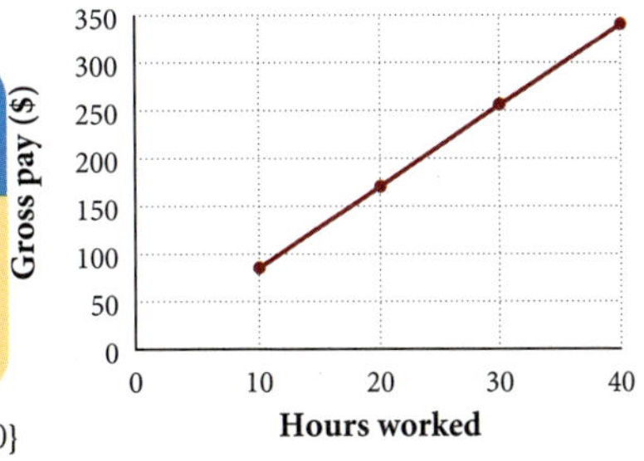

d. Domain = $\{x \mid 10 \leq x \leq 40\}$; Range = $\{y \mid 85 \leq y \leq 340\}$
e. Minimum = $85; Maximum = $340

41. a.

Time (sec)	Function Rule	Distance (ft)
t	$h = 16t - 16t^2$	h
0	$h = 16(0) - 16(0)^2$	0
0.1	$h = 16(0.1) - 16(0.1)^2$	1.44
0.2	$h = 16(0.2) - 16(0.2)^2$	2.56
0.3	$h = 16(0.3) - 16(0.3)^2$	3.36
0.4	$h = 16(0.4) - 16(0.4)^2$	3.84
0.5	$h = 16(0.5) - 16(0.5)^2$	4
0.6	$h = 16(0.6) - 16(0.6)^2$	3.84
0.7	$h = 16(0.7) - 16(0.7)^2$	3.36
0.8	$h = 16(0.8) - 16(0.8)^2$	2.56
0.9	$h = 16(0.9) - 16(0.9)^2$	1.44
1	$h = 16(1) - 16(1)^2$	0

b. Domain = $\{t \mid 0 \leq t \leq 1\}$; Range = $\{h \mid 0 \leq h \leq 4\}$
c.

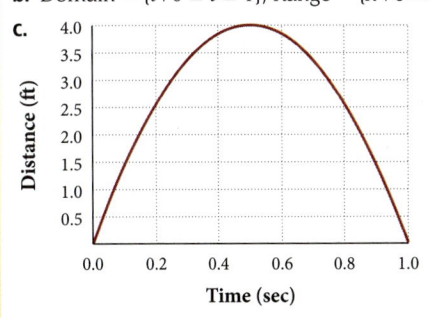

43. a. $A = \pi r^2, 0 \leq r \leq 3$

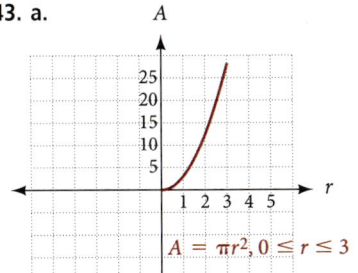

b. Domain = $\{r \mid 0 \leq r \leq 3\}$
Range = $\{A \mid 0 \leq A \leq 9\pi\}$

45. a. Yes
b. Domain = $\{t \mid 0 \leq t \leq 6\}$; Range = $\{h \mid 0 \leq h \leq 60\}$
c. $t = 3$
d. $h = 60$
e. $t = 6$

47. a. III **b.** I **c.** II **d.** IV   **49.** 10   **51.** -14 **53.** 1   **55.** -3 **57.** $-\frac{6}{5}$ **59.** $-\frac{7}{640}$ **61.** 150   **63.** 113   **65.** -9 **67. a.** 6 **b.** 7.5   **69. a.** 27 **b.** 6

71. Domain = All real numbers
Range = {y | y ≤ 5}
A Function

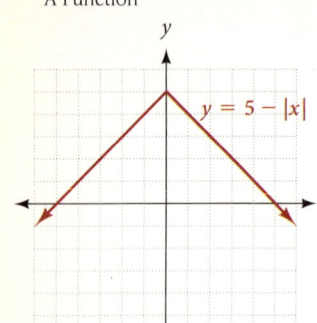

73. Domain = {x | x ≥ 3}
Range = All real numbers
Not a Function

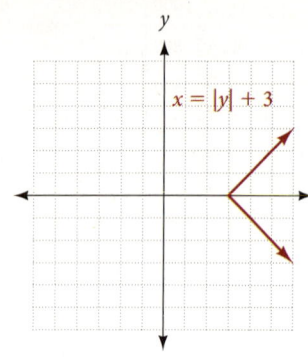

75. Domain = {x | −4 ≤ x ≤ 4}
Range = {y | −4 ≤ y ≤ 4}
Not a Function

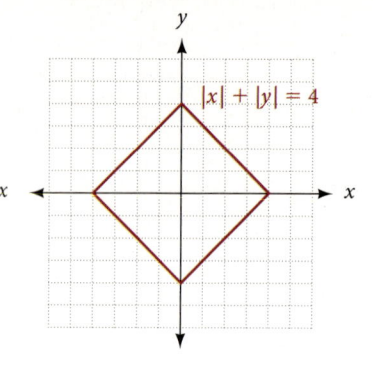

Problem Set 4.3

1. −1 **3.** −11 **5.** 2 **7.** 4 **9.** 35 **11.** −13 **13.** 1 **15.** −9 **17.** 8 **19.** 19 **21.** 16 **23.** 0 **25.** $3a^2 - 4a + 1$
27. 4 **29.** 0 **31.** 2 **33.** −8 **35.** −1 **37.** $2a^2 - 8$ **39.** $2b^2 - 8$ **41.** 0 **43.** −2 **45.** −3
47.   **49.** x = 4   **51.**

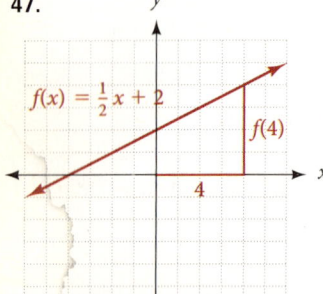

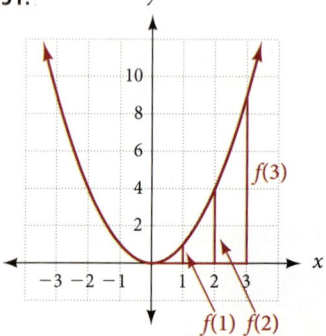

53. V(3) = 300, the painting is worth $300 in 3 years; V(6) = 600, the painting is worth $600 in 6 years. **55. a.** False **b.** False **c.** False
57. a. True **b.** True **c.** True **d.** False **e.** True
59. a. $5,625 **b.** $1,500 **c.** {t | 0 ≤ t ≤ 5} **d.**
e. {V(t) | 1,500 ≤ V(t) ≤ 18,000}
f. about 2.42 years

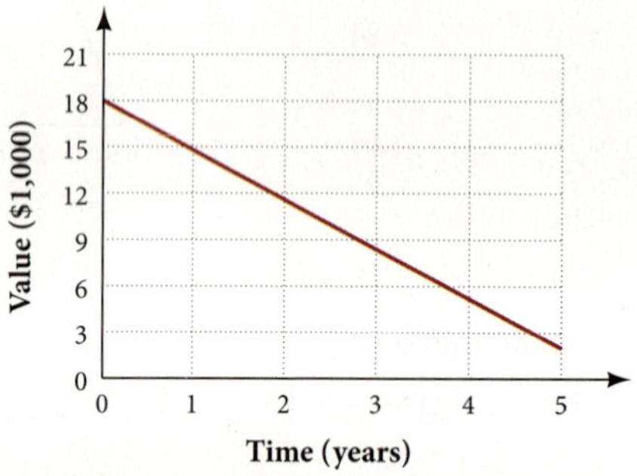

61. $-0.1x^2 + 27x - 500$ **63.** $6x^2 - 2x - 4$ **65.** $2x^2 + 8x + 8$ **67.** $0.6M - 42$ **69.** $4x^2 - 7x + 3$ **71. a.** 2 **b.** 0 **c.** 1 **d.** 4
73. a.

Weight (ounces)	0.6	1.0	1.1	2.5	3.6	4.8	5.0	5.3
Cost ($)	0.88	0.88	1.05	1.22	1.39	1.56	1.56	1.73

b. More than 2 ounces, but not more than 3 ounces; 2 < x ≤ 3
c. Domain: {x | 0 < x ≤ 6}
d. Range: {c | c = 0.88, 1.05, 1.22, 1.39, 1.56, 1.73}

Problem Set 4.4

1. $6x + 2$ **3.** $-2x + 8$ **5.** $4x - 7$ **7.** $3x^2 + x - 2$ **9.** $-2x + 3$ **11.** $9x^3 - 15x^2$ **13.** $\dfrac{3x^2}{3x-5}$ **15.** $\dfrac{3x-5}{3x^2}$
17. $3x^2 + 4x - 7$ **19.** 15 **21.** 98 **23.** $\dfrac{3}{2}$ **25.** 1 **27.** 40 **29.** 147 **31. a.** 81 **b.** 29 **c.** $(x+4)^2$ **d.** $x^2 + 4$
33. a. −2 **b.** −1 **c.** $4x^2 + 12x - 1$ **35.** $(f \circ g)(x) = 5\left[\dfrac{x+4}{5}\right] - 4 = x + 4 - 4 = x$; $(g \circ f)(x) = \dfrac{(5x-4)+4}{5} = \dfrac{5x}{5} = x$ **37.** 6 **39.** 2
41. 3 **43.** −8 **45.** 6 **47.** 5 **49.** 3 **51.** −6
53. a. $R(x) = 11.5x - 0.05x^2$ **b.** $C(x) = 2x + 200$ **c.** $P(x) = -0.05x^2 + 9.5x - 200$ **d.** $\overline{C}(x) = 2 + \dfrac{200}{x}$
55. a. $M(x) = 220 - x$ **b.** M(24) = 196 **c.** 142 **d.** 135 **e.** 128 **57.** 196 **59.** 4 **61.** 1.6 **63.** 3 **65.** 2,400

Chapter 4 Review

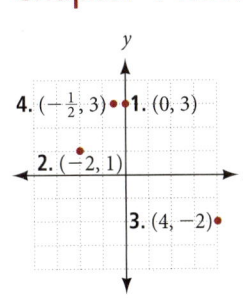

4. $(-\frac{1}{2}, 3)$ 1. $(0, 3)$
2. $(-2, 1)$
3. $(4, -2)$

5. yes 6. no 7. yes 8. no
9. Domain = {0, 1, −1}, Range = {3, 2}, A function
10. Domain = {−1, 0}, Range = {3, 4, 2}, Not a function
11. Domain = { $x \mid -2 \leq x \leq 2$ }, Range = { $y \mid -2 \leq y \leq 2$ }, A function
12. Domain = { $x \mid x \geq -2$ }, Range = All real numbers, Not a function
13. Domain = {2, 3, 4}; Range = {4, 3, 2}; a function
14. Domain = {6, −4, −2}; Range = {3, 0}; a function
15. 0 16. 1 17. 1 18. $3a + 2$ 19. 1 20. 31

Chapter 4 Cumulative Review

1. −25 2. −6 3. 74 4. 144 5. 36 6. 25 7. 27 8. 27 9. 1 10. −9 11. $\frac{7}{9}$ 12. $\frac{7}{9}$ 13. $\frac{1}{10}$ 14. $\frac{2}{21}$
15. −2 16. $-\frac{8}{5}$ 17. $6x - 4y$ 18. $10x + 9y$ 19. 15 20. $\frac{7}{12}$ 21. $\frac{5}{9}$ 22. −2 23. $\frac{11}{2}, -\frac{5}{2}$ 24. ∅ 25. $y = -3x - 1$
26. $y = \frac{3}{2}x - 5$ 27. −1 28. 7 29. 9 30. $2x$ 31. Domain = {−1, 2, 3}; Range = {−1, 3}; a function

Chapter 4 Test

1. $(1, \frac{1}{2})$ 2. $(-\frac{3}{2}, -\frac{5}{2})$ 3. $(2, -2)$ 4. $(-2, 2)$ 5. yes 6. no 7. no 8. yes 9. Domain = {−2, −3}; Range = {0, 1}
10. Domain = All real numbers; Range = { $y \mid y \geq -9$ }; a function 11. Domain = {0, 1, 2}; Range = {0, 3, 5}; a function
12. Domain = All real numbers; Range = All real numbers; a function 13. Domain = { $x \mid -4 \leq x \leq 4$ }; Range = { $y \mid -1 \leq y \leq 3$ }
14. Domain = { $x \mid -3 \leq x \leq 3$ }; Range = { $y \mid -4 \leq y \leq 4$ } 15. 11 16. −4 17. 8 18. 4 19. 5 20. 0 21. 3 22. 0

Chapter 5

Problem Set 5.1

1. 4 3. 5 5. 3 7. 2, 3 9. 3, −5 11. 2 13. 11 15. 3 17. 5 19. 12 21. 4 23. 8 25. 7 27. 11
29. 12 31. 11 33. −9, 9 35. 7 37. 6 39. −3, 5 41. 11 43. $x = y = 35, z = 145$ 45. $x = y = 28, z = 152$
47. $x = 31, y = 62, z = 118$ 49. $x = y = 24, z = 156$

Problem Set 5.2

1. 120° 3. 65° 5. 78° 7. $m \angle 2 = 60°$ 9. $m \angle 1 = 144°; m \angle 6 = 36°$ 11. parallel 13. not enough information
15. not parallel 17. $x = 17$ 19. $x = 10$ 21. $x = 5$ 23. $x = 41$ 25. $x = -3, 11$ 27. $x = 8$ 29. $x = 4, -\frac{3}{2}$
31. $x = 13, -1$ 33. $x = 1, 6$ 35. $x = 6$ 37. $x = 9$ 39. $x = 12, 1$ 41. $x = -2, 9$ 43. $x = 12$

Problem Set 5.3

1. right 3. acute 5. equilateral 7. 22.5°, 45°, 112.5° 9. 53°, 90° 11. 80°, 60°, 40° 13. $x = \pm 8$ 15. $x = 9$
17. $x = 3, -3$ 19. $x = 12$ 21. $x = 12$ 23. $x = 15$ 25. $x = 10$ 27. $x = -2\sqrt{7}, 2\sqrt{7}$ 29. $x = 8$ 31. $x = 32$
33. $x = 44$ 35. $x = 45$ 37. $x = 11$ 39. $x = 25$ 41. $x = 40$ 43. $x = 4$ 45. $x = 36$ 47. $x = 4$
49. $x = 10, m \angle A = 40°, m \angle B = 60°, m \angle C = 80°$ 51. $x = 5, m \angle A = 40°, m \angle B = 65°, m \angle C = 75°$
53. $x = 15, m \angle A = 36°, m \angle B = 80°, m \angle C = 64°$ 55. $x = 12, m \angle A = 12°, m \angle B = 132°, m \angle C = 36°$ 57. $x = 12, m \angle A = 55°, m \angle B = 47°$
59. $x = 5, m \angle A = 63°, m \angle B = 52°$ 61. $x = 6, m \angle A = 24°, m \angle B = 53°$ 63. $x = 5, m \angle A = 80°, m \angle B = 84°$

Problem Set 5.4

1. a. $\angle F$, b. $\angle C$, c. $\overline{FE}$, d. $\overline{AC}$ 3. a. $\angle E$, b. $\angle B$, c. $\overline{AB}$, d. $\overline{DC}$ 5. $\triangle CDB$ 7. $\triangle KLM$ 9. $\overline{GI}$ transitive 11. $\overline{CD}$ reflexive
13. $x = 2, AC = DF = 16$ 15. $x = 3, AC = DF = 5$ 17. $x = 8, AB = DE = 25$ 19. SAS 21. Not Congruent 23. SAS
25. $\triangle ABC \cong \triangle DEF$, SAS; $x = 3$ 27. $\triangle ABC \cong \triangle DEF$, ASA; $x = 5$ 29. $\triangle ABC \cong \triangle DEF$, AAS; $x = 17$ 31. $\triangle ABC \cong \triangle DEF$, SAS; $x = 5$
33. $\triangle ABC \cong \triangle DEC$, SAS; $x = 10, y = 15$ 35. $\triangle ABC \cong \triangle DEC$, SAS; $x = 20, y = 5$ 37. $\triangle ABC \cong \triangle DEC$, SAS; $x = 16, y = 12$
39. $\triangle ABC \cong \triangle DEC$, ASA; $x = 5, y = 18$

Problem Set 5.5

1. 360°   **3.** 540°   **5.** 1440°   **7.** 13   **9.** 13.5   **11.** 63   **13.** $OD = 3.5; m\angle BCD = 80°; AC = 28$
15. $m\angle BOC = 145°; m\angle ODC = 82°; OD = 8.5$ **17.** $AC = 18; m\angle ODA = 22°; m\angle ABC = 112°$ **19.** FG   **21.** ∠OHG   **23.** ∠HEF
25. HG   **27.** $WO \cong XO \cong YO$ **29.** $XY \cong YZ \cong WZ$ **31.** $XY \cong YZ \cong ZW$ **33.** ASA, $x = 8, y = 7$ **35.** ASA, $x = 5, y = 3$
37. AAS, $x = 6, y = 9$ **39.** ASA, $x = 12, y = 17$ **41.** AAS, $x = 4, y = 10$ **43.** ASA, $x = 12, y = 7$

Problem Set 5.6

1. 10   **3.** $\frac{12}{5}$ **5.** $\frac{3}{2}$ **7.** $\frac{10}{9}$ **9.** 7   **11.** 14   **13.** 18   **15.** 6   **17.** 40   **19.** $h = 9$ **21.** $y = 14$ **23.** $x = 12$
25. $a = 25$ **27.** $y = 32$
29.   **31.**

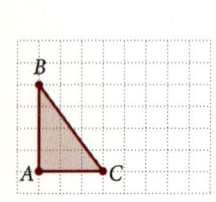

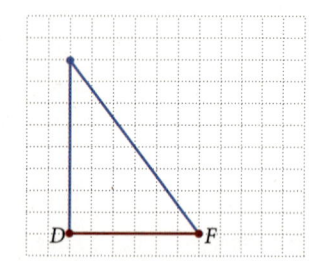

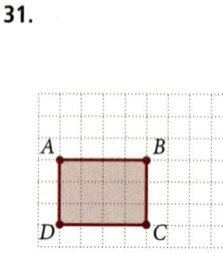

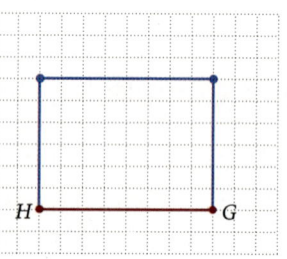

33. $x = 6$ **35.** $x = 8$ **37.** $x = 5$ **39.** $x = 6$ **41.** 45 in.   **43.** 960 pixels   **45.** 177 feet

Problem Set 5.7

1. 5   **3.** 15   **5.** 5   **7.** $7\sqrt{2}$ **9.** 5   **11.** 8   **13.** $10\sqrt{3}$ **15.** $3\sqrt{2}$ **17.** $4\sqrt{5}$ **19.** $2\sqrt{5}$ **21.** $7\sqrt{3}$
23. $x = 3, AB = 5, AC = 4, BC = 3$ **25.** $x = 3, AB = 10, AC = 8, BC = 6$ **27.** $x = 12, AB = 25, AC = 24, BC = 7$
29. $x = 9, \overline{AB} = 41, \overline{AC} = 40, \overline{BC} = 9$ **31.** Shortest side 4, third side $4\sqrt{3}$ **33.** Shortest side $\frac{6}{\sqrt{3}} = 2\sqrt{3}$, longest side $4\sqrt{3}$
35. 40 ft   **37.** 101.6 ft² = $\frac{176}{3}\sqrt{3}$ **39.** $\frac{4\sqrt{2}}{5}$ **41.** 8   **43.** $\frac{4}{\sqrt{2}} = 2\sqrt{2}$ **45.** 1,414 ft
47. a. $\sqrt{2}$ inches   **b.** $\sqrt{3}$ inches   **49. a.** $x\sqrt{2}$ **b.** $x\sqrt{3}$

Problem Set 5.8

1. $\sin A = \frac{4}{5}, \cos A = \frac{3}{5}, \tan A = \frac{4}{3}, \cot A = \frac{3}{4}, \sec A = \frac{5}{3}, \csc A = \frac{5}{4}$
3. $\sin A = \frac{2}{\sqrt{5}}, \cos A = \frac{1}{\sqrt{5}}, \tan A = 2, \cot A = \frac{1}{2}, \sec A = \sqrt{5}, \csc A = \frac{\sqrt{5}}{2}$
5. $\sin A = \frac{2}{3}, \cos A = \frac{\sqrt{5}}{3}, \tan A = \frac{2}{\sqrt{5}}, \cot A = \frac{\sqrt{5}}{2}, \sec A = \frac{3}{\sqrt{5}}, \csc A = \frac{3}{2}$
7. $\sin A = \frac{5}{6}, \cos A = \frac{\sqrt{11}}{6}, \tan A = \frac{5}{\sqrt{11}}, \sin B = \frac{\sqrt{11}}{6}, \cos B = \frac{5}{6}, \tan B = \frac{\sqrt{11}}{5}$
9. $\sin A = \frac{1}{\sqrt{2}}, \cos A = \frac{1}{\sqrt{2}}, \tan A = 1, \sin B = \frac{1}{\sqrt{2}}, \cos B = \frac{1}{\sqrt{2}}, \tan B = 1$
11. $\sin A = \frac{3}{5}, \cos A = \frac{4}{5}, \tan A = \frac{3}{4}, \sin B = \frac{4}{5}, \cos B = \frac{3}{5}, \tan B = \frac{4}{3}$
13. $\sin A = \frac{\sqrt{3}}{2}, \cos A = \frac{1}{2}, \tan A = \sqrt{3}, \sin B = \frac{1}{2}, \cos B = \frac{\sqrt{3}}{2}, \tan B = \frac{1}{\sqrt{3}}$
15. $x \approx 10.1$ **17.** $x = 13.2$ **19.** $x = 17.8$ **21.** $x = 10.8$ **23.** $B = (4, 3) \sin A = \frac{3}{5}, \cos A = \frac{4}{5}, \tan A = \frac{3}{4}$ **25.** 82°   **27.** $90° - x$
29. x   **31.**

x	sin x	csc x
30°	$\frac{1}{2}$	2
45°	$\frac{1}{\sqrt{2}}$	$\sqrt{2}$
60°	$\frac{\sqrt{3}}{2}$	$\frac{2}{\sqrt{3}}$

33. 2   **35.** $\frac{2+\sqrt{3}}{2}$ **37.** 0   **39.** $4 + 2\sqrt{3}$ **41.** 1   **43.** $-\frac{3\sqrt{3}}{2}$ **45.** $\sqrt{2}$
47. $\frac{2}{\sqrt{3}}$ **49.** 1   **51.** $\sqrt{2}$ **53.** $\sin A = 0.6, \sin B = 0.8, \cos A = 0.8, \cos B = 0.6$
55. $\sin A = 0.96, \sin B = 0.28, \cos A = 0.28, \cos B = 0.96$ **57.** $\sin \theta = \frac{1}{\sqrt{3}}, \cos \theta = \frac{\sqrt{2}}{\sqrt{3}}$
59. $\sin \theta = \frac{1}{\sqrt{3}}, \cos \theta = \frac{\sqrt{2}}{\sqrt{3}}$

Problem Set 5.9

1. 10 ft   **3.** 39 m   **5.** 2.13 ft   **7.** 8.535 yd   **9.** 32°   **11.** 30°   **13.** 59.20°   **15.** $B = 65°, a = 10$ m, $b = 22$ m
17. $B = 57.4°, b = 67.9$ inches, $c = 80.6$ inches   **19.** $B = 79° 18', a = 1.121$ cm, $c = 6.037$ cm   **21.** $A = 14°, a = 1.4$ ft, $b = 5.6$ ft
23. $A = 63° 30', a = 650$ mm, $c = 726$ mm   **25.** $A = 66.55°, b = 2.356$ mi, $c = 5.921$ mi   **27.** $A = 23°, B = 67°, c = 95$ ft
29. $A = 42.8°, B = 47.2°, b = 2.97$ cm   **31.** $A = 61.42°, B = 28.58°, a = 22.41$ inches   **33.** 49°   **35.** 11   **37.** 36
39. $x = 79, h = 40$ **41.** 42°   **43.** $x = 7.4, y = 6.2$ **45.** $h = 18, x = 11$ **47.** 15   **49.** 35.3°   **51.** 35.3°   **53.** 30.7 ft
55. 201.5 ft   **57.** 39 cm, 69°   **59.** 78.4°   **61.** 39 ft   **63.** 36.6°

Problem Set 5.10

1. Width = 8 mm, Perimeter = 26 mm   3. Width = 6 in., Area = 36 in²   5. Rectangle; 54 cm²   7. Square; 20.25 cm²
9. RE = 4 mm; RT = 16 mm   11. x = 4; Area = 44 cm²; Perimeter = 30 cm   13. x = 7; Area = 400 cm²; Perimeter = 80 cm
15. x = 2; Area = 49 cm²; Perimeter = 28 cm   17. $15\frac{3}{4}$ in.   19. 76 in.   21. 168 ft
23. Perimeter = 18.28 in; Area = 22.28 in²   23. Perimeter = 25.12 in; Area = 50.24 in²   27. 178.5 in.   29. 37.68 in.
31. 24,492 mi   33. 124 tiles   35. 1.18 in²   37. 8,509 mm²
39. The area increases from 25 ft² to 49 ft², which is an increase of 24 ft².   41. 22.6 cm

Chapter 5 Review

1. $x = -4$ 2. $x = 5$ 3. $x = 6, m\angle 1 = m\angle 2 = 79°$ 4. $x = 7, \angle ABD = 58°, \angle DBC = 32°$ 5. $x = 8$ 6. $x = 9$ 7. $x = 6$
8. $x = 9$ 9. $x = 16$ 10. $x = 8$ 11. $ABC \cong DEF$, SAS, $x = 8$ 12. ASA, $x = 12, y = 10$ 13. $x = 6$ 14. 5.4 ft   15. 6.4 ft
16. $x = 7.3$ 17. $x = 7.2$ 18. $x = 9.4$ 19. $x = 8.5$ 20. $x = 4$, Area = 120   21. Perimeter = 42.84 in., Area = 7.74 in²

Chapter 5 Cumulative Review

1. 27   2. -7 3. 28   4. 50   5. 18   6. 8   7. 30   8. 51   9. 27   10. 20   11. $\frac{9}{13}$ 12. $\frac{7}{11}$ 13. $\frac{1}{6}$ 14. $\frac{65}{72}$
15. -3 16. 0   17. $10x - 12y$ 18. $2x + 3y$ 19. 16   20. 3   21. $\frac{5}{6}$ 22. 6   23. $2, -\frac{10}{3}$ 24. $-\frac{2}{5}, -10$ 25. $y = -\frac{2}{3}x + 6$
26. $y = 6x + 1$ 27. -2 28. -1 29. 16   30. $x^2 - 4x + 4$ 31. Domain = {$-2, 6$}; Range = {3, 1, 7}; not a function
32. $x = 12, 125°, 55°$ 33. $x = 14, 45°, 90°$ 34. $x = 3$ 35. SAS; $x = 9, \angle A = \angle D = 31°$

Chapter 5 Test

1. $x = -1, \frac{3}{2}$ 2. $x = 9$ 3. $x = 17$ 4. $x = 5$ 5. $x = 23$ 6. $x = -\frac{1}{2}, 2$ 7. $x = 4$ 8. $x = 18$ 9. SAS, $x = 4, y = 9$
10. AAS, $x = 2, y = 4$ 11. $x = 5$ 12. $x = 3$ 13. $x = 17.5$ 14. P = 15.97 in., A = 8.43 in.²   15. $x = 4$, P = 44 m, A = 117 m²

# Solutions to Selected Practice Problems

Solutions to most of the practice problems are shown here. Before you look here to see where you have made a mistake, you should try the problem you are working on twice. If you do not get the correct answer the second time you work the problem, then the solution here should show you where you went wrong.

Chapter 1

Section 1.1

2. $\frac{3}{7} \cdot \frac{2}{5} = \frac{3 \cdot 2}{7 \cdot 5} = \frac{6}{35}$ 3. $9 \cdot \frac{1}{4} = \frac{9}{1} \cdot \frac{1}{4} = \frac{9}{4}$ 4. $\left(\frac{3}{4}\right)^3 = \frac{3}{4} \cdot \frac{3}{4} \cdot \frac{3}{4} = \frac{27}{64}$ 5. The reciprocal of 5 is $\frac{1}{5}$ because $5 \cdot \frac{1}{5} = \frac{5}{1} \cdot \frac{1}{5} = \frac{5}{5} = 1$.

6. The reciprocal of $\frac{1}{4}$ is 4 because $\frac{1}{4} \cdot 4 = \frac{1}{4} \cdot \frac{4}{1} = \frac{4}{4} = 1$.   7. The reciprocal of $\frac{3}{4}$ is $\frac{4}{3}$ because $\frac{3}{4} \cdot \frac{4}{3} = \frac{12}{12} = 1$.

8. The reciprocal of x is $\frac{1}{x}$ because $x \cdot \frac{1}{x} = \frac{x}{1} \cdot \frac{1}{x} = \frac{x}{x} = 1$.

17. $5 + (7 + y) = (5 + 7) + y$
    $= 12 + y$

18. $3(2x) = (3 \cdot 2)x$
    $= 6x$

19. $\frac{1}{3}(3a) = \left(\frac{1}{3} \cdot 3\right)a$
    $= a$

20. $5\left(\frac{1}{5}x\right) = \left(5 \cdot \frac{1}{5}\right)x$
    $= 1x$
    $= x$

21. $9\left(\frac{2}{3}x\right) = \left(9 \cdot \frac{2}{3}\right)x$
    $= 6x$

22. $7(6x + 8) = 7(6x) + 7(8)$
    $= 42x + 56$

23. $9(7x + 11y) = 9(7x) + 9(11y)$
    $= 63x + 99y$

24. $\frac{1}{3}(3x + 6) = \frac{1}{3}(3x) + \frac{1}{3}(6)$
    $= x + 2$

25. $4(5y + 2) + 8 = 4(5y) + 4(2) + 8$
    $= 20y + 8 + 8$
    $= 20y + 16$

26. $a\left(2 - \frac{1}{a}\right) = a \cdot 2 - a \cdot \frac{1}{a} = 2a - 1$

27. $5\left(\frac{1}{5}x + 8\right) = 5 \cdot \frac{1}{5}x + 5 \cdot 8 = x + 40$

28. $12\left(\frac{2}{3}x + \frac{3}{4}y\right) = 12 \cdot \frac{2}{3}x + 12 \cdot \frac{3}{4}y = 8x + 9y$

29. $4x + 9x = (4 + 9)x$
    $= 13x$

30. $y + 5y = (1 + 5)y$
    $= 6y$

31. $\frac{3}{14} + \frac{7}{30} = \frac{3}{2 \cdot 7} + \frac{7}{2 \cdot 3 \cdot 5}$
    $= \frac{3 \cdot 5}{3 \cdot 5} \cdot \frac{3}{2 \cdot 7} + \frac{7}{2 \cdot 3 \cdot 5} \cdot \frac{7}{7}$
    $= \frac{45}{210} + \frac{49}{210}$
    $= \frac{94}{210}$
    $= \frac{47}{105}$

32. $10x + 3 + 7x + 12 = (10x + 7x) + (3 + 12)$
    $= (10 + 7)x + (3 + 12)$
    $= 17x + 15$

33. $9 + 5(4y + 8) + 10y = 9 + 20y + 40 + 10y$
    $= (20y + 10y) + (9 + 40)$
    $= 30y + 49$

Section 1.2

2. $7 - 4 = 7 + (-4) = 3$ 3. $-6 - 3 = -6 + (-3) = -9$ 4. $8 - (-2) = 8 + 2 = 10$ 5. $-4 - (-6) = -4 + 6 = 2$
6. $-3 - (-5) = -3 + 5 = 2$ 7. $(-7 - 4) + (-8) = -11 + (-8) = -19$ 10. $(-6 - 4)(8 - 12) = (-10)(-4) = 40$
11. $3 - 7(9 - 5) - 2 = 3 - 7(4) - 2 = 3 - 28 - 2 = -25 - 2 = -27$
12. $3(5 - 8)^3 - 2(-1 - 1)^2 = 3(-3)^3 - 2(-2)^2 = 3(-27) - 2(4) = -89$

13. $2\left(\frac{x}{2}\right) = 2\left(\frac{1}{2}x\right)$
    $= \left(2 \cdot \frac{1}{2}\right)x$
    $= x$

14. $2\left(\frac{x}{2} - 3\right) = 2 \cdot \frac{x}{2} - 2 \cdot 3$
    $= x - 6$

15. $\frac{-6 - 6}{-5 - 3} = \frac{-12}{-8}$
    $= \frac{3}{2}$

16. $\frac{5(-6) + 3(-2)}{4(-3) + 3} = \frac{-30 + (-6)}{-12 + 3}$
    $= \frac{-36}{-9}$
    $= 4$

17. $\frac{3^3 - 4^3}{3^2 + 4^2} = \frac{27 - 64}{9 + 16}$
    $= \frac{-37}{25}$
    $= -\frac{37}{25}$

18. $2(5y - 1) - y = 10y - 2 - y$
    $= 9y - 2$

Solutions to Selected Practice Problems S-1

19. $6 - 2(5x + 1) + 4x = 6 - 10x - 2 + 4x$
$= -6x + 4$

20. $4(3a + 1) - (7a - 6) = 12a + 4 - 7a + 6$
$= 5a + 10$

21. $\dfrac{3}{5} \div \dfrac{6}{7} = \dfrac{3}{5} \cdot \dfrac{7}{6}$
$= \dfrac{21}{30}$
$= \dfrac{7}{10}$

22. $12 \div \dfrac{3}{4} = \dfrac{12}{1} \cdot \dfrac{4}{3}$
$= \dfrac{48}{3}$
$= 16$

23. $-\dfrac{5}{6} \div 10 = -\dfrac{5}{6} \cdot \dfrac{1}{10}$
$= -\dfrac{5}{60}$
$= -\dfrac{1}{12}$

24. a. $7[4(-1) - 6] = 7(-4 - 6)$
$= 7(-10)$
$= -70$
b. $28(-1) - 42 = -28 - 42$
$= -70$
c. $28(-1) - 6 = -28 - 6$
$= -34$

Section 1.3

4. $(2.5 \times 10^6)(1.4 \times 10^2) = (2.5)(1.4) \times (10^6)(10^2)$
$= 3.5 \times 10^{6+2}$
$= 3.5 \times 10^8$

5. $(2{,}200{,}000)(0.00015) = (2.2 \times 10^6)(1.5 \times 10^{-4})$
$= (2.2)(1.5) \times (10^6)(10^{-4})$
$= 3.3 \times 10^{6+(-4)}$
$= 3.3 \times 10^2$

6. $\dfrac{6 \times 10^5}{2 \times 10^{-4}} = \dfrac{6}{2} \times \dfrac{10^5}{10^{-4}}$
$= 3.0 \times 10^{5-(-4)}$
$= 3.0 \times 10^{5+4}$
$= 3.0 \times 10^9$

7. $\dfrac{0.0038}{19{,}000{,}000} = \dfrac{3.8 \times 10^{-3}}{1.9 \times 10^7}$
$= \dfrac{3.8}{1.9} \times \dfrac{10^{-3}}{10^7}$
$= 2.0 \times 10^{-3-7}$
$= 2.0 \times 10^{-10}$

8. $\dfrac{(6.8 \times 10^{-4})(3.9 \times 10^2)}{7.8 \times 10^{-6}} = \dfrac{(6.8)(3.9)}{(7.8)} \times \dfrac{(10^{-4})(10^2)}{10^{-6}}$
$= 3.4 \times 10^{-4+2-(-6)}$
$= 3.4 \times 10^4$

9. $\dfrac{(0.000035)(45{,}000)}{0.000075} = \dfrac{(3.5 \times 10^{-5})(4.5 \times 10^4)}{7.5 \times 10^{-5}}$
$= \dfrac{(3.5)(4.5)}{(7.5)} \times \dfrac{(10^{-5})(10^4)}{10^{-5}}$
$= 2.1 \times 10^{-5+4-(-5)}$
$= 2.1 \times 10^4$

Section 1.4

1. $4(3) - 2 \stackrel{?}{=} 10$
$12 - 2 = 10$
$10 = 10$

2. $3(5) + 1 \stackrel{?}{=} 16$
$15 + 1 = 16$
$16 = 16$
$4(5) - 6 \stackrel{?}{=} 14$
$20 - 6 = 14$
$14 = 14$

3. $\dfrac{2}{3}x + 4 = -8$
$\dfrac{2}{3}x + 4 - \mathbf{4} = -8 - \mathbf{4}$
$\dfrac{2}{3}x = -12$
$\dfrac{\mathbf{3}}{\mathbf{2}} \cdot \dfrac{2}{3}x = \dfrac{\mathbf{3}}{\mathbf{2}}(-12)$
$x = -18$

4. $3a - 3 = -5a + 9$
$3a + \mathbf{5a} - 3 = -5a + \mathbf{5a} + 9$
$8a - 3 = 9$
$8a - 3 + \mathbf{3} = 9 + \mathbf{3}$
$8a = 12$
$\dfrac{\mathbf{1}}{\mathbf{8}} \cdot 8a = \dfrac{\mathbf{1}}{\mathbf{8}} \cdot 12$
$a = \dfrac{12}{8} = \dfrac{3}{2}$

5. a. $-x = \dfrac{2}{3}$
$-1(-x) = -1\left(\dfrac{2}{3}\right)$
$x = -\dfrac{2}{3}$
b. $-y = -4$
$-1(-y) = -1(-4)$
$y = 4$

6. $\dfrac{3}{5}x + \dfrac{1}{3} = -\dfrac{5}{6}$
$\dfrac{3}{5}x + \dfrac{1}{3} + \left(-\dfrac{1}{3}\right) = -\dfrac{5}{6} + \left(-\dfrac{1}{3}\right)$
$\dfrac{3}{5}x = -\dfrac{7}{6}$
$\dfrac{5}{3} \cdot \dfrac{3}{5}x = \dfrac{5}{3}\left(-\dfrac{7}{6}\right)$
$x = -\dfrac{35}{18}$

Or eliminating the fractions in the beginning we can solve it this way:
$30 \cdot \dfrac{3}{5}x + 30 \cdot \dfrac{1}{3} = 30\left(-\dfrac{5}{6}\right)$
$18x + 10 = -25$
$18x = -35$
$x = -\dfrac{35}{18}$

7. Method 1
Working with the decimals.
$$0.08x + 0.10(8{,}000 - x) = 680$$
$$0.08x + 800 - 0.10x = 680$$
$$-0.02x + 800 = 680$$
$$-0.02x + 800 + (-800) = 680 + (-800)$$
$$-0.02x = -120$$
$$x = \frac{-120}{-0.02} = 6{,}000$$

Method 2
Eliminating the decimals by multiplying each side by 100.
$$100(0.08x) + 100(0.10)(8{,}000 - x) = 100(680)$$
$$8x + 10(8{,}000 - x) = 68{,}000$$
$$8x + 80{,}000 - 10x = 68{,}000$$
$$-2x + 80{,}000 = 68{,}000$$
$$-2x = -12{,}000$$
$$x = \frac{-12{,}000}{-2}$$
$$= 6{,}000$$

8.
$$2(5y - 3) + 4 = 9y - 6$$
$$10y - 6 + 4 = 9y - 6$$
$$10y - 2 = 9y - 6$$
$$10y + (-9y) - 2 = 9y + (-9y) - 6$$
$$y - 2 = -6$$
$$y - 2 + 2 = -6 + 2$$
$$y = -4$$

9.
$$6 - 2(5x - 1) + 4x = 20$$
$$6 - 10x + 2 + 4x = 20$$
$$-6x + 8 = 20$$
$$-6x + 8 + (-8) = 20 + (-8)$$
$$-6x = 12$$
$$-\frac{1}{6}(-6x) = -\frac{1}{6}(12)$$
$$x = -2$$

10.
$$3(5x + 1) = 10 + 15x$$
$$15x + 3 = 10 + 15x$$
$$15x + 3 + (-3) = 10 + 15x + (-3)$$
$$15x = 7 + 15x$$
$$15x + (-15x) = 7 + 15x + (-15x)$$
$$0 = 7$$
no solution

11.
$$-4 + 8x = 2(4x - 2)$$
$$-4 + 8x = 8x - 4$$
$$-4 + 8x + 4 = 8x - 4 + 4$$
$$8x = 8x$$
$$8x + (-8x) = 8x + (-8x)$$
$$0 = 0$$
identity

Section 1.5

1.
$$2x + 2(4x - 10) = 12.5$$
$$2x + 8x - 20 = 12.5$$
$$10x - 20 = 12.5$$
$$10x = 32.5$$
$$x = 3.25$$
The length is $x = 3.25$ feet; the width is $4x - 10 = 4(3.25) - 10 = 3$ feet.

2.
$$x + 0.0725x = 23{,}466.30$$
$$1.0725x = 23{,}466.30$$
$$x = \frac{23{,}466.30}{1.0725}$$
$$= 21{,}880$$
The price of the car is $21,880.00.

3.

	Dollars at 8%	Dollars at 10%	Total
Number	x	$8{,}000 - x$	8,000
Interest	$0.08x$	$0.10(8{,}000 - x)$	680

$$0.08x + 0.10(8{,}000 - x) = 680$$
$$8x + 10(8{,}000 - x) = 68{,}000$$
$$8x + 80{,}000 - 10x = 68{,}000$$
$$-2x + 80{,}000 = 68{,}000$$
$$-2x = -12{,}000$$
$$x = 6{,}000$$
$6,000 at 8% and $2,000 at 10%

4.

Width	Length	Area (in²)
1	$l = 7 - 1 = 6$	6
2	$l = 7 - 2 = 5$	10
3	$l = 7 - 3 = 4$	12
4	$l = 7 - 4 = 3$	12
5	$l = 7 - 5 = 2$	10
6	$l = 7 - 6 = 1$	6

Section 2.1

1. $x^5 \cdot x^4 = (x \cdot x \cdot x \cdot x \cdot x)(x \cdot x \cdot x \cdot x) = x^9$

2. $(4^5)^2 = 4^5 \cdot 4^5 = 4^{10}$

3. $(2x)^3 = 2^3 x^3 = 8x^3$

4. $(-2x^3)(4x^5) = (-2 \cdot 4)(x^3 \cdot x^5) = -8x^8$

5. $(-3y^3)^3(2y^6) = (-27y^9)(2y^6) = -54y^{15}$

6. $(a^2)^3(a^3 b^4)^2(b^5)^2 = a^6 \cdot a^6 \cdot b^8 \cdot b^{10} = a^{12}b^{18}$

7. $3^{-2} = \dfrac{1}{3^2} = \dfrac{1}{9}$

8. $(-5)^{-3} = \dfrac{1}{(-5)^3} = -\dfrac{1}{125}$

9. $\left(\dfrac{2}{3}\right)^{-2} = \left(\dfrac{3}{2}\right)^2 = \dfrac{9}{4}$

10.
a. $\dfrac{3^7}{3^4} = 3^{7-4} = 3^3 = 27$

b. $\dfrac{x^3}{x^{10}} = x^{3-10} = x^{-7} = \dfrac{1}{x^7}$

c. $\dfrac{a^5}{a^{-7}} = a^{5-(-7)} = a^{12}$

d. $\dfrac{m^{-3}}{m^{-5}} = m^{-3-(-5)} = m^{-3+5} = m^2$

11.
a. $(3xy^5)^0 = 1$

b. $(3xy^5)^1 = 3xy^5$

12. $\dfrac{(x^2)^{-4}(x^3)^6}{(x^{-3})^5} = \dfrac{x^{-8}x^{18}}{x^{-15}} = \dfrac{x^{10}}{x^{-15}} = x^{10-(-15)} = x^{25}$

13. $\dfrac{18a^7b^{-4}}{36a^2b^{-8}} = \dfrac{18}{36} \cdot \dfrac{a^7}{a^2} \cdot \dfrac{b^{-4}}{b^{-8}} = \dfrac{1}{2} \cdot a^5 \cdot b^4 = \dfrac{a^5b^4}{2}$

14. $\dfrac{(3x^{-2}y^7)^2}{(x^5y^{-2})^{-4}} = \dfrac{9x^{-4}y^{14}}{x^{-20}y^8} = 9x^{16}y^6$

Section 2.2

6. $(3x^2 + 2x - 5) + (2x^2 - 7x + 3) = (3x^2 + 2x^2) + (2x - 7x) + (-5 + 3)$
$= (3 + 2)x^2 + (2 - 7)x + (-5 + 3)$
$= 5x^2 - 5x - 2$

7. $x^3 + 7x^2 + 3x + 2$
$-3x^3 - 2x^2 + 3x - 1$
$\overline{-2x^3 + 5x^2 + 6x + 1}$

8. $(4x^2 - 2x + 7) - (7x^2 - 3x + 1) = 4x^2 - 2x + 7 - 7x^2 + 3x - 1$
$= -3x^2 + x + 6$

9. $(7x - 4) - (3x + 5) = 7x - 4 - 3x - 5$
$= 4x - 9$

10. $2x - 4[6 - (5x + 3)]$
$= 2x - 4(6 - 5x - 3)$
$= 2x - 4(-5x + 3)$
$= 2x + 20x - 12$
$= 22x - 12$

11. $(9x - 4) - [(2x + 5) - (x + 3)]$
$= 9x - 4 - (2x + 5 - x - 3)$
$= 9x - 4 - (x + 2)$
$= 9x - 4 - x - 2$
$= 8x - 6$

12. $2(-2)^3 - 3(-2)^2 + 4(-2) - 8$
$= 2(-8) - 3(4) + 4(-2) - 8$
$= -16 - 12 - 8 - 8$
$= -44$

13. $P = -500 + 27x - 0.1x^2$
$= -500 + 27(170) - 0.1(170)^2$
$= -500 + 27(170) - 0.1(28,900)$
$= -500 + 4{,}590 - 2{,}890$
$= \$1{,}200$

Section 2.3

1. $5x^2(3x^2 - 4x + 2) = 5x^2(3x^2) + 5x^2(-4x) + 5x^2(2)$
$= 15x^4 - 20x^3 + 10x^2$

2. $(4x - 1)(x + 3) = (4x - 1)x + (4x - 1)3$
$= 4x^2 - x + 12x - 3$
$= 4x^2 + 11x - 3$

3. $x^2 - 5x + 6$
$3x + 2$
$\overline{3x^3 - 15x^2 + 18x}$
$2x^2 - 10x + 12$
$\overline{3x^3 - 13x^2 + 8x + 12}$

4. $(2a - 3b)(5a - b) = 10a^2 - 2ab - 15ab + 3b^2$
$ \text{F} \text{O} \text{I} \text{L}$
$= 10a^2 - 17ab + 3b^2$

5. $(6 - 3t)(2 + 5t) = 12 + 30t - 6t - 15t^2$
$\text{F}\text{O}\text{I}\text{L}$
$= 12 + 24t - 15t^2$

6. $\left(3x + \dfrac{1}{4}\right)\left(4x - \dfrac{1}{3}\right) = 12x^2 - x + x - \dfrac{1}{12}$
$\text{F}\text{O}\text{I}\text{L}$
$= 12x^2 - \dfrac{1}{12}$

7. $(a^4 - 2)(a^4 + 5) = a^8 + 5a^4 - 2a^4 - 10$
$\text{F}\text{O}\text{I}\text{L}$
$= a^8 + 3a^4 - 10$

8. $(7x - 2)(3y + 8) = 21xy + 56x - 6y - 16$
$\text{F}\text{O}\text{I}\text{L}$

9. $(3x - 2)^2 = (3x - 2)(3x - 2)$
$= 9x^2 - 6x - 6x + 4$
$= 9x^2 - 12x + 4$

10. $(x - y)^2 = x^2 - (2)xy + y^2$
$= x^2 - 2xy + y^2$

11. $(4t + 5)^2 = (4t)^2 + 2(4t)(5) + 5^2$
$= 16t^2 + 40t + 25$

12. $(5x - 3y)^2 = (5x)^2 - 2(5x)(3y) + (3y)^2$
$= 25x^2 - 30xy + 9y^2$

13. $(6 - a^4)^2 = 6^2 - 2(6)(a^4) + (a^4)^2$
$= 36 - 12a^4 + a^8$

14. $(4x - 3)(4x + 3) = 16x^2 + 12x - 12x - 9$
$= 16x^2 - 9$

Section 2.4

1. $15a^7 - 25a^5 + 30a^3$
$= 5a^3(3a^4) - 5a^3(5a^2) + 5a^3(6)$
$= 5a^3(3a^4 - 5a^2 + 6)$

2. $12x^4y^5 - 9x^3y^4 - 15x^5y^3$
$= 3x^3y^3(4xy^2) - 3x^3y^3(3y) - 3x^3y^3(5x^2)$
$= 3x^3y^3(4xy^2 - 3y - 5x^2)$

3. $4(a + b)^4 - 6(a + b)^3 + 16(a + b)^2$
$= 2(a + b)^2 \cdot 2(a + b)^2 - 2(a + b)^2 \cdot 3(a + b) + 2(a + b)^2 \cdot 8$
$= 2(a + b)^2[2(a + b)^2 - 3(a + b) + 8]$

4. $ab^3 + b^3 + 6a + 6 = b^3(a + 1) + 6(a + 1)$
$= (a + 1)(b^3 + 6)$

5. $15 - 3y^2 - 5x^2 + x^2y^2 = 3(5 - y^2) - x^2(5 - y^2)$
   $= (5 - y^2)(3 - x^2)$

6. $x^3 + 5x^2 + 3x + 15 = x^2(x + 5) + 3(x + 5)$
   $= (x + 5)(x^2 + 3)$

Section 2.5

5. $(y + 2)^2 + 8(y + 2) + 16 = [(y + 2) + 4]^2$
   $= (y + 6)^2$

6. $27x^2 - 36x + 12 = 3(9x^2 - 12x + 4)$
   $= 3(3x - 2)^2$

7. $x^2 - 16 = x^2 - 4^2 = (x + 4)(x - 4)$

8. $64 - t^2 = 8^2 - t^2 = (8 + t)(8 - t)$

9. $25x^2 - 36y^2 = (5x)^2 - (6y)^2 = (5x + 6y)(5x - 6y)$

10. $9x^6 - 1 = (3x^3)^2 - 1^2 = (3x^3 + 1)(3x^3 - 1)$

11. $x^2 - \dfrac{25}{64} = x^2 - \left(\dfrac{5}{8}\right)^2 = \left(x + \dfrac{5}{8}\right)\left(x - \dfrac{5}{8}\right)$

12. $x^4 - 81 = (x^2)^2 - 9^2$
    $= (x^2 - 9)(x^2 + 9)$
    $= (x - 3)(x + 3)(x^2 + 9)$

13. $(x - 4)^2 - 9 = [(x - 4) - 3][(x - 4) + 3]$
    $= (x - 7)(x - 1)$

14. $x^2 - 6x + 9 - y^2 = (x - 3)^2 - y^2$
    $= (x - 3 + y)(x - 3 - y)$

15. $x^3 + 5x^2 - 4x - 20 = x^2(x + 5) - 4(x + 5)$
    $= (x + 5)(x^2 - 4)$
    $= (x + 5)(x - 2)(x + 2)$

16. We use the distributive property to expand the product:
$$(x - 3)(x^2 + 3x + 9) = x^3 + 3x^2 + 9x$$
$$+ \quad -3x^2 - 9x - 27$$
$$= x^3 \qquad\qquad - 27$$

17. The first term is the cube of 3 and the second term is the cube of x. Therefore,
$$27 + x^3 = 3^3 + x^3 = (3 + x)[3^2 - 3(x) + x^2]$$
$$= (3 + x)(9 - 3x + x^2)$$

18. $8x^3 + y^3 = (2x)^3 + y^3$
    $= (2x + y)[(2x)^2 - (2x)(y) + y^2]$
    $= (2x + y)(4x^2 - 2xy + y^2)$

19. $a^3 - \dfrac{1}{27} = a^3 - \left(\dfrac{1}{3}\right)^3 = \left(a - \dfrac{1}{3}\right)\left[a^2 + a\left(\dfrac{1}{3}\right) + \left(\dfrac{1}{3}\right)^2\right]$
    $= \left(a - \dfrac{1}{3}\right)\left(a^2 + \dfrac{1}{3}a + \dfrac{1}{9}\right)$

20. $x^6 - 1 = (x^3)^2 - 1^2 = (x^3 + 1)(x^3 - 1)$
    $= (x + 1)(x^2 - x + 1)(x - 1)(x^2 + x + 1)$

Section 2.6

1. The leading coefficient is 1. We need two numbers whose sum is -1 and whose product is -12. The numbers are -4 and $+3$.
$$x^2 - x - 12 = (x - 4)(x + 3)$$

2. We need two numbers whose product is $-15y^2$ and whose sum is $+2y$. The numbers are $+5y$ and $-3y$.
$$x^2 + 2xy - 15y^2 = (x + 5y)(x - 3y)$$

3. Since there is no pair of integers whose product is 1 and whose sum is 1, the trinomial $x^2 + x + 1$ is not factorable. We say it is a prime polynomial.

4. $5x^2 + 25x + 30 = 5(x^2 + 5x + 6)$
   $= 5(x + 3)(x + 2)$

5. We list all possible factors along with their products as follows:

Possible Factors	First Term	Middle Term	Last Term
$(3x + 2)(x - 1)$	$3x^2$	$-x$	-2
$(3x - 2)(x + 1)$	$3x^2$	$+x$	-2
$(3x + 1)(x - 2)$	$3x^2$	$-5x$	-2
$(3x - 1)(x + 2)$	$3x^2$	$+5x$	-2

The first line has the correct middle term: $\quad 3x^2 - x - 2 = (3x + 2)(x - 1)$

6. $15x^4 + x^2 - 2 = (5x^2 + 2)(3x^2 - 1)$

7. $3x^2(x - 2) - 7x(x - 2) + 2(x - 2)$
$= (x - 2)(3x^2 - 7x + 2)$
$= (x - 2)(3x - 1)(x - 2)$
$= (x - 2)^2(3x - 1)$

8. $ac = -24$
$b = -5$
Two numbers whose product is -24 and whose sum is -5 are -8 and 3.
$8x^2 - 5x - 3 = 8x^2 - 8x + 3x - 3$
$= 8x(x - 1) + 3(x - 1)$
$= (x - 1)(8x + 3)$

9. $ac = 24$
$b = 11$
Two numbers whose product is 24 and whose sum is 11 are 8 and 3.
$4x^2 + 11x + 6 = 4x^2 + 8x + 3x + 6$
$= 4x(x + 2) + 3(x + 2)$
$= (x + 2)(4x + 3)$

10. $ac = -60$
$b = -4$
Two numbers whose product is -60 and whose sum is -4 are -10 and 6.
$4x^2 - 4x - 15 = 4x^2 - 10x + 6x - 15$
$= 2x(2x - 5) + 3(2x - 5)$
$= (2x - 5)(2x + 3)$

Section 2.7

1. $3x^8 - 27x^6 = 3x^6(x^2 - 9)$
$= 3x^6(x + 3)(x - 3)$

2. $4x^4 + 40x^3 + 100x^2 = 4x^2(x^2 + 10x + 25)$
$= 4x^2(x + 5)^2$

3. $y^4 + 36y^2 = y^2(y^2 + 36)$

4. $6x^2 - x - 15 = (3x - 5)(2x + 3)$

5. $3x^5 - 81x^2 = 3x^2(x^3 - 27)$
$= 3x^2(x - 3)(x^2 + 3x + 9)$

6. $3a^2b^3 + 6a^2b^2 - 3a^2b = 3a^2b(b^2 + 2b - 1)$

7. $x^2 - 10x + 25 - b^2 = (x - 5)^2 - b^2$
$= (x - 5 + b)(x - 5 - b)$

Section 2.8

1. $x^2 - x - 6 = 0$
$(x - 3)(x + 2) = 0$
$x - 3 = 0$ or $x + 2 = 0$
$x = 3$ or $x = -2$

2. $6 \cdot \dfrac{1}{2}x^3 = 6 \cdot \dfrac{5}{6}x^2 + 6 \cdot \dfrac{1}{3}x$
$3x^3 = 5x^2 + 2x$
$3x^3 - 5x^2 - 2x = 0$
$x(3x^2 - 5x - 2) = 0$
$x(3x + 1)(x - 2) = 0$
$x = 0$ or $3x + 1 = 0$ or $x - 2 = 0$
$x = 0$ or $x = -\dfrac{1}{3}$ or $x = 2$

3. $100x^2 = 500x$
$100x^2 - 500x = 0$
$100x(x - 5) = 0$
$100x = 0$ or $x - 5 = 0$
$x = 0$ or $x = 5$

4. $(x + 1)(x + 2) = 12$
$x^2 + 3x + 2 = 12$
$x^2 + 3x - 10 = 0$
$(x + 5)(x - 2) = 0$
$x + 5 = 0$ or $x - 2 = 0$
$x = -5$ or $x = 2$

5. $x^3 + 5x^2 - 4x - 20 = 0$
$x^2(x + 5) - 4(x + 5) = 0$
$(x + 5)(x^2 - 4) = 0$
$(x + 5)(x + 2)(x - 2) = 0$
$x + 5 = 0$ or $x + 2 = 0$ or $x - 2 = 0$
$x = -5$ or $x = -2$ or $x = 2$

6. Let $x =$ one integer and $x + 3 =$ the other.
$x^2 + (x + 3)^2 = 29$
$x^2 + x^2 + 6x + 9 = 29$
$2x^2 + 6x + 9 = 29$
$2x^2 + 6x - 20 = 0$
$x^2 + 3x - 10 = 0$
$(x + 5)(x - 2) = 0$
$x + 5 = 0$ or $x - 2 = 0$
$x = -5$ or $x = 2$
$x + 3 = -5 + 3$ $x + 3 = 2 + 3$
$= -2$ $= 5$
The two integers are -5 and -2 or 2 and 5.

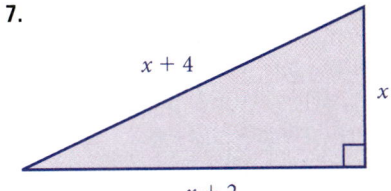

7.

$(x+4)^2 = (x+2)^2 + x^2$
$x^2 + 8x + 16 = x^2 + 4x + 4 + x^2$
$x^2 + 8x + 16 = 2x^2 + 4x + 4$
$0 = x^2 - 4x - 12$
$0 = (x-6)(x+2)$
$x - 6 = 0$ or $x + 2 = 0$
$x = 6$ or $x = -2$

Since the length of a side cannot be negative, the shortest side is 6. The other two sides are $6 + 2 = 8$ and $6 + 4 = 10$.

8.

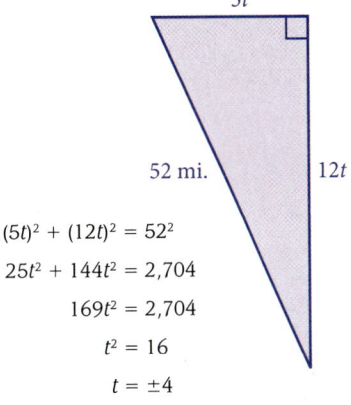

$(5t)^2 + (12t)^2 = 52^2$
$25t^2 + 144t^2 = 2{,}704$
$169t^2 = 2{,}704$
$t^2 = 16$
$t = \pm 4$

Since time can't be negative, the answer is 4 hours.

9. If $v = 64$ and $h = 64$ then $h = vt - 16t^2$ becomes $64 = 64t - 16t^2$
$16t^2 - 64t + 64 = 0$
$t^2 - 4t + 4 = 0$
$(t-2)^2 = 0$
$t = 2$ seconds

10. $R = (1{,}300 - 100p)p$
If $R = 3{,}600$, then
$3{,}600 = (1{,}300 - 100p)p$
$3{,}600 = 1{,}300p - 100p^2$
$100p^2 - 1{,}300p + 3{,}600 = 0$
$p^2 - 13p + 36 = 0$
$(p-4)(p-9) = 0$
$p - 4 = 0$ or $p - 9 = 0$
$p = 4$ or $p = 9$

The price can be set at either $4 or $9.

Section 3.1

1. $\dfrac{x^2 - 9}{x + 3} = \dfrac{(x+3)(x-3)}{x+3}$
$= x - 3$

2. $\dfrac{y^2 - y - 6}{y^2 - 4} = \dfrac{(y-3)(y+2)}{(y-2)(y+2)}$
$= \dfrac{y-3}{y-2}$

3. $\dfrac{3a^3 + 3}{6a^2 - 6a + 6} = \dfrac{3(a^3+1)}{6(a^2-a+1)}$
$= \dfrac{3(a+1)(a^2-a+1)}{6(a^2-a+1)}$
$= \dfrac{a+1}{2}$

4. $\dfrac{x^2 + 4x + ax + 4a}{x^2 + ax + 4x + 4a} = \dfrac{x(x+4) + a(x+4)}{x(x+a) + 4(x+a)}$
$= \dfrac{(x+4)(x+a)}{(x+a)(x+4)}$
$= 1$

5. When $a = 7$ and $b = 4$ the expression $\dfrac{a-b}{b-a}$ becomes
$\dfrac{7-4}{4-7} = \dfrac{3}{-3} = -1$

6. $\dfrac{7-x}{x^2 - 49} = \dfrac{-1(x-7)}{(x+7)(x-7)} = \dfrac{-1}{x+7}$

7. $C = 165(3.14) = 518$ feet to the nearest foot $r = \dfrac{d}{t} = \dfrac{518}{40} = 13.0$ feet per second

Section 3.2

1. $\dfrac{3}{4} \cdot \dfrac{12}{27} = \dfrac{3 \cdot 12}{4 \cdot 27}$
$= \dfrac{3 \cdot 2 \cdot 2 \cdot 3}{2 \cdot 2 \cdot 3 \cdot 3 \cdot 3}$
$= \dfrac{1}{3}$

2. $\dfrac{6x^4}{4y^9} \cdot \dfrac{12y^5}{3x^2} = \dfrac{\overset{2}{6} \cdot \overset{3}{12} \, x^4 y^5}{\overset{}{3} \cdot \overset{}{4} x^2 y^9}$
$= \dfrac{6x^2}{y^4}$

3. $\dfrac{x+5}{x^2-25} \cdot \dfrac{x-5}{x^2 - 10x + 25} = \dfrac{(x+5)(x-5)}{(x+5)(x-5)(x-5)^2}$
$= \dfrac{1}{(x-5)^2}$

4. $\dfrac{3y^2 - 3y}{3y - 12} \cdot \dfrac{y^2 - 2y - 8}{y^2 + 3y + 2} = \dfrac{3y(y-1)(y-4)(y+2)}{3(y-4)(y+1)(y+2)}$
$= \dfrac{y(y-1)}{y+1}$

5. $\dfrac{5}{9} \div \dfrac{10}{27} = \dfrac{5}{9} \cdot \dfrac{27}{10}$
$= \dfrac{5 \cdot 3 \cdot 3 \cdot 3}{3 \cdot 3 \cdot 2 \cdot 5}$
$= \dfrac{3}{2}$

6. $\dfrac{9x^4}{4y^3} \div \dfrac{3x^2}{8y^5} = \dfrac{9x^4}{4y^3} \cdot \dfrac{8y^5}{3x^2}$
$= \dfrac{\overset{3}{9} \cdot \overset{2}{8} x^4 y^5}{\overset{}{3} \cdot \overset{}{4} x^2 y^3}$
$= 6x^2 y^2$

7. $\dfrac{xy^2 - y^3}{x^2 - y^2} \div \dfrac{x^3 + y^3}{x^2 + 2xy + y^2} = \dfrac{xy^2 - y^3}{x^2 - y^2} \cdot \dfrac{x^2 + 2xy + y^2}{x^3 + y^3}$

$= \dfrac{y^2\cancel{(x-y)}\cancel{(x+y)}(x+y)}{\cancel{(x+y)}\cancel{(x-y)}\cancel{(x+y)}(x^2 - xy + y^2)}$

$= \dfrac{y^2}{x^2 - xy + y^2}$

8. $\dfrac{a^2 + 3a - 4}{a - 4} \cdot \dfrac{a + 3}{a^2 - 4a + 3} \div \dfrac{a + 1}{a^2 - 2a - 3}$

$= \dfrac{(a^2 + 3a - 4)(a + 3)(a^2 - 2a - 3)}{(a - 4)(a^2 - 4a + 3)(a + 1)}$

$= \dfrac{(a + 4)\cancel{(a-1)}(a + 3)\cancel{(a-3)}\cancel{(a+1)}}{(a - 4)\cancel{(a-3)}\cancel{(a-1)}\cancel{(a+1)}}$

$= \dfrac{(a + 4)(a + 3)}{a - 4}$

9. $\dfrac{xa + xb - ya - yb}{xa + 2x + ya + 2y} \cdot \dfrac{xa + 2x + ya + 2y}{xa + xb + ya + yb}$

$= \dfrac{x(a + b) - y(a + b)}{x(a + 2) + y(a + 2)} \cdot \dfrac{x(a + 2) + y(a + 2)}{x(a + b) + y(a + b)}$

$= \dfrac{\cancel{(a+b)}(x - y)\cancel{(a+2)}\cancel{(x+y)}}{\cancel{(a+2)}(x + y)\cancel{(a+b)}\cancel{(x+y)}}$

$= \dfrac{x - y}{x + y}$

10. $(5x^2 - 45) \cdot \dfrac{3}{5x - 15} = \dfrac{5x^2 - 45}{1} \cdot \dfrac{3}{5x - 15}$

$= \dfrac{\cancel{5}(x + 3)\cancel{(x-3)}3}{\cancel{5}\cancel{(x-3)}}$

$= 3(x + 3)$

Section 3.3

1. $\dfrac{3}{8} + \dfrac{1}{8} = \dfrac{3 + 1}{8}$

$= \dfrac{4}{8}$

$= \dfrac{1}{2}$

2. $\dfrac{x}{x^2 - 9} + \dfrac{3}{x^2 - 9} = \dfrac{x + 3}{x^2 - 9}$

$= \dfrac{\cancel{x+3}}{\cancel{(x+3)}(x - 3)}$

$= \dfrac{1}{x - 3}$

3. $\dfrac{2x - 7}{x - 2} - \dfrac{x - 5}{x - 2} = \dfrac{2x - 7 - (x - 5)}{x - 2}$

$= \dfrac{2x - 7 - x + 5}{x - 2}$

$= \dfrac{x - 2}{x - 2}$

$= 1$

4. $\dfrac{3}{10} + \dfrac{11}{42} = \dfrac{3}{2 \cdot 5} + \dfrac{11}{2 \cdot 3 \cdot 7}$

$= \dfrac{3}{2 \cdot 5} \cdot \dfrac{\mathbf{3 \cdot 7}}{\mathbf{3 \cdot 7}} + \dfrac{11}{2 \cdot 3 \cdot 7} \cdot \dfrac{\mathbf{5}}{\mathbf{5}}$

$= \dfrac{63}{2 \cdot 3 \cdot 5 \cdot 7} + \dfrac{55}{2 \cdot 3 \cdot 5 \cdot 7}$

$= \dfrac{118}{2 \cdot 3 \cdot 5 \cdot 7}$

$= \dfrac{\cancel{2} \cdot 59}{\cancel{2} \cdot 3 \cdot 5 \cdot 7}$

$= \dfrac{59}{105}$

5. $\dfrac{-3}{x^2 - 2x - 8} + \dfrac{4}{x^2 - 16} = \dfrac{-3}{(x - 4)(x + 2)} + \dfrac{4}{(x - 4)(x + 4)}$

$= \dfrac{-3}{(x - 4)(x + 2)} \cdot \dfrac{\mathbf{x + 4}}{\mathbf{x + 4}} + \dfrac{4}{(x - 4)(x + 4)} \cdot \dfrac{\mathbf{x + 2}}{\mathbf{x + 2}}$

$= \dfrac{-3x - 12 + 4x + 8}{(x - 4)(x + 4)(x + 2)}$

$= \dfrac{\cancel{x - 4}}{\cancel{(x - 4)}(x + 4)(x + 2)}$

$= \dfrac{1}{(x + 4)(x + 2)}$

6. $\dfrac{x - 4}{2x - 6} + \dfrac{3}{x^2 - 9} = \dfrac{x - 4}{2(x - 3)} + \dfrac{3}{(x + 3)(x - 3)}$

$= \dfrac{x - 4}{2(x - 3)} \cdot \dfrac{\mathbf{x + 3}}{\mathbf{x + 3}} + \dfrac{3}{(x + 3)(x - 3)} \cdot \dfrac{\mathbf{2}}{\mathbf{2}}$

$= \dfrac{(x - 4)(x + 3) + 3 \cdot 2}{2(x + 3)(x - 3)}$

$= \dfrac{x^2 - x - 12 + 6}{2(x + 3)(x - 3)}$

$= \dfrac{x^2 - x - 6}{2(x + 3)(x - 3)}$

$= \dfrac{\cancel{(x - 3)}(x + 2)}{2(x + 3)\cancel{(x - 3)}}$

$= \dfrac{x + 2}{2(x + 3)}$

7. $\dfrac{2x - 4}{x^2 + 5x + 4} - \dfrac{x - 4}{x^2 + 6x + 8} = \dfrac{2x - 4}{(x + 4)(x + 1)} - \dfrac{x - 4}{(x + 4)(x + 2)}$

$= \dfrac{2x - 4}{(x + 4)(x + 1)} \cdot \dfrac{\mathbf{x + 2}}{\mathbf{x + 2}} - \dfrac{x - 4}{(x + 4)(x + 2)} \cdot \dfrac{\mathbf{x + 1}}{\mathbf{x + 1}}$

$= \dfrac{(2x - 4)(x + 2) - (x - 4)(x + 1)}{(x + 4)(x + 1)(x + 2)}$

$= \dfrac{(2x^2 - 8) - (x^2 - 3x - 4)}{(x + 4)(x + 1)(x + 2)}$

$= \dfrac{x^2 + 3x - 4}{(x + 4)(x + 1)(x + 2)}$

$= \dfrac{\cancel{(x + 4)}(x - 1)}{\cancel{(x + 4)}(x + 1)(x + 2)}$

$= \dfrac{x - 1}{(x + 1)(x + 2)}$

8. $\dfrac{x^2}{x-4} + \dfrac{x+12}{4-x} = \dfrac{x^2}{x-4} + \dfrac{x+12}{4-x} \cdot \dfrac{-1}{-1}$

$= \dfrac{x^2}{x-4} + \dfrac{-x-12}{x-4}$

$= \dfrac{x^2 - x - 12}{x-4}$

$= \dfrac{(x-4)(x+3)}{x-4}$

$= x + 3$

9. $2 + \dfrac{25}{5x-1} = \dfrac{2}{1} + \dfrac{25}{5x-1}$

$= \dfrac{2}{1} \cdot \dfrac{5x-1}{5x-1} + \dfrac{25}{5x-1}$

$= \dfrac{10x - 2 + 25}{5x-1}$

$= \dfrac{10x + 23}{5x-1}$

10. $\dfrac{1}{x} + \dfrac{1}{3x} = \dfrac{3}{3x} + \dfrac{1}{3x} = \dfrac{4}{3x}$

Section 3.4

1. $\dfrac{x}{3} + 1 = \dfrac{1}{2}$ LCD = 6

$\mathbf{6}\left(\dfrac{x}{3} + 1\right) = \mathbf{6} \cdot \dfrac{1}{2}$

$6 \cdot \dfrac{x}{3} + 6 \cdot 1 = 6 \cdot \dfrac{1}{2}$

$2x + 6 = 3$

$2x = -3$

$x = -\dfrac{3}{2}$

2. $\dfrac{2}{a+5} = \dfrac{1}{3}$

LCD = $3(a+5)$

$\mathbf{3(a+5)} \cdot \dfrac{2}{a+5} = \mathbf{3(a+5)} \cdot \dfrac{1}{3}$

$6 = a + 5$

$1 = a$

3. $\dfrac{x}{x+1} - \dfrac{1}{2} = \dfrac{-1}{x+1}$ LCD = $2(x+1)$

$\mathbf{2(x+1)}\left(\dfrac{x}{x+1} - \dfrac{1}{2}\right) = \mathbf{2(x+1)} \cdot \dfrac{-1}{x+1}$

$2(x+1) \cdot \dfrac{x}{x+1} - 2(x+1) \cdot \dfrac{1}{2} = 2(x+1) \cdot \dfrac{-1}{x+1}$

$2x - (x+1) = 2(-1)$

$2x - x - 1 = -2$

$x - 1 = -2$

$x = -1$

The only possible solution is $x = -1$, but when $x = -1$, the original equation has two undefined terms. There is no solution to the equation.

4. $\dfrac{x}{x^2 - 9} - \dfrac{1}{x+3} = \dfrac{1}{4x - 12}$

$\dfrac{x}{(x+3)(x-3)} - \dfrac{1}{x+3} = \dfrac{1}{4(x-3)}$ LCD = $4(x+3)(x-3)$

$4(x+3)(x-3) \cdot \dfrac{x}{(x+3)(x-3)} - 4(x+3)(x-3) \cdot \dfrac{1}{x+3} = 4(x+3)(x-3) \cdot \dfrac{1}{4(x-3)}$

$4x - 4(x-3) = x + 3$

$4x - 4x + 12 = x + 3$

$12 = x + 3$

$9 = x$

5. $1 - \dfrac{2}{x} = \dfrac{8}{x^2}$ LCD = x^2

$\mathbf{x^2}\left(1 - \dfrac{2}{x}\right) = \mathbf{x^2} \cdot \dfrac{8}{x^2}$

$x^2 \cdot 1 - x^2 \cdot \dfrac{2}{x} = x^2 \cdot \dfrac{8}{x^2}$

$x^2 - 2x = 8$

$x^2 - 2x - 8 = 0$

$(x-4)(x+2) = 0$

$x - 4 = 0$ or $x + 2 = 0$

$x = 4$ or $x = -2$

6. $\dfrac{y+1}{3(y+4)} = \dfrac{8}{(y+4)(y-4)}$

The LCD is $3(y+4)(y-4)$.
Multiplying each side by the LCD gives us

$(y+1)(y-4) = 8 \cdot 3$

$y^2 - 3y - 4 = 24$

$y^2 - 3y - 28 = 0$

$(y-7)(y+4) = 0$

$y = 7$ or $y = -4$

The only solution is 7 because the original equation is undefined when y is -4.

7. $x = \dfrac{y+2}{y-1}$

$x(y-1) = y + 2$

$xy - x = y + 2$

$xy - y = x + 2$

$y(x-1) = x + 2$

$y = \dfrac{x+2}{x-1}$

8. $\dfrac{1}{a} = \dfrac{1}{x} + \dfrac{1}{b}$

$axb \cdot \dfrac{1}{a} = axb \cdot \dfrac{1}{x} + axb \cdot \dfrac{1}{b}$

$xb = ab + ax$

$xb - ax = ab$

$x(b - a) = ab$

$x = \dfrac{ab}{b-a}$

Section 3.5

1. $\sqrt{18} = \sqrt{9 \cdot 2}$
$= \sqrt{9}\sqrt{2}$
$= 3\sqrt{2}$

2. $\sqrt{50x^2y^3} = \sqrt{25x^2y^2 \cdot 2y}$
$= \sqrt{25x^2y^2}\sqrt{2y}$
$= 5xy\sqrt{2y}$

3. $\sqrt[3]{54a^4b^3} = \sqrt[3]{27a^3b^3 \cdot 2a}$
$= \sqrt[3]{27a^3b^3}\sqrt[3]{2a}$
$= 3ab\sqrt[3]{2a}$

4. $\sqrt{75x^5y^8} = \sqrt{25x^4y^8 \cdot 3x}$
$= \sqrt{25x^4y^8}\sqrt{3x}$
$= 5x^2y^4\sqrt{3x}$

5. $\sqrt[4]{48a^8b^5c^4} = \sqrt[4]{16a^8b^4c^4 \cdot 3b}$
$= 2a^2bc\sqrt[4]{3b}$

6. $\sqrt{\dfrac{5}{9}} = \dfrac{\sqrt{5}}{\sqrt{9}}$
$= \dfrac{\sqrt{5}}{3}$

7. $\sqrt{\dfrac{2}{3}} = \dfrac{\sqrt{2}}{\sqrt{3}}$
$= \dfrac{\sqrt{2}}{\sqrt{3}} \cdot \dfrac{\sqrt{3}}{\sqrt{3}}$
$= \dfrac{\sqrt{6}}{3}$

8. $\dfrac{5}{\sqrt{2}} = \dfrac{5}{\sqrt{2}} \cdot \dfrac{\sqrt{2}}{\sqrt{2}}$
$= \dfrac{5\sqrt{2}}{2}$

9. $\dfrac{3\sqrt{5x}}{\sqrt{2y}} = \dfrac{3\sqrt{5x}}{\sqrt{2y}} \cdot \dfrac{\sqrt{2y}}{\sqrt{2y}}$
$= \dfrac{3\sqrt{10xy}}{2y}$

10. $\dfrac{5}{\sqrt[3]{9}} = \dfrac{5}{\sqrt[3]{3^2}}$
$= \dfrac{5}{\sqrt[3]{3^2}} \cdot \dfrac{\sqrt[3]{3}}{\sqrt[3]{3}}$
$= \dfrac{5\sqrt[3]{3}}{\sqrt[3]{3^3}}$
$= \dfrac{5\sqrt[3]{3}}{3}$

11. $\sqrt{\dfrac{48x^3y^4}{7z}} = \dfrac{\sqrt{48x^3y^4}}{\sqrt{7z}}$
$= \dfrac{\sqrt{16x^2y^4}\sqrt{3x}}{\sqrt{7z}}$
$= \dfrac{4xy^2\sqrt{3x}}{\sqrt{7z}}$
$= \dfrac{4xy^2\sqrt{3x}}{\sqrt{7z}} \cdot \dfrac{\sqrt{7z}}{\sqrt{7z}}$
$= \dfrac{4xy^2\sqrt{21xz}}{7z}$

12. $\sqrt{16x^2} = 4|x|$

13. $\sqrt{25x^3} = 5|x|\sqrt{x}$

14. $\sqrt{x^2 + 10x + 25}$
$= \sqrt{(x+5)^2}$
$= |x+5|$

15. $\sqrt{2x^3 + 7x^2}$
$= \sqrt{x^2(2x+7)}$
$= |x|\sqrt{2x+7}$

16. $\sqrt[3]{(-3)^3}$
$= \sqrt[3]{-27}$
$= -3$

17. $\sqrt[3]{(-1)^3}$
$= \sqrt[3]{-1}$
$= -1$

Section 3.6

1. $6x^2 + 7x + 2 = 0$
$a = 6, b = 7, c = 2$
$x = \dfrac{-7 \pm \sqrt{7^2 - 4(6)(2)}}{2(6)}$
$= \dfrac{-7 \pm \sqrt{49 - 48}}{12}$
$= \dfrac{-7 \pm \sqrt{1}}{12}$
$= \dfrac{-7 \pm 1}{12}$
$x = \dfrac{-7 + 1}{12}$ or $x = \dfrac{-7 - 1}{12}$
$x = -\dfrac{1}{2}$ or $x = -\dfrac{2}{3}$

2. $\dfrac{x^2}{2} + x = \dfrac{1}{3}$ LCD = 6
$6 \cdot \dfrac{x^2}{2} + 6 \cdot x = 6 \cdot \dfrac{1}{3}$
$3x^2 + 6x = 2$
$3x^2 + 6x - 2 = 0$
$a = 3, b = 6, c = -2$
$x = \dfrac{-6 \pm \sqrt{6^2 - 4(3)(-2)}}{2(3)}$
$= \dfrac{-6 \pm \sqrt{36 + 24}}{6}$
$= \dfrac{-6 \pm \sqrt{60}}{6}$
$= \dfrac{-6 \pm 2\sqrt{15}}{6}$
$= \dfrac{2(-3 \pm \sqrt{15})}{2 \cdot 3}$
$= \dfrac{-3 \pm \sqrt{15}}{3}$

3.
$$12 = 32t - 16t^2$$
$$16t^2 - 32t + 12 = 0$$
$$4t^2 - 8t + 3 = 0 \quad \text{Divide by 4}$$
$$t = \frac{8 \pm \sqrt{64 - 4(4)(3)}}{2(4)}$$
$$= \frac{8 \pm \sqrt{16}}{8}$$
$$= \frac{8 \pm 4}{8}$$
$$t = \frac{8 + 4}{8} = \frac{12}{8} = \frac{3}{2} \quad \text{or} \quad t = \frac{8 - 4}{8} = \frac{4}{8} = \frac{1}{2}$$

Note: Since the solutions are rational numbers, we could have solved the equation by factoring.

4.
$$P = \$1{,}320$$
$$P = -500 + 27x - 0.1x^2$$
$$1{,}320 = -500 + 27x - 0.1x^2$$
$$0.1x^2 - 27x + 1{,}820 = 0$$
$$x = \frac{27 \pm \sqrt{(-27)^2 - 4(0.1)(1{,}820)}}{(2)(0.1)}$$
$$= \frac{27 \pm \sqrt{729 - 728}}{0.2}$$
$$= \frac{27 \pm \sqrt{1}}{0.2}$$
$$= \frac{27 + 1}{0.2} \quad \text{or} \quad = \frac{27 - 1}{0.2}$$
$$= \frac{28}{0.2} \quad \text{or} \quad = \frac{26}{0.2}$$
$$x = 140 \quad \text{or} \quad x = 130$$

Section 4.1

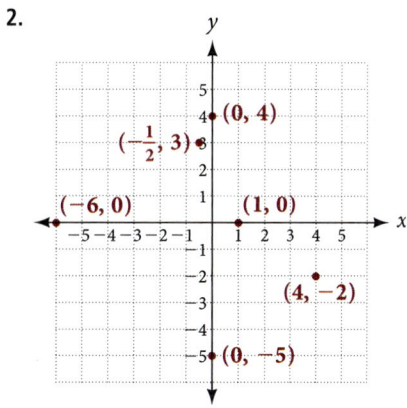

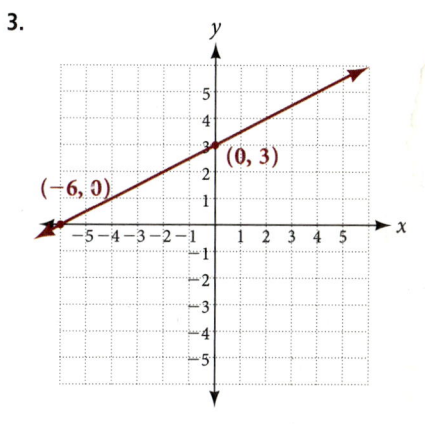

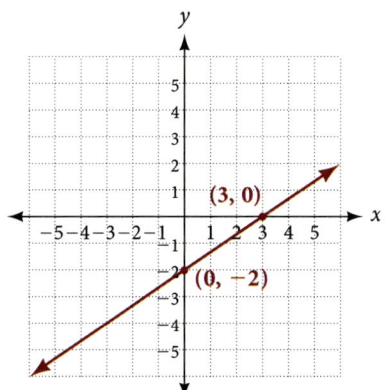

1. Points plotted: $(-1, 3)$, $(1, 3)$, $(-1, -3)$, $(1, -3)$

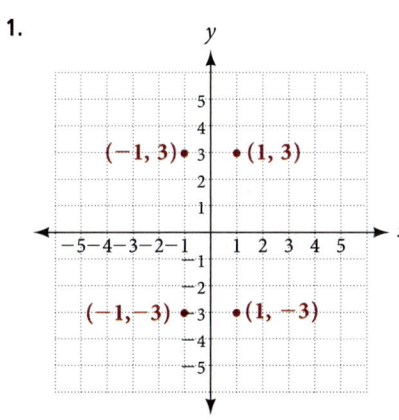

2. Points plotted: $(0, 4)$, $\left(-\tfrac{1}{2}, 3\right)$, $(-6, 0)$, $(1, 0)$, $(4, -2)$, $(0, -5)$

3. Line through $(-6, 0)$ and $(0, 3)$

4. Line through $(3, 0)$ and $(0, -2)$

5. Horizontal line $y = 4$ and vertical line $x = -1$

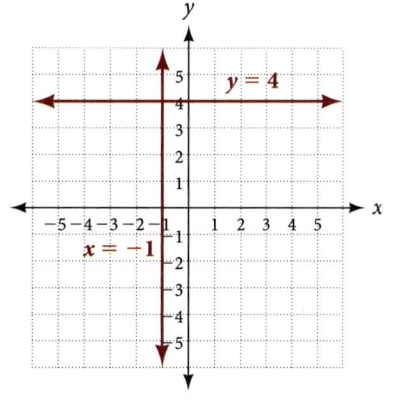

Section 4.2

1.

x	y
0	0
10	80
20	160
30	240
40	320

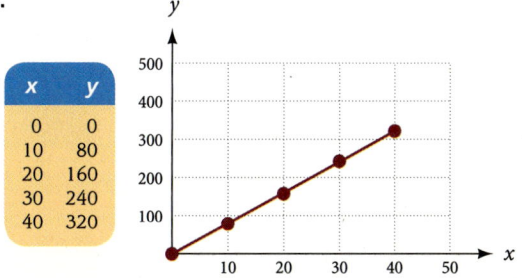

4.

t	h
0	0
$\tfrac{1}{2}$	20
1	32
$\tfrac{3}{2}$	36
2	32
$\tfrac{5}{2}$	20
3	0

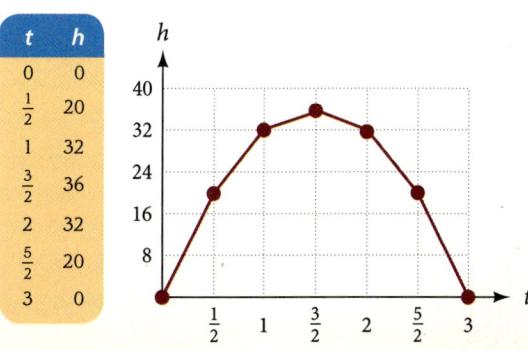

5.

x	y
5	−3
0	−2
−3	−1
−4	0
−3	1
0	2
5	3

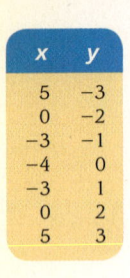

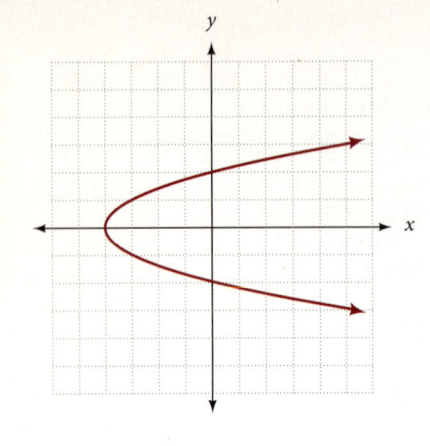

6.

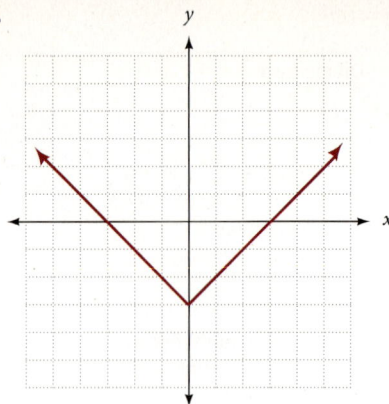

Section 5.1

1.
$\overline{AB} = \overline{BC}$
$5x = 9x - 8$
$8 = 4x$
$x = 2$

Therefore, $\overline{AB} = 5x$, and $\overline{BC} = 9x - 8$
$= 5(2)$ $= 9(2) - 8$
$= 10$ $= 10$

2.
$m\angle ABC = m\angle CBD$
$6x + 4 = 9x - 11$
$15 = 3x$
$x = 5$

Therefore, $m\angle ABC = 6x + 4$, and $m\angle CBD = 9x - 11$
$= 6(5) + 4$ $= 9(5) - 11$
$= 34$ $= 34$

Section 5.2

9. $m\angle 8 = m\angle 4$, Corresponding angles
$m\angle 8 = m\angle 9$, given
Therefore, $m\angle 9 = m\angle 4$, transitive property

Section 5.3

2. Since $\overline{AC} = \overline{CB}$,
$m\angle CAB = m\angle CBA$, which are both $(5x + 7)°$.

The sum of the interior angles is 180°, so we have
$(5x + 7) + (5x + 7) + (7x - 4) = 180$
$17x + 10 = 180$
$17x = 170$
$x = 10$

Section 5.6

1. First term = 2, Second term = 3, Third term = 6, Fourth term = 9.
The means are 3 and 6; the extremes are 2 and 9.

2. a. $5 \cdot 18 = 6 \cdot 15$
$90 = 90$

b. $13 \cdot 3 = 39 \cdot 1$
$39 = 39$

c. $\frac{2}{3} \cdot 5 = \frac{3}{5} \cdot 2$
$\frac{10}{3} = \frac{10}{3}$

d. $0.12 \cdot 3 = 0.18 \cdot 2$
$0.36 = 0.36$

7.

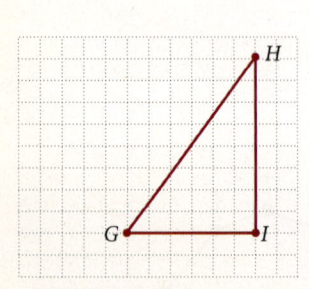

Section 5.8

3. a. $\frac{2 \tan 45°}{1 + \tan^2 45°} = \frac{2(1)}{1 + (1)^2} = \frac{2}{2} = 1$ **b.** $\cos^2 30° - \sin^2 30° = \left(\frac{\sqrt{3}}{2}\right)^2 - \left(\frac{1}{2}\right)^2 = \frac{3}{4} - \frac{1}{4} = \frac{1}{2}$

Index

A

Absolute value, 4
Acute angles, 306
Acute triangle, 327
Adding real numbers, 15
Addition and subtraction
   of polynomials, 90
Addition Property of Equality, 39
Addition
   of real numbers, 15
   of polynomials, 90
   Property of Equality, 39
   with different denominators, 182
   with same denominator, 181
   Associative Property of, 6
   Commutative Property of, 6
Additive Identity Property, 10
Additive Inverse Property, 10
Algebra with functions, 283
Algebraic Expressions
   finding the value of, 23
Alternate Exterior Angle Theorem, 319
Alternate Interior Angle Theorem, 318
Angle bisector, 305
Angle Theorems
   alternate exterior, 319
   alternate interior, 318
   consecutive interior, 319
   corresponding, 318
Angles, 304
   acute, 306
   complementary, 306
   congruent, 309
   exterior, 316
   interior, 306
   obtuse, 306
   right, 306
   sides of, 304
   straight, 306
   supplementary, 306
   vertical, 308
Area, 399
Associative Property of Addition, 6
Associative Property of Multiplication, 6
Axes, 238

B

Bisector
   angle, 305
Blueprint for problem solving, 53

C

Circle, 400
Circumference, 400
Cofunction Theorem, 378
Collinear, 303
Commutative Property of Addition, 6
Commutative Property of Multiplication, 6
Complementary angles, 306
Composition of functions, 284
Congruent, 309
Congruent triangles, 337
   tests for, 339
Consecutive Interior Angle Theorem, 319
Constant term, 43
Coplanar, 303
Corresponding Angle Theorem, 318
Cosecant, 377
Cosine, 377
Cotangent, 377

D

Degree measure, 305
Degree of a polynomial, 90
Diameter, 400
Difference, 16
Difference of two squares, 102
Distributive Property, 7
Division
   with fractions, 21
   with rational expressions, 171
   with real numbers, 18
   with the number 0, 22
Domain of the function, 254

E

Endpoint, 303
Equations with rational expressions, 193
Equality
   Addition Property of, 39
Equilateral triangle, 327
Equations
   equivalent, 39
   rational, 193
   linear, one variable, 39
   linear, two variables, 239
   solving by factoring, 136
Exponents
   properties of, 77
Exterior angles, 316

F

Factoring
   by grouping, 108
   by trial and error, 122
   trinomials with a leading coefficient of 1, 121
   when the lead coefficient is not 1, 122
   strategy to factor a polynomial, 129
FOIL Method, 100
Formulas, 196
Fractions, 3
   addition of, 181, 182
   dividing, 21
   multiplication, 3
   subtraction, 181, 182
Function maps, 270
Function notation, 269
Functions, 253
   algebra with, 283
   composition of, 284
   domain of, 254
   maps, 270
   notation, 269
   range of, 254
Fundamental Property of Proportions, 356

G

Greatest common factor, 107

H

Horizontal and vertical lines, 241
Horizontal axis, 237
Hypotenuse, 327

I

Identity, 44
Intercepts, 240
Interior angles, 316
Interior angles of a polygon, 347
Isosceles triangle, 327

L

Least common denominator, 182
Line, 303
   horizontal, 241
   vertical, 241
   graph, 237
   segment, 303
   parallel, 315
   vertical, 241
Linear equation in one variable, 39
Linear equation in two variables, 239
Linear equations, 239
   in one variable, 39
   in two variables, 239
Line graphs, 237
Line segment, 303

M

Midpoint, 304
Monomial, 89
Multiplication, 3
   associative property of, 6
   commutative property of, 6
   with fractions, 3
   with polynomials, 99
   with rational expressions, 171
   with real numbers, 17
Multiplication Property of Equality, 40
Multiplicative Identity Property, 10
Multiplicative Inverse Property, 10

O

Obtuse angles, 306
Obtuse triangle, 327
Opposites, 3
Ordered pairs, 237
Order of operations, 19
Origin, 238

P

Paired data, 237
Parallel lines, 315
   Theorem, 318
Parallelograms, 347, 349
Perfect square trinomials, 113
Perimeter, 399
Perpendicular, 321
Plane, 303
Plane geometry, 303
Point, 303
Polygon, 399
   interior angles of, 347
Polynomial, 89
   addition and subtraction of, 90
   degree of, 90
   multiplying, 99
   factoring strategy, 129
Polynomials and function notation, 92
Postulate AAS, 340
Postulate ASA, 339
Postulate SAS, 340
Postulate SSS, 340
Product, 17
Properties of exponents, 77
Properties of radicals, 369
Property 1 for Radicals, 203
Property 2 for Radicals, 203
Property 3 for Exponents, 78
Property 3 for Radicals, 207
Property 4 for Exponents, 78
Property 5 for Exponents, 79
Property 6 for Exponents, 79
Property 7 for Exponents, 80

Index

Proportion, 355
    fundamental property of, 356
    terms of, 355
Pythagorean Theorem, 365

Q

Quadratic
    equation, 135
    formula, 215
    standard form, 135
    Theorem, 215
Quotient, 18

R

Radical notation, 203
Radicals, 203
    properties of, 369
    notation, 203
Radius, 400
Range of a function, 254
Rational expressions, 162
    dividing, 171
    equations with, 193
    multiplying, 171
Rationalizing the denominator, 205
Ratios, 161
Ray, 304
Real numbers
    adding, 15
    dividing, 18
    multiplying, 17
    subtracting, 16
Reciprocals, 4
Rectangular coordinate system, 235, 238
Reducing to lowest terms, 159
Reflexive properties in triangles, 338
Relation, 255
Right angles, 306
Right triangle, 327
    solving, 385
Right triangle trigonometry, 377

S

Scalene, 327
Scatter diagrams, 237
Scientific notation, 31
Secant, 377
Sides of the angle, 304
Significant digits, 385
Similar figures, 358
Similar terms, 90
Similar triangles, 357
Simplified form for radical expressions, 203, 368
Simplifying expressions, 20
Sine, 377
Solution set, 39
Solving equations by factoring, 136
Solving linear equations in one variable, 43
Solving right triangles, 385

Square of a binomial, 101
Square root of a perfect square, 206
Standard form for quadratic equations, 135
Straight angles, 306
Subtracting real numbers, 16
Subtraction
    of polynomials, 90
    with different denominators, 182
    with same denominator, 181
Sum, 15
Sum and difference of two cubes, 115
Supplementary angles, 306
Symmetric properties in triangles, 338

T

Table building, 55
Tangent, 377
Term
    constant, 43
    similar, 90
    of a proportion, 355
    variable, 43
Terms of a proportion, 355
Tests for congruent triangles, 339
The difference of two squares, 114
Theorems for parallel lines, 318
Theorems for vertical angles, 309
Transitive properties in triangles, 338
Transitive Property, 321
Transversal, 315
Triangles, 327
    acute, 327
    congruent, 337
    equilateral, 327
    isosceles, 327
    obtuse, 327
    reflexive properties of, 338
    right, 327
    similar, 357
    symmetric properties of, 338
    transitive properties in, 338
Trigonometric functions, 377

V

Variable term, 43
Vertex, 304
Vertical angles, 308
    Theorems, 309
Vertical axis, 237
Vertical lines, 241
Vertical Line Test, 258

X

x-axis, 238
x-coordinate, 237
x-intercept, 240

Y

y-axis, 238
y-coordinate, 237
y-intercept, 240

Z

Zero-Factor Property, 135